CCNA Voice 640-461

Official Cert Guide

Jeremy Cioara, CCIE No. 11727

Mike Valentine, CCNA, CCNP, CCVP

Cisco Press

800 East 96th Street

Indianapolis, IN 46240

CCNA Voice 640-461 Official Cert Guide

Jeremy Cioara, CCIE No. 11727; Mike Valentine, CCNA, CCNP, CCVP

Published by:
Cisco Press
800 East 96th Street
Indianapolis, IN 46240 USA

Printed in the United States of America

Fourth Printing: November 2012

Library of Congress Cataloging-in-Publication Data

Cioara, Jeremy.

CCNA voice official exam certification guide / Jeremy Cioara, Mike Valentine.

p. cm.

"CCNA voice 640-461."

ISBN 978-1-58720-417-3 (hardcover w/cd) 1. Internet telephony—Examinations—Study guides. I. Valentine, Michael, 1966- II. Title.

TK5105.8865.C523 2012

004.69'5076—dc23

2011024500

ISBN-10: 1-58720-417-7

ISBN-13: 978-1-58720-417-3

Warning and Disclaimer

This book is designed to provide information about the 640-461 ICOMM exam certification exam. Every effort has been made to make this book as complete and as accurate as possible, but no warranty or fitness is implied.

The information is provided on an "as is" basis. The authors, Cisco Press, and Cisco Systems, Inc., shall have neither liability nor responsibility to any person or entity with respect to any loss or damages arising from the information contained in this book or from the use of the discs or programs that may accompany it.

The opinions expressed in this book belong to the authors and are not necessarily those of Cisco Systems, Inc.

Trademark Acknowledgments

All terms mentioned in this book that are known to be trademarks or service marks have been appropriately capitalized. Cisco Press or Cisco Systems, Inc., cannot attest to the accuracy of this information. Use of a term in this book should not be regarded as affecting the validity of any trademark or service mark.

Corporate and Government Sales

The publisher offers excellent discounts on this book when ordered in quantity for bulk purchases or special sales, which may include electronic versions and/or custom covers and content particular to your business, training goals, marketing focus, and branding interests. For more information, please contact:

U.S. Corporate and Government Sales
1-800-382-3419 corpsales@pearsontechgroup.com

For sales outside the United States, please contact:
International Sales
international@pearsoned.com

Feedback Information

At Cisco Press, our goal is to create in-depth technical books of the highest quality and value. Each book is crafted with care and precision, undergoing rigorous development that involves the unique expertise of members from the professional technical community.

Readers' feedback is a natural continuation of this process. If you have any comments regarding how we could improve the quality of this book, or otherwise alter it to better suit your needs, you can contact us through e-mail at feedback@ciscopress.com. Please make sure to include the book title and ISBN in your message.

We greatly appreciate your assistance.

Publisher: Paul Boger

Associate Publisher: Dave Dusthimer

Cisco Representative: Anthony Wolfenden

Cisco Press Program Manager: Jeff Brady

Executive Editor: Brett Bartow

Managing Editor: Sandra Schroeder

Development Editor: Deadline Driven Publishing

Project Editor: Mandie Frank

Copy Editor: Sheri Cain

Technical Editors: Brion Washington, John Swartz

Editorial Assistant: Vanessa Evans

Designer: Gary Adair

Composition: Mark Shirar

Indexer: Brad Herriman

Proofreader: Water Crest Publishing

About the Authors

Jeremy D. Cioara, CCIE No. 11727, works in many facets of the Cisco networking realm. As an author, he has written multiple books for Cisco Press and Exam Cram. As an instructor, he teaches at Interface Technical Training (www.interfacett.com) in Phoenix, Arizona. Likewise, Jeremy has recorded many E-Learning titles at CBTNuggets (www.cbtnuggets.com). Finally, Jeremy is the CIO of AdTEC Networks and works as a network consultant, focusing on Cisco network and VoIP implementations. Jeremy also casually blogs about Cisco topics at Tekcert (www.tekcert.com) in his "free time." Thankfully, he is married to the Certified Best Wife in the World (CBWW), who helps him manage his time and priorities and prevents him from getting an enormous Cisco logo tattooed across his chest.

Michael Valentine has 15 years of experience in the IT field, specializing in network design and installation. Currently, he is a Cisco trainer with Skyline Advanced Technology Services and specializes in Cisco Unified Communications, CCNA, and CCNP classes. His accessible, humorous, and effective teaching style has demystified Cisco for hundreds of students since he began teaching in 2002. Mike holds a Bachelor of Arts degree from the University of British Columbia and currently holds CCNA, CCDA, CCNP, CCVP, and CCSI No. 31461 certifications. Mike has developed courseware and labs for Cisco and its training partners. Mike is the coauthor of *CCNA Exam Cram (Exam 640-802)*, Third Edition (Que 2008), authored the *CCNA Voice Quick Reference Guide*, and has served as technical editor and contributor on several Cisco Press titles.

About the Technical Reviewers

Brion S. Washington, CCNA, is a senior voice engineer consultant in Atlanta, GA. He has more than ten years of Cisco experience, with the last five years dedicated to VoIP; he has worked with all the Cisco VoIP products. Brion has done many large projects involving VoIP, from complete network design, implementation, and the last level of escalation. He is currently finishing up his CCVP.

John Swartz, CCIE No. 4426, is the founder of Boson Software and training, 3DSNMP, Purple Penguin and Inner Four. Currently focused on cloud technologies using the VBLOCK Infrastructure Platform by VCE.

He is also focused on mobile technology his company has published over 500 apps for the iPhone. John created the original Cisco Press CCNA Network simulator, the Boson Netsim, and numerous practice tests. He has been a Cisco instructor for 12 years, starting with basic courses and now teaching Unified Computing, Nexus switching and other data center technology. John lives in Florida with his wife and three kids.

Dedication

From Jeremy D. Cioara:

This book is dedicated to you. Yes...the person reading this *right now*. No, I'm not being cheesy, I'm serious! The only real way people are truly successful and fulfilled in this career is to love what they're doing. Because of that, I put much effort (within grammatical boundaries) into not just communicating technical mumbo jumbo—hey! Microsoft Word didn't correct that! Who knew "mumbo jumbo" was a real word?—but making it fun and interesting to read. I hope this book sparks something in you that blooms into an interesting, fun, and fulfilling career.

(In case you're curious, dictionary.com defines "mumbo jumbo" as *senseless or pretentious language, usually designed to obscure an issue, confuse a listener, or the like*. It also says that mumbo jumbo is a masked man who combats evil in the western Sudan. I don't think either of these was my intention...)

From Mike Valentine:

This book is dedicated to my wife Liana, without whose unflinching support, it might never have happened. You and me, love.

In memory of my Dad.

Acknowledgments

Jeremy D. Cioara: When you go see a movie, ever notice how the credits roll for about 5 minutes with hundreds of names? It's the same with this book. There are probably hundreds of names you'll never see that had some part in making this book possible. My thanks goes to all of them!

Personally, I give thanks to Jesus Christ who is…well, everything! Without Christ, my world of color quickly fades to a dull, boring grey. Thanks to my wife, who tirelessly homeschools our three kiddos and puts up with my countless Matrix analogies to explain anything under the sun. Finally, thanks to Interface Technical Training (www.interfacett.com), CBTNuggets (www.cbtnuggets.com), and Pearson (www.pearson.com) for allowing me to communicate my love for all things networking to people everywhere.

Mike Valentine: In fear of forgetting someone, let me try to list all the people who helped make this book happen:

Brett Bartow: For asking, answering, and adapting. Thank you, sir.

Jeremy Cioara: For trusting me with all the hard stuff…kidding, man.

Dayna, Ginny, Chris, and all the unknown soldiers at Cisco Press: They tempered, refined, redrew, and otherwise helped create what you are holding. Professionals, all; I salute them.

Toby Sauer, Dave Schulz, and Dave Bateman: My colleagues at Skyline and, most importantly, my good friends; for their opinions, their commiseration and support, and for making me a better instructor and author. Thank you, my friends. (Please go buy their books, too; you will not regret it.)

Andy de Maria: Thank you for your empathy, flexibility, and your trust.

Ed Misely: A good friend and terrifyingly capable technical resource, for his assistance with my labs.

My family: Thank you so much for your support, patience, your love, and your belief in me.

The readers and posters on the Cisco Learning Community: For your early input and support. Here it is, finally. I sincerely hope you enjoy it.

Contents at a Glance

Contents

Icons Used in This Book

Communication Server

PC

PC with Software

Sun Workstation

Macintosh

Terminal

ISDN/Frame Relay Switch

Token Ring

Laptop

File Server

Web Server

Ciscoworks Workstation

ATM Switch

Modem

Gateway

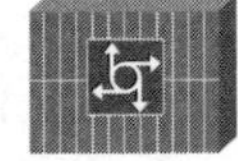
Access Server

IBM Mainframe

Front End Processor

Cluster Controller

Multilayer Switch without Text

Printer

Router

Bridge

Hub

DSU/CSU

FDDI

Catalyst Switch

Network Cloud

Line: Ethernet

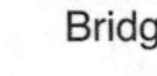
Line: Serial

Line: Circuit-Switched

Phone

IP Phone

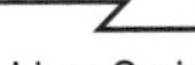
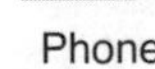

Repeater

PBX Switch

File Server

Cisco Unified Communications 500 Series for Small Business

Cisco Unity Express

Cisco Unified Communication Manager

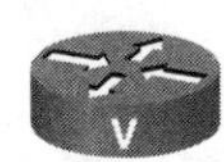

Voice-Enabled Router

Voice-Enabled Workgroup Switch

Legacy PBX

Multilayer Switch without Text

Unified Personal Communicator (UPC)

Command Syntax Conventions

The conventions used to present command syntax in this book are the same conventions used in the IOS Command Reference. The Command Reference describes these conventions as follows:

- **Boldface** indicates commands and keywords that are entered literally as shown. In actual configuration examples and output (not general command syntax), boldface indicates commands that are manually input by the user (such as a **show** command).
- *Italics* indicate arguments for which you supply actual values.
- Vertical bars (|) separate alternative, mutually exclusive elements.
- Square brackets [ ] indicate optional elements.
- Braces { } indicate a required choice.
- Braces within brackets [{ }] indicate a required choice within an optional element.

Introduction

Welcome to the world of CCNA Voice! As technology continues to evolve, the realm of voice, which was traditionally kept completely separate from data, has now begun to merge with the data network. This brings together two different worlds of people: data technicians—historically accustomed to working with routers, switches, servers, and the like—and voice technicians, historically accustomed to working with PBX systems, digital handsets, and trunk lines. Regardless of your background, one of the primary goals of the new CCNA Voice certification is to bridge these two worlds together.

In June 2008, Cisco announced new CCNA specialties, including CCNA Security, CCNA Wireless, and CCNA Voice. These certifications, released ten years after the initial CCNA, represent Cisco's growth into new and emerging industries. Certification candidates can now specialize in specific areas of study. Figure I-1 shows the basic organization of the certifications and exams used to achieve your CCNA Voice certification.

Figure I-1 *Cisco Certifications and CCNA Voice Certification Path*

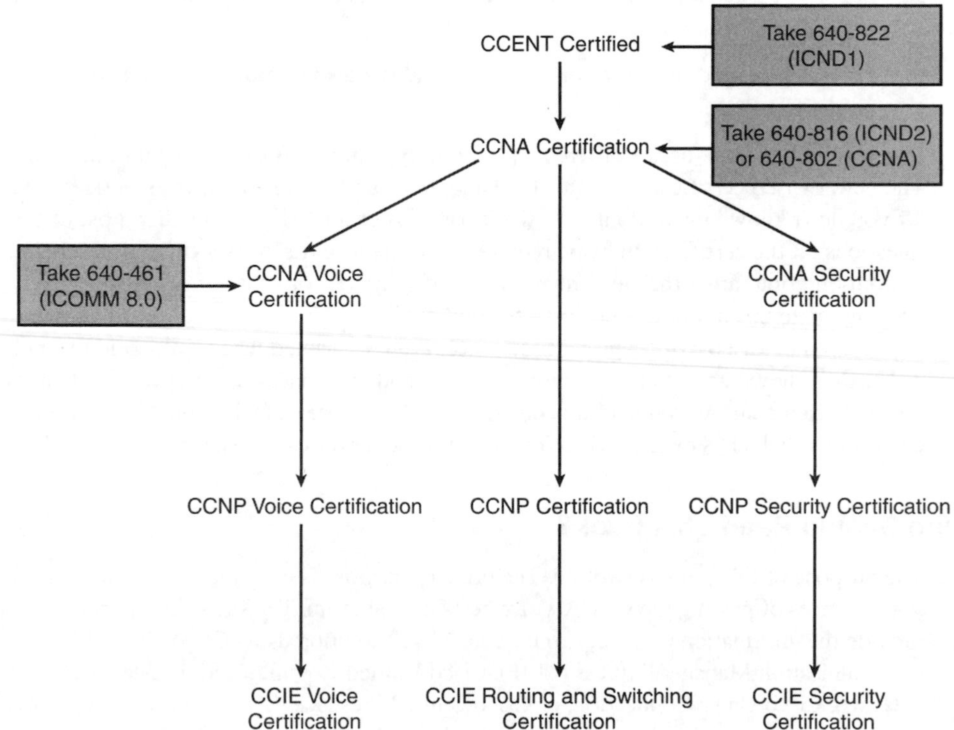

As you can see from Figure I-1, a traditional CCNA certification is a prerequisite before you venture into the CCNA Voice certification.

Goals and Methods

The most important and somewhat obvious goal of this book is to help you pass the Implementing Introducing Cisco Voice and Unified Communications Administration v8.0 (ICOMM 8.0) exam (640-461). In fact, if the primary objective of this book were different, the book's title would be misleading. The methods used in this book help you pass the ICOMM 8.0 exam and make you much more knowledgeable about how to do your job.

This book uses several key methodologies to help you discover the exam topics that you need to review in more depth, to help you fully understand and remember those details, and to help you prove to yourself that you have retained your knowledge of those topics. So, this book does not try to help you pass by memorization, but helps you truly learn and understand the topics. The CCNA Voice exam is the foundation for many of the Cisco professional certifications, and it would be a disservice to you if this book did not help you truly learn the material. Therefore, this book helps you pass the CCNA Voice exam by using the following methods:

- Helping you discover which test topics you have not mastered
- Providing explanations and information to fill in your knowledge gaps
- Supplying exercises and scenarios that enhance your ability to recall and deduce the answers to test questions
- Providing practice exercises on the topics and the testing process via test questions on the CD-ROM

In addition, this book uses a different style from typical certification-preparation books. The newer Cisco certification exams have adopted a style of testing that essentially says, "If you don't know how to do it, you won't pass this exam." This means that most of the questions on the certification exam require you to deduce the answer through reasoning or configuration rather than just memorizing facts, figures, or syntax from a book. To accommodate this newer testing style, the authors have written this book as a real-world explanation of Cisco VoIP topics. Most concepts are explained using real-world examples rather than showing tables full of syntax options and explanations, which are freely available on Cisco.com. As you read this book, you definitely get a feeling of, "This is how I can *do* this, "which is exactly what you need for the newer Cisco exams.

Who Should Read This Book?

The purpose of this book is twofold. The primary purpose is to tremendously increase your chances of passing the CCNA Voice certification exam. The secondary purpose is to provide the information necessary to manage a VoIP solution using Cisco Unified Communication Manager Express (CME), Cisco Unified Communications Manager (CUCM), Cisco Unity Connection, or Cisco Unified Presence. Cisco's new exam approach provides an avenue to write the book with both a real-world and certification-study approach at the same time. As you read this book and study the configuration examples and exam tips, you have a true sense of understanding how you could deploy a VoIP system, while at the same time feeling equipped to pass the CCNA Voice certification exam.

Strategies for Exam Preparation

Strategies for exam preparation will vary depending on your existing skills, knowledge, and equipment available. Of course, the ideal exam preparation would consist of building a small voice lab with a Cisco Integrated Services Router, virtualized lab versions of CUCM, Unity Connection, and Presence servers, a switch, and a few IP Phones, which you could then use to work through the configurations as you read this book. However, not everyone has access to this equipment, so the next best step you can take is to read the chapters and jot down notes with key concepts or configurations on a separate notepad. Each chapter begins with a "Do I Know This Already?" quiz, which is designed to give you a good idea of the chapter's content and your current understanding of it. In some cases, you might already know most of or all the information covered in a given chapter.

After you read the book, look at the current exam objectives for the CCNA Voice exam listed on Cisco.com (www.cisco.com/certification). If there are any areas shown in the certification exam outline that you would still like to study, find those sections in the book and review them.

When you feel confident in your skills, attempt the practice exam included on the CD with this book. As you work through the practice exam, note the areas where you lack confidence and review those concepts or configurations in the book. After you have reviewed the areas, work through the practice exam a second time and rate your skills. Keep in mind that the more you work through the practice exam, the more familiar the questions will become, so the practice exam will become a less accurate judge of your skills.

After you work through the practice exam a second time and feel confident with your skills, schedule the real ICOMM 8.0 (640-461) exam through Vue (www.vue.com). You should typically take the exam within a week from when you consider yourself ready to take the exam, so that the information is fresh in your mind.

Keep in mind that Cisco exams are very difficult. Even if you have a solid grasp of the information, many other factors play into the testing environment (stress, time constraints, and so on). If you pass the exam on the first attempt, fantastic! If not, know that this commonly happens. The next time you attempt the exam, you will have a major advantage: You already experienced the exam first-hand. Although future exams may have different questions, the topics and general "feel" of the exam remain the same. Take some time to study areas from the book where you felt weak on the exam. Retaking the exam the same or following day from your first attempt is a little aggressive; instead, schedule to retake it within a week, while you are still familiar with the content.

640-461 ICOMM 8.0 Exam Topics

Table I-1 lists the exam topics for the 640-461 ICOMM 8.0 exam. This table also lists the book parts in which each exam topic is covered.

Table I-1 *640-461 ICOMM 8.0 Exam Topics*

Chapter Where Topic Is Covered	Exam Topic
Describe the characteristics of a Cisco Unified Communications solution	
Chapter 2	Describe the Cisco Unified Communications components and their functions
Chapter 2	Describe call signaling and media flows
Chapter 6	Describe quality implications of a VoIP network
Provision end users and associated devices	
Chapter 5, Chapter 9	Describe user creation options for Cisco Unified Communications Manager and Cisco Unified Communications Manager Express
Chapter 9	Create or modify user accounts for Cisco Unified Communications Manager
Chapter 5	Create or modify user accounts for Cisco Unified Communications Manager Express using the GUI
Chapter 9	Create or modify endpoints for Cisco Unified Communications Manager
Chapter 5	Create or modify endpoints for Cisco Unified Communications Manager Express using the GUI
Chapter 6, Chapter 10	Describe how calling privileges function and how calling privileges impact system features
Chapter 5, Chapter 9	Create or modify directory numbers
Chapter 7, Chapter 11, Chapter 12	Enable user features and related calling privileges for extension mobility, call coverage, intercom, native presence, and unified mobility remote destination configuration
Chapter 14	Enable end users for Cisco Unified Presence
Chapter 7, Chapter 11, Chapter 12	Verify user features are operational
Configure voice messaging and presence	
Chapter 13	Describe user creation options for voice messaging
Chapter 13	Create or modify user accounts for Cisco Unity Connection
Chapter 14	Describe Cisco Unified Presence
Chapter 14	Configure Cisco Unified Presence

Table I-1 *640-461 ICOMM 8.0 Exam Topics*

Chapter Where Topic Is Covered	Exam Topic
Maintain Cisco Unified Communications system	
Chapter 16	Generate CDR and CMR reports
Chapter 16	Generate capacity reports
Chapter 16	Generate usage reports
Chapter 16	Generate RTMT reports to monitor system activities
Chapter 17	Monitor voicemail usage
Chapter 16	Remove unassigned directory numbers
Chapter 16	Perform manual system backup
Provide end user support	
Chapter 15, Chapter 16	Verify PSTN connectivity
Chapter 15, Chapter 16	Define fault domains using information gathered from end user
Chapter 15, Chapter 16	Troubleshoot endpoint issues
Chapter 17	Identify voicemail issues and resolve issues related to user mailboxes
Chapter 15, Chapter 16	Describe causes and symptoms of call quality issues
Chapter 5, Chapter 9	Reset single devices
Chapter 11	Describe how to use phone applications

How This Book Is Organized

Although this book could be read cover-to-cover, it is designed to be flexible and allow you to easily move between chapters and sections of chapters to cover just the material that you need more work with. If you do intend to read all the chapters, the order in the book is an excellent sequence to use.

The core chapters, Chapters 1 through 17, cover the following topics:

- **Chapter 1,** "Traditional Voice Versus Unified Voice." This chapter discusses what would be known as the traditional telephony world. It begins where the telephone system originally started: analog connectivity. It then moves into the realm of digital connections and considerations and concludes the traditional voice discussion with the primary pieces that you need to know from the public switched telephone network (PSTN). Chapter 1 then moves into the unified voice realm, discussing the benefits of VoIP, the process of coding and decoding audio, digital signal processors (DSP), and the core VoIP protocols.

- **Chapter 2**, "Understanding the Pieces of Cisco Unified Communications." This chapter primarily focuses on the components of a Cisco VoIP network. By breaking down the voice infrastructure into four distinct areas, each component can be categorized and described. These components include endpoints, call processing agents, applications, and network infrastructure devices.

- **Chapter 3**, "Understanding the Cisco IP Phone Concepts and Registration." This chapter discusses the preparation and base configuration of the LAN infrastructure to support VoIP devices. This preparation includes support for Power over Ethernet (PoE), voice VLANs, a properly configured DHCP scope for VoIP devices, and the Network Time Protocol (NTP).

- **Chapter 4**, "Getting Familiar with CME Administration." This chapter familiarizes you with Cisco Unified Communication Manager Express (CME) administration by unpacking the two primary administrative interfaces of CME: command-line and the Cisco Configuration Professional (CCP) GUI.

- **Chapter 5**, "Managing Endpoint and End Users with CME." This chapter focuses on the process to create and assign directory numbers (DN) and user accounts to Cisco IP Phones. The chapter walks through these configurations in both the command-line and CCP interfaces.

- **Chapter 6**, "Understanding the CME Dial-Plan." Now that the internal VoIP network is operational through the CME configuration, this chapter examines connections to the outside world through the PSTN or over an IP network. Concepts covered in this chapter include the configuration of physical voice port characteristics, dial peers, digit manipulation, class of restriction (COR), and quality of service (QoS).

- **Chapter 7**, "Configuring Cisco Unified CME Voice Productivity Features." This chapter examines feature after feature supported by the CME router. By the time you're done with this chapter, you'll understand how to configure features such as intercom, paging, Call Park and pickup, and many others.

- **Chapter 8**, "Administrator and End-User Interfaces." This chapter introduces the administration interfaces for CUCM, CUC, and CUP. From the administrative GUI for each application to the common Unified Serviceability interface, disaster recovery, and CLI, the fundamentals of navigation and configuration are laid out in a clear and logical sequence.

- **Chapter 9**, "Managing Endpoints and End Users in CUCM." The configuration and management of users and phones is covered in this chapter, including integration with LDAP.

- **Chapter 10**, "Understanding CUCM Dial-Plan Elements and Interactions." The guts of the call-routing system in CUCM are explained with simplicity and clarity. Call flows in different deployments and under different conditions of use and failure (including CAC and AAR) are demonstrated and compared, and the great mystery of partitions and calling search spaces (CSS) is revealed for the simple truth it really is.

- **Chapter 11**, "Enabling Telephony Features with CUCM." A small but excellent sample of the billions* (*approximately) of features available in CUCM, including Extension Mobility and call coverage.
- **Chapter 12**, "Enabling Mobility Features in CUCM." A step-by-step guide to enabling some of the most popular and powerful features in CUCM: Mobile Connect and Mobile Voice Access.
- **Chapter 13**, "Voicemail Integration with Cisco Unity Connection." The power, stability and wealth of features available in CUC are examined, followed by a look at the configuration of user accounts and their mail boxes.
- **Chapter 14**, "Enabling Cisco Unified Presence Support." The capabilities, features, and basic configuration of the CUP server and clients are covered, giving an introduction to one of the most powerful additions to the Unified Communications capabilities of any business.
- **Chapter 15**, "Common CME Management and Troubleshooting Issues." This chapter takes the CME concepts you learned and builds them into troubleshooting scenarios. The chapter begins by discussing a general troubleshooting process you can employ for any technical troubleshooting situation, then walks through many common CME troubleshooting situations dealing with IP phone registration. The chapter concludes by discussing dial-plan and QoS troubleshooting methods.
- **Chapter 16**, "Management and Troubleshooting of Cisco Unified Communications Manager." This chapter reviews the tools available to administrators to assist in the care and feeding of their CUCM servers. From the myriad of built-in reporting tools to the power of the RTMT, the administrator is introduced to his arsenal of tools to monitor the health and performance of the system.
- **Chapter 17**, "Monitoring Cisco Unity Connection." The wealth of built-in reporting and monitoring tools for CUC are reviewed in this chapter.

In addition to the 17 main chapters, this book includes tools to help you verify that you are prepared to take the exam. Chapter 18, "Final Preparation," includes guidelines that you can follow in the final days before the exam. Also, the CD-ROM includes quiz questions and memory tables that you can work through to verify your knowledge of the subject matter.

This chapter includes the following topics:

- **Where It All Began: Analog Connections:** This section discusses the simplest type of modern voice communication: analog connections.
- **The Evolution: Digital Connections:** Modern businesses quickly outgrow analog circuits. This section discusses the process of converting analog voice into digital signals and using digital circuits to send multiple calls over a single line.
- **Understanding the PSTN:** Just about all voice circuits currently terminate at the world's largest voice network, the PSTN. This section discusses the components of the PSTN, focusing specifically on PBX and Key Systems, and the methods used to connect to the PSTN.
- **The New Yet Not-So-New Frontier:** VoIP: Voice has been converted to digital format for decades; however, putting that digital content in a packet is a relatively new adventure. This section discusses the core concepts behind VoIP, including the coding/decoding (codec) process, DSPs, and the protocols used to deliver audio.

CHAPTER 1

Traditional Voice Versus Unified Voice

Welcome to the world of voice. No, not VoIP; you're not there yet. Before you can enter that world, a foundation must be laid. The traditional telephony network has been in place since the early 1900s, and it is not going to disappear overnight. Thus, until then, the new VoIP networks must integrate with traditional telephony networks. To perform this integration, you must have a basic understanding of the world of traditional voice telephony. This chapter walks you through the foundations of the public switched telephone network (PSTN), private branch exchange (PBX) systems, and analog and digital circuitry.

"Do I Know This Already?" Quiz

The "Do I Know This Already?" quiz allows you to assess whether you should read this entire chapter or simply jump to the "Exam Preparation Tasks" section for review. If you are in doubt, read the entire chapter. Table 1-1 outlines the major headings in this chapter and the corresponding "Do I Know This Already?" quiz questions. You can find the answers in Appendix A, "Answers Appendix."

Table 1-1 *"Do I Know This Already?" Foundation Topics Section-to-Question Mapping*

Foundation Topics Section	Questions Covered in This Section
Where It All Began: Analog Connections	1–3
The Evolution: Digital Connections	4–9
Understanding the PSTN	10
The New Yet Not-So-New Frontier: VoIP	11–13

1. Analog phones connected to the PSTN typically use which of the following signal types?

   a. Loop start

   b. Ground start

   c. CAS

   d. CCS

2. Which of the following issues is prevented by using ground start signaling?
   a. Echo
   b. Glare
   c. Reflexive transmissions
   d. Mirrored communication
3. Which of the following signaling types represents supervisory signaling?
   a. Off-hook signal
   b. Dial tone
   c. DTMF
   d. Congestion
4. What are two disadvantages of using analog connectivity?
   a. Conversion complexity
   b. Signal quality
   c. Limited calls per line
   d. Lack of common voice services
5. Which of the following systems allows you to send multiple voice calls over a single digital circuit by dividing the calls into specific time slots?
   a. MUX
   b. DE-MUX
   c. TDM
   d. TCP
6. When using T1 CAS signaling, which bits are used to transmit signaling information within each voice channel?
   a. First bit of each frame
   b. Last bit of each frame
   c. Second and third bits of every third frame
   d. Eighth bit of every sixth frame
7. How large is each T1 frame sent over a digital CAS connection?
   a. 8 bits
   b. 24 bits
   c. 80 bits
   d. 193 bits

8. Which of the following time slots are used for T1 and E1 signaling when using CCS connections? (Choose two.)

   a. Time slot 1
   b. Time slot 16
   c. Time slot 17
   d. Time slot 23
   e. Time slot 24

9. Which of the following standards created by the ITU designates international numbering plans for devices connected to the PSTN?

   a. ITU-T
   b. E.164
   c. ITU-161
   d. T-161

10. What frequency range is accurately reproduced by the Nyquist theorem on the PSTN?

    a. 200–9000 Hz
    b. 300–3400 Hz
    c. 300–4000 Hz
    d. 20–20,000 Hz

11. What amount of bandwidth is consumed by the audio payload of G.729a?

    a. 4.3 kbps
    b. 6.3 kbps
    c. 8 kbps
    d. 16 kbps

12. Which of the following are high-complexity codecs? (Choose two.)

    a. G.711 µ-law
    b. G.729
    c. G.729a
    d. iLBC

Foundation Topics

Where It All Began: Analog Connections

In 1877, Thomas Edison created a brilliant device known as a phonograph, which is shown in Figure 1-1.

Figure 1-1 *Replica of Edison's Phonograph*

This device was able to record sounds by pressing a needle into a cylinder covered with tinfoil on a rhythmic basis as a person spoke into a sound-collecting horn. The phonograph could then play back this sound by moving the needle at a steady speed back over the indentions made in the tinfoil. This "archaic" form of recording is one representation of an analog signal.

An analog signal uses a property of the device that captures the audio signal to convey audio information. In the case of Edison's phonograph, the property was the various indentions in tinfoil. In today's world, where everything is connected through some form of cabling, electric currents are used to send analog signals. When you speak into an analog phone, the sounds that come out of your mouth are converted into electricity. The volume and pitch that you use when speaking result in different variations of electrical current. Electrical voltage, frequency, current, and charge are all used in some combination to convey the properties of your voice. Figure 1-2 illustrates perhaps a more familiar view of using electrical signals to capture the properties of voice.

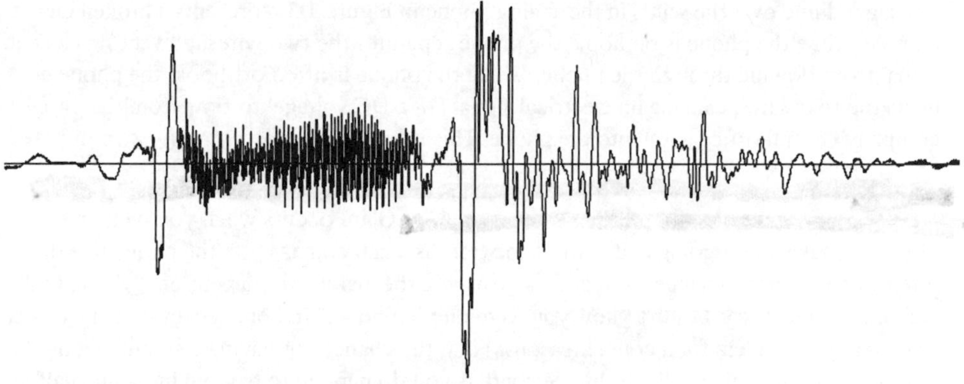

Figure 1-2 *Electrical Analog Waveform of Human Speech*

Note: The analog waveform shown in Figure 1-2 is from me (Jeremy) saying, "Hello."

Analog phone lines use the properties of electricity to convey changes in voice over cabling. Of course, there is far more than just voice to send over the phone lines. The analog phones you use at home must convey many different types of signaling, too. Signaling includes messages such as dial tone, dialed digits, busy signals, and so on. These signaling types are discussed in just a moment. For now, let's look at the cabling used to make analog connections function.

Each analog circuit is composed of a pair of wires. One wire is the ground, or positive side of the connection (often called the tip). The other wire is the battery, or negative side of the connection (often called the ring). You'll commonly hear phone technicians talk about these wires as the "tip and ring." These two wires are what power the analog phone and allow it to function, just like the wires that connect your car battery to the car. Figure 1-3 illustrates the connections of the tip and ring wire to your analog phone.

Key Topic

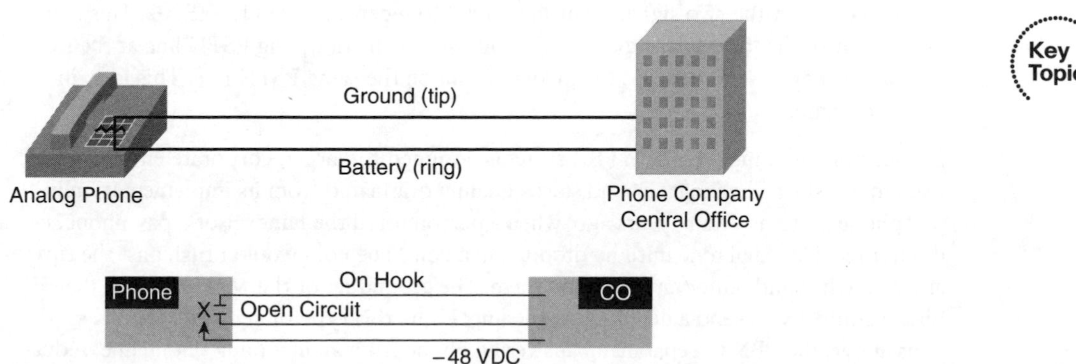

Figure 1-3 *Connections of the Ground and Battery Wires to an Analog Phone*

The jagged line over the wires in the analog phone in Figure 1-3 represents a broken circuit. Anytime the phone is on hook, the phone separates the two wires, preventing electric signal from flowing through the phone. When the phone is lifted off hook, the phone connects the two wires, causing an electrical signal (48V DC voltage) to flow from the phone company central office (CO) into the phone. This is known as *loop start signaling*.

Loop start signaling is the typical signaling type used in home environments. Loop start signaling is susceptible to a problem known as glare. Glare occurs when you pick up the phone to make an outgoing call at the same time as a call comes in on the phone line before the phone has a chance to ring. This gives you the awkward moment of, "Uhhh...Oh! Hello, Bob! I'm sorry, I didn't know you were on the phone." In home environments, this is not usually a problem for a couple reasons. First, the chances of having a simultaneous outgoing and incoming call are slim. Second, if you do happen to have an incoming call, it's always meant for your house (unless the caller dialed the wrong number).

In business environments, glare can become a significant problem because of the large number of employees and high call volume. For example, a corporation may have a key system (which allows it to run its own, internal phone system) with five analog trunks to the PSTN, as shown in Figure 1-4.

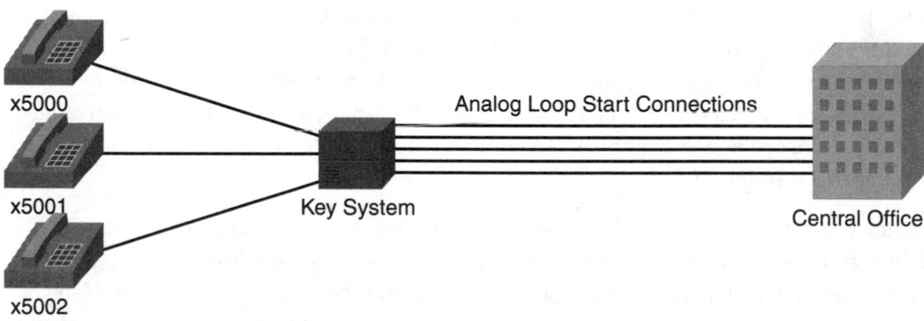

Figure 1-4 *Illustration of Glare*

If a call comes in for x5002 at the same time as x5000 picks up the phone, the key system will connect the two signals, causing x5000 to receive the call for x5002. This happens because the loop start signal from x5000 seizes the outgoing PSTN line at the same time as the key system receives the incoming call on the same PSTN line. This is an instance of glare.

Because of glare, most modern PBX systems designed for larger, corporate environments use ground start signaling. Ground start signaling originated from its implementation in pay phone systems. Many years ago, when a person lifted the handset of a pay phone, he did not receive a dial tone until he dropped in a coin. The coin would brush past the tip and ring wires and temporarily ground them. The grounding of the wires signaled the phone company to send a dial tone on the line. Using this type of signaling in PBX systems allows the PBX to separate an answering phone from an incoming phone line, reducing the problem of glare. To receive a dial tone from the CO, the PBX must send a ground signal on the wires. This intentionally signals to the telephone CO that an outgoing call is going to happen, whereas using the loop start method of signaling just connects the wires to receive an incoming call or place an outgoing call.

Tip: Many other types of signaling exist in the analog world. These include supervisory signaling (on hook, off hook, ringing), informational signaling (dial tone, busy, ringback, and so on), and address signaling (dual-tone multifrequency (DTMF) and Pulse). These are discussed in detail as part of the CVOICE certification series (for more information, see www.ciscopress.com/bookstore/product.asp?isbn=1587055546).

The Evolution: Digital Connections

Analog signaling was a massive improvement over tin cans and string, but still posed plenty of problems of their own. First, an analog electrical signal experiences degradation (signal fading) over long distances. To increase the distance the analog signal could travel, the phone company had to install repeaters (shown in Figure 1-5) to regenerate the signal as it became weak.

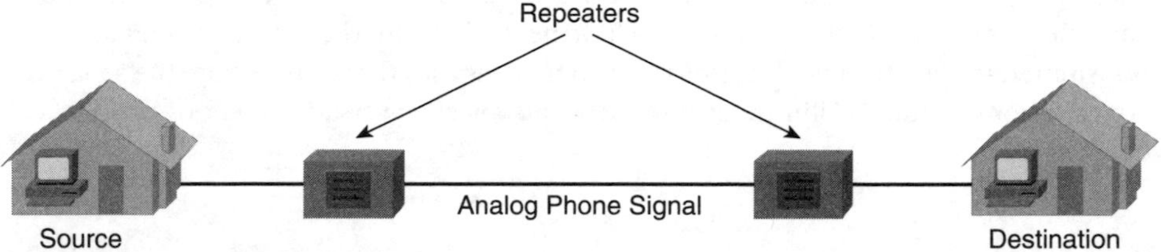

Figure 1-5 *Analog Signal Repeaters*

Unfortunately, as the analog signal was regenerated, the repeater device was unable to differentiate between the voice traveling over the wire and line noise. Each time the repeater regenerated the voice, it also amplified the line noise. Thus, the more times a phone company regenerated a signal, the more distorted and difficult to understand the signal became.

The second difficulty encountered with analog connections was the sheer number of wires the phone company had to run to support a large geographical area or a business with a large number of phones. Because each phone required two wires, the bundles of wire became massive and difficult to maintain (imagine the hassle of a single pair of wires in the bundle breaking). A solution to send multiple calls over a single wire was needed. A digital connection is that solution.

Moving from Analog to Digital

Simply put, digital signals use numbers to represent levels of voice instead of a combination of electrical signals. When someone talks about "digitizing voice," they are speaking of the process of changing analog voice signals into a series of numbers (shown in Figure 1-6) that you can use to put the voice back together at the other end of the line.

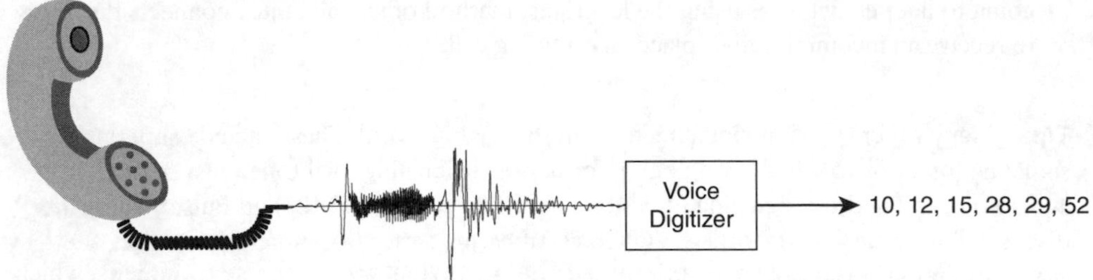

Figure 1-6 *Converting Analog to Digital Signals*

Essentially, each number sent represents a sound that someone made while speaking into a telephone handset. Today's network devices can easily transmit a numeric value any distance a cable can run without any degradation or line noise, which solves the signal degradation issues faced by analog phone connections. However, could digital signaling resolve the huge number of wires required by multiple analog connections? Indeed, it can.

Digital voice uses a technology known as time-division multiplexing (TDM). TDM allows voice networks to carry multiple conversations at the same time over a single, four-wire path. Because the multiple conversations have been digitized, the numeric values are transmitted in specific time slots (thus, the "time division") that differentiate the separate conversations. Figure 1-7 illustrates three separate voice conversations sent over a digital connection.

Key Topic

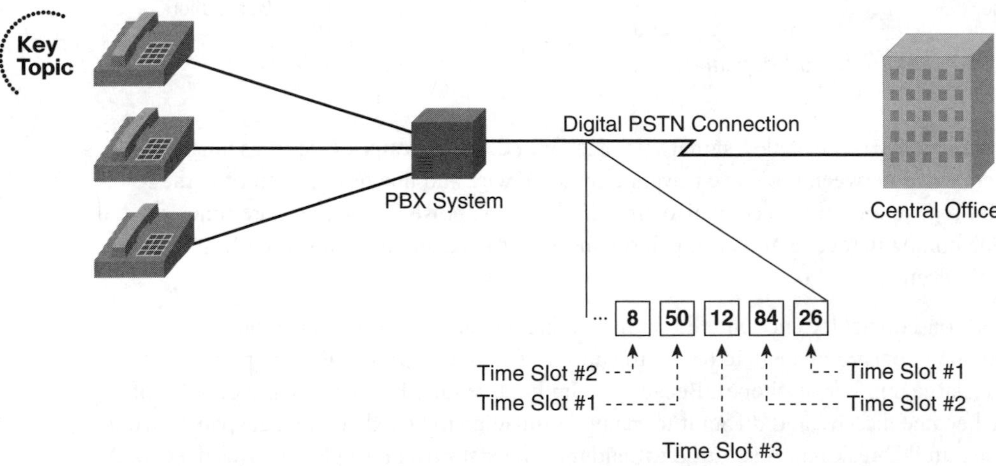

Figure 1-7 *Time-Division Multiplexing Voice Channels*

Notice each of the voice conversations in Figure 1-7 has been digitized and assigned a numeric value and transmitted over the digital PSTN connection. Based on the time the voice data was sent, the PSTN carrier is able to distinguish and reassemble the voice conversations.

Note: Although the values in each time slot are shown in decimal in Figure 1-10, they are actually transmitted and interpreted in binary.

Corporations use digital voice connections to the PSTN as T1 circuits in the United States, Canada, and Japan. A T1 circuit is built from 24 separate 64-kbps channels known as a digital signal 0 (DS0). Each one of these channels is able to support a single voice call. Corporations in areas outside the United States, Canada, and Japan use E1 circuits, which allow you to use up to 30 DS0s for voice calls.

Although digital technology solves the problems of signal degradation and the inability to send multiple calls over a single line that occur in analog technology, it creates a new issue: signaling. With analog circuits, supervisory signals were passed by connecting the tip and ring wires together. The phone company generated informational and address signals through specific frequencies of electricity. By solving the problems associated with analog signaling, digital signaling also removed the typical signaling capabilities. To solve this, two primary styles of signaling were created for digital circuits:

- **Channel associated signaling (CAS):** Signaling information is transmitted using the same bandwidth as the voice.
- **Common channel signaling (CCS):** Signaling information is transmitted using a separate, dedicated signaling channel.

The following sections discuss these two styles of signaling.

Channel Associated Signaling

T1 digital connections that use CAS actually "steal" binary bits that would typically have been used to communicate voice information and use them for signaling. Initially, this may seem crazy; if you take the binary bits that are used to resynthesize the voice, won't the voice quality drop significantly? Although the voice quality does drop some, the number of binary bits stolen for signaling information is small enough that the change in voice quality is not noticeable.

Note: Because T1 CAS steals bits from the voice channel to transfer signaling information, it is often called *robbed bit signaling (RBS).*

The voice device running the T1 line uses the eighth bit on every sixth sample in each T1 channel (DS0). Figure 1-8 illustrates this concept.

As you can see from Figure 1-8, the 24 channels of the digital T1 circuit carry only voice data for the first five frames that they send. On the sixth frame (marked with an S in Figure 1-8), the eighth bit (also called the least significant bit) is stolen for the voice devices to transmit signaling information. This process occurs for every sixth frame after this (12th, 18th, 24th, and so on). This stolen bit relays the signaling information for each respective DS0 channel. For example, the bits stolen from the third DS0 channel relay the signaling information only for that channel.

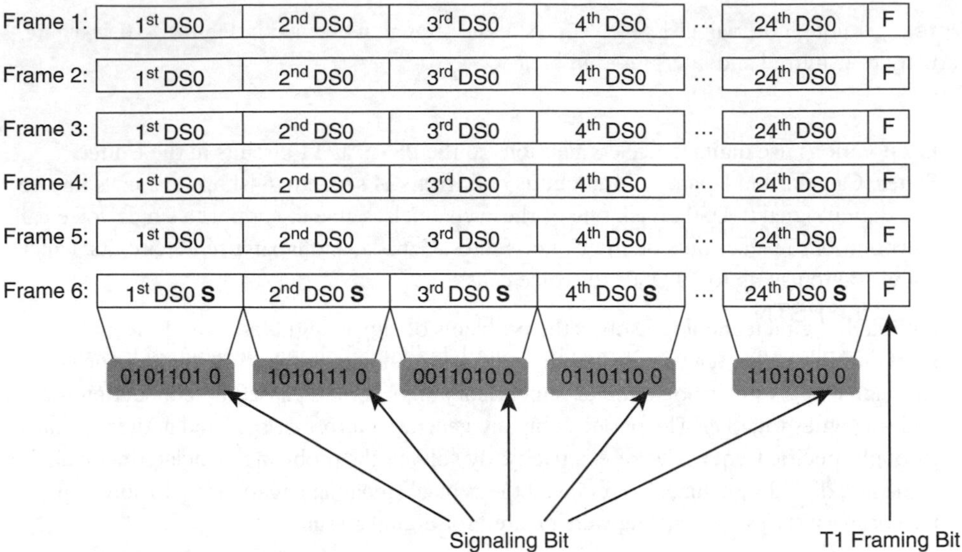

Figure 1-8 *CAS T1 Signaling Bits*

Common Channel Signaling

CCS dedicates one of the DS0 channels from a T1 or E1 link for signaling information. This is often called out-of-band signaling because the signaling traffic is sent completely separate from the voice traffic. As a result, a T1 connection using CCS has only 23 usable DS0s for voice. Because CCS dedicates a full channel of the circuit for signaling, the "stolen bit" method of signaling using ABCD bits is no longer necessary. Rather, a full signaling protocol sends the necessary information for all voice channels. The most popular signaling protocol used is Q.931, which is the signaling protocol used for ISDN circuits.

CCS is the most popular connection used between voice systems worldwide because it offers more flexibility with signaling messages, more bandwidth for the voice bearer channels, and higher security (because the signaling is not embedded in the voice channel). CCS also allows PBX vendors to communicate proprietary messages (and features) between their PBX systems using ISDN signaling, whereas CAS does not offer any of these capabilities.

Tip: When using CCS configurations with T1 lines, the 24th time slot is always the signaling channel. When using CCS configurations with E1 lines, the 17th time slot is always the signaling channel.

Note: Although ISDN is the most popular protocol used with CCS configurations, CCS can use other protocols. For example, telephone companies use the Signaling System 7 (SS7) protocol (described later) with CCS configurations to communicate between COs.

Understanding the PSTN

All the signaling standards and communication methods discussed in the previous section typically focus around the connection to one massive voice network, known as the PSTN. If you have ever made a call from a home telephone, you have experienced the results of the traditional telephony network. This network is not unlike many of the data networks of today. Its primary purpose is to establish worldwide pathways to allow people to easily connect, converse, and disconnect.

Pieces of the PSTN

When the phone system was originally created, individual phones were wired together to allow people to communicate. If you wanted to connect with more than one person, you needed multiple phones. As you can imagine, this solution was short lived as a more scalable system was found. The modern PSTN is now a worldwide network (much like the Internet), built from the following pieces, as shown in Figure 1-9:

- **Analog telephone:** Able to connect directly to the PSTN and is the most common device on the PSTN. Converts audio into electrical signals.
- **Local loop:** The link between the customer premises (such as a home or business) and the telecommunications service provider.
- **CO switch:** Provides services to the devices on the local loop. These services include signaling, digit collection, call routing, setup, and teardown.
- **Trunk:** Provides a connection between switches. These switches could be CO or private.

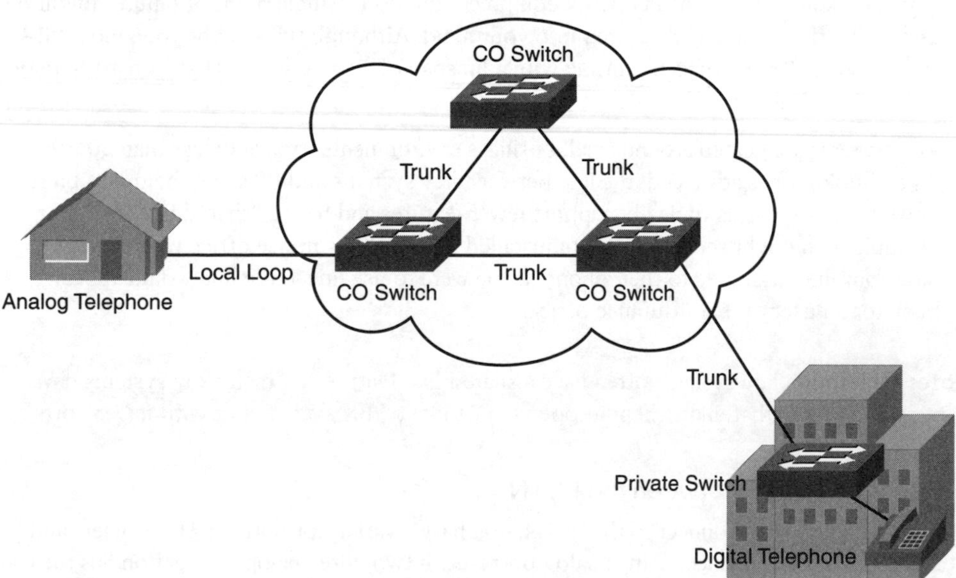

Figure 1-9 *PSTN Components*

- **Private switch:** Allows a business to operate a "miniature PSTN" inside its company. This provides efficiency and cost savings because each phone in the company does not require a direct connection to the CO switch.
- **Digital telephone:** Typically connects to a PBX system. Converts audio into binary 1s and 0s, which allows more efficient communication than analog.

Many believe that the PSTN will eventually be absorbed into the Internet. Although this may be true, advances must be made on the Internet to ensure proper quality of service (QoS) guarantees for voice calls.

Understanding PBX and Key Systems

Many businesses have hundreds or even thousands of phones they support in the organization. If the company purchases a direct PSTN connection for each one of these phones, the cost would be astronomical. Instead, most organizations choose to use a PBX or key system internally to manage in-house phones. These systems allow internal users to make phone calls inside the office without using any PSTN resources. Calls to the PSTN forward out the company's PSTN trunk link.

When you first look at a PBX system, it looks like a large box full of cards. Each card has a specific function:

- **Line cards:** Provide the connection between telephone handsets and the PBX system.
- **Trunk cards:** Provide connections from the PBX system to the PSTN or other PBX systems.
- **Control complex:** Provides the intelligence behind the PBX system; all call setup, routing, and management functions are contained in the control complex.

If you look at a PBX from a network equipment mindset, "single point of failure" might be one of the first thoughts that jump into your mind. Although this may be true, most PBX systems offer 99.999 percent uptime with a lifespan of 7 to 10 years. That's a hard statistic to beat in just about any industry.

Key systems are geared around small business environments (typically less than 50 users). As technology has advanced, the line between key systems and PBXs has begun to blur; however, key systems typically support fewer features and have a "shared line" feel. For example, you might see a key system installed in a small insurance office where users all have four lines assigned to their phone. If Joe were to use line 1, the line would appear busy for all users at the insurance office.

> **Note:** Although key systems often have a shared-line feature set, many key systems have numerous features that allow them to operate just like a PBX system, but with fewer ports.

Connections to and Between the PSTN

When you want to connect to the PSTN, you have a variety of options. Home users and small offices can connect using analog ports. Each two-wire analog connection has the capability to support a single call. For home users, a single, analog connection to the PSTN may be sufficient. For small offices, the number of incoming analog connections directly relates to the office size and average call volume. As businesses grow, you can consolidate the multiple analog connections into one or more digital T1 or E1 connections, as shown in Figure 1-10.

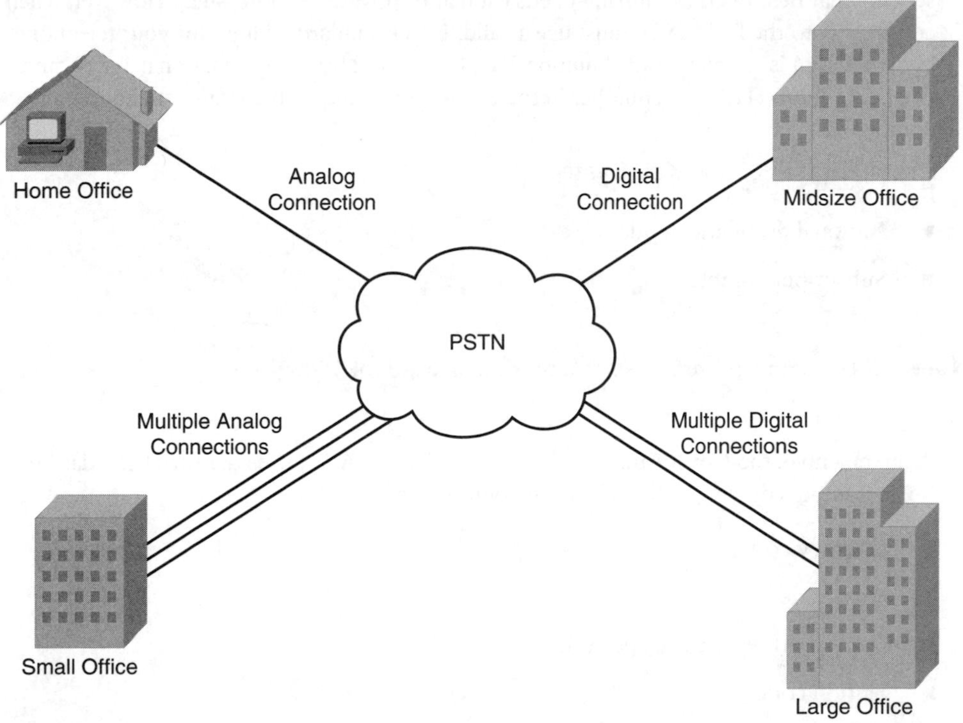

Figure 1-10 *Connections to the PSTN*

In the PSTN lies a network of networks, similar to the Internet, which connects offices from multiple telephony providers together into a massive worldwide network. For all the telephony providers of the world to communicate together, a common signaling protocol must be used, similar to the way TCP/IP operates in the data realm. The voice signaling protocol used around the world is SS7.

SS7 is an out-of-band (CCS-style) signaling method used to communicate call setup, routing, billing, and informational messages between telephone company COs around the world. When a user makes a call, the first CO to receive the call performs an SS7 lookup to locate the number. Once the destination is found, SS7 is responsible for routing the call through the voice network to the destination and providing all informational signaling (such as ring back) to the calling device.

Note: SS7 is primarily a telephony service provider technology. You do not typically directly interface with the SS7 protocol from a telephony customer perspective.

PSTN Numbering Plans

Just as data networks use IP addressing to organize and locate resources, voice networks use a numbering plan to organize and locate telephones all around the world. Organizations managing their own internal telephony systems can develop any internal number scheme that best fits the company needs (similar to private IP addressing). However, when connecting to the PSTN, you must use a valid, E.164 standard address for your telephone system. E.164 is an international numbering plan created by the International Telecommunication Union (ITU). Each number in the E.164 numbering plan contains the following components:

- Country code
- National destination code
- Subscriber number

Note: E.164 numbers are limited to a maximum length of 15 digits.

As an example, the North American Numbering Plan (NANP) uses the E.164 standard to break numbers down into the following components:

- Country code
- Area code
- Central office or exchange code
- Station code

For example, the NANP number 1-602-555-1212 breaks down as shown in Figure 1-11.

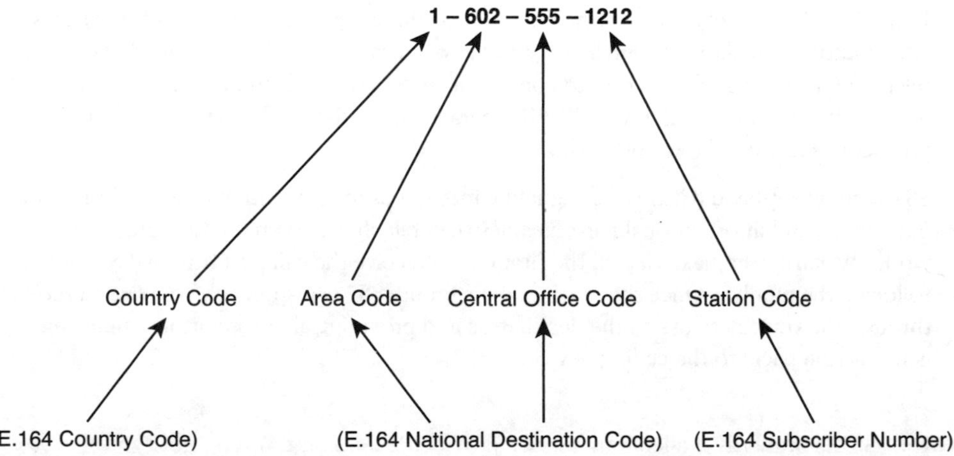

Figure 1-11 *NANP Phone Number Example*

Even though the NANP defines specific categories of numbers that the E.164 standard does not include, the number still falls under the three broad categories, also shown in Figure 1-11.

The New Yet Not-So-New Frontier: VoIP

When I (Jeremy) first started exploring VoIP technology, I often talked excitedly with people about the amazing process of converting spoken voice into packets. Many of these people replied, "Ah, those processes have been happening for a long time!" Indeed, it is true: Everything we discussed thus far deals with taking spoken voice (or analog data) and converting it into binary 1s and 0s (digital data). Digitizing voice is now considered "old school."

So, what's so new about VoIP? Here it is: taking those "old school" 1s and 0s and placing them into a data packet with IP addressing information in the headers. You can then take that VoIP packet and send it across the data network at your office. But, is it that simple? Not necessarily. Our concerns now turn to ensuring that the packet gets to its destination in time (QoS), choosing the proper coding and decoding (codec) methods, making sure that the VoIP packet doesn't fall into the wrong hands (encryption), and a plethora of other concerns. However, these topics will unfold in their due time—for now, take a moment to simply enjoy walking into the "new frontier" of VoIP!

VoIP: Why It Is a Big Deal for Businesses

When many people first learn about VoIP, they commonly say, "So, we are sending voice over data cables instead of voice cables...what is so big about that?" It seems like the biggest benefit is saving cabling costs, nothing more. After you dig deeper into the ramifications of running voice over data networks, you begin to uncover many business benefits that were previously untapped.

The business benefits of VoIP include the following:

- **Reduced cost of communicating:** Instead of relying on expensive tie lines or toll charges to communicate between offices, VoIP allows you to forward calls over WAN connections.
- **Reduced cost of cabling:** VoIP deployments typically cut cabling costs in half by running a single Ethernet connection instead of both voice and data cables. (This cost savings is most realized in newly constructed offices.)
- **Seamless voice networks:** Because data networks connect offices, mobile workers, and telecommuters, VoIP naturally inherits this property. The voice traffic is crossing "your network" (relatively speaking) rather than exiting to the PSTN. This also provides centralized control of all voice devices attached to the network and a consistent dial-plan. For example, all users can dial each other using four-digit extensions, even though many of them may be scattered around the world.
- **Take your phone with you:** Cost estimates for moves, adds, and changes (MAC) to a traditional PBX system range from $55 to $295 per MAC. With VoIP phone systems, this cost is virtually eliminated. In addition, IP phones are becoming increasingly plug-and-play within the local offices, allowing moves with little to no

reconfiguration of the voice network. In addition, when combined with a VPN configuration, users can take IP phones home with them and retain their work extension.

- **IP SoftPhones:** SoftPhones represent an ideal example of the possibilities when combining voice and data networks. Users can now plug a headset into their laptop or desktop and allow it to act as their phone. SoftPhones are becoming increasingly more integrated with other applications such as e-mail contact lists, instant messenger, and video telephony.
- **Unified e-mail, voicemail, fax:** All messaging can be sent to a user's e-mail inbox. This allows users to get all messages in one place and easily reply, forward, or archive messages.
- **Increased productivity:** VoIP extensions can forward to ring multiple devices before forwarding to voicemail. This eliminates the "phone tag" game.
- **Feature-rich communications:** Because voice, data, and video networks have combined, users can initiate phone calls that communicate with or invoke other applications from the voice or data network to add additional benefits to a VoIP call. For example, calls flowing into a call center can automatically pull up customer records based on caller ID information or trigger a video stream for one or more of the callers.
- **Open, compatible standards:** In the same way that you can network Apple, Dell, and IBM PCs together, you can now connect devices from different telephony vendors together. Although this benefit has yet to be fully realized, this will allow businesses to choose the best equipment for their network, regardless of the manufacturer.

The Process of Converting Voice to Packets

Long ago, Dr. Harry Nyquist (and many others) created a process that allows equipment to convert analog signals (flowing waveforms) into digital format (1s and 0s). It is important to understand this process, because it will guide your configuration of VoIP audio sample sizes, Digital Signal Processor (DSP) resources, and codecs.

The origin of the digital conversion process (which fed many of the developments discussed earlier) takes us back to the 1920s, a far throw from our VoIP world. The Bell Systems Corporation tried to find a way to deploy more voice circuits with less wire, because analog voice technology required one pair of wires for each voice line. For organizations that required many voice circuits, this meant running bundles of cable. After plenty of research, Nyquist found that he could accurately reconstruct audio streams by taking samples that numbered twice the highest audio frequency used in the audio.

Here is how it breaks down. Audio frequencies vary based on the volume, pitch, and so on that comprise the sound. Here are a few key facts:

- The average human ear is able to hear frequencies from 20–20,000 Hz.
- Human speech uses frequencies from 200–9,000 Hz.
- Telephone channels typically transmit frequencies from 300–3,400 Hz.
- The Nyquist theorem is able to reproduce frequencies from 300–4,000 Hz.

Now, you might think, "If human speech uses frequencies between 200–9,000 Hz and the normal telephone channel only transmits frequencies from 300–3,400 Hz, how can you understand human conversation over the phone?" That's a great question! Studies have found that telephone equipment can accurately transmit understandable human conversation by sending only a limited range of frequencies. The telephone channel frequency range (300–3,400 Hz) gives you enough sound quality to identify the remote caller and sense their mood. The telephone channel frequency range does not send the full spectrum of human voice inflection and lowers the actual quality of the audio. For example, if you've ever listened to talk radio, you can always tell the difference in quality between the radio host and the telephone caller.

Nyquist believed that you can accurately reproduce an audio signal by sampling at twice the highest frequency. Because he was after audio frequencies from 300–4,000 Hz, it would mean sampling 8,000 times (2 * 4000) every second. So, what's a sample? A sample is a numeric value. More specifically, in the voice realm, a sample is a numeric value that consumes a single byte of information. As Figure 1-12 illustrates, during the process of sampling, the sampling device puts an analog waveform against a Y-axis lined with numeric values.

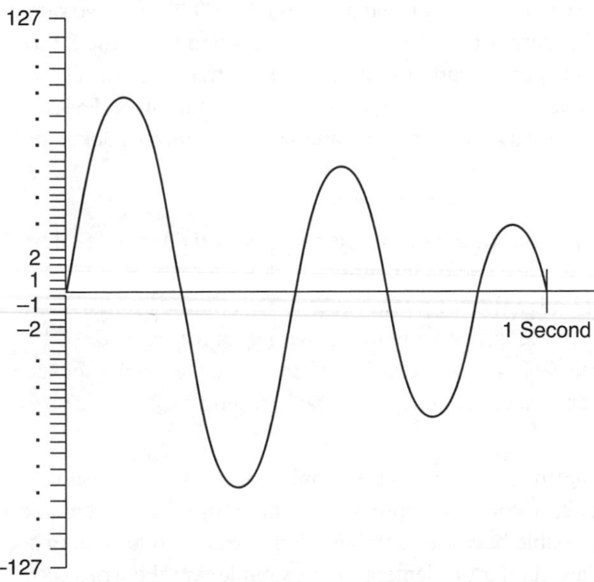

Figure 1-12 *Converting Analog Voice Signals to Digital*

This process of converting the analog wave into digital, numeric values is known as quantization. Because 1 byte of information can represent only values 0–255, the quantization of the voice scale is limited to values measuring a maximum peak of +127 and a maximum low of –127. Notice in Figure 1-12 that the 127 positive and negative values are not evenly spaced. This is by design. To achieve a more accurate numeric value (and thus, a more accurate reconstructed signal at the other end), the amplitude values more common to voice are tightly packed with numeric values, whereas the "fringe amplitudes" on the high and low end of the spectrum are more spaced apart.

The sampling device breaks the 8 binary bits in each byte into two components: a positive/negative indicator and the numeric representation. As shown in Figure 1-13, the first bit indicates positive or negative, and the remaining seven bits represent the actual numeric value.

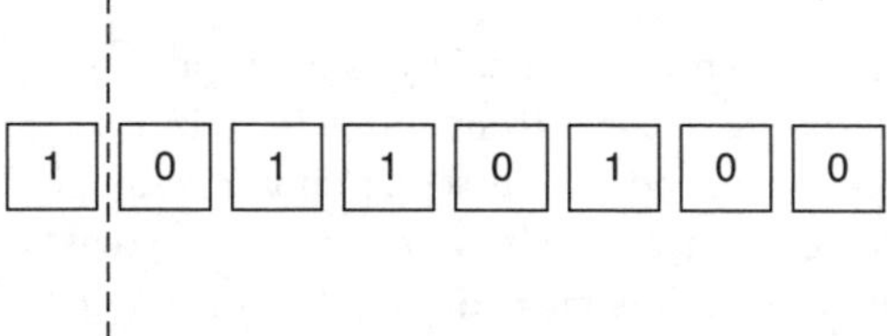

Figure 1-13 *Encoding Voice into Binary Values*

Because the first bit in Figure 1-13 is a 1, you read the number as positive. The remaining seven bits represent the number 52. This is the digital value used for one voice sample. Now, remember, the Nyquist theorem dictates that you need to take 8,000 of those samples every single second. Doing the math, figure 8,000 samples a second times the 8 bits in each sample, and you get 64,000 bits per second. It's no coincidence that uncompressed audio (including the G.711 audio codec) consumes 64 kbps. Once the sampling device assigns numeric values to all these analog signals, a router can place them into a packet and send them across a network.

Note: There are two forms of the G.711 codec: µ-law (used primary in the United States and Japan) and a-law (used everywhere else). The quantization method described in the preceding paragraph represents G.711 a-law. G.711 µ-law codes in exactly the opposite way. If you were to take all the 1 bits in Figure 1-13 and make them 0s and take all the 0 bits and make them 1s, you would have the G.711 µ-law equivalent. Yes, it doesn't make sense to code it that way, but who said things we do in the United States should make sense?

The last and optional step in the digitization process is to apply compression measures. Advanced codecs, such as G.729, allow you to compress the number of samples sent and thus use less bandwidth. This is possible because sampling human voice 8,000 times a second produces many samples that are similar or identical. For example, say the word "cow" out loud to yourself (provided you are in a relatively private area). That takes about a second to say, right? If not, say it slower until it does. Now, listen to the sounds you are making. There's the distinguished "k" sound that starts the word, then you have the "ahhhhhh" sound in the middle, followed by the "wa" sound at the end. If you were to break that into 8,000 individual samples, chances are most of them would sound the same.

The process G.729 (and most other compressed codecs) uses to compress this audio is to send a sound sample once and simply tell the remote device to continue playing that sound for a certain time interval. This is often described as "building a codebook" of the human voice traveling between the two endpoints. Using this process, G.729 is able to reduce bandwidth down to 8 kbps for each call; a fairly massive reduction in bandwidth.

Unfortunately, chopping the amount of bandwidth down comes with a price. Quality is usually impacted by the compression process. Early on in the voice digitization years, the powers that be created a measurement system known as a Mean Opinion Score (MOS) to rate the quality of the various voice codecs. The test used to rate the quality of voice is simple: A listener listens to a caller say the sentence, "Nowadays, a chicken leg is a rare dish," and rates the clarity of this sentence on a scale of 1–5. Table 1-2 shows how each audio codec fared in MOS testing.

Table 1-2 *Audio Codec Bandwidth and MOS Values*

Key Topic

Codec	Bandwidth Consumed	MOS
G.711	64 kbps	4.1
Internet Low Bitrate Codec (iLBC)	15.2 kbps	4.1
G.729	8 kbps	3.92
G.726	32 kbps	3.85
G.729a	8 kbps	3.7
G.728	16 kbps	3.61

Table 1-2 leads into a much-needed discussion about audio coder/decoders (codecs). You can use quite a few different audio codecs on your network, each geared for different purposes and environments. For example, some codecs are geared specifically for military environments where audio is sent through satellite link and bandwidth is at a premium. These codecs sacrifice audio quality to achieve very streamlined transmissions. Other codecs are designed to meet the need for quality.

If you stay in the Cisco realm for long, you will hear two codecs continually repeated: G.711 and G.729. This is because Cisco designed all its IP phones with the ability to code in either of these two formats. G.711 is the "common ground" between all VoIP devices. For example, if a Cisco IP phone is attempting to communicate with an Avaya IP phone, they may support different compressed codecs, but can at least agree on G.711 when communicating.

Note: G.729 comes in two different variants: G.729a (annex A) and G.729b (annex B). G.729a sacrifices some audio quality to achieve a much more processor-efficient coding process. G.729b introduces support for Voice Activity Detection (VAD), which makes voice transmissions more efficient. You learn more about these variants in the following section.

Role of Digital Signal Processors

Cisco designed its routers with one primary purpose in mind: routing. Moving packets between one location and another is not a processor-intensive task, thus Cisco routers are not equipped with the kind of memory and processing resources typical PCs are equipped with. For example, from a router's perspective, having 256 MB of RAM is quite a bit. From a PC's perspective, 256 MB will barely help you survive the Microsoft Windows boot process.

Moving into the realm of VoIP, the network now requires the router to convert loads of voice into digitized, packetized transmissions. This task would easily overwhelm the resources you have on the router. This is where DSPs come into play. DSPs offload the processing responsibility for voice-related tasks from the processor of the router. This is similar to the idea of purchasing an expensive video card for a PC to offload the video processing responsibility from the PC's processor.

Specifically, a DSP is a chip that performs all the sampling, encoding, and compression functions on audio coming into your router. If you were to equip your router with voice interface cards (VIC), allowing it to connect to the PSTN or analog devices, but did not equip your router with DSPs, the interfaces would be worthless. The interfaces would be able to actively connect to the legacy voice networks, but would not have the power to convert any voice into packetized form.

DSPs typically come in chips to install in your Cisco router that look like old memory SIMMs, as shown in Figure 1-14.

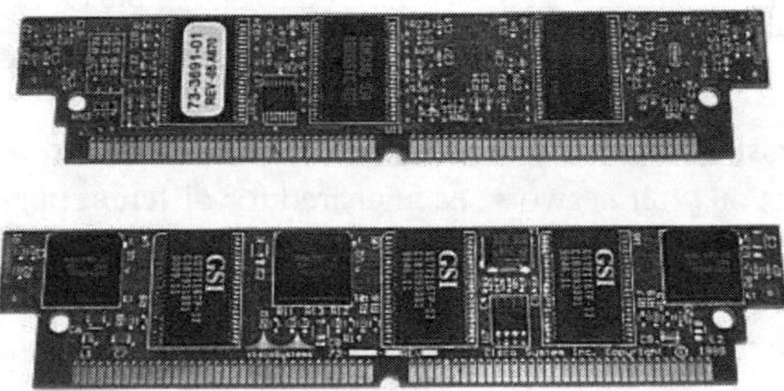

Figure 1-14 *DSP Chip*

Some Cisco routers can also have DSPs embedded on the motherboard or added in riser cards. Above all, it is important for you to add the necessary number of DSPs to your router to support the number of active voice call, conferencing, and transcoding (converting one codec to another) sessions you plan to support.

Tip: Cisco provides a DSP calculator that provides the number of DSP chips you need to purchase based on the voice network you are supporting. This tool can be found at www.cisco.com/web/applicat/dsprecal/index.html (Cisco.com login required). Keep in mind that a growing network will always require more DSP resources. It is usually best to pack the router full with as many DSP resources as you can fit in it; you're going to need them!

You can add DSP chips either directly to a router's motherboard (if the router supports this) or to the network modules you add to the router to support voice cards. Cisco bundles these DSP chips into packet voice DSP modules (PVDM), which resemble the old memory SIMMs (refer to Figure 1-14). Based on the DSP requirements given by the Cisco DSP calculator, you can then purchase one or more of the following PVDMs:

- **PVDM2-8:** Provides .5 DSP chip
- **PVDM2-16:** Provides 1 DSP chip
- **PVDM2-32:** Provides 2 DSP chips
- **PVDM2-48:** Provides 3 DSP chips
- **PVDM2-64:** Provides 4 DSP chips

Alas, not all codecs are created equal. Some codecs consume more DSP resources to pass through the audio conversion process than other codecs consume. Table 1-3 shows the codecs considered medium and high complexity.

Table 1-3 *Medium- and High-Complexity Codecs*

Medium Complexity	High Complexity
G.711 (a-law and µ-law)	G.728
G.726	G.723
G.729a, G.729ab	G.729, G.729b
—	iLBC

Generally speaking, the DSP resources are able to handle roughly double the number of medium-complexity calls per DSP as high-complexity calls.

Note: Newer DSP chips are capable of handling calls more efficiently and can handle more high-complexity calls per chip than older DSP hardware. To find the exact amount of calls per DSP, use the Cisco DSP calculator tool mentioned in the previous tip.

Understanding RTP and RTCP

When you walk into the VoIP world, you encounter a whole new host of protocol standards. Think of the Real-time Transport Protocol (RTP) and Real-time Transport Control Protocol (RTCP) as the protocols of voice. RTP operates at the transport layer of the OSI model on top of UDP. Having two transport layer protocols is odd, but that's exactly what is happening here. UDP provides the services it always does: port numbers (that is, session multiplexing) and header checksums (which ensure that the header information does not become corrupted). RTP adds time stamps and sequence numbers to the header information. This allows the remote device to put the packets back in order when it receives them at the remote end (function of the sequence number) and use a buffer to remove jitter (slight delays) between the packets to give a smooth audio playout (function of the time stamp). Figure 1-15 represents the RTP header information contained in a packet.

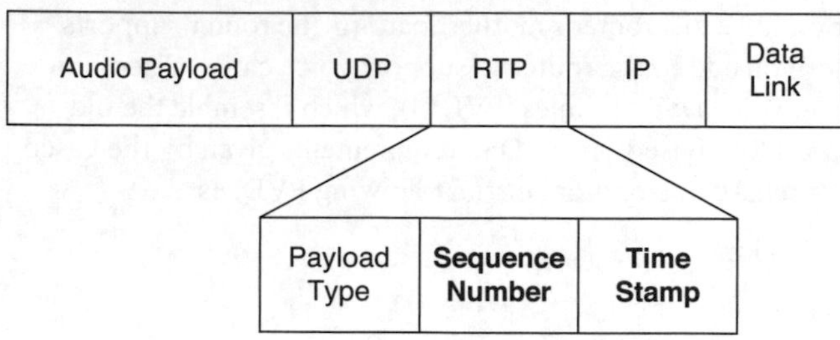

Figure 1-15 *RTP Header Information*

The Payload Type field in the RTP header is used to designate what type of RTP is in use. You can use RTP for audio or video purposes.

Once two devices attempt to establish an audio session, RTP engages and chooses a random, even UDP port number from 16,384 to 32,767 for each RTP stream. Keep in mind that RTP streams are one way. If you are having a two-way conversation, the devices establish dual RTP streams, one in each direction. The audio stream stays on the initially chosen port for the duration of the audio session. (The devices do not dynamically change ports during a phone call.)

At the time the devices establish the call, RTCP also engages. Although this protocol sounds important, its primary job is statistics reporting. It delivers statistics between the two devices participating in the call, which include:

- Packet count
- Packet delay
- Packet loss
- Jitter (delay variations)

Although this information is useful, it is not nearly as critical as the actual RTP audio streams. Keep this in mind when you configure QoS settings.

As the devices establish the call, the RTP audio streams use an even UDP port from 16,384 to 32,767, as previously discussed. RTCP creates a separate session over UDP between the two devices by using an odd-numbered port from the same range. Throughout the call duration, the devices send RTCP packets at least once every 5 seconds. The Cisco Unified Communication Manager Express (CME) router can log and report this information, which allows you to determine the issues that are causing call problems (such as poor audio, call disconnects, and so on) on the network.

Note: RTCP uses the odd-numbered port following the RTP port. For example, if the RTP audio uses port 17,654, the RTCP port for the session will be 17,655.

Exam Preparation Tasks

Review All the Key Topics

Review the most important topics in the chapter, noted with the key topics icon in the outer margin of the page. Table 1-4 lists and describes these key topics and identifies the page numbers on which each is found.

Table 1-4 *Key Topics for Chapter 1*

Key Topic Element	Description	Page Number
Figure 1-3	Illustrates the wired connections to an analog phone	7
List	Two methods used to deliver signaling with digital circuits	11
Figure 1-7	Illustrates TDM	12
Tip	Specific signaling time slot for T1 and E1 circuits using CCS	12
List	Components of the PSTN	13
Table 1-2	Common audio codecs, bandwidth consumption, and MOS rating	21
Note	PVDM ratings	21
Text	RTP concepts and port ranges	24

Definitions of Key Terms

Define the following key terms from this chapter, and check your answers in the Glossary:

analog signal, loop start signaling, ground start signaling, glare, time-division multiplexing (TDM), channel associated signaling (CAS), common channel signaling (CCS), robbed bit signaling (RBS), Q.931, local loop, private branch exchange (PBX), key system, Signaling System 7 (SS7), E.164, quantization, Nyquist theorem, mean opinion score (MOS), G.711, G.726, G.728, G.729, Real-time Transport Protocol (RTP), Real-time Transport Control Protocol (RTCP)

- **Did Someone Say Unified?:** Cisco is moving full-steam ahead with an initiative to unify voice, video, and data networks into a single infrastructure. This section overviews the core four Cisco unified applications to support VoIP networks.
- **Understanding Cisco Unified Communications Manager Express:** Communications Manager Express (CME) breaks the typical "data device" paradigm by allowing you to run an entire VoIP system using a Cisco router. This section discusses the target market segment for CME, key CME features, and communication between CME and Cisco IP Phones.
- **Understanding Cisco Unified Communications Manager:** Cisco Unified Communications Manager (CUCM), Cisco's flagship call management application, is used in thousands of businesses worldwide. This section details key CUCM features, CUCM database architecture, and CUCM-to-Cisco IP Phone interaction.
- **Understanding Cisco Unity Connection:** A telephony network just isn't complete without voicemail! Cisco Unity Connection provides this voicemail functionality along with many other capabilities discussed in this section.
- **Understanding Cisco Unified Presence:** Unified Presence truly blends the voice and data realms by giving visibility to a telephony user's status and supporting enterprise instant messaging applications (such as Cisco Unified Personal Communicator). This section highlights key features and placement of Cisco Unified Presence.

Understanding the Pieces of Cisco Unified Communications

Have you ever heard the saying, "You don't know what you don't know?" With the speed of technology development constantly increasing, the chances of this being true for you increase as well. Unfortunately, that can lead to missed opportunities to implement some really great features (at best) or increase your company's productivity and efficiency (at worst). This chapter resolves that issue (at least, as it relates to the core Cisco Unified products). What are all these servers and software? What do they do? How can they benefit my company? These are all questions we will unpack as you read this chapter.

"Do I Know This Already?" Quiz

The "Do I Know This Already?" quiz allows you to assess whether you should read this entire chapter or simply jump to the "Exam Preparation Tasks" section for review. If you are in doubt, read the entire chapter. Table 2-1 outlines the major headings in this chapter and the corresponding "Do I Know This Already?" quiz questions. You can find the answers in Appendix A, "Answers Appendix."

Table 2-1 *"Do I Know This Already?" Foundation Topics Section-to-Question Mapping*

Foundation Topics Section	Questions Covered in This Section
Understanding Cisco Unified Communications Manager Express	1–5
Understanding Cisco Unified Communications Manager	6–9
Understanding Cisco Unity Connection	10–11
Understanding Cisco Unified Presence	12

1. Which of the following products would support integrated FXS ports?
   a. Cisco Unified Communications Manager Express
   b. Cisco Unified Communications Manager
   c. Cisco Unified Presence
   d. Cisco Unity Connection

2. Which of the following signaling methods can CME use for endpoint control? (Choose two.)

   a. SCCP
   b. SIP
   c. H.323
   d. MGCP

3. Even though Cisco designed CME for small office deployments, you can connect through a SIP trunk to a larger VoIP infrastructure, such as CUCM.

   a. True
   b. False

4. Two users are talking on Cisco IP Phones to each other in the same office. Midway through the call, an administrator reboots the CME router. What happens to the current call?

   a. The call is immediately disconnected.
   b. The call disconnects once the TCP keepalive between the Cisco IP Phones and the CME router fails.
   c. The call remains active; however, no supplemental features (such as hold, transfer, etc...) are available for the remainder of the call.
   d. The call remains active, and the Cisco IP Phones enable SIP Proxy mode for RTP.

5. Which of the following equipment allows the CME router to convert analog audio into VoIP packets?

   a. CPU/Processor
   b. Digital Signal Processor
   c. SIP CODEC conversion
   d. CODEC IOS enablement

6. Each Unity Express module supports a different number of ports. What does the port value indicate?

   a. Number of concurrent calls the module supports
   b. Number of voicemail boxes the module supports
   c. Number of imported mailboxes the module supports
   d. Amount of storage space the module supports

7. Which of the following key features supported by CUCM does CUCM Business Edition *not* support?

   a. SIP/SCCP endpoint control
   b. Appliance-based OS
   c. Gateway control
   d. Redundancy

8. If a CUCM Publisher server fails, which of the following events occur?

a. A CUCM Subscriber takes over Publisher functions until the original Publisher server preempts the role.

b. A CUCM Subscriber takes over Publisher functions permanently; if the original Publisher server returns, it is demoted to a Subscriber role.

c. Administrative access to the CUCM database becomes read-only; user-facing features are writable to the Subscriber.

d. Outside calling is disabled until the CUCM Publisher returns to the network.

9. How many redundant CUCM servers does a Cisco IP Phone support for failover purposes (not including SRST devices)?

a. 2

b. 3

c. 4

d. 6

10. How many call processing servers does Cisco support adding in a CUCM cluster?

a. 2

b. 3

c. 6

d. 8

e. 9

11. When run on a single server, Cisco Unity Connection can support up to 20,000 mailboxes. When combining two Cisco Unity Connection servers in an active/active redundancy pair, how many mailboxes are supported?

a. 20,000

b. 35,000

c. 40,000

d. 45,000

12. Which of the following governs the number of concurrent calls supported by a Cisco Unity Connection server?

a. Trunk Ports

b. Voicemail Ports

c. SIP Proxy Server

d. CUCM SCCP Hunt Group

13. Cisco Unified Presence supplies which of the following capabilities? (Choose two.)

a. Cisco Unified Personal Communicator

b. Enterprise IM Solution

c. Visible status indicators of IP Phone users

d. Multipoint audio and HD video

Foundation Topics

Did Someone Say Unified?

Just glancing through the Cisco product suite for VoIP, it appears that Cisco is trying to say something. The resounding theme again and again is "unified" (with "collaboration" coming in at a close second). When you peel back the glossy marketing surface, you find that there is a lot more to "unified" than just VoIP. This topic crosses boundaries and brings together all communication into one, seamless framework. The interaction we experience today was only seen in the science-fiction movies decades ago: rooms full of corporate strategists interacting with a partner company half way around the world through huge flat-panel monitors surrounding the conference desk. Virtual workgroups comprised of telecommuting individuals sharing whiteboards, documentation, and project plans in real time; a "road warrior" sales manager leading the video-streamed team meeting from a mobile device in his car (while safely pulled over on the side of the road, of course). Yes, we've definitely come a long way from dial-up modems and bulletin board systems (BBS).

The Cisco unified strategy encompasses all electronic communication types: voice, video, and data. Nonetheless, most engineers who hear the phrase "Unified Communications" relate it directly to the Cisco VoIP products, which is what you're here to learn. The Cisco Unified Communications products break down into four core solutions:

- Cisco Unified Communications Manager Express
- Cisco Unified Communications Manager
- Cisco Unity Connection
- Cisco Unified Presence

Keep in mind that these are the "core" solutions, because you can add many additional applications to expand the features and functionalities of the system. For example, the Cisco Unified Contact Center platforms allows you to add call-center capabilities to your network, such as skills-based call routing, call queuing, live monitoring of conversations, and so on. Cisco Unified MeetingPlace adds enhanced conference-call capabilities, document collaboration, and training platforms. The list goes on and on; it's all impressive technology, but not core to the system.

Our goal in the rest of this book is to give you the information you need to become proficient using each core Unified Communications platforms. Keep in mind that "proficient" does not equate to "expert level." Just as the Cisco CCNA certification was created to lay the foundation, technology needed to manage day-to-day operations in a Cisco-based organization (and hopefully lead you into the CCNP certification), the CCNA Voice certification will also lay this same foundation as it relates to VoIP (and hopefully lead you into CCNP Voice certification). As you go through this book, you will be amazed at just how much technology you learn, and you'll be amazed when you see just how much further the technology goes.

Well, let's start off by discussing one of my (Jeremy's) favorite Unified Communications platforms: Cisco Unified Communication Manager Express (CME).

Note: Years ago, Cisco Unified Communications Manager (CUCM) and Cisco Unified Communications Manager Express (CME) used to be called Cisco CallManager (CCM) and Cisco CallManager Express (CME). Years later, the CallManager name lives on because it's just simpler to say. So, if you ever hear someone say, "Let's get CallManager set up!," it probably wouldn't be the best thing for you to say, "Did you know Cisco renamed that to Cisco Unified Communications Manager?"

Understanding Cisco Unified Communications Manager Express

The amount a Cisco router is able to accomplish is simply amazing: routing tables, routing protocols, security with access lists, Network Address Translation (NAT)...the list goes on and on. But, what if someone told you that the Cisco router you've been using for years could also support your IP telephony network? That was the goal with the Cisco Unified Communications Manager Express (CME) platform. Your Cisco Integrated Service Router (ISR) platform not only routes data, but terminates analog and digital voice circuits (such as FXO, FXS, and T1 ports, which are discussed later), supports VoIP endpoints (such as Cisco IP Phones), and even handles the more advanced features, such as conference calling, video telephony, and automatic call distribution (ACD). Depending on the platform you use, CME can scale to support up to 450 IP Phones, which makes it a good solution for small and even some midsize businesses.

Although Cisco designed CME 8.X to operate on the ISR Generation 2 (G2) platforms, you can also run it on the original ISR platforms (such as the 1800, 2800, or 3800 series), provided you have the proper IOS upgrade. Cisco integrates CME into the IOS software itself, based on the feature set you download. Since the release of IOS 15, a Product Authorization Key (PAK) is needed to activate the CME feature set. Table 2-2 shows the current ISR G2 platforms available and the maximum number of phones supported on each platform.

Table 2-2 *Unified CME Supported ISR G2 Platforms*

Platform	Maximum Phones
Cisco 2901 ISR G2	35
Cisco 2911 ISR G2	50
Cisco 2921 ISR G2	100
Cisco 2951 ISR G2	150
Cisco 3925 ISR G2	250
Cisco 3945 ISR G2	350
Cisco 3925E ISR G2	400
Cisco 3945E ISR G2	450

Keep in mind that these figures are current at the time of writing, but they change all the time as hardware platforms upgrade and become more efficient.

CME Key Features

Because CME runs on a Cisco router, it has the unique advantage of acting as an all-in-one device for controlling Cisco IP Phones and trunking to the public switched telephone network (PSTN) through various connections. The following are the key features supported by CME:

- **Call processing and device control:** As mentioned previously, CME acts as the all-in-one call control device. It handles the signaling to the endpoints, call routing, call termination, and call features.
- **Command-line or GUI-based configuration:** Because Cisco integrated CME directly into the IOS, you have the full flexibility of command-line configuration. You can also use a GUI utility, such as the Cisco Configuration Professional (CCP), to interface with a point-and-click style of configuration.
- **Local directory service:** The CME router can house a local database of users you can use for authentication in the IP Telephony (IPT) network.
- **Computer Telephony Integration (CTI) support:** CTI allows the IPT network to integrate with the applications running on the data network. For example, you could use the Cisco Unified CallConnector to make calls directly from your Microsoft Outlook contact list.
- **Trunking to other VoIP systems:** Although CME can run as a standalone deployment interfacing directly with the PSTN, it can also integrate with other VoIP deployments. For example, you could use CME for a small, 40-user office and have it connect directly over your data network to the corporate headquarters supported by a full Cisco Unified Communications Manager (CUCM) cluster of servers.
- **Direct integration with Cisco Unity Express (CUE):** CUE, which runs through a module installed in a Cisco router, can provide voicemail services to the IP Phones supported by CME.

Although it's not a key feature of CME, keep in mind that by using a CME platform, you have an amazing opportunity to unify the support staff on your network. Rather than having one technician who focuses on maintaining the Cisco network and another technician who focuses on maintaining the phone system, you now have one individual capable of handling everything.

CME Interaction with Cisco IP Phones

Although there is much to be said about the process a Cisco IP Phone uses to contact and register with a CME router (which is discussed in Chapter 3, "Understanding the Cisco IP Phone Concepts and Registration"), let's focus on the roll of CME after a phone has completed registration. This gives you an idea of how the CME router interacts with the Cisco IP Phones and route calls across the data network. This discussion also lays the foundation for how the other call-management applications support the IPT network. First, let's follow the VoIP flow shown in Figure 2-1.

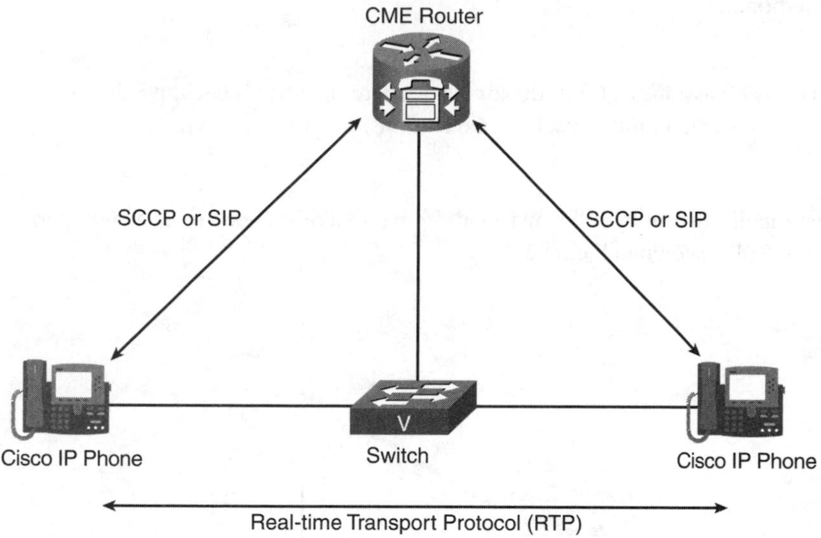

Figure 2-1 *CME Call Flow for On-Network Cisco IP Phones*

Essentially, the relationship between CME and the Cisco IP Phones is similar to the relationship between a mainframe and dumb terminals. CME controls virtually every action performed at the Cisco IP Phones. For example, if a user picks up the handset, an off-hook state is sent from the Cisco IP Phone to the CME router using either the Skinny Client Control Protocol (SCCP) or the Session Initiation Protocol (SIP). We discuss the differences between these protocols in Chapter 3, but in a nutshell, SCCP and SIP are both signaling protocols that allow the call-management platform (CME, in this case) to communicate with and control an IP Phone. As the user begins to dial digits, each digit is sent to the CME router (again, via SCCP or SIP). After the user completely dials the phone number of the other Cisco IP Phone shown in Figure 2-1, CME sends some signaling messages causing the phone to ring. After the user answers the ringing phone, CME connects the IP Phones directly and steps out of the communication stream. The phones now communicate directly using the Real-time Transport Protocol (RTP), which handles the actual audio stream between the devices.

The fact that CME steps out of the middle of the RTP stream and allows the IP phones to communicate directly is fantastic because of two primary reasons. First, it eliminates the CME router as a point of failure. After CME establishes the RTP stream between the IP phones, it can crash, reboot, or catch fire, and the conversation between the two endpoints continues unhindered (provided the fire did not also burn up the switch). The other benefit is that the CME router does not become a bottleneck for the RTP stream. If the links to the CME router became saturated or the router ran out of resources, RTP packets can drop, causing the call quality to degrade. Keep in mind that we're only talking about the RTP stream, which contains the audio of the call. All the phone features (such as hold, transfer, conference, and so on) are still managed using SCCP or SIP, so those would not be available until the CME router came back online. Likewise, after the users disconnect from the call, the phone would not be available to place or receive any calls until the CME router returned online.

> **Note:** This discussion assumes a CME deployment where there is no backup call-management device. This is common for smaller organizations.

Let's broaden the discussion by following a call from a Cisco IP Phone to an analog phone attached to the PSTN shown in Figure 2-2.

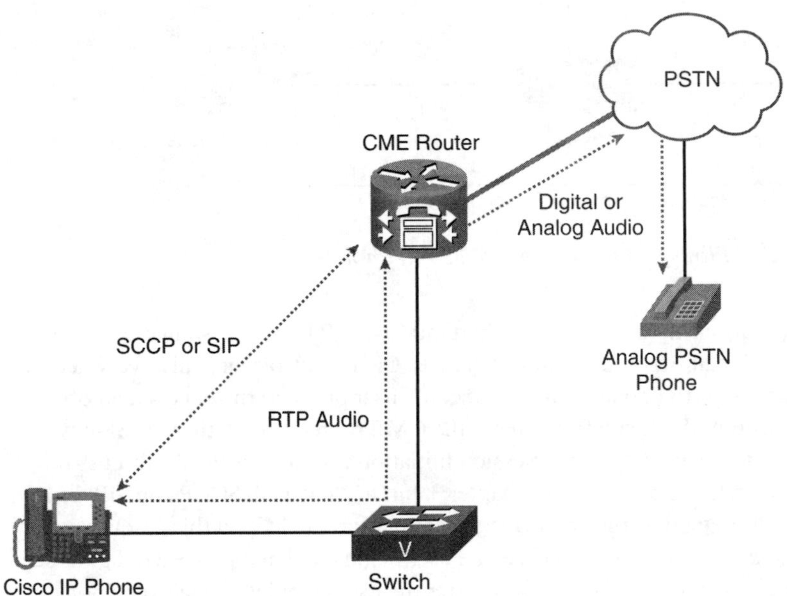

Figure 2-2 *CME Call Flow for Calls to the PSTN*

As the user of the Cisco IP Phone picks up the phone to place the call, all the communication is handled by SCCP or SIP. After the user finishes dialing the phone number of the PSTN phone, the CME router realizes (due to its dial-plan) that the call needs to exit out a PSTN-connected interface. CME now assumes the role of voice gateway and signals to the PSTN to establish the call on behalf of the Cisco IP Phone. Keep in mind that this is the "old school" signaling discussed in Chapter 1 since the CME router is attached to the PSTN using a digital (T1/E1) or analog (FXO) trunk. Once the audio for the call connects, the CME router assumes the role of converting between VoIP audio and PSTN audio. Because this conversion is processor intensive, the CME router is equipped with Digital Signal Processors (DSP), which are simply additional "mini processors" dedicated to voice functions.

> **Note:** Many VoIP deployments now connect to the PSTN using an Internet Telephony Service Provider (ITSP) rather than a traditional TSP. In this case, the CME router relays the voice call over a SIP trunk rather than an analog or digital circuit.

The CME router has now established two different "legs" of the call: one to the PSTN and the other to the Cisco IP Phone. It stands in the middle, independently handling signaling from both sides in two different formats. Unlike the first example, the CME router is now a critical piece of the ongoing RTP audio flow. If it were to fail in the middle of the call, the audio for the call would also fail.

A Match Made in Heaven: CME and CUE

CME is a fantastic, all-in-one call processing device, provided someone is there to answer the phone. However, just like the full-blown CUCM, there are no built-in voicemail capabilities. All that changes when CUE steps onto the stage. CUE is a voicemail system that you can install into your CME router in one of two form factors: Internal Services Module (ISM) or Service Module (SM). The ISM form factor installs internal to the CME router and uses solely flash memory for storage. The full SM form factor installs externally to the router and uses a hard disk for storage. Because of this, it can hold roughly ten times as much voicemail as the ISM form factor.

Cisco designed ISM and SM for the ISR G2 series of routers (1900, 2900, and 3900). They are upgrades from the previous Advanced Integration Modules (AIM) and Network Modules (NM), shown in Figure 2-3 and Figure 2-4, which were used in the first-generation ISR routers (1800, 2800, and 3800).

Figure 2-3 *CUE AIM*

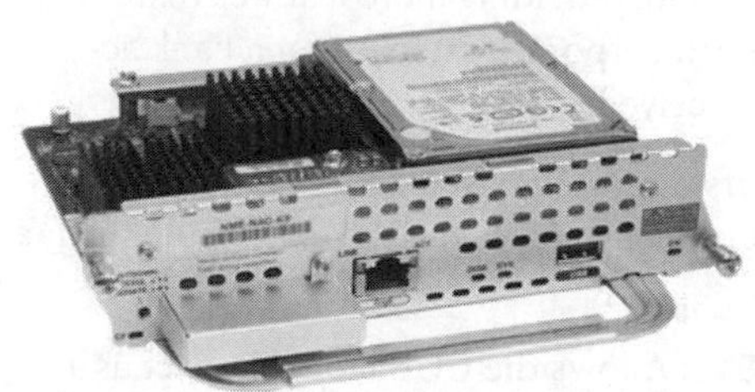

Figure 2-4 *CUE NM*

Each module runs its own Linux-based operating system (OS) that you can access and manage from the Cisco IOS of the CME router. Table 2-3 shows the current hardware limitations of each module type.

Table 2-3 *Cisco Unity Express Modules*

Platform	Hardware	Included Ports	Maximum Ports	Maximum Mailboxes	Storage (Hours)
Cisco 2800/3800 (ISR Generation 1)	AIM2-CUE-K9	6	6	65	14
	NME-CUE	8	24	275	300
Cisco 2900/3900 (ISR Generation 2)	ISM-SRE-300-K9	2	10	100	60
	SM-SRE-700-K9	4	32	300	600

The Included Ports and Maximum Ports shown in Table 2-3 represent how many concurrent sessions the CUE module can handle (essentially, how many people can retrieve or leave voicemail at the same time). You can increase the ports by purchasing licenses.

Working with the CUE module is a little bizarre at first. Although you install the CUE module into the CME router, it still runs its own, independent OS. After you access it from the IOS command line, you will find yourself in a completely new OS that looks and feels like the IOS but has a completely different set of commands. Most administrators use the CUE command-line access just long enough to get the CUE web-based GUI running and then handle all administration from there.

Despite the small size of CUE, it has some powerful features to support your IPT network:

- **Voicemail:** The core feature provided by CUE. The type of CUE module you choose sets the maximum mailboxes and storage space used for this feature (refer to Table 2-2).
- **Auto-attendant:** The automated voice everyone loves to hear! An auto-attendant might replace an organization's operator and allows dial by name, dial by extension, and some basic menu capabilities.
- **Interactive voice response (IVR) system:** CUE includes basic IVR capabilities that allow callers to move through a menu system, providing more features than the auto-attendant. For example, a user can query a corporate database for a bank account balance or the date of a last payment received.
- **Native T.37 fax processing:** CME supports the T.37 fax standard that allows it to receive and process faxes as TIFF e-mail attachments. CUE can then distribute the fax to a user's mailbox.
- **Survivable Remote Site Voicemail (SRSV):** Allows the CUE module to act as a backup to a primary voicemail server (perhaps running the full Unity Connection) if the IP phones are not able to reach the primary voicemail server. This feature

integrates with Survivable Remote Site Telephony (SRST), which allows the CME router to act as a backup to a primary call-management system (perhaps CUCM).

- **Standards-based:** All signaling between the CME router and CUE module use the SIP protocol, which is becoming the universal standard for voice signaling.

Understanding Cisco Unified Communications Manager

When Cisco first released CallManager 2.4, it ran as an application on the Microsoft Windows NT 4.0 OS supported by the Internet Information Server (IIS) web server. You could even install it on a laptop to give yourself a battery backup power supply in case of a power outage.

Note: Cisco does not recommend installing CallManager on a laptop to give yourself a battery backup power supply.

Needless to say, Cisco has come a long way as CUCM moves through the 8.X versions and beyond. It now runs as a Linux appliance on certified hardware platforms and acts as the powerful call processing component of the Cisco Unified Communications solution. Think of CUCM as the "director" behind any large organization's Cisco IPT solution. It provides the core device control, call routing, permissions, features, and connectivity to outside applications. The scope of control handled by CUCM is so large, Cisco created two certification exams for it in the CCNP Voice certification track (CIPT1 and CIPT2). In a nutshell, the importance of CUCM to Cisco VoIP is monumental. With that being said, don't let the massive size of CUCM to overwhelm you. Cisco has done a fantastic job at allowing you to manage nearly every CUCM option through a crisp, well-designed web-based GUI.

CUCM Key Features

Although CUCM supports numerous capabilities, here are some of the most important:

- **Full support for audio and video telephony:** The core feature provided by CUCM. In the same way CME acts as the director of a small organization, CUCM supports audio and video calls for midsize to enterprise class corporations.
- **Appliance-based operation:** Modern CUCM versions run as an appliance, which means the underlying operating system is hardened (secured) and inaccessible.
- **Redundant server cluster:** As the saying goes, "One is none, two is one." CUCM supports redundant servers configured in a cluster relationship. The clustering capabilities replicate both database information (containing static data such as directory numbers and route plans) and real-time information (containing dynamic data, such as active calls). CUCM clusters can scale to 30,000 IP phones (SCCP or SIP in unsecure mode) or 27,000 IP phones (SCCP or SIP in secure mode).
- **Intercluster and voice gateway control and communication:** Even though a CUCM cluster has a limit of 30,000 IP phones, you can create as many clusters as you like (with up to 30,000 IP phones each) and connect them together using intercluster trunk connections. In addition to using Intercluster trunk links to call outside of your

own cluster, CUCM can also connect to voice gateways (such as a Cisco router), which can connect to various other voice networks (such as the PSTN or legacy PBX systems).

- **Built-in Disaster Recovery System (DRS):** As a built-in feature, the CUCM DRS service allows you to back up the CUCM database (and any additional files you'd prefer) to a network device or over Secure FTP (SFTP).
- **VMWare Virtualization Support:** If you love VMWare, you'll love the fact that CUCM 8.0 is supported in a VMWare ESXi environment. This brings all the high availability and scalability benefits of virtualization to your CUCM deployment.
- **Directory service support or integration:** VoIP networks can use network user accounts for a variety of purposes (phone control, attendant console control, and so on). CUCM has the capability to be its own directory server to hold user accounts or it can integrate into an existing corporate directory structure (such as Microsoft Active Directory) and pull user account information from there.

Note: Cisco does produce a single-server CUCM solution known as CUCM Business Edition. This single server provides CUCM, Unified Mobility, and Unity Connection for up to 500 IP Phones at an aggressive price. The catch? CUCM Business Edition does not support clustering (server redundancy).

CUCM Database Replication and Interacting with Cisco IP Phones

Even though CUCM can scale to a massive size, it interacts with Cisco IP Phones in a similar manner to CME...just bigger! The first thing you'll want to catch is that CUCM works in a cluster relationship. When most people think of a server cluster, they think of multiple servers with mirrored configurations that assume the identity of each other in the case of failure. That is *not* the functionality of a CUCM server cluster. Instead, think of a CUCM server cluster as multiple, individual servers with their own, unique configuration working for the common good of the IP Phones. The CUCM cluster relationship includes two types of communication:

- **CUCM Database Relationship:** The CUCM IBM Informix database includes all the "static" data of the cluster (directory numbers, route plan, calling permissions, and so on). This data replicates to all the servers in the cluster.
- **CUCM Runtime Data:** Just like the name sounds, the runtime data encompasses anything that happens in "real time" in the CUCM cluster. For example, when a device registers with a CUCM server, it communicates to all the other servers that it now "owns" that IP phone. CUCM uses a method designed specifically by Cisco for this type of communication: Intracluster Communication Signaling (ICCS).

Although the CUCM runtime data is important for functionality, there's simply not too much to it. All the servers in the CUCM cluster form TCP sessions to each other for ICCS communication (TCP ports 8002 to 8004). Then, anytime something of interest happens (such as a phone registering, a call initiation, a call disconnect, and so on), the servers inform each other of the event. All this ICCS communication takes virtually no additional configuration on your part (other than adding the CUCM server to the cluster).

On the other hand, the CUCM database relationship has a little more substance to it. It is not any more difficult to manage than ICCS (it just works behind the scenes), but it does have some key concepts you'll want to know. First off, the CUCM IBM Informix database uses a one-way replication method where a server known as the Publisher holds the master copy of the database. All changes to the database (with some minor exceptions) occur on the Publisher server and replicate to the Subscribers. Figure 2-5 illustrates this relationship.

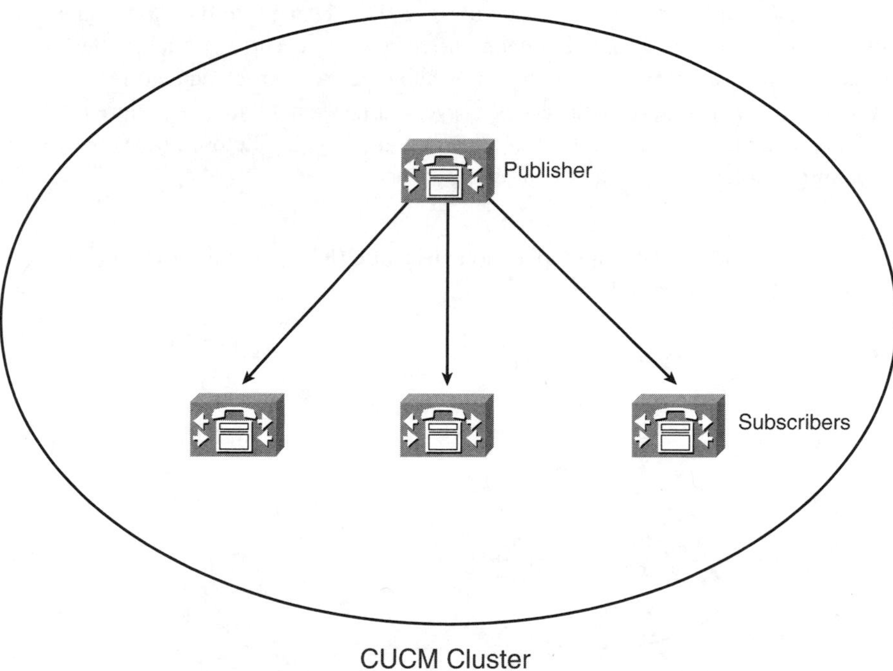

Figure 2-5 *CUCM Database Relationship*

Each CUCM cluster supports a single Publisher and up to eight Subscribers. Think of the Subscribers as the "workhorses" of the IPT network. These servers are providing dial tone, receiving digits, routing calls, and streaming music on hold. However, the Publisher typically performs only two primary functions: It maintains the database and serves TFTP requests. Because the Publisher serves such a critical role in maintaining the only writable copy of the CUCM database, it is usually kept from all the heavier work. The TFTP requests come from the Cisco IP Phones during their boot process (covered in Chapter 3).

Tip: In smaller environments (500 IP phones or less), it's perfectly fine to use the Publisher for call processing and database management. However, once you exceed that number, it's generally a best practice to pull the Publisher out of call processing and leave that work to the Subscribers. Likewise, once you exceed 1,250 users, Cisco recommends moving the TFTP server role to a dedicated server.

Note: I'm sure at this point, many of you have a look of concern because of this question: "If the CUCM Publisher has the only writable copy of the database, what happens if it fails?" If the Publisher fails, the CUCM cluster goes into a type of "locked configuration" mode. You can no longer make changes to the database (such as adding a new IP phone, changing the route plan, modifying a music on hold selection, and so on). The only exception to this is the user-facing features. These include functions such as forwarding your phone, enabling the message waiting light, pressing the Do Not Disturb button, and many others. The CUCM Subscribers are able to write these changes to their local database, replicate them to the other Subscribers in the cluster, and eventually back to the Publisher when you have restored connectivity. By allowing the user-facing features to write to the Subscriber database, the users will never know when a Publisher failure has occurred! This capability emerged in CUCM version 5 and beyond. Before this version, a Publisher failure absolutely impacted the user experience.

Once you understand the database replication piece of CUCM, the call flow is nearly identical to CME, as shown in Figure 2-6.

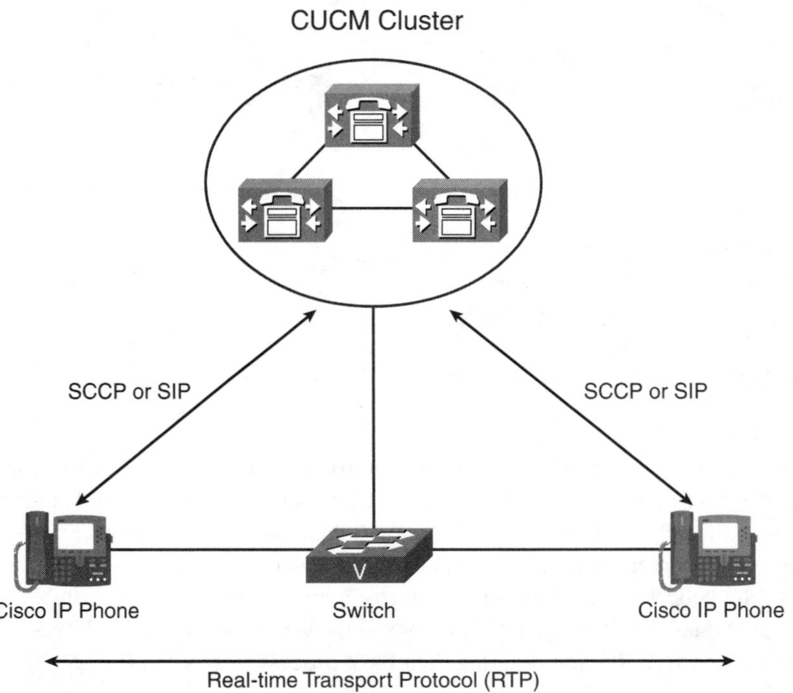

Figure 2-6 *CUCM Call Processing*

As with CME, the primary CUCM server used by the Cisco IP Phone receives the SCCP or SIP off-hook message and responds appropriately in a stimulus/response fashion until the call is placed. One difference is that the IP phones can now use redundant servers (as opposed to the single CME router). Based on your configuration, a phone can use a list of up to three redundant CUCM servers for failover purposes. If the primary server is down, it uses the secondary. If the secondary is down, it uses the tertiary. Because your CUCM cluster can support up to nine call processing servers (one Publisher, up to eight Subscribers), you can manually load balance the IP phones by assigning different primary, secondary, and tertiary CUCM servers to different groups of phones.

Note: A common mistake when first learning CUCM functions is to confuse the primary server with the Publisher database role. Remember, the main goal of the Publisher is to maintain the writable copy of the database. It will most likely *not* be the primary server of an IP phone (except in very small environments). One of the Subscribers will be the IP phone's primary server.

Understanding Cisco Unity Connection

Long before the Cisco "unified" product campaign, there was the Cisco Unity product. Although most people identify Cisco Unity as "the voicemail solution" for your VoIP network, Cisco designed it to be much more. The term "Unity" related to messages: voice messages for sure, but also e-mail messages, fax messages, instant messages...you get the idea. The goal of Unity was to make any message retrievable from any voice-enabled device or application. For example, a caller could leave a voicemail that you could retrieve from an e-mail client inbox. You could listen to your e-mails from a mobile device or have them faxed to an offsite fax machine. Essentially, regardless of how a message was left, you could retrieve it using a variety of clients.

Just as the original Cisco CallManager software, the original Cisco Unity ran on a Microsoft Windows OS and used the Microsoft Exchange e-mail server solution as a message store for voicemail. Cisco introduced Unity Connection as a smaller alternative using the same appliance-based model as CUCM years after the original Cisco Unity release. As time passed, support for Cisco Unity Connection grew so much that it surpassed the original Windows-based Unity platform and is now the more popular and scalable solution. Table 2-4 depicts the various voicemail solutions you can use for your IPT network and their scalability limits.

Table 2-4 *Cisco Voice Messaging Systems Comparison*

Platform	Maximum Mailboxes	Platform	Redundancy
Cisco Unity Express	300	Router	Not supported
CUCM Business Edition	500	Appliance	Not supported
Cisco Unity	15,000 per server	Windows Server	Active/Passive
Cisco Unity Connection	20,000 per server	Appliance	Active/Active

In addition to supporting more mailboxes, Cisco Unity Connection now supports features (such as personal call transfer rules and speech recognition) that are not available in the other Cisco voice-messaging products.

Note: The Redundancy column shown in Table 2-4 shows the type of failover supported by the voicemail solutions. CUE and CUCM Business Edition do not support any failover. The original Cisco Unity supported Active/Passive, which meant a backup server would sit idle until the primary server failed. Cisco Unity Connection supports Active/Active, allowing the redundant servers to load balance the mailboxes.

Cisco Unity Connection Key Features

The following are some of the notable features of Cisco Unity Connection:

- **Proven appliance-based platform:** Cisco Unity Connection is built on top of the same stable, hardened, appliance-based operating system as CUCM. (These two software products even use the same installation DVD.)
- **Up to 20,000 mailboxes per server:** Cisco Unity Connection scales to a massive size per server. Even though Unity Connection supports a single-server configuration, most organizations will opt for a high availability pair of servers.
- **Access voicemails from anywhere:** Carrying on the original dream of Cisco Unity, Cisco Unity Connection allows voicemail retrieval from phone, e-mail, web browser, mobile devices, and instant-messenger platforms.
- **LDAP directory server integration:** Similar to CUCM, Cisco Unity Connection can integrate with an existing corporate directory (such as Microsoft Active Directory) to avoid creating a duplicate user database.
- **Microsoft Exchange support:** Cisco Unity Connection can integrate with an existing Microsoft Exchange deployment to enable fantastic features, such as different call treatment based on your Exchange calendar, e-mail text-to-speech (hear your emails read to you from a phone), manage Exchange calendar (accept, decline, cancel, and so on) from a phone, and so on.
- **Voice Profile for Internet Mail (VPIM) support:** VPIM is a standard allowing voicemail servers to integrate together to exchanging voicemails (and other messaging).
- **Active/active high availability:** Cisco Unity Connection uses a Publisher/Subscriber IBM Informix database scheme just like CUCM between a pair of servers. The pair of servers can support up to 20,000 mailboxes in a redundant fashion. Both servers can accept client requests (giving it the active/active redundancy). Typically, the largest Cisco Unity Connection server can support up to 250 voicemail ports (essentially allowing 250 people to check their voicemail at a time). By creating a high-availability pair, you can now support 500 voicemail ports.

Note: If one of the servers in a Unity Connection high-availability pair fails, all 20,000 mailboxes are still available, but the number of voicemail ports is reduced to the maximum supported by the single server.

Cisco Unity Connection and CUCM Interaction

Outside of CUCM Business Edition, where CUCM and Unity Connection reside on a single server, Cisco Unity Connection operates independently from CUCM. As a matter of fact, you can run Cisco Unity Connection as a voicemail server for other non-Cisco VoIP deployments or even PBX systems; it is not tied to only the CUCM product. Because of this, there is not a simple "click this button in CUCM to magically integrate with Cisco Unity Connection." Instead, you set up Cisco Unity Connection as an outside system that CUCM can communicate with using the SCCP or SIP signaling protocols. Figure 2-7 illustrates this connection.

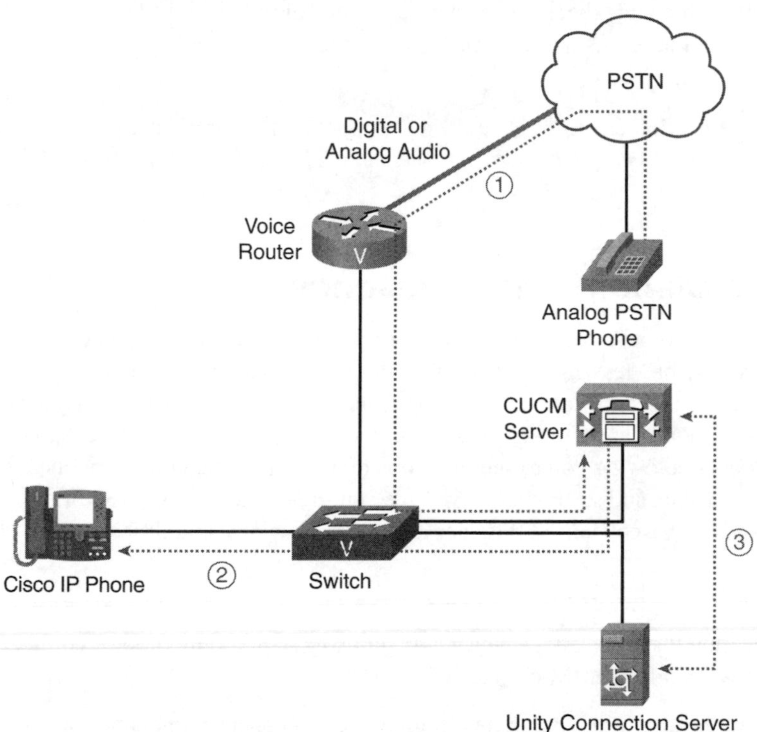

Figure 2-7 *CUCM and Cisco Unity Connection Communication*

Although you can dive into much detail regarding the communication between CUCM and Unity Connection, let's start by breaking down the high-level diagram shown in Figure 2-7:

1. An incoming call from the PSTN arrives at the voice router. The dial-plan on the voice router causes it to route the incoming call to the CUCM server.
2. CUCM receives the call and directs it to the appropriate IP phone (using SCCP or SIP). If someone does not answer the IP phone or diverts the call to voicemail, CUCM forwards the call to the preconfigured extension that reaches the Unity Connection server.

3. CUCM transfers the call (once again using SCCP or SIP) to the Unity Connection server. The extension of the originally called phone is contained in the signaling messages, which allows Unity Connection to send the call to the correct voicemail box.

After the caller leaves a message on the voicemail server, Cisco Unity Connection then "makes a call" (via SCCP or SIP) back to a Message Waiting Indicator (MWI) extension on the CUCM server. CUCM then lights the voicemail indicator on the Cisco IP Phone, alerting the user that they have a voice message waiting. All this interaction between CUCM and Unity Connection is done using voicemail ports, which are licensed features. The more voicemail port licenses you purchase for the Unity Connection server, the more concurrent communication it supports. Be sure to purchase enough licenses! You should consider calls to the auto-attendant, checking voicemail, leaving voicemail, and MWI communication in the number of supported voicemail ports.

Note: Cisco Unity Connection can fully integrate and support CME deployments. Many organizations use a centralized Unity Connection voicemail cluster to support many CME-based remote offices.

Understanding Cisco Unified Presence

Cisco Unified Presence promotes an awareness of the VoIP and data networks. Many folks commonly use Instant Messenger (IM) clients to communicate. In these applications, you are able to see the status of a user, gauging whether they are available, busy, or offline before you begin to chat with them. Cisco Unified Presence stretches this capability to the voice network, allowing you to see the status of a user (are they on the phone, off the phone, not available, and so on) before you pick up the phone to dial. In addition to this core functionality, Cisco Unified Presence adds the following capabilities to your voice network:

- **Enterprise instant messaging:** Unified Presence incorporates the Jabber Extensible Communication Platform (XCP), which is an industry standard method of communicating between different IM clients.
- **Message compliance:** Many industries require strict compliance guidelines on instant messenger communication. Cisco Presence supports logging functionality for all types of IM communication (even conversations encrypted with Transport Layer Security [TLS]).
- **Interdomain federation:** Unfortunately, this feature has nothing to do with Star Trek; however, it has everything to do with connectivity. Using Interdomain federation connections from Unified Presence, you can connect your organization to other domains, such as Google Talk or WebEx Connect, thus giving you worldwide reach.
- **Jabber XCP extensibility:** XCP allows you to extend Unified Presence into virtually any area of the data or voice network. XCP can allow features such as peer-to-peer file sharing, application sharing, video-conference systems, and so on. XCP integrates with nearly any infrastructure, such as directory services, databases, and web portals.

- **Secure messaging:** Applications integrating into Unified Presence can use IPSec or TLS standards to encrypt and secure all communication.

Cisco Unified Personal Communicator

You'll hardly find any documentation available about Cisco Unified Presence without seeing a mention of Cisco Unified Personal Communicator. Once you understand the purpose of Personal Communicator, you'll understand why. This single-software application brings together frequently used services in a single location: soft phone, presence, instant messaging, visual voicemail, employee directory, communication history, video, and web conferencing. When you initially see Cisco Unified Personal Communicator, it looks like another IM client (shown in Figure 2-8); under the hood, it's so much more.

Original source: http://www.cisco.com/en/US/prod/collateral/voicesw/ps6789/ps6836/ps6844/data_sheet_c78-609335.html

Figure 2-8 *Cisco Unified Personal Communicator*

At its core, Personal Communicator serves as an IM client, supporting peer-to-peer chat, multiuser chat, and persistent chat. Persistent chat uses the idea of "rooms," which allow users to join existing chat sessions and see the conversation history (rather than simply the new messages appearing after they have joined). Personal Communicator uses LDAP authentication to log users into the IM client, for user searches, and to add contacts. Because Personal Communicator uses the Cisco Unified Presence server, you are able to see not only the status of a user as it relates to IM conversations, but you can also see the status of their phone (off hook, on hook, and so on).

From Personal Communicator, you can start both voice and video calls with video resolutions able to reach High Definition (HD) quality. Personal Communicator also acts as a full softphone, allowing audio only calls from other audio devices (IP phones, softphones, and so on) in the network. If you combine Personal Computer with one of the multipoint video conference solutions offered by Cisco (such as Cisco Unified MeetingPlace or Cisco Unified Video Conferencing), you are even able to run video-based conference calls with multiple video sources!

Exam Preparation Tasks

Review All the Key Topics

Review the most important topics in the chapter, noted with the key topics icon in the outer margin of the page. Table 2-5 lists and describes these key topics and identifies the page numbers on which each is found.

Table 2-5 *Key Topics for Chapter 2*

Key Topic Element	Description	Page Number
Text	CME interaction with IP phones; functions of SCCP and SIP	33
List	Key features of CUCM	37
Text	Understanding the CUCM Publisher and Subscriber roles and relationships	39
Note	Understanding the difference between a primary CUCM server and the Publisher server	41
List	Method used to route incoming calls to Cisco Unity Connection server from CUCM	43

Definitions of Key Terms

Define the following key terms from this chapter, and check your answers in the Glossary:

Cisco Unified Communications Manager (CUCM), Cisco Unified Communications Manager Express (CME), Cisco Unity Connection, Cisco Unified Presence, automatic call distribution (ACD), Skinny Client Control Protocol (SCCP), Session Initiation Protocol (SIP), Real-Time Transport Protocol (RTP), Internet Telephony Service Provider (ITSP), digital signal processors (DSP), Cisco Unity Express (CUE), interactive voice response (IVR), CUCM Publisher, CUCM Subscriber, CUCM Business Edition, Intracluster Communication Signaling (ICCS), Voice Profile for Internet Mail (VPIM), Cisco Unified Personal Communicator

This chapter includes the following topics:

- **Connecting and Powering Cisco IP Phones:** To provide a centralized power system, the Cisco IP Phones must receive their power from a centralized source using PoE. This section discusses the different options for PoE and the selection criterion of each.
- **VLAN Concepts and Configuration:** VLANs allow you to break the switched network into logical pieces to provide management and security boundaries between the voice and data network. This section discusses the concepts and configuration behind VLAN.
- **Understanding Cisco IP Phone Boot Process:** This section discusses the foundations of the Cisco IP Phone boot process. Understanding this process is critical to troubleshooting issues with the IP Telephony system.
- **Configuring a Router-Based DHCP Server:** This section discusses configuring a Cisco router as a DHCP server for your network.
- **Setting the Clock of a Cisco Device with NTP:** Because a VoIP network heavily depends on accurate time, the sole focus of this section is keeping the clocks accurate on Cisco devices by using NTP.
- **IP Phone Registration:** Once the Cisco IP Phone receives all its network configuration settings, it is ready to speak to a call processing agent. This section describes the process and protocols that make it happen.

CHAPTER 3

Understanding the Cisco IP Phone Concepts and Registration

You walk into the new corporate headquarters for Fizzmo Corp. On the top of each desk is a Cisco 7945G IP Phone, glowing with a full-color display and two line instances. Smiling, courteous agents are busy taking phone calls from callers excited to purchase the latest Fizzmo wares. Samantha (located in the north corner) is checking her visual voicemail, while Emilio (located in the south hall) is getting the latest weather report through an XML IP phone service.

How did we get here? How do you take a newly constructed building and transform it into a bustling call center? That's what this chapter is all about. We walk through the key concepts and technologies used to build a Cisco VoIP network. By the time you are done with this chapter, you will have all the conceptual knowledge you need to have in place before you can move into the installation and configuration of the Cisco VoIP system.

"Do I Know This Already?" Quiz

The "Do I Know This Already?" quiz allows you to assess whether you should read this entire chapter or simply jump to the "Exam Preparation Tasks" section for review. If you are in doubt, read the entire chapter. Table 3-1 outlines the major headings in this chapter and the corresponding "Do I Know This Already?" quiz questions. You can find the answers in Appendix A, "Answers Appendix."

Table 3-1 *"Do I Know This Already?" Foundation Topics Section-to-Question Mapping*

Foundation Topics Section	Questions Covered in This Section
Connecting and Powering Cisco IP Phones	1–2
VLAN Concepts and Configuration	3–8
Understanding Cisco IP Phone Boot Process	9
Configuring a Router-Based DHCP Server	10
Setting the Clock of a Cisco Device with NTP	11
IP Phone Registration	12

1. Which of the following is an industry standard used for powering devices using an Ethernet cable?
   - **a.** Cisco Inline Power
   - **b.** 802.1Q
   - **c.** 802.3af
   - **d.** Local power brick
2. Which of the following are valid methods for powering a Cisco IP Phone? (Select all that apply.)
   - **a.** Power brick
   - **b.** Crossover coupler
   - **c.** PoE
   - **d.** Using pins 1, 2, 3, and 4
3. Which of the following terms are synonymous with a VLAN? (Choose two.)
   - **a.** IP subnet
   - **b.** Port security
   - **c.** Broadcast domain
   - **d.** Collision domain
4. Which of the following trunking protocols would be used to connect a Cisco switch to a non-Cisco switch device?
   - **a.** VTP
   - **b.** 802.3af
   - **c.** 802.1Q
   - **d.** ISL
5. How should you configure a port supporting voice and data VLANs that is connected to a Cisco IP Phone?
   - **a.** Access
   - **b.** Trunk
   - **c.** Dynamic
   - **d.** Dynamic Desired
6. How does a device attached to a Cisco IP Phone send data to the switch?
   - **a.** As tagged (using the voice VLAN)
   - **b.** As untagged
   - **c.** As tagged (using the data VLAN)
   - **d.** As tagged (using the CoS value)

7. Which of the following commands should you use to configure a port for a voice VLAN 12?

   a. switchport mode voice vlan 12
   b. switchport trunk voice vlan 12
   c. switchport voice vlan 12
   d. switchport vlan 12 voice

8. Which of the following commands would you use to forward DHCP requests from an interface connected to the 172.16.1.0/24 subnet to a DHCP server with the IP address 172.16.100.100?

   a. forward-protocol 172.16.1.0 255.255.255.0 172.16.100.100
   b. forward-protocol dhcp 172.16.1.0 255.255.255.0 172.16.100.100
   c. ip helper-address 172.16.1.0 172.16.100.100
   d. ip helper-address 172.16.100.100

9. How does the Cisco switch communicate voice VLAN information after a Cisco IP Phone has received PoE and started the boot process?

   a. Through CDP
   b. Using 802.1Q
   c. Using the proprietary ISL protocol
   d. Voice VLAN information must be statically entered on the Cisco IP Phone.

10. Which DHCP option provides the IP address of a TFTP server to a Cisco IP Phone?

    a. Option 10
    b. Option 15
    c. Option 150
    d. Option 290

11. Which of the following NTP stratum numbers would be considered the best?

    a. Stratum 0
    b. Stratum 1
    c. Stratum 2
    d. Stratum 3

12. Which of the following protocols could be used for Cisco IP Phone registration? (Choose two.)

    a. SCCP
    b. SIP
    c. DHCP
    d. H.323

Foundation Topics

Connecting and Powering Cisco IP Phones

Before we can get to the point of plugging in phones and having happy users placing and receiving calls, we must first lay the foundational infrastructure of the network. This includes technologies such as Power over Ethernet (PoE), voice VLANs, and Dynamic Host Configuration Protocol (DHCP). The network diagram shown in Figure 3-1 represents the placement of these technologies. As you read this chapter, each section will act as a building block to reach this goal. The first item that must be in place is power for the Cisco IP Phones.

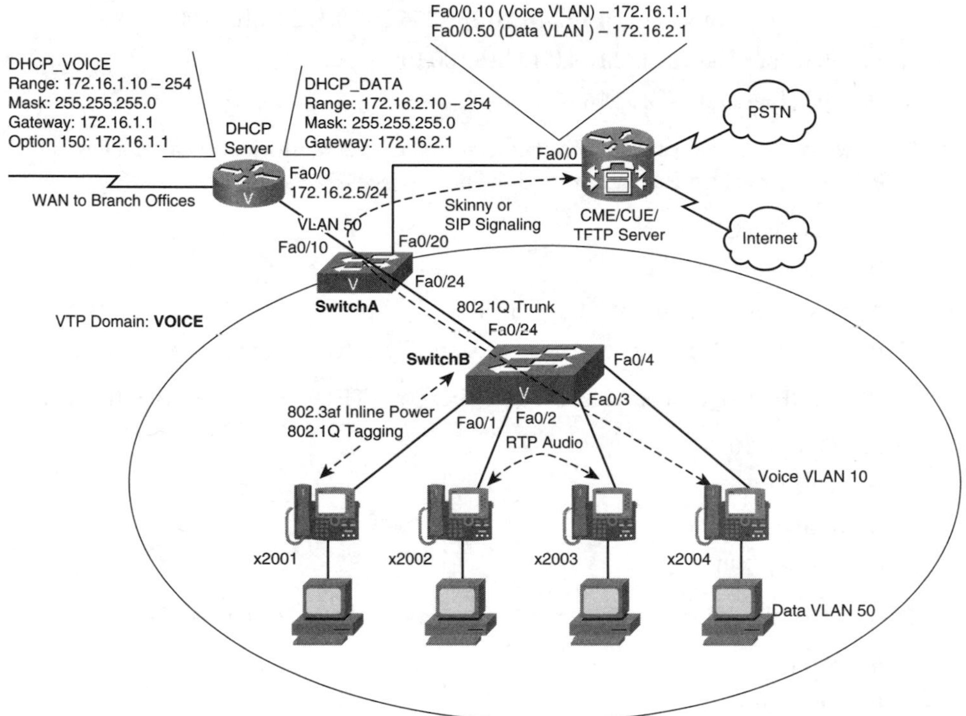

Figure 3-1 *VoIP Network*

Cisco IP Phones connect to switches just like any other network device (such as PCs, IP-based printers, and so on). Depending on the model of IP phone you are using, it may also have a built-in switch. Figure 3-2 illustrates the connections on the back of a Cisco 7960 IP Phone.

The ports shown in Figure 3-2 are as follows:

- **RS232:** Connects to a expansion module (such as a 7914, 7915, or 7916)
- **10/100 SW:** Used to connect the IP phone to the network
- **10/100 PC:** Used to connect a co-located PC (or other network device) to the IP Phone

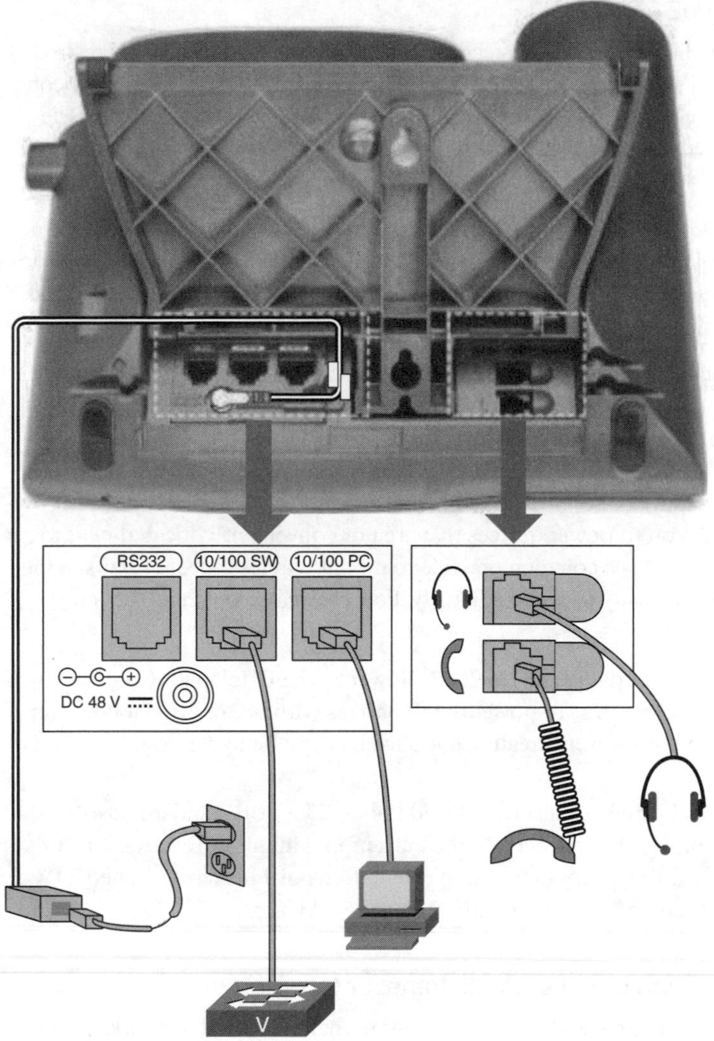

Figure 3-2 *Cisco IP Phone Ethernet Connections*

After you physically connect the IP phone to the network, it needs to receive power in some way. There are three potential sources of power in a Cisco VoIP network:

- Cisco Catalyst Switch PoE (Cisco prestandard or 802.3af power)
- Power Patch Panel PoE (Cisco prestandard or 802.3af power)
- Cisco IP Phone Power Brick (wall power)

Let's dig deeper into each one of these power sources.

Cisco Catalyst Switch PoE

If you were to create an Ethernet cable (Category 5 or 6), you would find that there are eight wires (four pairs of wires) to crimp into an RJ-45 connector on each end of the connection. Further study reveals that only four of the wires are used to transmit data. The other four remain unused and idle...until now.

The terms inline power and PoE describe two methods you can use to send electricity over the unused Ethernet wires to power a connected device. There is now a variety of devices that can attach solely to an Ethernet cable and receive all the power they need to operate. In addition to Cisco IP Phones, other common PoE devices include wireless access points and video surveillance equipment.

Powering devices through an Ethernet cable offers many advantages over using a local power supply. First, you have a centralized point of power distribution. Many users expect the phone system to continue to work even if the power is out in the company offices. By using PoE, you can connect the switch powering the IP phones to an uninterruptible power supply (UPS) instead of placing a UPS at the location of each IP phone. PoE also enables you to power devices that are not conveniently located next to a power outlet. For example, it is a common practice to mount wireless access points in the ceiling, where power is not easily accessible. Finally, PoE eliminates much of the "cord clutter" at employees' desks.

PoE became an official standard (802.3af) in 2003. However, the IP telephony industry was quickly developing long before this. To power the IP phones without an official PoE standard, some proprietary methods were created, one such method being Cisco Inline Power.

Note: The IEEE standards body has recently created the 802.3at PoE standard (also called PoE Plus), the goal of which is to increase the current maximum PoE wattage from 15.4W to 25.5W. In addition, some proprietary implementations of PoE have reached 51W of power by using all four pairs of wire in the Ethernet cable.

Powering the IP Phone Using a Power Patch Panel or Coupler

Many companies already have a significant investment in their switched network. To upgrade all switches to support PoE would be a significant expense. These organizations may choose to install intermediary devices, such as a patch panel, that are able to inject PoE on the line. The physical layout for this design is demonstrated in Figure 3-3.

By using the power patch panel, you still gain the advantage of centralized power and backup without requiring switch upgrades.

Note: Keep in mind that Cisco switches must also provide quality of service (QoS) and voice VLAN support capabilities, which may require switch hardware upgrades. Be sure your switch supports these features before you consider a power patch panel solution.

Inline PoE injectors provide a low-cost PoE solution for single devices (one device per coupler). These are typically used to support wireless access points or other "single spot" PoE solutions. Using inline PoE couplers for a large IP Phone network would make a mess of your wiring infrastructure and exhaust your supply of electrical outlets (because each inline PoE coupler requires a dedicated plug).

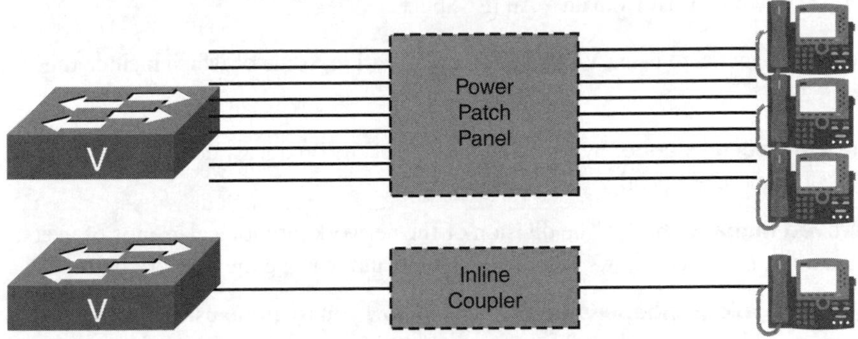

Figure 3-3 *Design for Power Patch Panels or Inline Couplers*

Powering the IP Phone with a Power Brick

Using a power brick to power a device is so simple that it warrants only brief mention. Thus, the reason for this section is primarily to mention that most Cisco IP Phones do not ship with power supplies. Cisco assumes most VoIP network deployments use PoE. If you have to choose between purchasing power bricks and upgrading your switch infrastructure, it's wise to check the prices of the power bricks. The average Cisco IP Phone power brick price is between $30–$40 USD. When pricing out a 48-switchport deployment, purchasing power bricks for all the IP phones may very well be in the same price range as upgrading the switch infrastructure.

Note: Some devices exceed the power capabilities of the 802.3af PoE standard. For example, when you add a sidecar module to a Cisco IP Phone (typically to support more line buttons), PoE connections can no longer support the device. These devices will need a power brick adapter.

VLAN Concepts and Configuration

After the IP phone has received power, it must determine its VLAN assignment. Because of security risks associated with having data and voice devices on the same network, Cisco recommends isolating IP phones in VLANs dedicated to voice devices. To understand how to implement this recommendation, let's first review a few key VLAN concepts.

VLAN Review

When VLANs were introduced a number of years ago, the concept was so radical and beneficial that it was immediately adopted into the industry. Nowadays, it is rare to find any reasonably sized network that is not using VLANs in some way.

VLANs allow you to break up switched environments into multiple broadcast domains. Here is the basic summary of a VLAN:

A VLAN = A Broadcast Domain = An IP Subnet

There are many benefits to using VLANs in an organization, some of which include the following:

- **Increased performance:** By reducing the size of the broadcast domain, network devices run more efficiently.
- **Improved manageability:** The division of the network into logical groups of users, applications, or servers allows you to understand and manage the network better.
- **Physical topology independence:** VLANs allow you to group users regardless of their physical location in the campus network. If departments grow or relocate to a new area of the network, you can simply change the VLAN on their new ports without making any physical network changes.
- **Increased security:** A VLAN boundary marks the end of a logical subnet. To reach other subnets (VLANs), you must pass through a routed (Layer 3) device. Any time you send traffic through a router, you have the opportunity to add filtering options (such as access lists) and other security measures.

VLAN Trunking/Tagging

VLANs are able to transcend individual switches, as shown in Figure 3-4.

If a member of VLAN_GRAY sends a broadcast message, it goes to all VLAN_GRAY ports on both switches. The same holds true for VLAN_WHITE. To accommodate this, the connection between the switches must carry traffic for multiple VLANs. This type of port is known as a trunk port.

Trunk ports are often called tagged ports because the switches send frames between each other with a VLAN "tag" in place. Figure 3-5 illustrates the following process:

1. HostA (in VLAN_GRAY) wants to send data to HostD (also in VLAN_GRAY). HostA transmits the data to SwitchA.
2. SwitchA receives the data and realizes that HostD is available through the FastEthernet 0/24 port (because HostD's MAC address has been learned on this port). Because FastEthernet 0/24 is configured as a trunk port, SwitchA puts the VLAN_GRAY tag in the IP header and sends the frame to SwitchB.
3. SwitchB processes the VLAN_GRAY tag because the FastEthernet 0/24 port is configured as a trunk. Before sending the frame to HostD, the VLAN_GRAY tag is removed from the header.
4. The tagless frame is sent to HostD.

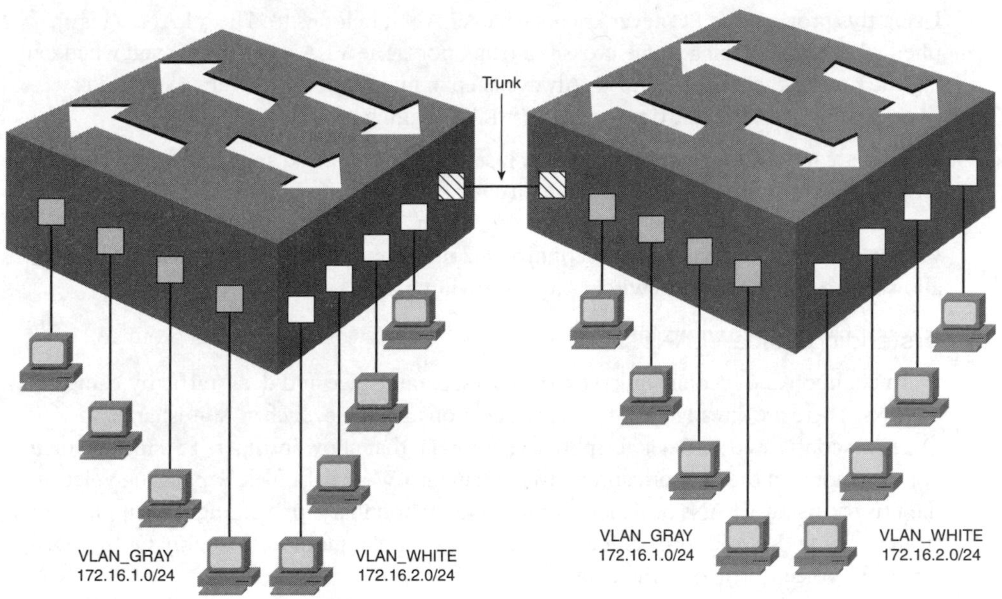

Figure 3-4 *VLANs Move Between Switches*

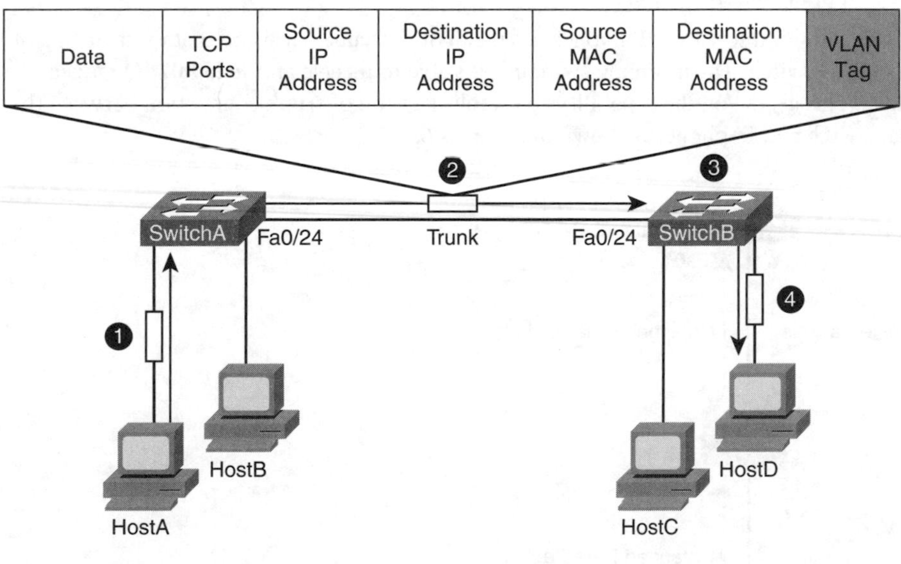

Key Topic

Figure 3-5 *VLAN Tags*

Using this process, the PC never knows what VLAN it belongs to. The VLAN tag is applied when the incoming frame crosses a trunk port. The VLAN tag is removed when exiting the port to the destination PC. Always keep in mind that VLANs are a switching concept; the PCs never participate in the VLAN tagging process.

VLANs are not a Cisco-only technology. Just about all managed switch vendors support VLANs. In order for VLANs to operate in a mixed-vendor environment, a common trunking or "tagging" language must exist between them. This language is known as 802.1Q. All vendors design their switches to recognize and understand the 802.1Q tag, which is what allows us to trunk between switches in any environment.

Understanding Voice VLANs

It is a common and recommended practice to separate voice and data traffic by using VLANs. There are already easy-to-use applications available, such as Wireshark and Voice Over Misconfigured Internet Telephones (VOMIT), that allow intruders to capture voice conversations on the network and convert them into WAV data files. Separating voice and data traffic using VLANs provides a solid security boundary, preventing data applications from reaching the voice traffic. It also gives you a simpler method to deploy QoS, prioritizing the voice traffic over the data.

One initial difficulty you can encounter when separating voice and data traffic is the fact that PCs are often connected to the network using the Ethernet port on the back of a Cisco IP Phone. Because you can assign a switchport to only a single VLAN, it initially seems impossible to separate voice and data traffic. That is, until you see that Cisco IP Phones support 802.1Q tagging.

The switch built into Cisco IP Phones has much of the same hardware that exists inside of a full Cisco switch. The incoming switchport is able to receive and send 802.1Q tagged packets. This gives you the capability to establish a type of trunk connection between the Cisco switch and IP phone, as shown in Figure 3-6.

Key Topic

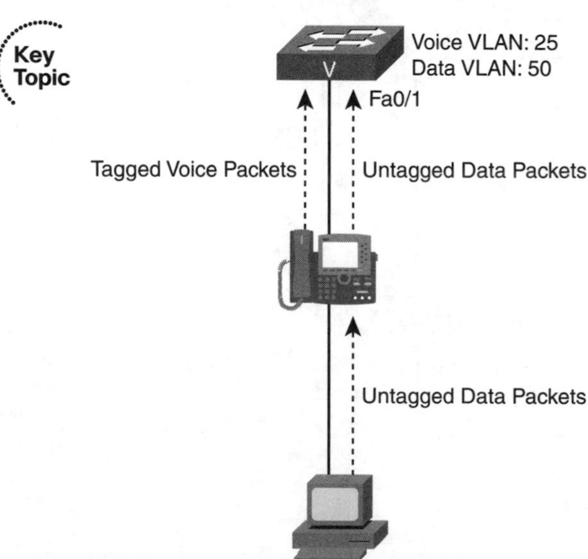

Figure 3-6 *Separating Voice and Data Traffic Using VLANs*

You might call the connection between the switch and IP phone a "mini-trunk" because a typical trunk passes a large number of VLANs (if not all VLANs). In this case, the IP phone tags its own packets with the correct voice VLAN (VLAN 25, in the case of Figure 3-6). Because the switch receives this traffic on a port supporting tagged packets (our mini-trunk), the switch can read the tag and place the data in the correct VLAN. The data packets pass through the IP phone and into the switch untagged. The switch assigns these untagged packets to whatever VLAN you have configured on the switchport for data traffic.

Note: Traditionally, a switchport on a Cisco switch that receives tagged packets is referred to as a trunk port. However, when you configure a switchport to connect to a Cisco IP Phone, you configure it as an access port (for the untagged data from the PC) while supporting tagged traffic from the IP phone. So, think of these ports as "access ports supporting tagged voice VLAN traffic."

VLAN Configuration

Configuring a Cisco switch to support Voice VLANs is a fairly simple process. First, you can add the VLANs to the switch, as shown in Example 3-1.

Example 3-1 *Adding and Verifying Data and Voice VLANs*

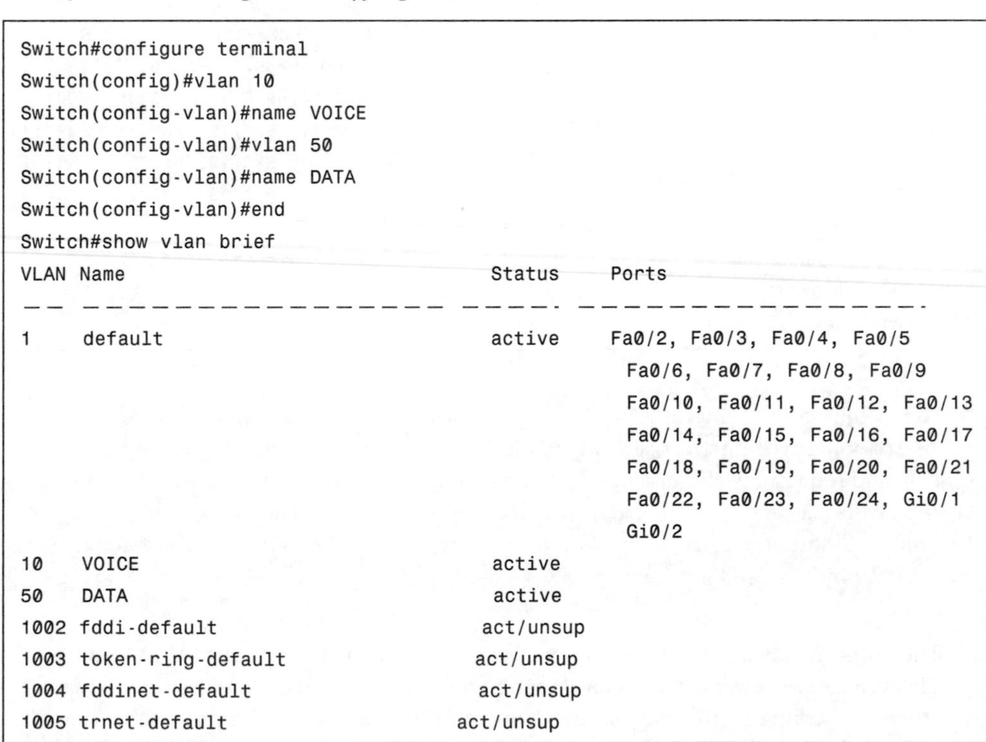

```
Switch#configure terminal
Switch(config)#vlan 10
Switch(config-vlan)#name VOICE
Switch(config-vlan)#vlan 50
Switch(config-vlan)#name DATA
Switch(config-vlan)#end
Switch#show vlan brief
VLAN Name                             Status    Ports
---- -------------------------------- --------- -------------------------------
1    default                          active    Fa0/2, Fa0/3, Fa0/4, Fa0/5
                                                Fa0/6, Fa0/7, Fa0/8, Fa0/9
                                                Fa0/10, Fa0/11, Fa0/12, Fa0/13
                                                Fa0/14, Fa0/15, Fa0/16, Fa0/17
                                                Fa0/18, Fa0/19, Fa0/20, Fa0/21
                                                Fa0/22, Fa0/23, Fa0/24, Gi0/1
                                                Gi0/2
10   VOICE                            active
50   DATA                             active
1002 fddi-default                     act/unsup
1003 token-ring-default               act/unsup
1004 fddinet-default                  act/unsup
1005 trnet-default                    act/unsup
```

Sure enough, VLANs 10 (VOICE) and 50 (DATA) now appear as valid VLANs on the switch. Now that the VLANs exist, you can assign the ports attaching to Cisco IP Phones (with PCs connected to the IP Phone) to the VLANs, as shown in Example 3-2.

Example 3-2 *Assigning Voice and Data VLANs*

```
Switch#configure terminal
Switch(config)#interface range fa0/2 - 24
Switch(config-if-range)#switchport mode access
Switch(config-if-range)#spanning-tree portfast
Switch(config-if-range)#switchport access vlan 50
Switch(config-if-range)#switchport voice vlan 10
Switch(config-if-range)#end
Switch#show vlan brief
VLAN Name                             Status    Ports
-- ------------------- ----- ------------------
1    default                          active    Gi0/1, Gi0/2
10   VOICE                            active    Fa0/2, Fa0/3, Fa0/4, Fa0/5
                                                 Fa0/6, Fa0/7, Fa0/8, Fa0/9
                                                 Fa0/10, Fa0/11, Fa0/12, Fa0/13
                                                 Fa0/14, Fa0/15, Fa0/16, Fa0/17
                                                 Fa0/18, Fa0/19, Fa0/20, Fa0/21
                                                 Fa0/22, Fa0/23, Fa0/24
50   DATA                             active    Fa0/2, Fa0/3, Fa0/4, Fa0/5
                                                 Fa0/6, Fa0/7, Fa0/8, Fa0/9
                                                 Fa0/10, Fa0/11, Fa0/12, Fa0/13
                                                 Fa0/14, Fa0/15, Fa0/16, Fa0/17
                                                 Fa0/18, Fa0/19, Fa0/20, Fa0/21
                                                 Fa0/22, Fa0/23, Fa0/24
1002 fddi-default                    act/unsup
1003 token-ring-default              act/unsup
1004 fddinet-default                 act/unsup
1005 trnet-default                   act/unsup
```

Note: When connecting Cisco IP Phones to a switch, you should also enable portfast (using **spanning-tree portfast**, as shown in Example 3-2), because the IP phones boot quickly and request a DHCP assigned address before a typical port with spanning-tree enabled would go active. Also, keep in mind that port Fa0/1 does not appear in the Example 3-2 output because it is configured as a trunk port (ports 2–24 are not considered trunks by Cisco IOS).

The ports are now configured to support a voice VLAN of 10 and a data VLAN of 50. This syntax is a newer form of configuration for IP Phone connections. In the "old days," you would configure the interface as a trunk port because the switch was establishing a trunking relationship between it and the IP phone. This was less secure because a hacker could remove the IP phone from the switchport and attach their own device (another managed switch or PC) and perform a VLAN-hopping attack. The more modern syntax configures the port as a "quasi-access port," because an attached PC will be able to access only VLAN 50. Only an attached Cisco IP Phone will be able to access the voice VLAN 10.

Note: Keep in mind that Cisco IP phones will be able to receive this voice VLAN configuration from the switch via CDP. After it receives the voice VLAN number, the IP Phone begins tagging its own packets. Non-Cisco IP Phones cannot understand CDP packets. This typically requires you to manually configure each of the non-Cisco IP Phones with its voice VLAN number from a local phone configuration window (on the IP phone).

Understanding the Cisco IP Phone Boot Process

Now that you learned about the VLAN architecture used with Cisco IP Phones, we can turn our attention to the IP Phones themselves. By understanding the IP Phone boot process, you can more fully understand how the Cisco IP Phone operates (which aids significantly in troubleshooting Cisco IP Phone issues). Here is the Cisco IP Phone boot process, start to finish:

1. The Cisco IP Phone connects to an Ethernet switchport. If the IP phone and switch support PoE, the IP phone receives power through either Cisco-proprietary PoE or 802.3af PoE.
2. As the Cisco IP Phone powers on, the Cisco switch delivers voice VLAN information to the IP phone using CDP as a delivery mechanism. The Cisco IP Phone now knows what VLAN it should use.
3. The Cisco IP Phone sends a DHCP request asking for an IP address on its voice VLAN.
4. The DHCP server responds with an IP address offer. When the Cisco IP Phone accepts the offer, it receives all the DHCP options that go along with the DHCP request. DHCP options include items such as default gateway, DNS server information, domain name information, and so on. In the case of Cisco IP Phones, a unique DHCP option is included, known as Option 150. This option directs the IP phone to a TFTP server. (You learn more about this in the upcoming section, "Configuring a Router-Based DHCP Server.")
5. After the Cisco IP Phone has the IP address of the TFTP server, it contacts the TFTP server and downloads its configuration file. Included in the configuration file is a list of valid call processing agents (such as Cisco Unified Communications Manager or Cisco Unified Communications Manager Express CME agents).
6. The Cisco IP Phone attempts to contact the first call processing server (the primary server) listed in its configuration file to register. If this fails, the IP phone moves to the next server in the configuration file. This process continues until the IP phone registers successfully or the list of call processing agents is exhausted.

Configuring a Router-Based DHCP Server

We currently made it to Step 4 in the preceding IP phone boot process. The phones in our network now need to receive IP address and TFTP server information. In the network design scenario used in this chapter, we use the WAN branch router as the DHCP server.

Using a router as a DHCP server is a somewhat common practice in smaller networks. Once you move into larger organizations, DHCP services are typically centralized onto server platforms. Either DHCP option is capable of sending TFTP server information to the IP phones.

Example 3-3 shows the syntax used to configure a WAN branch router as a DHCP server.

Example 3-3 *Configuring Router-Based DHCP Services*

```
WAN_RTR#configure terminal
WAN_RTR(config)#ip dhcp excluded-address 172.16.1.1 172.16.1.9
WAN_RTR(config)#ip dhcp excluded-address 172.16.2.1 172.16.2.9
WAN_RTR(config)#ip dhcp pool DATA_SCOPE
WAN_RTR(dhcp-config)#network 172.16.2.0 255.255.255.0
WAN_RTR(dhcp-config)#default-router 172.16.2.1
WAN_RTR(dhcp-config)#dns-server 4.2.2.2
WAN_RTR(dhcp-config)#exit
WAN_RTR(config)#ip dhcp pool VOICE_SCOPE
WAN_RTR(dhcp-config)#network 172.16.1.0 255.255.255.0
WAN_RTR(dhcp-config)#default-router 172.16.1.1
WAN_RTR(dhcp-config)#option 150 ip 172.16.1.1
WAN_RTR(dhcp-config)#dns-server 4.2.2.2
```

Note: This example uses a Cisco router as a DHCP server. I (Jeremy) took this approach because using a router as a DHCP server is simple and stable. That being said, most people use a Windows server or some other centralized device for DHCP services. Even Cisco Unified Communications Manager includes DHCP server capabilities. In these cases, you typically need to configure an **ip helper-address** *<central DHCP server IP address>* to forward DHCP requests to the central DHCP server for the voice VLAN devices.

The way in which Cisco routers approach DHCP configurations is slightly different from how many other DHCP servers do so. Most DHCP servers allow you to specify a range of IP addresses that you would like to hand out to clients. Cisco routers take the opposite approach: you first specify a range of addresses that you do not want to hand out to clients (using the **ip dhcp excluded-address** syntax from global configuration mode). Configuring the excluded addresses before you configure the DHCP pools ensures that the Cisco router does not accidentally hand out IP addresses before you have a chance to exclude them from the range. The DHCP service on the router will begin handing out IP addresses from the first nonexcluded IP address in the network range. In Example 3-3, this is 172.16.1.10 for the voice scope and 172.16.2.10 for the data scope.

Tip: Notice a DNS server of 4.2.2.2 is assigned to both the data and voice devices. This is a well-known, open DNS server on the Internet. This IP address works fantastically to test connectivity and DNS services in new network deployments because it is such a simple IP address to remember.

Also notice that the VOICE_SCOPE DHCP pool includes the option 150 syntax. This creates the custom TFTP server option to be handed out to the Cisco IP Phones along with their IP address information. In this case, the TFTP server of the IP phones is the same as the default gateway because we use the CME router as a call processing agent. As mentioned in the section, "Understanding the Cisco IP Phone Boot Process," the TFTP server holds the configuration files for the phones. When you configure a Cisco IP Phone in Cisco Unified Communications Manager (CUCM) or CME, an XML configuration file is generated and stored on a TFTP server. These CML configuration files have a filename format of SEP<*IP Phone MAC Address*>.cnf.xml and contain a base configuration for the IP phone (specifying language settings, URLs, and so on). Most importantly, these XML files contain a list of up to three CUCM server or CME IP addresses the Cisco IP Phone uses for registration. After the IP phone receives the XML file, it attempts to register with the first CUCM or CME server listed in the file. If it is unable to reach that server, it moves down to the next until the list is exhausted (at which point the IP phone reboots and tries it all over again).

Note: If the Cisco IP Phone has not yet been configured in CUCM or CME (no SEP<*MAC*>.cnf.xml file exists on the TFTP server), the IP Phone requests a file named XMLDefault.cnf.xml. This is a base configuration file typically used for a feature called Auto-Registration (allowing phones to register without being configured).

Tip: Many people often wonder the meaning of SEP at the beginning of the configuration filename. SEP stands for Selsius Ethernet Phone. Selsius was the name of the company Cisco acquired when they first began manufacturing VoIP technology.

Setting the Clock of a Cisco Device with NTP

The final task to prepare the network infrastructure to support a Cisco VoIP network is to set the time. Having an accurate time on Cisco devices is important for many reasons. Here is a quick list of just some of the reasons why you want an accurate clock on your network devices:

- It allows Cisco IP Phones to display the correct date and time to your users.
- It assigns the correct date and time to voicemail tags.
- It gives accurate times on Call Detail Records (CDR), which are used to track calls on the network.
- It plays an integral part in multiple security features on all Cisco devices.
- It tags logged messages on routers and switches with accurate time information.

When Cisco devices boot, many of them default their date and time to noon on March 1, 1993. You have two options in setting the clock: manually, using the **clock set** command from the privileged EXEC mode, or automatically, using the Network Time Protocol (NTP).

Devices setting the clock using NTP always have a more accurate time clock than a manually set clock. Likewise, all the NTP devices on your network will have the exact same time. These advantages make NTP the preferred clock-setting method. The accuracy of the clock on your device depends on the stratum number of the NTP server. A stratum 1 time server is one that has a radio or atomic clock directly attached. The device that receives its time from this server via NTP is considered a stratum 2 device. The device that receives its time from this stratum 2 device via NTP is considered a stratum 3 device, and so on. There are many publicly accessible stratum 2 and 3 (and even some stratum 1) devices on the Internet.

Note: You can obtain a list of publicly accessible NTP servers at www.ntp.org.

After you obtain one or more NTP servers to use, you can configure NTP support on your Cisco devices by using the syntax in Example 3-4.

Example 3-4 *Configuring a Cisco Router to Receive Time via NTP*

```
WAN_RTR#configure terminal
WAN_RTR(config)#ntp server 64.209.210.20
WAN_RTR(config)#clock timezone ARIZONA -7
```

The first command, **ntp server <ip address>**, configures your Cisco device to use the specified NTP server; 64.209.210.20 is one of many publicly accessible NTP servers. If this is the only command you enter, your clock on your device will set itself to the Universal Time Coordinated (UTC) time zone. To accurately adjust the time zone for your device, use the **clock timezone <name> <hours>** command. The previous syntax example set the time zone for Arizona to –7 hours from UTC.

Now that we configured the router to synchronize with an NTP server, we can verify the NTP associations and the current time and date using the commands shown in Example 3-5.

Example 3-5 *Verifying NTP Configurations*

```
WAN_RTR#show ntp associations
      address           ref clock      st  when  poll reach  delay  offset    disp
*~64.209.210.20     138.23.180.126     3    14    64  377    65.5    2.84      7.6
 * master (synced), # master (unsynced), + selected, - candidate, ~ configured
WAN_RTR#show clock
11:25:48.542 CA1_DST Mon Dec 13 2010
```

The key information from the **show ntp associations** command is just to the left of the configured NTP server address. The asterisk indicates that your Cisco device has synchronized with this server. You can configure multiple NTP sources for redundancy, but the Cisco device will only choose one master NTP server to use at a time.

After you configure the Cisco router to synchronize with an NTP server, you can configure it to provide date and time information to a CUCM server, which can then provide that date and time information to the Cisco IP Phones in your network. To allow other devices (such as a CUCM server) to pull date and time information from a Cisco router using NTP, use the **ntp master** <*stratum number*> command from global configuration mode. For example, entering **ntp master 4** instructs the Cisco router to deliver date and time information to requesting clients, marking it with a stratum number of 4.

Note: Example 3-4 illustrates configuring a Cisco router to support NTP. This is necessary if you are supporting a Cisco IP Telephony network using Communication Manager Express (CME). If you were using a full CUCM solution, you'd configure NTP on the CUCM server.

IP Phone Registration

Now that the Cisco IP Phone has gone through the complete process, it is ready to register with the call-management system (CME or CUCM). Before we discuss this final step, keep in mind what the phone has gone through up to this point:

1. The phone has received Power over Ethernet (PoE) from the switch.
2. The phone has received VLAN information from switch via CDP.
3. The phone has received IP information from the DHCP server (including Option 150).
4. The phone has downloaded its configuration file from the TFTP server.

The Cisco IP Phone is now looking at a list of up to three call processing servers (depending on how many you have configured) that it found in the configuration file it retrieved from the TFTP server. The phone tries to register with the first call processing server. If that fails, it continues down the list it received from the TFTP server until the phone makes it through all the listed call processing servers (at which point it reboots if it finds no servers online).

Key Topic

If the IP phone finds an active server in the list, it goes through the registration process using either the Skinny Client Control Protocol (SCCP) or Session Initiation Protocol (SIP). The protocol the phone uses depends on the firmware it is using. Today, most Cisco IP Phones use the SCCP, which is Cisco proprietary. However, as the SIP protocol matures, widespread support continues to grow. Because SIP is an industry standard, using it across your network provides benefits such as vendor neutrality and inter-vendor operation.

Note: The SIP standard is moving so quickly, by the time you read this, SCCP may not be the most popular protocol for Cisco IP Telephony networks. SCCP will most likely take its place in the proprietary protocol history books (which contain other items, such as the InterSwitch Link [ISL] trunking protocol and the Cisco original inline power method).

Regardless of the protocol used, the registration process is simple: The Cisco IP Phone contacts the call processing server and identifies itself by its MAC address. The call processing server looks at its database and sends the operating configuration to the phone. The operating configuration is different than the settings found in the configuration XML file located on the TFTP server. The TFTP server configuration is "base level settings," including items such as device language, firmware version, call processing server IP addresses, port numbers, and so on. The operating configuration contains items such as directory/line numbers, ring tones, softkey layout (on-screen buttons), and so on. Although the TFTP server configuration is sent using the TFTP protocol, the operating configuration is sent using SIP or SCCP.

These protocols (SIP or SCCP) are then used for the vast majority of the phone functionality following the registration. For example, as soon as a user picks up the handset of the phone, it sends a SCCP or SIP message to the call processing server indicating an off-hook condition. The server quickly replies with a SCCP or SIP message to play dial tone and collect digits. As the user dials, digits are transmitted to the call processing server using SCCP or SIP; call progress tones, such as ringback or busy, are delivered from the call processing server to the phone using SCCP or SIP. Hopefully, you get the idea: The Cisco IP Phone and call processing server have a dumb terminal and mainframe style of relationship, and the "language of love" between them is SCCP or SIP.

Exam Preparation Tasks

Review All the Key Topics

Review the most important topics in the chapter, noted with the key topics icon in the outer margin of the page. Table 3-2 lists and describes these key topics and identifies the page number on which each is found.

Table 3-2 *Key Topics for Chapter 3*

Key Topic Element	Description	Page Number
Figure 3-5	Trunking tag concepts	56
Figure 3-6	Separating voice and data traffic using VLANs	58
Examples 3-1 and 3-2	Configuring voice and data VLANs	59-60
Note	CDP delivers Voice VLAN information	59
Text	Cisco phones receive DHCP Option 150 to download an .xml configuration file via TFTP.	63
Text	Two primary signaling protocols to Cisco IP Phones are SIP and SCCP.	65

Definitions of Key Terms

Define the following key terms from this chapter, and check your answers in the Glossary:

802.3af Power over Ethernet (PoE), Cisco Inline Power, Cisco Discovery Protocol (CDP), virtual LAN (VLAN), trunking, 802.1Q, Dynamic Trunking Protocol (DTP), Skinny Client Control Protocol (SCCP), Session Initiation Protocol (SIP), Network Time Protocol (NTP)

- **Managing CME Using the Command Line:** Because CME runs on an IOS-based platform, an understanding of working with the CLI is essential. This section discusses the essentials of command-line interaction and the most common uses of the command line with CME.
- **Managing CME Using a Graphic User Interface:** Over the last few years, Cisco has released many different GUI-based utilities for configuring and interacting with Cisco routers and switches. Cisco Configuration Professional (CCP) has evolved to become the most fully functional all-in-one graphic interface for managing your CME router. This section discusses the CCP interface and common configuration areas.

CHAPTER 4

Getting Familiar with CME Administration

If you access Google and search for Cisco Unified Communication Manager Express (CME) configurations, you'll be overwhelmed with many useful syntax examples and configuration screenshots. However, before you can implement those configurations, you need to be familiar with the foundation administration of the CME router. This chapter discusses the methods of CME configuration: command line and GUI.

"Do I Know This Already?" Quiz

The "Do I Know This Already?" quiz allows you to assess whether you should read this entire chapter or simply jump to the "Exam Preparation Tasks" section for review. If you are in doubt, read the entire chapter. Table 4-1 outlines the major headings in this chapter and the corresponding "Do I Know This Already?" quiz questions. You can find the answers in Appendix A, "Answers Appendix."

Table 4-1 *"Do I Know This Already?" Foundation Topics Section-to-Question Mapping*

Foundation Topics Section	Questions Covered in This Section
Managing CME Using the Command Line	1–3
Managing CME Using the Graphic User Interface (GUI)	4–7

1. Which of the following represent valid command-line interfaces that you can use to manage a CME router? (Choose three.)

   a. Console connection

   b. Cisco ASDM

   c. Cisco SDM

   d. Telnet

   e. SSH

2. Which of the following configuration modes represents the area where you can apply core CME configurations that affect the entire telephony system?
   a. Dial-peer configuration
   b. Telephony-service configuration
   c. Global configuration
   d. Voice port configuration
3. Which of the following **show** commands enable you to see all the Cisco IP Phones that have registered with the CME router?
   a. show ip phone registered
   b. show ephone registered
   c. show voip registered
   d. show voice registered
4. What type of compressed archive can you download from Cisco to reinstall all support files for the integrated GUI back into the router flash?
   a. .TAR file
   b. .ZIP file
   c. .GZ file
   d. .RAR file
5. Which version of CCP installs into the flash of a Cisco router?
   a. There is only one CCP version, which cannot install into the router flash.
   b. CCP integrates with the SDM utility in the router's flash.
   c. CCP Lite.
   d. CCP Express.
6. Which of the following is not necessary to configure the CME router to support CCP?
   a. Enabling write operations to the router's flash
   b. Creating a level 15 user account
   c. Enabling HTTP services on the router
   d. Configuring Telnet or SSH for local logins
7. How does the communication change by selecting the Connect Securely checkbox when adding a device to a CCP community?
   a. You are prompted for the router's enable secret password rather than the VTY line password.
   b. Only level 15 user accounts are permitted to access the device.
   c. CCP uses HTTPS/SSH to connect to the device.
   d. CCP permits only communication over an IPSec managed connection.

Foundation Topics

Managing CME Using the Command Line

As you read this paragraph, you identify yourself as one of two types of people. The first type of person sees the "Managing CME Using the Command Line" title and thinks, "Fantastic! I love working in the command line—it's so flexible and customizable. Let's get started!" The second type of person sees the same title and thinks, "I thought we moved away from Microsoft DOS years ago. If you want to edit an autoexec.bat file, be my guest. I'll opt for the point-and-click GUI." If these two types of people met, they would probably not enjoy each other's company and would most likely end up throwing computer equipment at each other.

Note: In most cases, the GUI-based administrator would win the initial confrontation; however, the command-line administrator would later unleash a Denial of Service (DoS) attack on the GUI-based administrator that would make up for any initial loss.

Thankfully, with the modern configuration tools supported by CME, both administrators can peacefully coexist. Command-line administration still remains the most flexible and supports all CME features. However, the GUI-based utilities, specifically the Cisco Configuration Professional (CCP), have evolved enough to support simple configuration and troubleshooting for the vast majority of CME features. In some cases, the configuration performed using CCP is much more efficient than the using command-line administration. With that said, troubleshooting typically lives in the command-line domain and each upcoming sections presents a variety of **show** or **debug** commands that you can use to verify or troubleshoot the operation of your CME router. To access the command-line interface (CLI) of the CME router, use one of three methods:

- **Console port:** This how you initially configure the Cisco router before anyone has assigned a management IP address to the device. You can access the console port using a serial interface on a desktop or laptop PC and a Cisco rollover cable with the appropriate adapters.
- **Telnet access:** Since the 1970s, people have used Telnet to manage a variety of command-line systems. The industry now considers Telnet to be an unsecure protocol because it transmits data in clear text.
- **SSH access:** Secure Shell (SSH) performs the same function as Telnet but secures communication with a heavy dose of encryption. All modern Cisco equipment supports SSH capabilities out of the box, whereas older Cisco equipment might need an IOS upgrade to a security feature set.

Note: The foundation IOS commands, such as **enable**, **configure terminal**, and **show**, are covered in the CCENT and CCNA certification guides, which are a prerequisite for the CCNA Voice certification. They are not covered here.

To support a majority of the VoIP configuration, Cisco developed a **telephony-service** configuration mode. You can access this mode from global configuration mode, as shown in Example 4-1.

Example 4-1 *Accessing Telephony Service Configuration*

```
CME_ROUTER#conf t
Enter configuration commands, one per line.  End with CNTL/Z.
CME_ROUTER(config)#telephony-service
CME_ROUTER(config-telephony)#?
Cisco Unified Communications Manager Express configuration commands.
For detailed documentation see:
http://www.cisco.com/en/US/products/sw/voicesw/ps4625/tsd_products_support_series_
  home.html

  after-hours              define after-hours patterns, date, etc
  application              The selected application
  authentication           Config CME authentication server
  auto                     Define dn range for auto assignment
  auto-reg-ephone          Enable Ephone Auto-Registration
  bulk-speed-dial          Bulk Speed dial config
  call-forward             Configure parameters for call forwarding
  call-park                Configure parameters for call park
  caller-id                Configure caller id parameters
  calling-number           Replace calling number with local for hairpin
  cnf-file                 Ephone CNF file config options
  codec                    Define default codec for CME service
  conference               Configure conference type for adhoc
  create                   create cnf for ethernet phone
  date-format              Set date format for IP Phone display
  <output omitted>
```

Although there are commands that move outside of the telephony-service configuration mode (especially the critical **dial-peer** configurations, which are discussed in Chapter 6, "Understanding the CME Dial-Plan"), Cisco keeps the core configurations centralized in one place.

As mentioned, most troubleshooting commands are performed from the CLI. Example 4-2 shows one of the most common verification and troubleshooting commands used with CME: **show ephone registered.** This command verifies the active phones registered with CME and the status of their lines.

Example 4-2 *show ephone registered Command Output*

```
CME_ROUTER#show ephone registered
ephone-1[0] Mac:0014.A89E.F845 TCP socket:[1] activeLine:0 REGISTERED in SCCP
ver 17/9
mediaActive:0 offhook:0 ringing:0 reset:0 reset_sent:0 paging 0 debug:0 caps:8
IP:172.30.100.40 36964 7970  keepalive 27761 max_line 8
button 1: dn 1  number 1005 CH1   IDLE         CH2   IDLE
Preferred Codec: g711ulaw
```

Managing CME Using a Graphic User Interface

Managing CME using a Graphic User Interface (GUI) offers a number of advantages, some of which can be seen on the surface, whereas others might not be as evident. First, many small offices use an all-in-one administrator whose knowledge is spread across many different technologies. To force this level of administrator to learn a complete command-line operating system to interact with CME is unrealistic. Similarly, some offices use consultants or contract network administrators to manage their network. Providing an easy-to-use GUI allows one of the more technically inclined users at the office to take care of the day-to-day administration (changing directory numbers, adding phones, and so on) without the involvement of dedicated IT staff. Finally, the point-and-click of a graphic interface can be more efficient at times than typing configuration commands.

Although Cisco has released multiple GUI management tools to configure CME over the years, two primary tools are used today: the integrated CME GUI and Cisco Configuration Professional (CCP). The integrated CME GUI is powered by HTML and JAR (Java) files loaded into the flash of the CME router. Typically, the CME router ships with these files preloaded by Cisco into the flash; however, you can also download a TAR package of files from the Cisco website (assuming you have a valid support contract) and extract the files into the flash of the router. With minimal command-line configuration (assigning an IP address and enabling the HTTP server), you can have the integrated CME GUI up and running in no time, as shown in Figure 4-1.

Although the integrated CME GUI is not "pretty" by today's standards, it is functional, which enables you to handle most core functions of CME: adding/changing phone configurations, modifying the dial-plan, configuring hunt groups, and so on. Keep in mind that this, too, runs from the flash of the CME router, so conserving space is more important in Cisco's mind than adding "gloss" to the utility.

Cisco focused the integrated CME GUI on configuring only the telephony aspects of the CME router. It created the CCP to configure *all* major aspects of the CME router. It enables simple (and often wizard-based) configuration of the router, firewall, Intrusion Prevention System (IPS), Virtual Private Network (VPN), Unified Communications, and common WAN and LAN features and configurations. You can download the latest version of the CCP software from the Cisco website free of charge. CCP is roughly a 200 MB installation on a local PC. After you install it on your local PC, you can manage any supported Cisco platform using the utility. (You do not need to install anything on the managed devices.)

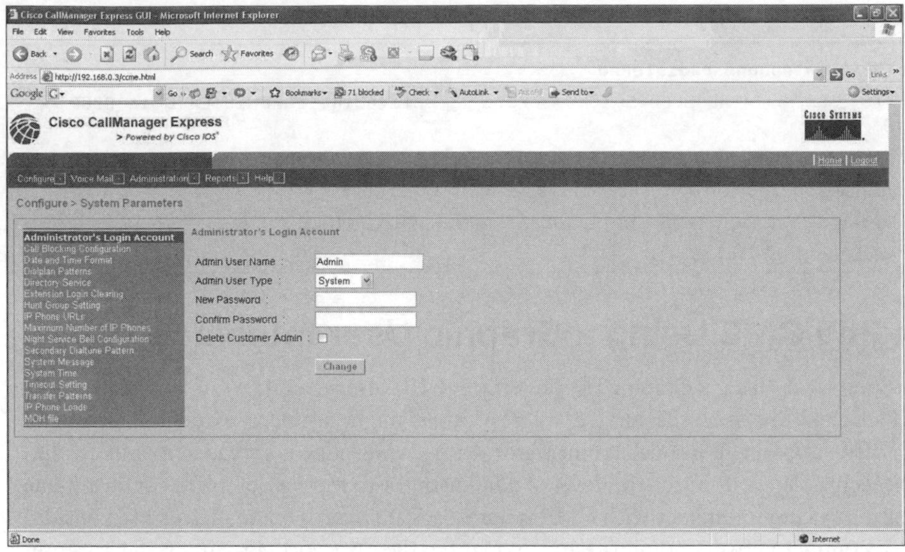

Figure 4-1 *Integrated CME GUI*

Note: Cisco created Cisco Configuration Professional Express (CCP Express), which is a similar GUI utility that is loaded into the flash of the router. CCP Express only focuses on configuring basic LAN/WAN connections, firewall, and Network Address Translation (NAT). It is not able to configure Unified Communications features.

When you initially open CCP, it prompts you to configure a community, as shown in Figure 4-2.

A community is a group of devices you want to manage using CCP. Before you can add a device, you must configure it with these four things:

- **Reachable IP address:** CCP must be able to reach the CME router on the IP address you specify.
- **Level-15 Username and Password:** Use this administrative account to manage the CME router.
- **Integrated HTTP Services:** HTTP services enable the CCP utility to discover the CME router.
- **Local Authentication for Telnet/SSH:** CCP logs into the CME router to apply configurations based on the GUI interaction.

Example 4-3 shows the configuration of a CME router to support CCP-based configuration.

Example 4-3 *Configuring the CME Router to Support CCP*

```
CME_Router#configure terminal
Enter configuration commands, one per line.  End with CNTL/Z.
CME_Router(config)#interface fastEthernet 0/0
```

```
CME_Router(config-if)#ip address 172.30.100.77 255.255.255.0
CME_Router(config-if)#no shutdown
CME_Router(config-if)#exit
CME_Router(config)#username Neo privilege 15 secret ci$co
CME_Router(config)#ip http server
CME_Router(config)#ip http secure-server
% Generating 1024 bit RSA keys, keys will be non-exportable...
*Jan 15 23:13:38.719: %SSH-5-ENABLED: SSH 1.99 has been enabled
*Jan 15 23:13:38.955: %PKI-4-NOAUTOSAVE: Configuration was modified.  Issue "write
memory" to save new certificate
CME_Router(config)#line vty 0 4
CME_Router(config-line)#login local
CME_Router(config-line)#transport input telnet ssh
CME_Router(config-line)#end
CME_Router#
```

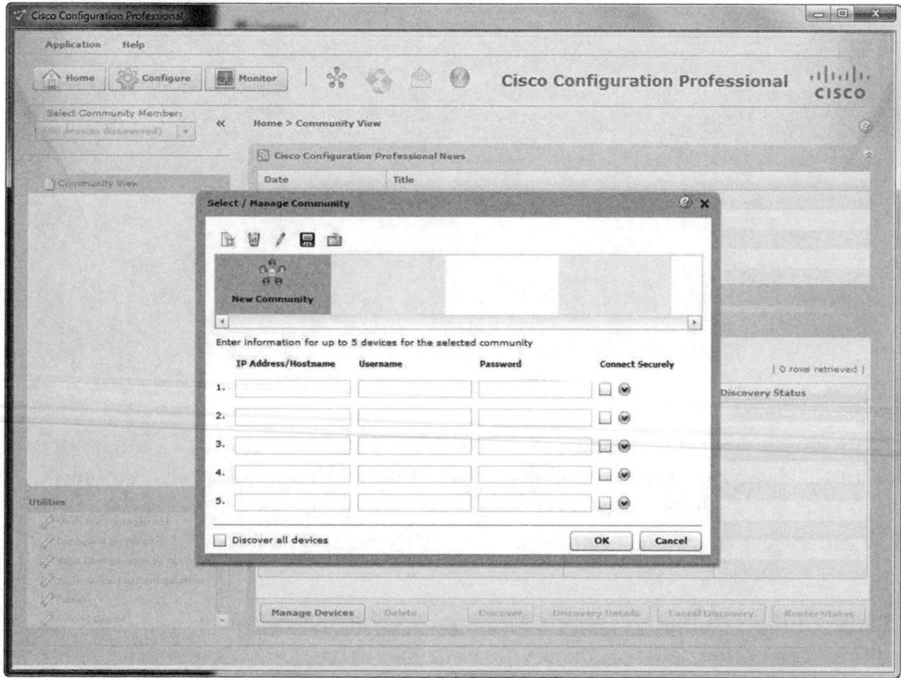

Figure 4-2 *Creating a CCP Community of Managed Devices*

By default, CCP attempts to connect to the router using Telnet and HTTP, which are both clear-text protocols. Of course, secure connections are always better. By simply checking the Connect Securely checkbox in CCP (as shown in Figure 4-3), it now uses SSH and HTTPS to connect to and configure the CME router.

After you connect to the CME router, CCP runs a discovery process, which identifies the router hardware, software, interfaces, and modules. After this process completes, you are able to configure the device. Although CCP has many configuration options available relating to routing and security, we primarily focus on the Unified Communications features, as shown in Figure 4-4.

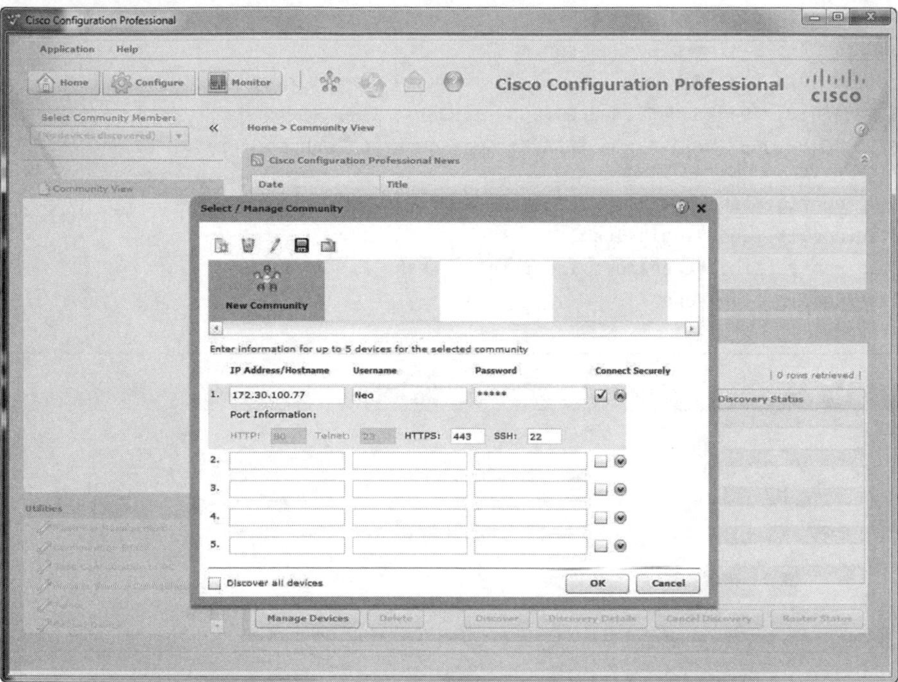

Figure 4-3 *Connecting to the CME Router Securely*

When you initially expand the Unified Communications folder under the Configuration section of CCP, you see only one available option: Unified Communications Features. Only one option appears because you have not yet configured the CME router to support CME features. As you can see from Figure 4-4, CCP enables you to configure the router in one of four ways:

- **Cisco Unified Communications Manager Express:** Most small businesses choose this option. CCP configures the router as a standalone CME system.
- **Gateway:** Enables you to use the CME router as a voice gateway. Voice gateways translate between dissimilar networks (such as your VoIP network connecting to the public switched telephone network [PSTN] over analog or digital trunks). Use this option if you are managing your Cisco phones with another system (such as Cisco Unity Connection Manager [CUCM]) and want to use this router only as a voice gateway.
- **Cisco Unified CME as SRST:** Enables the IP Phones to use the CME router as a failover device should they lose connectivity with the primary call-management system.
- **None:** Disables all Unified Communications capabilities of the CME router.

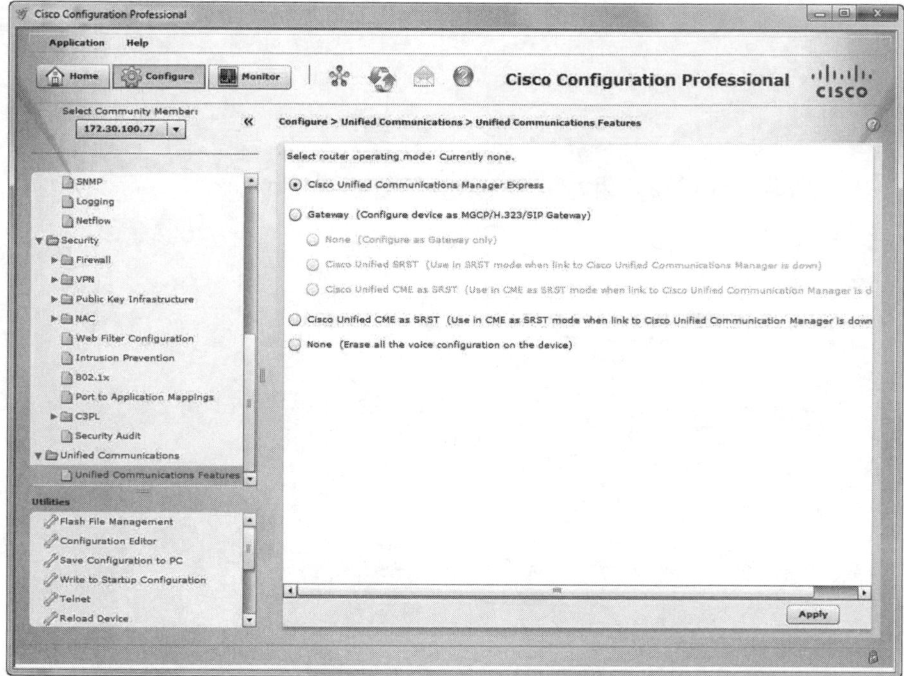

Figure 4-4 *Unified Communications Initial Configuration*

Note: If you choose the Gateway option shown in Figure 4-4, you have the ability to select three suboptions. None configures the router as a voice gateway only (does not support IP Phones). Cisco Unified SRST and Cisco Unified CME as SRST enable the IP Phones to use the CME router as a failover device should they lose connectivity with the primary call-management system. These topics are discussed in the CIPT1 and CIPT2 CCNP Voice titles.

One of my (Jeremy) favorite CCP features is the configuration confirmation screen (shown in Figure 4-5) that appears after you choose to apply a configuration.

Not only does the confirmation screen act as an "are you sure?" confirmation that you indeed want to apply the commands, but CCP also shows all the commands it will submit to the CME router. This is fantastic for two reasons: First, it enables you to verify that CCP is performing the configuration you want. Second, it enables you to learn the commands that CCP uses to configure the CME router. After you click the Deliver button (refer to Figure 4-5), CCP sends the commands to the CME router using either Telnet or SSH (depending on the type of connection you chose when you created the device community in CCP).

After CCP applies the initial CME configuration to the router, the Unified Communications menu refreshes to display additional configuration options, as shown in Figure 4-6.

These areas are what the upcoming chapters discuss, so fasten your seatbelts and move on to Chapter 5, "Managing Endpoint and End Users with CME."

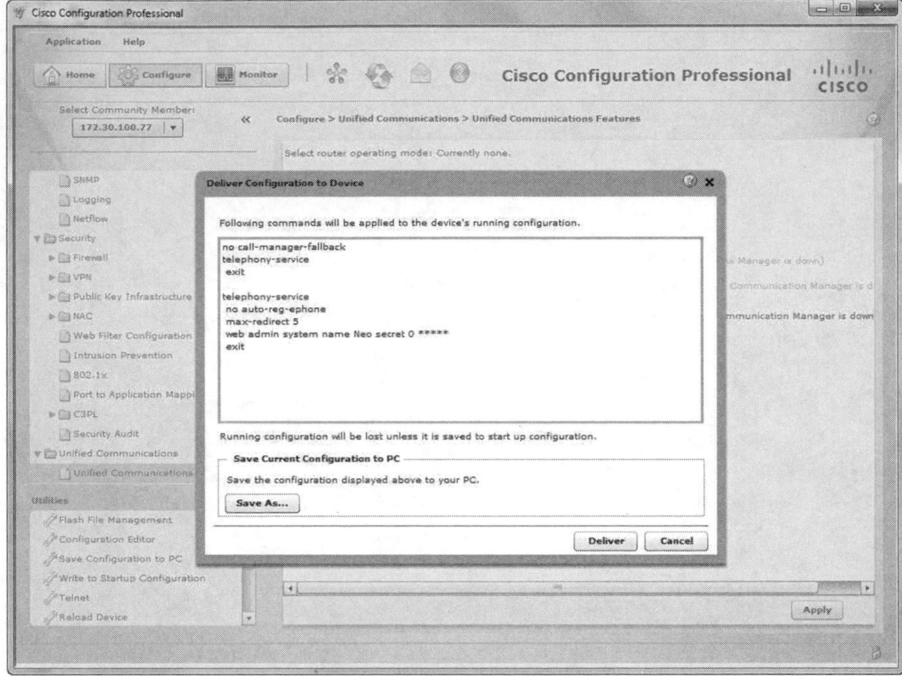

Figure 4-5 *CCP Configuration Confirmation Window*

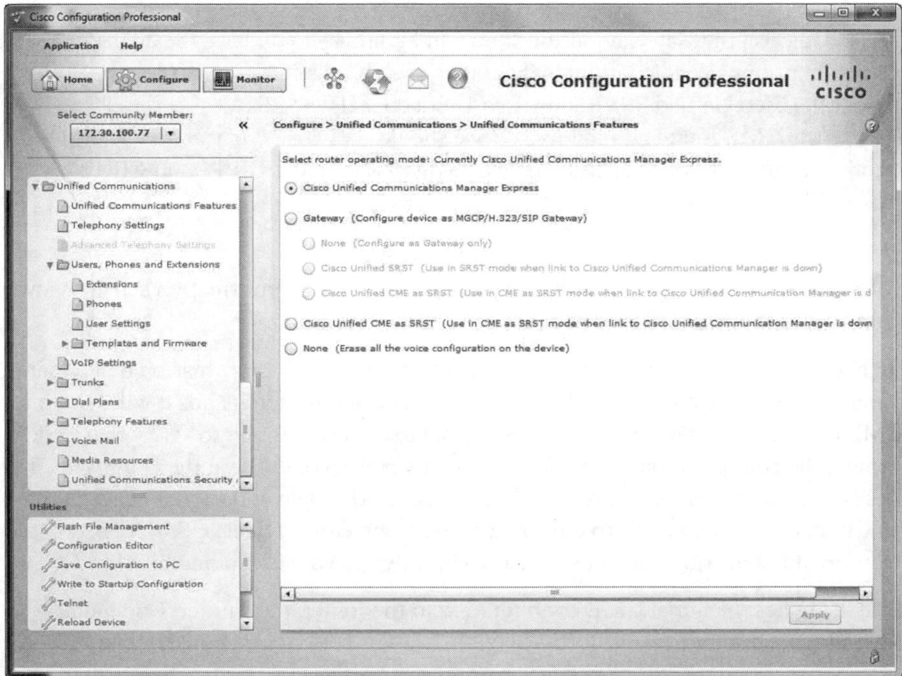

Figure 4-6 *CCP Unified Communications Configuration Options*

Exam Preparation Tasks

Review All the Key Topics

Review the most important topics in the chapter, noted with the key topics icon in the outer margin of the page. Table 4-2 lists and describes these key topics and identifies the page numbers on which each is found.

Table 4-2 *Key Topics for Chapter 4*

Key Topic Element	Description	Page Number
Example 4-1	Accessing the telephony-service configuration of a CME router	72
List	Requirements for supporting connection from CCP	74
List	The modes of VoIP configuration supported by CCP	76

- **Ensuring the Foundation:** This section explores the configuration foundation that must be in place to allow for a working IP telephony environment.
- **Ephone and Ephone-DN—The Keys to Ringing Phones:** Configuring Cisco IP Phones and directory numbers from the CLI is the most flexible method you can use to manage your end devices. This section walks through the ephone (IP phone) and ephone-dn (directory number) setup.
- **Adding Directory Numbers, Phones, and Users with CCP:** For the not-so-command-line-at-heart, Cisco enables you to configure the same telephony components using a point-and-click GUI. This section walks through the configuration of the core telephony components using CCP.

CHAPTER 5

Managing Endpoints and End Users with CME

Up to this point, you have seen the key concepts of traditional voice and VoIP, understood the core pieces that make up the Cisco Unified Communications suite of products, and correctly implemented the underlying network foundation for VoIP. Now, we've reached our moment of glory: It's time to make some phones ring! This chapter focuses on the most common tasks performed by a Cisco Unified Communication Manager Express (CME) administrator: IP phone provisioning and maintenance. The topics covered will take you all the way from the initial installation of a new IP phone to the day-to-day maintenance of IP phones and end users on the VoIP network.

"Do I Know This Already?" Quiz

The "Do I Know This Already?" quiz allows you to assess whether you should read this entire chapter or simply jump to the "Exam Preparation Tasks" section for review. If you are in doubt, read the entire chapter. Table 5-1 outlines the major headings in this chapter and the corresponding "Do I Know This Already?" quiz questions. You can find the answers in Appendix A, "Answers Appendix."

Table 5-1 *"Do I Know This Already?" Foundation Topics Section-to-Question Mapping*

Foundation Topics Section	Questions Covered in This Section
Ensuring the Foundation	1–2
Ephone and Ephone-DN—The Keys to Ringing Phones	3–6
Adding Directory Numbers, Phones, and Users with CCP	7–10

1. If a Cisco IP Phone not yet configured in CME contacts the TFTP server, which file will be sent?
   a. XMLDefault.cnf.xml
   b. 000000000000.xml
   c. A file named after the MAC address of the IP phone
   d. None of the above

2. What minimum set of telephony-service commands does CME require before it is prepared to accept IP phone registrations? (Choose three.)

   a. ip source-address
   b. no shutdown
   c. ephone
   d. ephone-dn
   e. max-ephones
   f. max-ephone-dn
   g. max-dn

3. Entering the command **ephone-dn 2** from global configuration mode creates what type of ephone-dn?

   a. Single line
   b. Dual line
   c. Quad line
   d. Octo line

4. What is the simplest method you can use to have an ephone-dn respond to an internal four-digit extension as well as a ten-digit DID number?

   a. Create incoming translation patterns assigned to the dial-peer.
   b. Use partitions and calling search spaces.
   c. Assign a secondary number using the "number" command.
   d. Assign a calling party transformation mask.

5. You want to assign the directory number 1000 (ephone-dn 5) to the third line button on a Cisco 7961 IP Phone (ephone 12). What syntax accomplishes this task?

   a. button 3:1000
   b. button 12:3:1000
   c. button 5:3:12
   d. button 3:5

6. You want to verify the line assignment and registration status of the IP phones currently supported by CME. Which of the following **show** commands provides this information?

   a. show run
   b. show ephone
   c. show voip ephone registration
   d. show voip registration

7. Before CCP enables you to configure any phones or extensions, what component must you configure?
   a. Dynamic routing protocol
   b. Static route to the CME system
   c. Telephony service
   d. Unity express voicemail
8. CCP automatically saves changes to flash as you configure them on the router.
   a. True
   b. False
9. Which of the following is the only required field when adding a user account using CCP?
   a. UserID
   b. First Name
   c. Last Name
   d. Password
10. How does CCP link phones and extensions together?
   a. By navigating to the Phone configuration page, you can assign extensions to phones.
   b. By creating a user account, you can link the user account, phone, and extension together.
   c. You can assign extensions to phones from the Extension configuration page.
   d. You must use the command-line ephone/ephone-dn syntax to link phones and extensions together.

Foundation Topics

Ensuring the Foundation

Chapter 3, "Understanding the Cisco IP Phone Concepts and Registration," and Chapter 4, "Getting Familiar with CME Administration," put configurations in place that are key to a working CME system. It is wise to take a high-level review of these concepts before you jump into the configuration of the IP phones.

Just about all the concepts discussed so far focus on the boot process of the Cisco IP Phone. The following list outlines the Cisco IP Phone boot process, which is illustrated in Figure 5-1:

1. The 802.3af PoE switch sends a small DC voltage on the Ethernet cable, detects an unpowered 802.3af device, and supplies power to the line.
2. The switch delivers voice VLAN information to the Cisco IP Phone using Cisco Discovery Protocol (CDP).
3. The IP Phone sends a Dynamic Host Configuration Protocol (DHCP) request on its voice VLAN. The DHCP server replies with IP addressing information, including DHCP Option 150, which directs the IP phone to the TFTP server.
4. The IP phone contacts the TFTP server and downloads its configuration file and firmware.
5. Based on the IP address listed in the configuration file, the IP phone contacts the call processing server (the CME router, in this case), which supports VoIP functions.

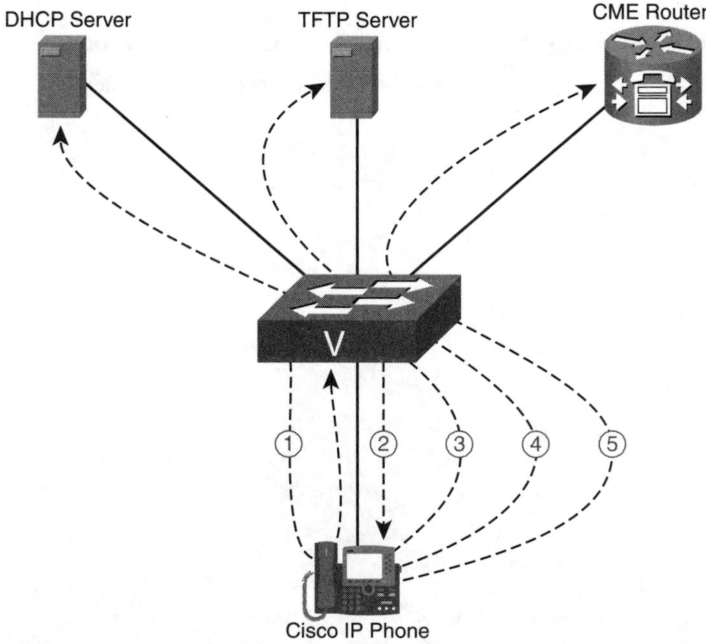

Figure 5-1 *Cisco IP Phone Boot Process*

Note: Figure 5-1 shows the DHCP server, TFTP server, and CME router as three separate devices. To save resources, smaller networks typically combine all three of these functions into one device. In this case, the CME router also acts as the DHCP and TFTP server for the network.

To meet the demands of this boot process, we put the following configuration in place over the previous two chapters:

- Voice VLAN
- DHCP services
- TFTP services

Voice VLAN

To separate the voice and data traffic, you must configure each port connecting to a Cisco IP Phone for a voice VLAN. In Example 5-1, the voice VLAN is 100 and the data VLAN (for the PC device attaching to the IP phone) is 200.

Example 5-1 *Configuring Voice VLANs*

```
Switch#configure terminal
Switch(config)#interface fa0/1
Switch(config-if)#switchport mode access
Switch(config-if)#switchport voice vlan 100
Switch(config-if)#switchport access vlan 200
Switch(config-if)#spanning-tree portfast
```

DHCP Services

After the IP phone receives its voice VLAN information, it begins sending DHCP requests. The configuration in Example 5-2 enables a Cisco router to become a DHCP server for the voice VLAN and deliver the needed IP address information to the IP phones.

Example 5-2 *Configuring a DHCP Scope on a Router*

```
ROUTER(config)#ip dhcp pool VOICE_SCOPE
ROUTER(dhcp-config)#network 172.16.1.0 255.255.255.0
ROUTER(dhcp-config)#default-router 172.16.1.1
ROUTER(dhcp-config)#option 150 172.16.1.1
ROUTER(dhcp-config)#dns-server 4.2.2.2
```

In addition to delivering the standard DHCP information (IP address, subnet mask, default gateway, and DNS), the DHCP server delivers phone Option 150, which gives the IP phone the IP address of the TFTP server.

TFTP Services

During the IP phone boot process, the IP phone contacts the TFTP server to download its configuration files. Until this point, you might wonder, "Where is this TFTP server? And what is in that configuration file?" I'm glad you asked. The TFTP server is what it has always been: a simple file store that serves files to unauthenticated clients on demand. Although you can install and configure a TFTP server on virtually any device, typically the Cisco router (in the case of CME) or the CUCM Publisher server handles the role of the TFTP server.

The TFTP server plays the role of "file server" in the IP telephony network. The IP phones download their configuration and firmware files from this server. The CME router generates these configuration files as you work through the initial configuration. For example, if you specify a new firmware load for a Cisco IP Phone to use, the CME router would modify a configuration file in flash (or an external TFTP server, if that's what you chose to use) to list the new firmware image. The next time the phone reboots, it receives the new configuration file, realizes its firmware is out of date, and contacts the TFTP server to download and apply the new firmware image.

If you have not entered any individual phone configurations, the only configuration file sent to the Cisco IP Phones is the XMLDefault.cnf.xml file. This file contains the IP address and port number used to connect to the call processing server (the CME router, in our case) and the names of the firmware file the IP phone should use. After the IP phone has this configuration file, it downloads the necessary firmware and contacts the CME router.

Although Cisco routers have the capability to act as a TFTP server, they were never designed to handle a large IP telephony network with many requests. Also, each model of Cisco IP Phone has a number of associated firmware files. Some of the high-end Cisco IP Phones use firmware files that can be over 40 MB in size! It doesn't take too many of these to fill up the router's flash memory. Because of this, using an external TFTP server to store the support files for the VoIP network is a good idea.

Note: In CME 4.0 and later, configuration and firmware files can be stored on an external TFTP server using the command **cnf-file location tftp://<ip address of TFTP server>** from telephony service configuration mode. This can save some valuable flash space on your router.

To use your router's flash as a TFTP server, you must make the files individually available for the IP phones to download. Example 5-3 illustrates this process for the 7940/7960 IP phone firmware files.

Example 5-3 *Configuring Router-Based TFTP Services for IP Phone Firmware Files*

```
CME_Voice# dir flash:/phone/7940-7960
Directory of flash:/phone/7940-7960/
   98  -rw-      129824  May 12 2008 21:33:56 -07:00  P00308000500.bin
   99  -rw-         458  May 12 2008 21:33:56 -07:00  P00308000500.loads
  100  -rw-      705536  May 12 2008 21:34:00 -07:00  P00308000500.sb2
  101  -rw-      130228  May 12 2008 21:34:00 -07:00  P00308000500.sbn
129996800 bytes total (28583936 bytes free)
CME_Voice# configure terminal
Enter configuration commands, one per line.  End with CNTL/Z.
CME_Voice(config)# tftp-server flash:/phone/7940-7960/P00308000500.bin alias
  P00308000500.bin
CME_Voice(config)# tftp-server flash:/phone/7940-7960/P00308000500.loads alias
  P00308000500.loads
CME_Voice(config)# tftp-server flash:/phone/7940-7960/P00308000500.sb2 alias
  P00308000500.sb2
CME_Voice(config)# tftp-server flash:/phone/7940-7960/P00308000500.sbn alias
  P00308000500.sbn
```

The alias syntax that follows the **tftp-server** command enables the firmware file to be requested simply by asking for the aliased filename. This is necessary in the newer CME versions, which organize the firmware files into subdirectories. The Cisco IP Phones do not know the full path to the firmware file; they ask only for the firmware filename.

Note: Be sure to make all the firmware files in each subdirectory available through TFTP. The Cisco IP Phones use most, if not all, of these files during the boot process.

The firmware files for the Cisco 7940 and 7960 IP Phones are now available through TFTP from the Cisco Unified CME router. You can repeat this process as many times as necessary to make the firmware files available for the additional models of IP phones you have on your network.

Note: You can download the latest firmware files for Cisco IP Phones from the Cisco website assuming you have a valid support contract with Cisco. These files come in a TAR archive, which you can extract to the flash of your router with a single command.

Base CME Configuration

For the CME router to prepare to support the Cisco IP Phones, you must configure some core telephony service information. The key items you must configure include the following:

- IP Source Address
- Max-DN
- Max-Ephones

Although you can configure many additional features, these three are necessary to get the phone system up and running. Example 5-4 illustrates this configuration.

Example 5-4 *Provisioning CME Phone and Directory Number Support*

```
CME_Voice(config)# telephony-service
CME_Voice(config-telephony)# ip source-address 172.16.1.1
CME_Voice(config-telephony)# max-ephones ?
  <1-30>  Maximum phones to support
CME_Voice(config-telephony)# max-ephones 24
CME_Voice(config-telephony)# max-dn ?
  <1-150>  Maximum directory numbers supported
  <cr>
CME_Voice(config-telephony)# max-dn 48
```

The max-ephones parameter configures the maximum number of IP phones the router will support, whereas the max-dn parameter specifies the maximum number of directory numbers. The **ip source-address** command enables the router to know which IP address will receive registration requests from the IP phones.

Note: The max-ephones and max-dn parameters directly affect how much memory the router reserves to support the CME service. Setting the value much higher than you actually need might reserve excessive resources on your router and impact other network services. In addition, the max-ephones parameter should not be any higher than the number of feature licenses you have purchased for your CME system.

At this point, CME is nearly ready. The Cisco IP Phone has come to the CME router and said, "I want to use you as my call processing device." The CME router receives this request and says, "Great! But, who are you?" The CME router delivers no additional configuration to the IP phone because you have yet to enter it. So, let's enter away! Welcome to the world of ephones and ephone-dn.

Ephone and Ephone-DN—The Keys to Ringing Phones

In the movie *The Matrix*, Thomas Anderson (also known as Neo) finds out that the world around him is not real, but rather a computer-generated environment. He soon becomes a part of the Resistance, who fights against the forces behind this computer-generated world. He meets up with Morpheus, who teaches him more about this new world. Let's pick up with their dialog here:

> Morpheus: "This is the construct. It's our loading program. We can load anything from clothing, to equipment, weapons, training simulations, anything we need."
>
> Neo: "Right now, we're inside a computer program?"
>
> Morpheus: "Is it really so hard to believe? Your clothes are different. The plugs in your arms and head are gone. Your hair is changed. Your appearance now is what we call residual self image. It is the mental projection of your digital self."

The concepts of ephone and ephone-dn are not too far beyond this. You must now configure the CME router to support IP phones. Think about this as your construct" "the mental projection of your digital phones." Each ephone you configure is a representation of the settings of a physical Cisco IP Phone sitting in your office somewhere. Each ephone-dn represents a directory number that you can assign to one or more ephones. Because ephone-dns are assigned to ephones, you should configure them first.

Note: Although watching *The Matrix* is not yet required for Cisco certification, I (Jeremy) believe about 90 percent of the concepts in Cisco can somehow be linked back to this movie, which I highly recommend.

Understanding and Configuring Ephone-DNs

An ephone-dn in its simplest form is just a directory number that can be assigned to one or more buttons on one or more Cisco IP Phones. You can configure each ephone-dn you create as either a single- or dual-line mode ephone-dn. Here's the difference:

- **Single-line ephone-dn:** In single-line mode, the ephone-dn is able to make or receive only one call at a time. If a call arrives on an ephone-dn where there is already an active call, the caller receives a busy signal.
- **Dual-line ephone-dn:** In dual-line mode, the ephone-dn is able to handle two simultaneous calls. This is useful for supporting features like call waiting, conference calling, and consultative transfers.

In most network environments, dual-line configurations are useful for user IP phones, whereas single-line configurations are useful for network functions (such as intercom or paging). Example 5-5 configures two ephone-dns: the first as a single-line and the second as a dual-line.

Tip: Newer IOS versions also support octo-line configuration, which enables eight calls per line! This configuration is typically used for receptionist phones, shared lines (where many people share the same extension), or as a conference resource. Assigning all phones the octo-line configuration uses excessive resources on your CME router and is not recommended!

Example 5-5 *Configuring ephone-dn*

```
CME_Voice# config t
Enter configuration commands, one per line.  End with CNTL/Z.
CME_Voice(config)# ephone-dn ?
  <1-150>   ephone-dn tag
CME_Voice(config)# ephone-dn 1
CME_Voice(config-ephone-dn)# number ?
  WORD  A sequence of digits - representing telephone number
CME_Voice(config-ephone-dn)# number 1000
CME_Voice(config-ephone-dn)# exit
CME_Voice(config)# ephone-dn 2 ?
```

```
  dual-line  dual-line DN (2 calls per line/button)
  <cr>
CME_Voice(config)# ephone-dn 2 dual-line
CME_Voice(config-ephone-dn)# number 1001
```

That's all there is to it! Notice the range of ephone-dn tags from the context-sensitive help is 1–150. This tag is a logical number, which is used when assigning the ephone-dn to an ephone. You can choose any ephone-dn tag from the range when creating the ephone-dn as long as the total number of ephone-dns does not exceed the number specified using the max-dn command.

Note: Creating more ephone-dns than you have specified using the **max-dn** command results in the following error message:

CME_Voice(config)# ephone-dn 49

dn tag 49 exceeds legal range 1 to max-dn 48

To correct this error, increase the max-dn value from telephony service configuration mode.

The number syntax (which is used to assign a directory number to an ephone-dn) also supports a secondary number value. For example, you can enter the following:

```
CME_Voice(config)#ephone-dn 2 dual-line
CME_Voice(config-ephone-dn)#number 1001 secondary 4805551001
```

This enables the ephone-dn to answer for multiple phone numbers. This can be used if you want an internal extension to be reachable if someone on the internal network dialed a four-digit extension or the full Public Switched Telephone Network (PSTN) Direct Inward Dial (DID) number.

Note: DID is a feature supported by PSTN carriers that enables internal extensions to be reached by PSTN callers directly without the need to route calls through a receptionist.

Understanding and Configuring Ephones

When you configure an ephone in CME, it represents the configuration applied directly to a single Cisco IP Phone or SoftPhone managed by the CME router. Just as with the ephone-dns, the max-ephone parameter directly impacts the number of ephones you are able to create and manage on a CME router. Example 5-6 adds an ephone to the CME router.

Example 5-6 *Creating an Ephone*

```
CME_Voice(config)# ephone ?
  <1-24>  Ethernet phone tag
CME_Voice(config)# ephone 1
CME_Voice(config-ephone)#
```

After you enter the command **ephone <tag>**, the CME router moves you into ephone configuration mode. Every command you enter after this directly affects the Cisco IP Phone matched to this ephone.

After initially creating the ephone, you need to logically link it to the physical IP phone it represents. The CME router uses the MAC address of a Cisco IP Phone for this purpose. There are three ways to find the MAC address of a Cisco IP Phone:

- **On the box of the Cisco IP Phone:** The box the Cisco IP Phone ships in has the MAC address of the phone on it next to a UPC code.
- **On the back of the Cisco IP Phone:** A sticker on the back side of the phone lists the MAC address of the device. The address appears next to a UPC code.
- **From the Settings menu of the Cisco IP Phone:** All Cisco IP Phones have a Settings button that enables you to manually configure various settings for the device. On most models of Cisco IP Phones, choosing Settings and then Network Configuration displays the MAC address of the phone on the LCD display.

Note: Having a UPC code containing the MAC address information of the IP phone is beneficial if you have a handheld UPC scanner (barcode scanner). You can set up the scanner to allow you to scan the MAC address of each device and then input the extension number into an Excel spreadsheet. You can then export this information for bulk entry of devices.

When you have the MAC address of the IP phone, you can enter it from ephone configuration mode, as shown in Example 5-7.

Example 5-7 *Assigning a MAC Address to an Ephone*

Key Topic

```
CME_Voice(config)# ephone 1
CME_Voice(config-ephone)# mac-address ?
  H.H.H  Mac address
  <cr>
CME_Voice(config-ephone)# mac-address 0014.1c48.e71a
```

After you enter the ephone MAC address information, you can verify the IP phone registration status by using the **show ephone** command, shown in Example 5-8.

Example 5-8 *Verifying Ephone Registration Status*

```
CME_Voice# show ephone
ephone-1 Mac:0014.1C48.E6D1 TCP socket:[2] activeLine:0 REGISTERED in SCCP ver 11
  and Server in ver 8
mediaActive:0 offhook:0 ringing:0 reset:0 reset_sent:0 paging 0 debug:0 caps:8
IP:172.30.60.31 52777 Telecaster 7960  keepalive 0 max_line 6
```

```
ephone-2 Mac:000C.2957.ACF5 TCP socket:[-1] activeLine:0 UNREGISTERED

mediaActive:0 offhook:0 ringing:0 reset:0 reset_sent:0 paging 0 debug:0 caps:0
IP:0.0.0.0 0 Unknown 0  keepalive 0 max_line 0
```

This **show ephone** output shows two configured ephones: one currently registered (ephone 1) with the IP address 172.30.60.31 and the other currently unregistered (ephone 2).

> **Tip:** If you ever see a phone's status shown as DECEASED in the show ephone output, the CME router has lost connectivity with the IP phone through a TCP keepalive failure. The UNREGISTERED status indicates the CME router closed the connection to the IP phone in a normal manner.

With the ephone linked to a physical Cisco IP Phone, you can now begin assigning the buttons to the ephone-dns.

Associating Ephones and Ephone-DNs

Linking ephones and ephone-dns is probably the most confusing section of the CME configuration because there are so many options. You can assign ephone-dns by using the button command from ephone configuration mode. The basic syntax of this command is as follows:

```
button <physical button> <separator> <ephone-dn>
```

Example 5-9 demonstrates the basic use of the **button** command.

Key Topic

Example 5-9 *Assigning Ephone-DN 2 to Button 1 on Ephone 1*

```
CME_Voice(config)# ephone 1
CME_Voice(config-ephone)# button 1:2
CME_Voice(config-ephone)# restart
```

The button 1:2 syntax assigns ephone-dn 2 (configured in Example 3-5) to button 1 of ephone 1. The colon (:) separator designates that this is a "normal ring" button assignment. That is, calls to 1001 (the number of ephone-dn 2) causes the IP phone to audibly ring and the light on the handset to blink. The restart syntax causes the phone to perform a warm reboot and redownload its configuration file from the TFTP server. Figure 5-2 illustrates what the physical IP phone looks like after making this assignment.

You can assign multiple lines to a phone by either entering multiple button commands, as shown in Example 5-10, or putting multiple entries on the same line, as shown in Example 5-11.

Example 5-10 *Assigning Buttons Using Multiple Commands*

```
CME_Voice(config)# ephone 1
CME_Voice(config-ephone)# button 1:2
CME_Voice(config-ephone)# button 2:1
```

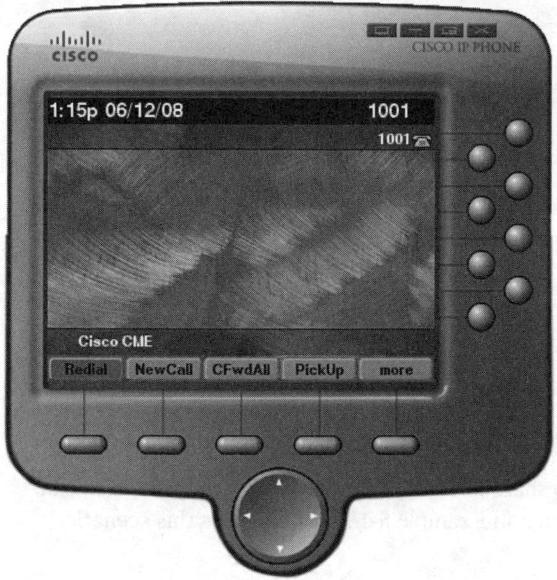

Figure 5-2 *Ephone 1 Following ephone-dn Assignment*

Example 5-11 *Assigning Buttons Using One Command*

```
CME_Voice(config)# ephone 1
CME_Voice(config-ephone)# button 1:2 2:1
```

The **show ephone** command is also used to verify button assignments, as shown in Example 5-12.

Example 5-12 *Verifying Button Assignments Using the* **show ephone** *Command*

```
CME_Voice# show ephone
ephone-1 Mac:0014.6A16.C2DA TCP socket:[5] activeLine:0 REGISTERED in SCCP ver 8
  and Server in ver 8
mediaActive:0 offhook:0 ringing:0 reset:0 reset_sent:0 paging 0 debug:0 caps:7
IP:172.30.60.32 14719 7912    keepalive 2701 max_line 2 dual-line
button 1: dn 2 number 1001 CH1   IDLE           CH2   IDLE
button 2: dn 1 number 1000 CH1   IDLE
```

Notice the last two lines of this **show** output. You are now able to verify the ephone-dns assigned to the ephone. As an added benefit, the output displays the actual extension as well. Notice the first button shows CH1 and CH2 as IDLE, whereas the second button just shows CH1 as IDLE. This is because ephone-dn 2 was configured with the dual-line syntax, which enables two active channels.

Let's wrap up this initial ephone/ephone-dn discussion with a final configuration example. In this example, a company wants to create five DNs for its three IP phones:

- IP phone 1 (normal employee)
  - Line 1: Directory number 1010
  - Line 2: Directory number 1015
- IP phone 2 (normal employee)
  - Line 1: Directory number 1011
  - Line 2: Directory number 1015
- IP phone 3 (receptionist)
  - Line 1: Directory number 1012
  - Line 2: Directory number 1013
  - Line 3: Directory number 1015

The company wants the employees to share directory number 1015 so any of them can answer incoming calls. The configuration in Example 5-13 accomplishes this scenario.

Example 5-13 *Multiple ephone-dn and ephone Configuration*

```
CME_Voice(config)# ephone-dn 10 dual-line
CME_Voice(config-ephone-dn)# number 1010
CME_Voice(config)# ephone-dn 11 dual-line
CME_Voice(config-ephone-dn)# number 1011
CME_Voice(config)# ephone-dn 12 dual-line
CME_Voice(config-ephone-dn)# number 1012
CME_Voice(config)# ephone-dn 13 dual-line
CME_Voice(config-ephone-dn)# number 1013
CME_Voice(config)# ephone-dn 15 dual-line
CME_Voice(config-ephone-dn)# number 1015
CME_Voice(config)# ephone 5
CME_Voice(config-ephone)# mac-address 00a0.932a.b34c
CME_Voice(config-ephone)# button 1:10 2:15
CME_Voice(config-ephone)# exit
CME_Voice(config)# ephone 6
CME_Voice(config-ephone)# mac-address 00a0.aa25.431b
CME_Voice(config-ephone)# button 1:11 2:15
CME_Voice(config-ephone)# exit
CME_Voice(config)# ephone 7
CME_Voice(config-ephone)# mac-address 00a0.a819.90a1
CME_Voice(config-ephone)# button 1:12
CME_Voice(config-ephone)# button 2:13
CME_Voice(config-ephone)# button 3:15
CME_Voice(config-ephone)# exit
```

> **Note:** The buttons for the receptionist phone also can be entered using the command **button 1:12 2:13 3:15**. Example 5-13 shows them entered on separate lines to make it a little easier to understand.

Adding Directory Numbers, Phones, and Users with CCP

For the not-so-command-line-at-heart, Cisco provides the Cisco Configuration Professional (CCP) GUI to configure ephones and ephone-dns using a point-and-click interface. Because the CCP keeps everything in order and formal, simply adding ephone-dns (called "Extensions" in CCP) and ephones (called "Phones" in CCP) is not possible. If you try to do it, CCP displays the error message shown in Figure 5-3.

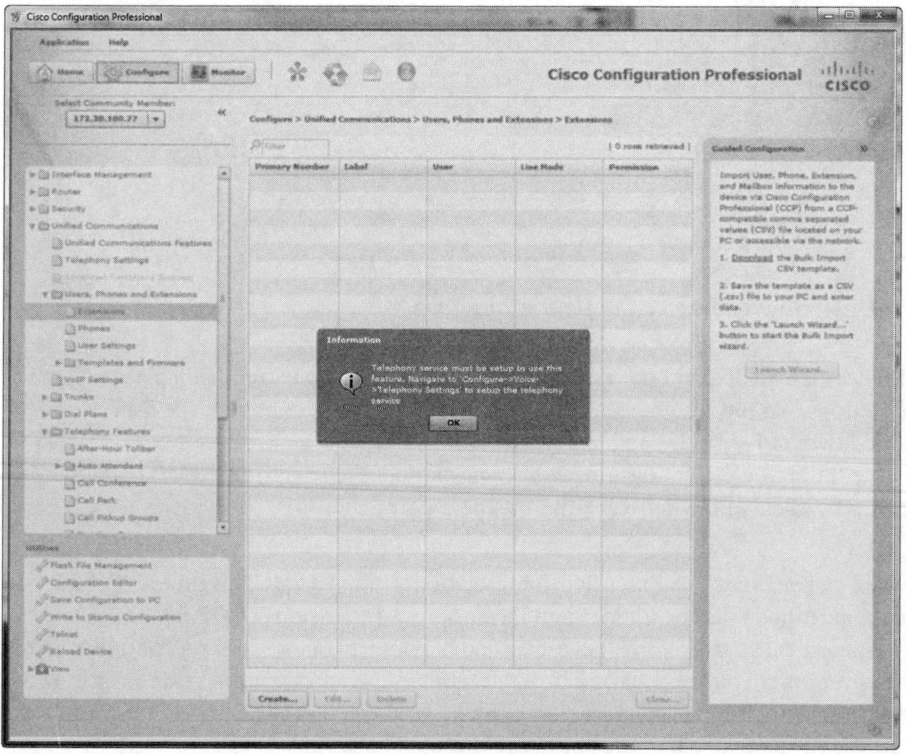

Figure 5-3 *CCP Forces Telephony Service Configuration*

The Telephony Service configuration is essentially the same configuration from the Base CME Configuration commands shown in Example 5-4. To access the Telephony Service configuration in CCP, access **Configure > Unified Communications > Telephony Settings.** CCP then provides a window to configure the max-ephones, max-dn, and ip source address commands (see Figure 5-4). The IOS CLI preview window confirms this.

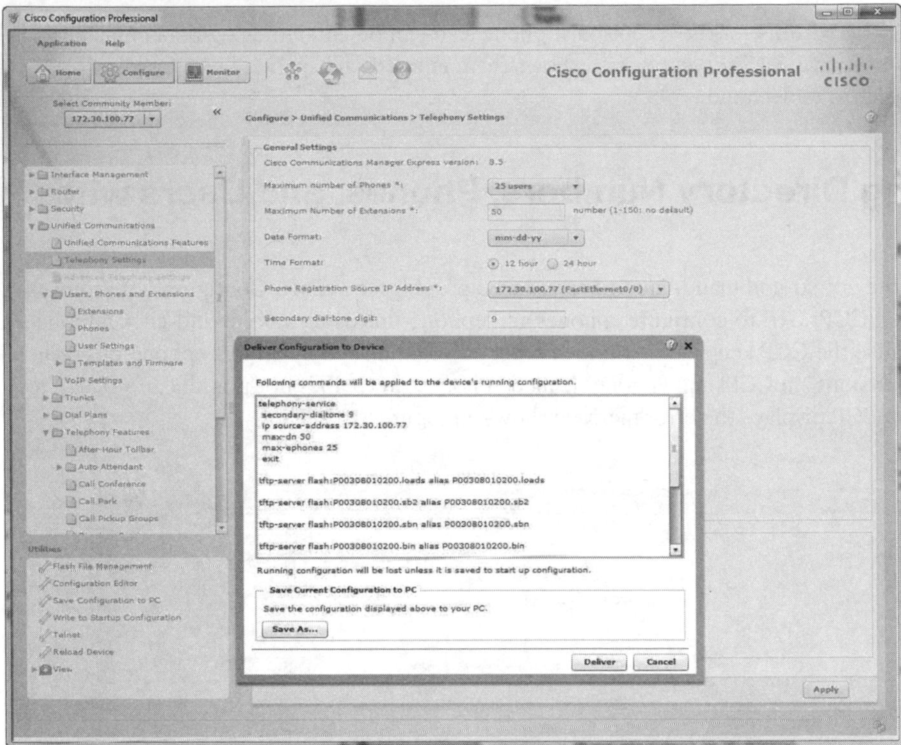

Figure 5-4 *Telephony Service Configuration from CCP*

> **Tip:** Sometimes, CCP might not refresh the CME configuration after making a change (such as configuring the base telephony service features). If you configure the telephony service as shown in Figure 5-4 and still receive the message stating that the telephony service must be set up (before configuring extensions or phones), refresh the CME configuration in CCP by clicking the Refresh button at the top of the utility.

Now, we can return to the Extension configuration within CCP (**Configure > Unified Communications > Users, Phones and Extensions > Extensions**). CCP does not immediately connect the extensions (ephone-dns) and phones (ephones); you can configure them in whatever order you feel most comfortable. For now, we start with the extensions: At the bottom of the CCP Extensions window is a button labeled Create. Clicking this causes the Create Extension dialog box to appear, as shown in Figure 5-5.

This window provides the same capabilities we had when we added the ephone-dn from a command-line interface (CLI) without knowing a lick of command-line syntax. Upon clicking the OK button, the CCP IOS CLI Preview displays the following syntax:

```
ephone-dn 1 dual-line
 number 1050 secondary 14805551050
 label Jeremy Cioara
 description Primary line
 exit
```

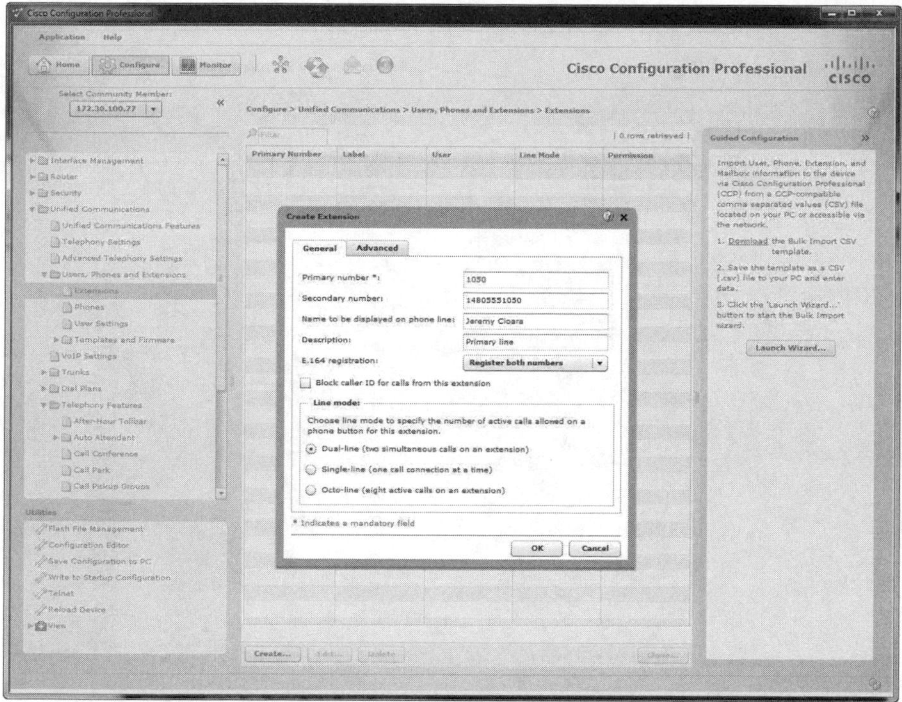

Figure 5-5 *Configuring Extensions in CCP*

As you add the extensions to CCP, they are stored on the CME running-configuration (don't forget to save when you're done!) and appear as valid extensions in CCP (see Figure 5-6).

Now, we can move on to adding the phones. Clicking the Phones CCP configuration (**Configure > Unified Communications > Users, Phones and Extensions > Phones**) reveals a window almost identical to the Extensions window. Clicking the Add button pops up a simple interface asking for two key items: the model of Cisco IP Phone you want to add and the associated MAC address (shown in Figure 5-7).

After you click the OK button, the CCP CLI Preview displays the following syntax:

```
ephone 1
 mac-address 0014.1C48.E71A
 type 7960
 auto-line
 exit
```

You can repeat this operation for as many phones as you want to add.

Tip: On the right pane of the CCP interface is a bulk import wizard. This exists both for the extensions and phones. If you format all the required data in a Comma Separated Value (CSV) file, you can perform a bulk import of both extensions and phones into the system (which takes much less time than adding them individually for a large group of devices).

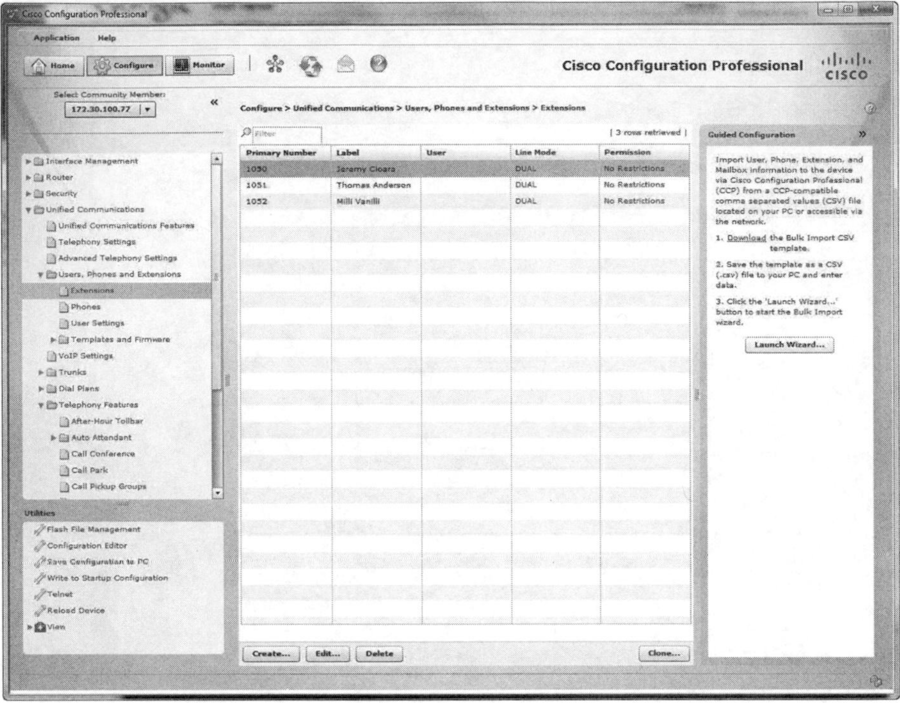

Figure 5-6 *CCP with Three Extensions Added*

Now that we added extensions and phones to the system, we can move into the final piece of the configuration: adding users. Unlike the pure command-line configuration of CME, which links ephone-dns to ephones, CCP ties the extensions and phones through a user account. By creating a user account, the end user can have the ability to manage his or her phone through a web interface. For now, let's get the phones working.

To add a user in CCP, navigate to **Configure > Unified Communications > Users, Phones and Extensions > User Settings**. After you are there, you can create a new user, as shown in Figure 5-8. The only required field is the User ID, which represents the username the user uses when logging on to the system. You can also configure these additional fields:

- First Name
- Last Name
- Display Name (used for Caller ID)
- Password (used to log on to the system)
- PIN (used for features like Extension Mobility)

This simple configuration window also contains the information CME uses to build a local directory of users. After you add these users to CME, the users of the VoIP system will be able to press the directories button on their IP phone, navigate to Local Directory, and search the database of users contained in CME.

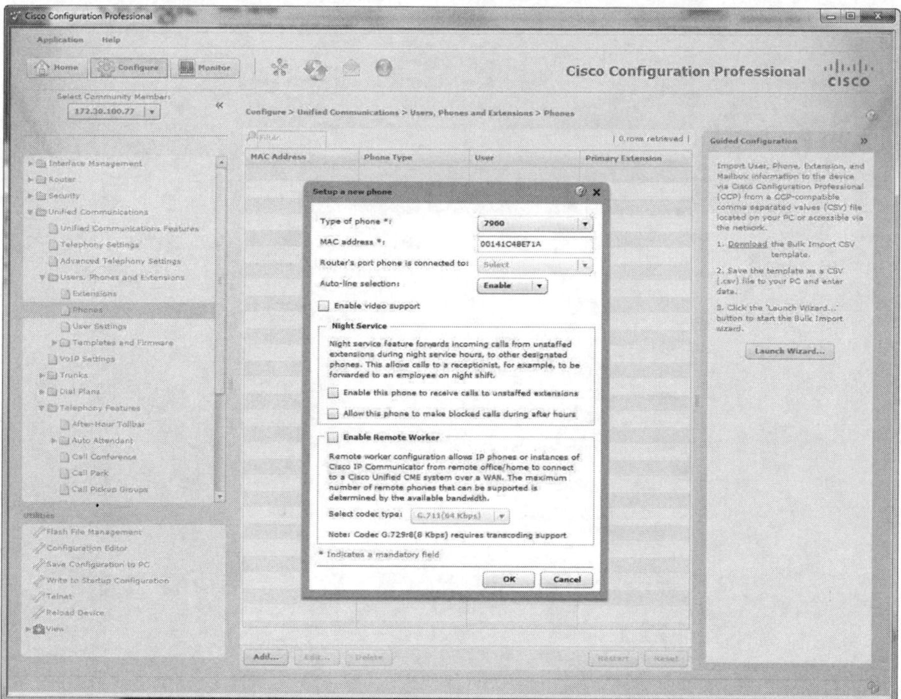

Figure 5-7 *Adding a Phone Using CCP*

After you enter the user information, you can click the Phones/Extensions tab at the top of the Create User window to associate the phone, extensions, and user account. Notice how CCP organizes this window in Figure 5-9. You first select the phone by referencing its MAC address. Based on the model you chose when you originally configured the phone, CCP displays an image of the phone model (definitely a nice touch by Cisco). Beneath the phone drop-down box, you can then choose which line you want to configure (because this is a Cisco 7960 IP Phone, you can configure up to six lines), the type of line (you most often choose a Regular line), and its ring type (normal ring, silent ring, beep ring, and so on). Near the bottom of the window, CCP displays the extensions you previously configured. In the case of Figure 5-9, I assigned extension 1050 to the ButtercupC user.

After we click the OK button in CCP to submit the configuration, it displays the following syntax:

```
ephone 1
 username ButtercupC password asdfasdf
 pin 12345
 button  1:1
 restart
 exit
ephone-dn 1
 name Buttercup Cioara
 exit
```

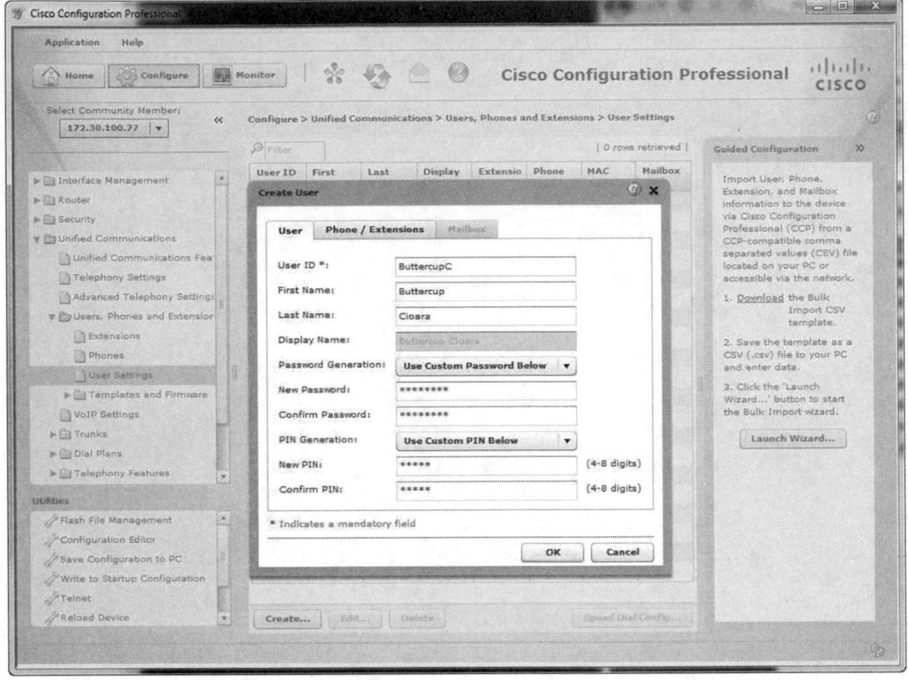

Figure 5-8 *Adding Users Using CCP*

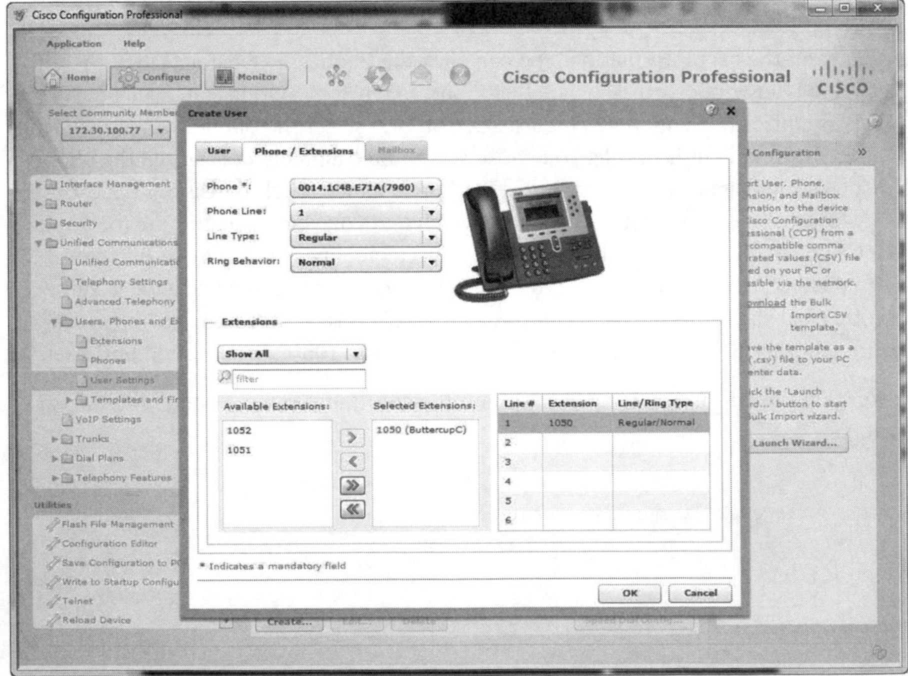

Figure 5-9 *Associating Phones, Extensions, and Users in CCP*

If you have not already connected the IP phone to the network, doing so now causes it to appear as a completely configured device.

Because nearly all verification and troubleshooting is typically a command-line realm, CCP provides a template set of **show** commands you are able to perform using a drop-down box. In addition, the CCP enables you to use the drop-down box as a "free form" field where you are able to type any custom show command you want. To access this area in CCP, navigate to **Configure > View > IOS Show Commands**. The View section is split apart from the normal configuration categories under the Utilities section of CCP. Figure 5-10 shows the output from the **show ephone register** command, which you can use to validate that the phone has indeed registered with CME.

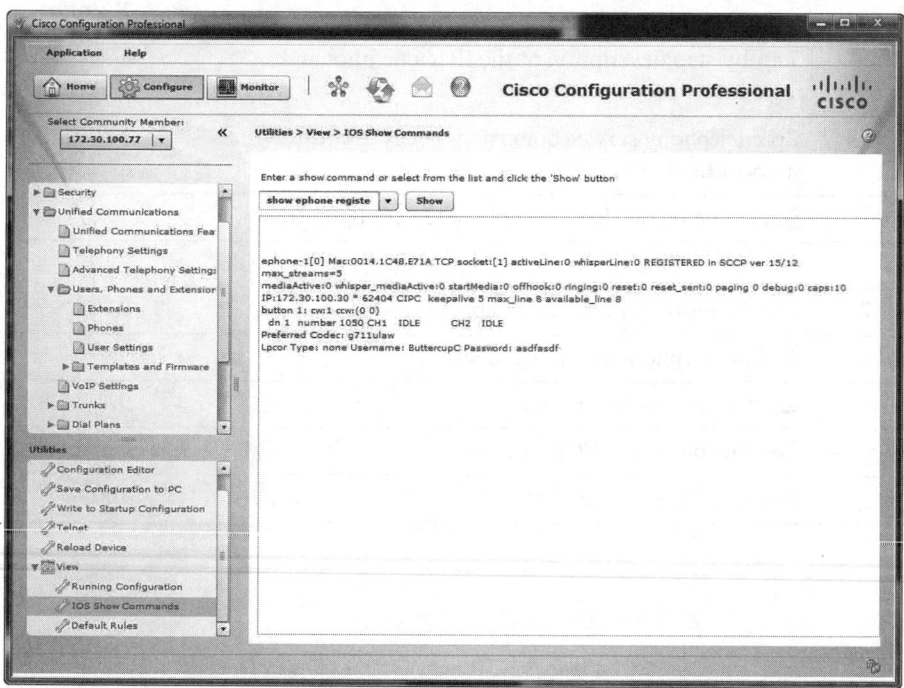

Figure 5-10 *Executing show Commands from CCP*

At this point, you configured CME to support a local IP telephony network within the office. Users are able to dial each other and receive calls. Eventually, someone in the office will want to make or receive an outside PSTN call, which will not be possible until you move on to Chapter 6, "Understanding the CME Dial-Plan."

Exam Preparation Tasks

Review All the Key Topics

Review the most important topics in the chapter, noted with the key topics icon in the outer margin of the page. Table 5-2 lists and describes these key topics and identifies the page numbers on which each is found.

Table 5-2 *Key Topics for Chapter 5*

Key Topic Element	Description	Page Number
List	Step-by-step description of the IP phone boot and registration process	84
List	Core telephony service commands necessary for CME to support IP phones	87
List	Single-line versus dual-line ephone-dn configurations	89
Example 5-5	Configuration process for an ephone-dn	89
Example 5-7	Configuration process for an ephone	91
Example 5-9	Assigning ephone-dns to ephones	92
Figure 5-5	Creating extensions in CCP	97
Figure 5-7	Creating phones in CCP	99
Figure 5-8	Creating users in CCP	100

- **Configuring Physical Voice Port Characteristics:** According to the OSI Model, the Physical layer is where it all begins. Similarly, some basic voice-port configuration items can prove essential to setting up your connections to the legacy voice environment. This section breaks down the key commands used in analog and digital voice connections.
- **Understanding and Configuring Dial-Peers:** If there were one most important topic for all things related to voice in CME, this would be it. Dial-peers define the "routing table" for your voice calls. This section discusses the configuration of both POTS and VoIP dial-peers through many practical examples.
- **Understanding Router Call Processing and Digit Manipulation:** Even the simplest of all VoIP environments need to modify dialed digits or caller ID information for incoming and outgoing calls. This section breaks down the core digit-manipulation commands and explores the flow of a typical CME-handled call.
- **Understanding and Implementing CME Class of Restriction (COR):** Just like you can use access lists to secure the data plane of your router, COR lists secure access to dialed numbers in your VoIP environment. This section discusses the concepts and configuration of COR lists.
- **Quality of Service:** Regardless of how beautiful and expansive your VoIP network, a voice network missing QoS is like confidential data lacking security—it won't last long. This section quickly overviews QoS and the configuration of AutoQoS on switches and routers.

CHAPTER 6

Understanding the CME Dial-Plan

Connecting a voice gateway to another voice network is similar to connecting a router to a data network: Plugging in the cable is the smallest part of the configuration! After the physical connections are in place, the configuration begins. Instead of routing tables, voice gateways build the logical dial-plan through a system of dial peers. This chapter explores the configuration and testing of dial peers in a voice environment.

"Do I Know This Already?" Quiz

The "Do I Know This Already?" quiz allows you to assess whether you should read this entire chapter or simply jump to the "Exam Preparation Tasks" section for review. If you are in doubt, read the entire chapter. Table 6-1 outlines the major headings in this chapter and the corresponding "Do I Know This Already?" quiz questions. You can find the answers in Appendix A, "Answers Appendix."

Table 6-1 *"Do I Know This Already?" Foundation Topics Section-to-Question Mapping*

Foundation Topics Section	Questions Covered in This Section
Configuring Physical Voice Port Characteristics	1–2
Understanding and Configuring Dial Peers	3–6
Understanding Router Call Processing and Digit Manipulation	7–9
Understanding and Implementing CME Class of Restriction	10
Quality of Service	11

1. Which of the following interface types would you use to connect an analog fax machine to the VoIP network?
    a. FXS
    b. FXO
    c. E&M
    d. BRI

2. Which of the following commands would you use to configure a T1 line to use channels 1 through 6 to connect to the PSTN using FXO loop start signaling?

   a. **pri-group 1-6 type fxo-loop-start**
   b. **pri-group 1 timeslots 1-6 type fxo-loop-start**
   c. **ds0-group 1-6 type fxo-loop-start**
   d. **ds0-group 1 timeslots 1-6 type fxo-loop-start**

3. You want to configure a dial peer to connect to a PBX system using a digital T1 CAS configuration. What type of dial peer would you create?

   a. Analog
   b. Digital
   c. POTS
   d. VoIP

4. You have the following configuration entered on your voice router:

```
dial-peer voice 99 pots
 destination-pattern 115.
 port 1/0/0
```

   A user dials the number 1159. What digits does the router send out the port 1/0/0?

   a. 1159
   b. 115
   c. 11
   d. 59
   e. 9

5. What is the default codec used by a VoIP dial peer?

   a. G.711 μ-law
   b. G.711 a-law
   c. G.723
   d. G.729

6. Which of the following destination patterns could you use to match any dialed number up to 32 digits in length? (Choose two.)

   a. .+
   b. [0-32]
   c. T
   d. &

7. After you create a translation rule, how is it applied?

a. To an interface

b. To a translation profile

c. Globally

d. To a dial peer

8. Which of the following digit manipulation commands will work under a VoIP dial peer?

a. **prefix**

b. **forward-digits**

c. **translation-profile**

d. **digit-strip**

9. What is the final method used by a router to match an inbound dial peer for incoming calls?

a. Using the **answer-address** command

b. Using **dial peer 0**

c. Using the **port** command

d. Using the **destination-pattern** command

10. If an ephone-dn lacks an incoming COR list and attempts to dial a dial peer assigned an outgoing COR list, what behavior occurs?

a. CME denies the call.

b. CME permits the call.

c. The call is rerouted to the next dial peer without an outgoing COR list.

d. CME disables the ephone-dn lacking an incoming COR list.

11. Which of the following is not an area you can use QoS to manage?

a. Packet jitter

b. Variable delay

c. Fixed delay

d. Router queuing

Foundation Topics

Configuring Physical Voice Port Characteristics

Before you can dive fully into the configuration of dial-plans using dial peers, you must first think about the physical characteristics of the voice ports on the router. Obviously, the voice ports plug into cables, which eventually connect to far-end devices. Beyond that, you can tune a few additional settings on the router to allow the voice ports to operate exactly to your specification. This section divides the discussion of these configurations into analog and digital forms.

Configuring Analog Voice Ports

Similar to Ethernet, when you connect a cable to an analog voice port on a router, it just works (provided a signal is coming from the other end). The router receives the electrical signals from the line and processes them normally. In addition to normal call processing, each interface type has a few settings you can tune to change the way it operates with the other end of the connection. This section describes configuration options for Foreign Exchange Station (FXS) ports and Foreign Exchange Office (FXO) ports.

Foreign Exchange Station Ports

FXS ports connect to end stations—that is, typical analog devices such as telephones, fax machines, and modems (shown in Figure 6-1).

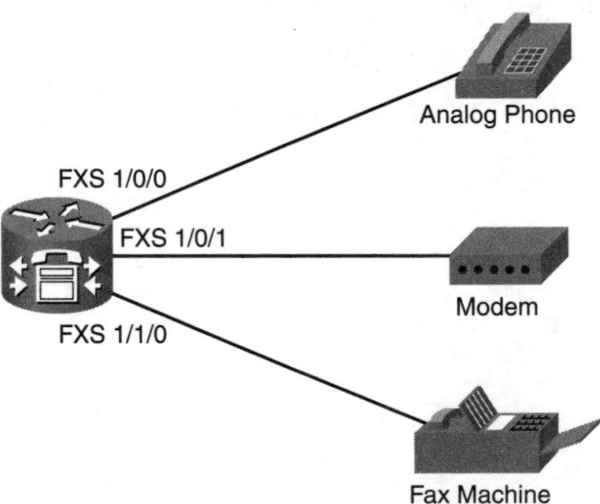

Figure 6-1 *FXS Port Connections*

When you are ready to configure your FXS voice ports, the best place to start is to find out what voice ports your router is equipped with. You can do this quickly by using the **show voice port summary** command, as shown in Example 6-1.

Example 6-1 *Identifying Voice Ports Using show voice port summary*

```
CME_Voice# show voice port summary
                                               IN       OUT
PORT           CH   SIG-TYPE    ADMIN OPER STATUS   STATUS   EC
============== == ============ ===== ==== ======== ======== ==
0/0/0          —   fxs-ls      up    dorm on-hook  idle     y
0/0/1          —   fxs-ls      up    dorm on-hook  idle     y
0/2/0          —   fxo-ls      up    dorm idle     on-hook  y
0/2/1          —   fxo-ls      up    dorm idle     on-hook  y
0/2/2          —   fxo-ls      up    dorm idle     on-hook  y
0/2/3          —   fxo-ls      up    dorm idle     on-hook  y
```

Note: If you are using your router for Cisco Unified Communication Manager Express (CME), each ephone-dn you configure shows up under the **show voice port summary** output as an EXFS port.

Based on the output from Example 6-1, you can see that this router is equipped with two FXS ports and four FXO ports.

FXS ports have three common areas of configuration:

- Signaling
- Call progress tones
- Caller ID information

You can use two types of signaling for analog FXS interfaces: ground start and loop start. The signal type dictates the method used by the attached device to signal that a phone is going off-hook. Table 6-2 briefly describes the differences between ground start and loop start signaling.

Table 6-2 *Comparing Ground Start and Loop Start*

Ground Start	Loop Start
Signals a new connection by grounding two of the wires in the cable temporarily	Signals by completing a circuit (by lifting the handset off-hook) and dropping the total DC voltage down on the line
Must be configured	Is the default
Typically used when connecting to PBX equipment	Typically used when connecting to analog devices, such as telephones, fax machines, and modems

You can use the syntax shown in Example 6-2 to set the signaling type on the voice port.

Example 6-2 *Configuring FXS Voice Port Signaling*

```
CME_Voice(config)# voice-port 0/0/0
CME_Voice(config-voiceport)# signal ?
  groundStart  Ground Start
  loopStart    Loop Start
CME_Voice(config-voiceport)# signal loopStart
```

If you have traveled to other countries, you probably noticed that phones sound different in different regions. Based on your geographical location, dial tones might be higher or lower and busy signals might be fast or slow. These are all considered call progress tones: audio signals that inform the caller how the call is progressing. By default, the FXS port of your router uses the call progress tones from the United States. If your router is serving another part of the world, use the command shown in Example 6-3 to adjust the call progress tones.

Example 6-3 *Adjusting Call Progress Tones*

```
CME_Voice(config)# voice-port 0/0/0
CME_Voice(config-voiceport)# cptone ?
  locale   2 letter ISO-3166 country code

AR Argentina              IN India                  PE Peru
AU Australia              ID Indonesia              PH Philippines
AT Austria                IE Ireland                PL Poland
BE Belgium                IL Israel                 PT Portugal
BR Brazil                 IT Italy                  RU Russian Federation
CA Canada                 JP Japan                  SA Saudi Arabia
CN China                  JO Jordan                 SG Singapore
CO Colombia               KE Kenya                  SK Slovakia
C1 Custom1                KR Korea Republic         SI Slovenia
C2 Custom2                KW Kuwait                 ZA South Africa
CY Cyprus                 LB Lebanon                ES Spain
CZ Czech Republic         LU Luxembourg             SE Sweden
DK Denmark                MY Malaysia               CH Switzerland
EG Egypt                  MX Mexico                 TW Taiwan
FI Finland                NP Nepal                  TH Thailand
FR France                 NL Netherlands            TR Turkey
DE Germany                NZ New Zealand            AE United Arab Emirates
GH Ghana                  NG Nigeria                GB United Kingdom
GR Greece                 NO Norway                 US United States
HK Hong Kong              OM Oman                   VE Venezuela
HU Hungary                PK Pakistan               ZW Zimbabwe
IS Iceland                PA Panama
```

Simply enter the two-digit country code to change the sound of all the progress tones on the device attached to the FXS port.

Finally, you can use the syntax shown in Example 6-4 to configure caller ID information for the device attached to the FXS port.

Example 6-4 *Configuring FXS Port Caller ID Information*

```
CME_Voice(config)# voice-port 0/0/0
CME_Voice(config-voiceport)# station-id name 3rd Floor Fax
CME_Voice(config-voiceport)# station-id number 5551000
```

This configuration allows other devices in your system to receive caller ID name and number information any time the device attached to the FXS port places a call to them.

Foreign Exchange Office Ports

FXO ports act as a trunk to the public switched telephone network (PSTN) central office (CO) or private branch exchange (PBX) systems, as shown in Figure 6-2.

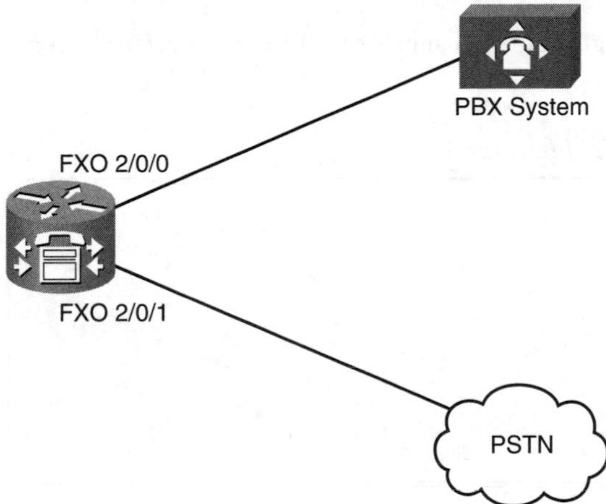

Figure 6-2 *FXO Port Connections*

FXO ports use many of the same commands as FXS ports, such as **signal** to set loop start or ground start signaling and **station-id** to set caller ID information. Two additional commands are of note:

- **dial-type**
- **ring number**

The **dial-type <dtmf/pulse>** command allows you to choose to use dual-tone multifrequency (DTMF) or pulse dialing. Yes, some areas of the world still require the use of pulse dialing (and rotary phones). If you are installing a voice network into one of these areas, this command is for you.

The **ring number <number>** command allows you to specify the number of rings that should pass before the router answers an incoming call to the FXO port. By default, this is set to one ring, which causes the router to answer an incoming call immediately. There might be instances where the FXO port of the router is attached to a loop of other devices (such as in a home office environment) and the user wants the other devices to have a chance to answer the call before the router picks up the line and processes it. In this case, you can set the ring number to a higher value.

Configuring Digital Voice Ports

Cisco provides digital T1 and E1 ports in the form of Voice and WAN Interface Cards (VWIC) for routers. These cards offer you the flexibility to configure them for a data connection or a voice connection. Unlike analog interfaces, you must configure digital interfaces before they will operate correctly, because the router does not know the type of network you will be using. As discussed in Chapter 1, "Traditional Voice Versus Unified Voice," two types of voice network configurations exist: T1/E1 channel associated signaling (CAS) or T1/E1 common channel signaling (CCS; commonly referred to as ISDN Primary Rate Interface [PRI]). The type of network to which you are connecting dictates the command you use to configure your VWIC card: **ds0-group** for T1/E1 CAS connections or **pri-group** for T1/E1 CCS connections.

Example 6-5 demonstrates how to configure all 24 channels of a T1 CAS interface to connect to a PSTN carrier.

Example 6-5 *Configuring a T1 CAS PSTN Interface*

```
CME_Voice# show controllers t1
T1 1/0 is down.
  Applique type is Channelized T1
  Cablelength is long gain36 0db
  Transmitter is sending remote alarm.
  Receiver has loss of signal.
  alarm-trigger is not set
  Soaking time: 3, Clearance time: 10
  AIS State:Clear  LOS State:Clear  LOF State:Clear
  Version info Firmware: 20050620, FPGA: 20, spm_count = 0
  Framing is SF, Line Code is AMI, Clock Source is Line.
   Current port master clock:local osc on this network module
  Data in current interval (215 seconds elapsed):
     0 Line Code Violations, 0 Path Code Violations
     0 Slip Secs, 0 Fr Loss Secs, 0 Line Err Secs, 0 Degraded Mins
     0 Errored Secs, 0 Bursty Err Secs, 0 Severely Err Secs, 215 Unavail Secs
CME_Voice# configure terminal
Enter configuration commands, one per line.  End with CNTL/Z.
CME_Voice(config)# controller t1 1/0
CME_Voice(config-controller)# framing ?
  esf  Extended Superframe
  sf   Superframe

CME_Voice(config-controller)# framing esf
CME_Voice(config-controller)# linecode ?
```

```
  ami   AMI encoding
  b8zs  B8ZS encoding
CME_Voice(config-controller)# linecode b8zs
CME_Voice(config-controller)# clock source ?
  free-running  Free Running Clock
  internal      Internal Clock
  line          Recovered Clock

CME_Voice(config-controller)# clock source line
CME_Voice(config-controller)# ds0-group ?
  <0-23>  Group Number
CME_Voice(config-controller)# ds0-group 1 ?
  timeslots  List of timeslots in the ds0-group
CME_Voice(config-controller)# ds0-group 1 timeslots ?
  <1-24>  List of T1 timeslots
CME_Voice(config-controller)# ds0-group 1 timeslots 1-24 ?
  type  Specify the type of signaling
  <cr>
CME_Voice(config-controller)# ds0-group 1 timeslots 1-24 type ?
  e&m-delay-dial        E & M Delay Dial
  e&m-fgd               E & M Type II FGD
  e&m-immediate-start   E & M Immediate Start
  e&m-lmr               E & M land mobil radio
  e&m-wink-start        E & M Wink Start
  ext-sig               External Signaling
  fgd-eana              FGD-EANA BOC side
  fxo-ground-start      FXO Ground Start
  fxo-loop-start        FXO Loop Start
  fxs-ground-start      FXS Ground Start
  fxs-loop-start        FXS Loop Start
  none                  Null Signalling for External Call Control
  <cr>
CME_Voice(config-controller)# ds0-group 1 timeslots 1-24 type fxo-loop-start ?
CME_Voice(config-controller)#^Z
CME_Voice# show voice port summary
                                        IN       OUT
PORT      CH  SIG-TYPE    ADMIN OPER STATUS   STATUS   EC
========= == ============ ===== ==== ======== ======== ==
1/0:1     01  fxo-ls      up    down idle     on-hook  y
1/0:1     02  fxo-ls      up    down idle     on-hook  y
1/0:1     03  fxo-ls      up    down idle     on-hook  y
1/0:1     04  fxo-ls      up    down idle     on-hook  y
1/0:1     05  fxo-ls      up    down idle     on-hook  y
1/0:1     06  fxo-ls      up    down idle     on-hook  y
1/0:1     07  fxo-ls      up    down idle     on-hook  y
```

```
1/0:1     08  fxo-ls       up    down idle      on-hook  y
1/0:1     09  fxo-ls       up    down idle      on-hook  y
1/0:1     10  fxo-ls       up    down idle      on-hook  y
1/0:1     11  fxo-ls       up    down idle      on-hook  y
1/0:1     12  fxo-ls       up    down idle      on-hook  y
1/0:1     13  fxo-ls       up    down idle      on-hook  y
1/0:1     14  fxo-ls       up    down idle      on-hook  y
1/0:1     15  fxo-ls       up    down idle      on-hook  y
1/0:1     16  fxo-ls       up    down idle      on-hook  y
1/0:1     17  fxo-ls       up    down idle      on-hook  y
1/0:1     18  fxo-ls       up    down idle      on-hook  y
1/0:1     19  fxo-ls       up    down idle      on-hook  y
1/0:1     20  fxo-ls       up    down idle      on-hook  y
1/0:1     21  fxo-ls       up    down idle      on-hook  y
1/0:1     22  fxo-ls       up    down idle      on-hook  y
1/0:1     23  fxo-ls       up    down idle      on-hook  y
1/0:1     24  fxo-ls       up    down idle      on-hook  y
```

There are many commands to discuss in Example 6-5, starting with the **show controllers t1** command. This command allows you to identify the T1 interfaces on your router. These interfaces do not appear in the **show ip interface brief** output, because the router does not know if you will configure the interface as a voice or data connection. After you identify the slot and port of your T1 interface, you can then configure the necessary **framing** and **linecode** commands. These commands let you change how the T1 or E1 interface formats the frames it sends to the service provider. Set these values based on the service provider to which you are connecting.

Note: If you are in the United States, most service providers use extended super frame (ESF) framing and B8ZS linecoding.

After you set the framing and linecoding, you can move into the clocking. The command **clock source line** instructs the router to receive its interface clocking from the service provider. If you are connecting to a PSTN carrier, this is the norm. If your router is connecting to a PBX system inside your company, you can enter the command **clock source internal**, which allows the router to provide clocking information to the PBX system.

Finally, the **ds0-group** command configures the line as a T1 CAS connection and allows you to enter the specific number of time slots you want to provision. In Example 6-5, all 24 time slots are provisioned under DS0 group 1. You can choose any value from 0 to 23 for the group number. This value acts as an identifier for the time slots you place into it. You can provision a single T1 line for many different purposes. For example, you could create DS0 group 5 with time slots 1–5 that connect to an onsite PBX system. You could then create DS0 group 6 using time slots 6–24 that connect to the PSTN (provided the PBX system and PSTN carrier are provisioned for these same time slot settings). Figure 6-3 illustrates the physical design of this network type.

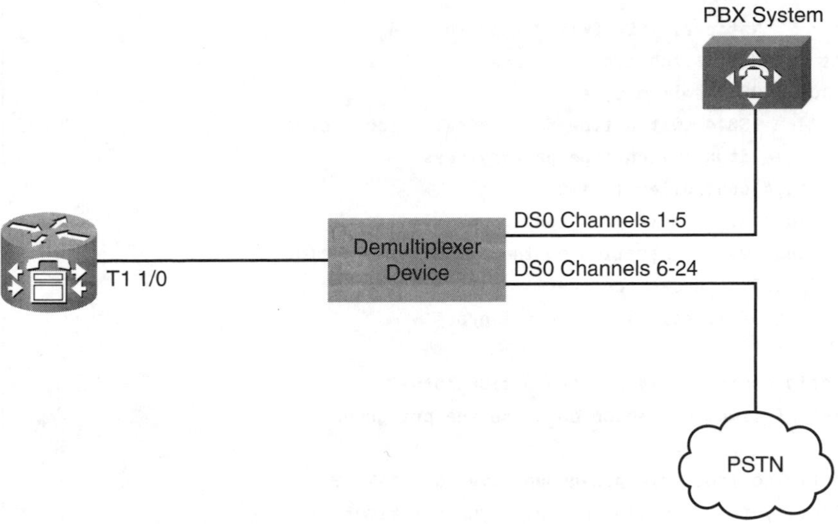

Figure 6-3 *Provisioning Multiple Connections with a Single T1 Interface*

Note: The demultiplexing device shown in Figure 6-3 allows you to break the single T1 interface into multiple interfaces with specific channel assignments.

Notice that the **ds0-group** command also allows you to set the signaling type. This gives you the ability to connect to many different network types. A PSTN carrier typically uses FXO loop start signaling over the T1 CAS connection (this might differ depending on your location and service provider). PBX systems often support one of the various Ear and Mouth (E&M) signaling types.

After you enter the **ds0-group** command, the router automatically creates a voice port for each time slot you provision, as you can see from the **show voice port summary** output in Example 6-5. The port is listed as 1/0:1 because 1/0 represents the physical interface and the additional 1 represents the DS0 group number. Make a note of this port identifier because you need it to configure the dial peers. Each port listed represents a different channel on the T1 interface.

The digital T1/E1 interface for a CCS (ISDN PRI) PSTN connection is configured using similar syntax as the CAS. Example 6-6 demonstrates a configuration that provisions all 24 time slots of a VWIC interface as a PRI PSTN connection.

Example 6-6 *Configuring a T1 CCS PSTN Interface*

```
CME_Voice(config)# isdn switch-type ?
  primary-4ess     Lucent 4ESS switch type for the U.S.
  primary-5ess     Lucent 5ESS switch type for the U.S.
  primary-dms100   Northern Telecom DMS-100 switch type for the U.S.
  primary-dpnss    DPNSS switch type for Europe
  primary-net5     NET5 switch type for UK, Europe, Asia and Australia
```

```
  primary-ni      National ISDN Switch type for the U.S.
  primary-ntt     NTT switch type for Japan
  primary-qsig    QSIG switch type
  primary-ts014   TS014 switch type for Australia (obsolete)
CME_Voice(config)# isdn switch-type primary-5ess
CME_Voice(config)# controller t1 1/0
CME_Voice(config-controller)# pri-group ?
  nfas_d     Specify the operation of the D-channel timeslot.
  service    Specify the service type
  timeslots  List of timeslots in the pri-group
  <cr>
CME_Voice(config-controller)# pri-group timeslots ?
  <1-24>  List of timeslots which comprise the pri-group

CME_Voice(config-controller)# pri-group timeslots 1-24 ?
  nfas_d   Specify the operation of the D-channel timeslot.
  service  Specify the service type
CME_Voice(config-controller)# pri-group timeslots 1-24
CME_Voice(config-controller)#^Z
CME_Voice# show voice port summary
                                          IN       OUT
PORT      CH   SIG-TYPE    ADMIN OPER STATUS   STATUS   EC
========= == ============ ===== ==== ======== ======== ==
1/0:23    01  isdn-voice  up    dorm none     none     y
1/0:23    02  isdn-voice  up    dorm none     none     y
1/0:23    03  isdn-voice  up    dorm none     none     y
1/0:23    04  isdn-voice  up    dorm none     none     y
1/0:23    05  isdn-voice  up    dorm none     none     y
1/0:23    06  isdn-voice  up    dorm none     none     y
1/0:23    07  isdn-voice  up    dorm none     none     y
1/0:23    08  isdn-voice  up    dorm none     none     y
1/0:23    09  isdn-voice  up    dorm none     none     y
1/0:23    10  isdn-voice  up    dorm none     none     y
1/0:23    11  isdn-voice  up    dorm none     none     y
1/0:23    12  isdn-voice  up    dorm none     none     y
1/0:23    13  isdn-voice  up    dorm none     none     y
1/0:23    14  isdn-voice  up    dorm none     none     y
1/0:23    15  isdn-voice  up    dorm none     none     y
1/0:23    16  isdn-voice  up    dorm none     none     y
1/0:23    17  isdn-voice  up    dorm none     none     y
1/0:23    18  isdn-voice  up    dorm none     none     y
1/0:23    19  isdn-voice  up    dorm none     none     y
1/0:23    20  isdn-voice  up    dorm none     none     y
1/0:23    21  isdn-voice  up    dorm none     none     y
1/0:23    22  isdn-voice  up    dorm none     none     y
1/0:23    23  isdn-voice  up    dorm none     none     y
```

When configuring a CCS connection, the first step is to set the ISDN switch type to match the type of switch your local service provider is using. Example 6-6 sets this to primary-5ess. After you configure the switch type, the router allows you to enter the **pri-group** command. This works identically to the **ds0-group** command in that it allows you to provision a specific number of time slots for use with the PSTN carrier. This command does not allow you to select a signaling type because the router assumes ISDN PRI signaling.

Note: This example assumes you have enough Digital Signal Processor (DSP) resources to support a full PRI connection. If your router does not have enough DSPs, it displays an error message when you try to create use the **pri-group** command. The error message says exactly how many channels the router can support.

After you enter the **pri-group** command, the router creates 24 ISDN voice ports that it will use for incoming and outgoing voice calls. This is verified with the **show voice port summary** command. Notice that the voice port is labeled with the identifier 1/0:23. This represents channel 23 (time slot 24) of the T1 ISDN PRI connection (channels are listed from 0–23, whereas time slots are listed 1–24), which is the dedicated signaling channel used to bring up the other 23 voice bearer channels.

Note: When using T1 interfaces, channel 23 (time slot 24) is always the signaling channel. When using E1 interfaces, channel 16 (time slot 17) is always the signaling channel.

As before, make a note of this port identifier for the ISDN circuit. The router requires you to identify this interface when configuring your dial peers.

Understanding and Configuring Dial Peers

When you initially entered the Cisco world, you probably learned about the concept of static routing. This method of routing allows you to manually enter destinations the router is able to reach on the data network. Dial peers use a similar concept to this; think of dial peers as static routes for your voice network. By default, the CME router only knows how to reach the ephone-dns you configure for the Cisco IP phones. You can connect the CME router to any number of FXS, FXO, or digital T1/E1 connections, but until you create dial peers for these connections, the router will not use them.

Dial peers define voice reachability information. Simply put, these are the phone numbers you can dial. For example, you might connect an analog phone to the FXS port of the router. As soon as you make the connection, the analog phone receives a dial tone and is able to place calls. However, no one will be able to call the analog phone because it does not yet have a phone number. Using a dial peer, you can assign one or more phone numbers to this analog device. Furthermore, dial peers allow you to use wildcards to define ranges of phone numbers. This is useful when you want to define large groups of numbers available from a destination such as a PBX system or PSTN connection.

You can create two primary types of dial peers:

- **Plain Old Telephone Service (POTS) dial peer:** Defines voice reachability information for any traditional voice connection (that is, any device connected to an FXS, FXO, E&M, or digital voice port).
- **Voice over IP (VoIP) dial peer:** Defines voice reachability information for any VoIP connection (that is, any device that is reachable through an IP address).

Figure 6-4 illustrates the placement of POTS and VoIP dial peers in a network.

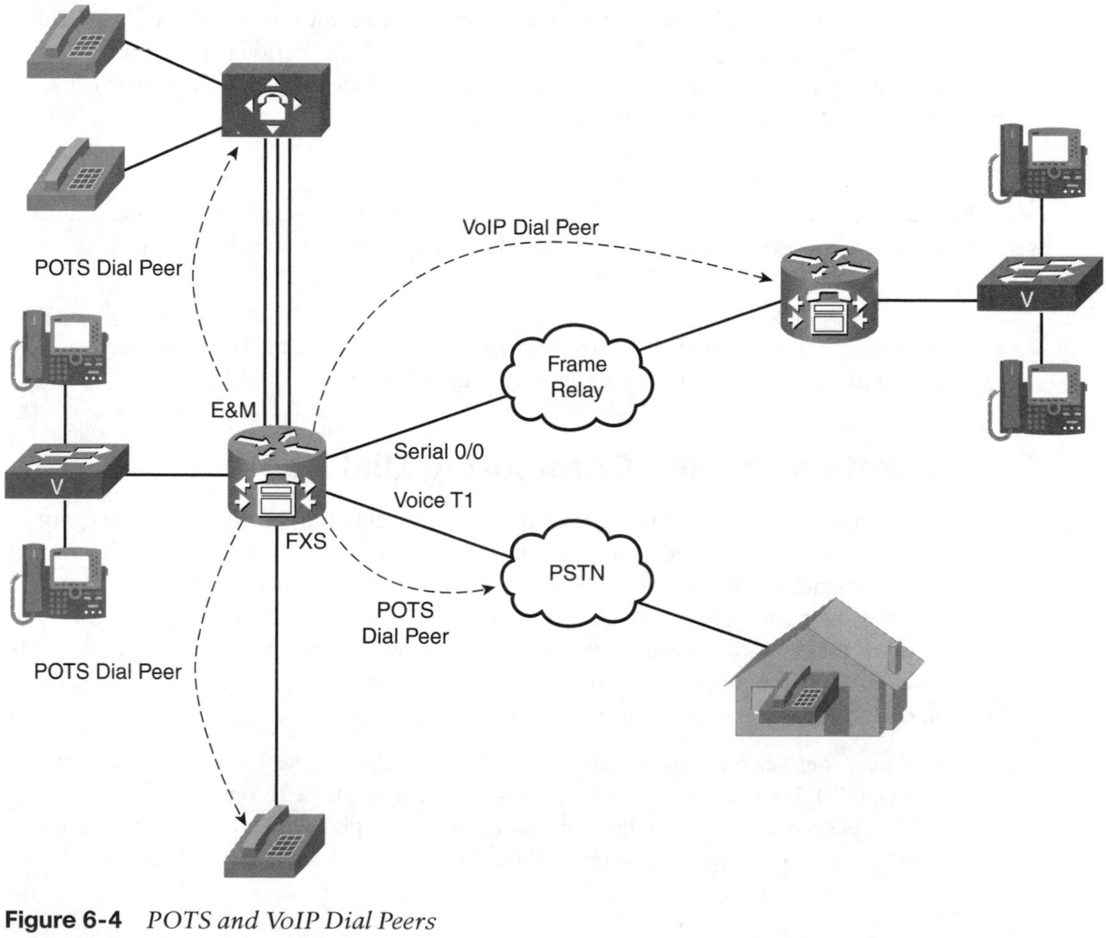

Figure 6-4 *POTS and VoIP Dial Peers*

Voice Call Legs

To accurately configure dial peers, you must first understand the concept of call legs. A call leg represents a connection to or from a voice gateway from a POTS or VoIP source. Figure 6-5 illustrates an example voice connection scenario.

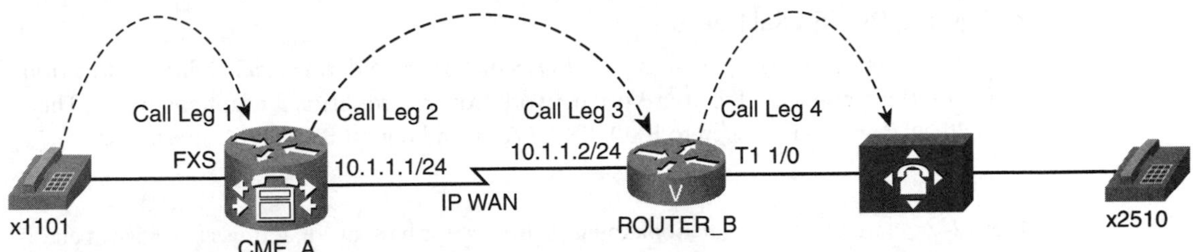

Figure 6-5 *Voice Connection Call Legs*

As illustrated in Figure 6-5, the phone on the left (extension 1101) makes a call to the phone on the right (extension 2510). For this call to pass through successfully, four call legs must exist:

- **Call leg 1:** The incoming POTS call leg from x1101 on CME_A.
- **Call leg 2:** The outgoing VoIP call leg from CME_A to ROUTER_B.
- **Call leg 3:** The incoming VoIP call leg on ROUTER_B from CME_A.
- **Call leg 4:** The outgoing POTS call leg to x2510 from ROUTER_B.

If the call was placed in the opposite direction (from x2510 to x1101), the same number of call legs would be needed, but in reverse. Thus, to provide a two-way calling environment that enables x1101 to call x2510 and vice versa, you need a total of eight call legs.

It is critical to understand the concept of call legs to properly configure the dial peers on your router. Each call leg identified in Figure 6-5 represents a dial peer that must exist on your router. These dial peers define not only the reachability information (phone numbers) for the devices, but also the path the audio must travel. From CME_A's perspective, it receives audio from x1101 on an FXS port (call leg 1). CME_A must then pass that voice information over the IP network to 10.1.1.2 (call leg 2). From ROUTER_B's perspective, it receives a call from x1101 on the IP WAN network (call leg 3). It must then take that call and pass it to the PBX system out the digital T1 1/0 interface (call leg 4).

As you can see from Figure 6-5, call legs are matched on the inbound and outbound direction. In the same way, you must configure dial peers to match voice traffic in both directions. In some cases, you can use a single dial peer for bidirectional traffic. For example, creating a single POTS dial peer for x1101 will match incoming and outgoing calls to x1101. At other times, you must create more than one dial peer for inbound and outbound traffic. For example, CME_A requires an outbound VoIP dial peer to send the call to ROUTER_B (10.1.1.2). ROUTER_B needs an inbound VoIP dial peer to receive the call from CME_A. As you see the multiple examples of dial peers in the upcoming sections, these concepts become clearer.

Note: Keep in mind, the engine behind all this is the DSP resources in the CME router. These workhorses convert the analog audio to digital (VoIP packets) and vice versa. If the CME routers did not have any DSP resources, this conversion would not be possible.

Configuring POTS Dial Peers

As previously mentioned, you can use POTS dial peers to define reachability information for anything connected to your VoIP network from the traditional telephony world. This includes devices connected to FXO, FXS, E&M, and digital BRI/T1/E1 interfaces.

Tip: If you are connecting to something that does not have an IP address (such as an analog phone, fax machine, PBX, or the PSTN), it is a POTS dial peer.

The network in Figure 6-6 demonstrates the configuration of POTS dial peers.

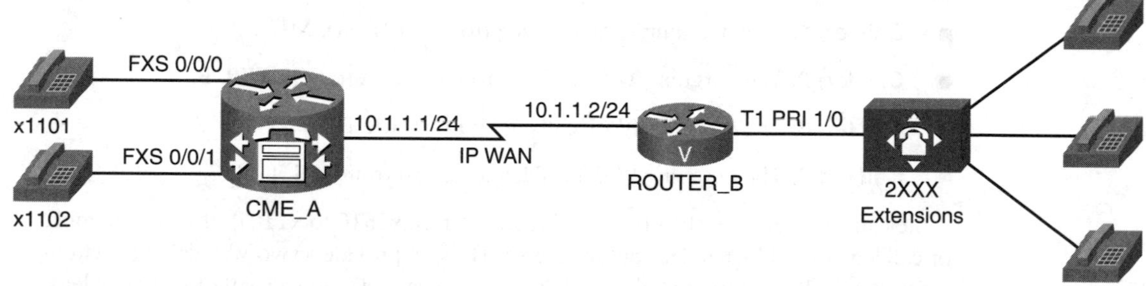

Figure 6-6 *Dial Peer Configuration Scenario*

The configuration begins with the CME_A router. To create POTS dial peers, you can use the syntax **dial-peer voice <tag> pots** from global configuration mode. The tag value can be any number you want (from 1 to 2,147,483,647), as long as it is unique on the router. Although this tag does not have any impact on the reachability information you assign to the devices, many administrators have a common practice of relating a dial peer tag value to the phone number of the device. Example 6-7 assigns the extensions shown in Figure 6-6 to the analog phones attached to the CME_A router's FXS ports.

Example 6-7 *Configuring POTS Dial Peers for FXS Ports*

```
CME_A(config)# dial-peer voice ?
  <1-2147483647>  Voice dial-peer tag
CME_A(config)# dial-peer voice 1101 ?
  mmoip  Multi Media Over IP
  pots   Telephony
```

```
  vofr   Voice over Frame Relay
  voip   Voice over IP
CME_A(config)# dial-peer voice 1101 pots
CME_A(config-dial-peer)# destination-pattern ?
  WORD  A sequence of digits - representing the prefix or full telephone number
CME_A(config-dial-peer)# destination-pattern 1101
CME_A(config-dial-peer)# port 0/0/0
CME_A(config-dial-peer)# exit
CME_A(config)# dial-peer voice 1102 pots
CME_A(config-dial-peer)# destination-pattern 1102
CME_A(config-dial-peer)# port 0/0/1
```

After you create the dial peer, you can then assign the phone number to the attached device(s) by using the **destination-pattern** and **port** commands. After you enter this configuration, you can place calls between the phones attached to the CME_A router. Before you place any calls, it is always best to verify the dial peer configuration.

Tip: After a dial peer is created, you no longer need to specify the type of dial peer when entering the configuration mode. For example, to create a VoIP dial peer with tag 50, you enter **dial-peer voice 50 voip.** If you want to return to the configuration mode later, you need only enter **dial-peer voice 50.** If you want to change the type of dial peer (VoIP or POTS), you must delete the dial peer (**no dial-peer voice 50**) and create it again.

The **show dial-peer voice** command (without the **summary** keyword) shows you all the dial peers on your router, but uses about a page of output for each dial peer. Although this information can be useful at times, the summary view, which is shown in Example 6-8, is usually much easier to digest. Notice at the bottom of the output are the dial peer tags 1101 and 1102, displayed as POTS dial peers with the proper destination pattern and port assignments. The other dial peers listed (with tags 20005–20014) are dial peers created by the CME routers for the ephone-dns configured in Chapter5, "Managing Endpoints and End Users with CME."

Example 6-8 *Verifying Dial Peers*

```
CME_A# show dial-peer voice summary
dial-peer hunt 0
             AD                                  PRE PASS                  OUT
TAG    TYPE  MIN  OPER PREFIX    DEST-PATTERN    FER THRU SESS-TARGET      STAT PORT
20005  pots  up   up             1500$           0                         50/0/20
20006  pots  up   up             1501$           0                         50/0/21
20007  pots  up   up             1502$           0                         50/0/22
20008  pots  up   up             1503$           0                         50/0/23
20009  pots  up   up             1504$           0                         50/0/24
```

```
20010  pots  up   up         1505$        0                 50/0/25
20011  pots  up   up         1506$        0                 50/0/26
20012  pots  up   up         1507$        0                 50/0/27
20013  pots  up   up         1508$        0                 50/0/28
20014  pots  up   up         1509$        0                 50/0/29
1101   pots  up   up         1101         0              up   0/0/0
1102   pots  up   up         1102         0              up   0/0/1
```

You can test the configuration by placing a call between the devices. Because this is a book, you will not actually hear the phones ring, so Example 6-9 shows a useful **debug** command that you can use to see the router process the dialed digits from the phone attached to the FXS port.

Example 6-9 *Using debug voip dialpeer to Analyze Digit Processing*

```
CME_A# debug voip dialpeer
voip dialpeer default debugging is on
.Jul  2 17:16:44.698: //-1/77671F238035/DPM/dpMatchPeersCore:
   Calling Number=, Called Number=1, Peer Info Type=DIALPEER_INFO_SPEECH
.Jul  2 17:16:44.698: //-1/77671F238035/DPM/dpMatchPeersCore:
   Match Rule=DP_MATCH_DEST; Called Number=1
.Jul  2 17:16:44.698: //-1/77671F238035/DPM/dpMatchPeersCore:
   Result=Partial Matches(1) after DP_MATCH_DEST
.Jul  2 17:16:44.702: //-1/77671F238035/DPM/dpMatchPeersMoreArg:
   Result=MORE_DIGITS_NEEDED(1)
.Jul  2 17:16:45.114: //-1/77671F238035/DPM/dpMatchPeersCore:
   Calling Number=, Called Number=11, Peer Info Type=DIALPEER_INFO_SPEECH
.Jul  2 17:16:45.114: //-1/77671F238035/DPM/dpMatchPeersCore:
   Match Rule=DP_MATCH_DEST; Called Number=11
.Jul  2 17:16:45.114: //-1/77671F238035/DPM/dpMatchPeersCore:
   Result=Partial Matches(1) after DP_MATCH_DEST
.Jul  2 17:16:45.114: //-1/77671F238035/DPM/dpMatchPeersMoreArg:
   Result=MORE_DIGITS_NEEDED(1)
.Jul  2 17:16:45.914: //-1/77671F238035/DPM/dpMatchPeersCore:
   Calling Number=, Called Number=110, Peer Info Type=DIALPEER_INFO_SPEECH
.Jul  2 17:16:45.914: //-1/77671F238035/DPM/dpMatchPeersCore:
   Match Rule=DP_MATCH_DEST; Called Number=110
.Jul  2 17:16:45.914: //-1/77671F238035/DPM/dpMatchPeersCore:
   Result=Partial Matches(1) after DP_MATCH_DEST
.Jul  2 17:16:45.914: //-1/77671F238035/DPM/dpMatchPeersMoreArg:
   Result=MORE_DIGITS_NEEDED(1)
.Jul  2 17:16:46.426: //-1/77671F238035/DPM/dpMatchPeersCore:
   Calling Number=, Called Number=1101, Peer Info Type=DIALPEER_INFO_SPEECH
.Jul  2 17:16:46.426: //-1/77671F238035/DPM/dpMatchPeersCore:
   Match Rule=DP_MATCH_DEST; Called Number=1101
```

```
.Jul  2 17:16:46.426: //-1/77671F238035/DPM/dpMatchPeersCore:
   Result=Success(0) after DP_MATCH_DEST
.Jul  2 17:16:46.426: //-1/77671F238035/DPM/dpMatchPeersMoreArg:
   Result=SUCCESS(0)
   List of Matched Outgoing Dial-peer(s):
     1: Dial-peer Tag=1101
```

Notice the highlighted output from the **debug** command in Example 6-9. This shows the router performing digit-by-digit call processing. As the attached phone dials each digit, the router processes that digit and attempts to find a match from among its dial peer configuration. For the first three dialed digits, the result is clear: more digits needed. After the caller dials the fourth digit, the router matches dial-peer tag 1101 and processes the call.

Now, we can turn our attention to the POTS dial peer configuration on ROUTER_B, which has a T1 PRI connection to a PBX system hosting 2XXX extensions (four-digit extensions beginning with the number 2). In the earlier section, "Configuring Digital Voice Ports," the physical characteristics of the T1 VWIC interface were configured to support T1 PRI connectivity (by using the **pri-group** command). When that command was entered, the router automatically created the voice port 1/0:23, which represented the signaling channel of the T1 PRI connection. Example 6-10 now configures the router to use this T1 PRI port anytime it receives a call for a 2XXX extension.

Example 6-10 *Configuring a POTS Dial Peer for a T1 Interface*

```
ROUTER_B(config)# dial-peer voice 2000 pots
ROUTER_B(config-dial-peer)# destination-pattern 2...
ROUTER_B(config-dial-peer)# no digit-strip
ROUTER_B(config-dial-peer)# port 1/0:23
```

It's that simple. Notice that you can use the "." wildcard to represent any dialed digit. This instructs the router to send all 2XXX extensions out port 1/0:23 (the T1 PRI interface). One additional command in this example brings up a big point of discussion: **no digit-strip**. This command prevents the router from automatically stripping dialed digits from this dial peer. Now, why would the router do that? Because of the POTS dial peer rule Cisco programmed into Cisco IOS. Here's the rule:

Digit-Stripping Rule of POTS Dial Peers:

The router automatically strips any explicitly defined digit from a POTS dial peer before forwarding the call.

An explicitly defined digit is any non-wildcard digit. In the case of Example 6-10, 2 is an explicitly defined digit. This rule is in place primarily to assist with stripping outside dialing codes before sending calls to the PSTN. For example, organizations commonly require users to dial 9 to access an outside line (often receiving a second dial tone after they have dialed 9). However, if you keep this access digit prepended to the dialed phone number, the PSTN carrier rejects the call. Therefore, if you create a POTS dial peer with the

destination-pattern 9....... command (for seven-digit dialing), the router automatically strips the explicitly defined 9 digit before sending the call to the PSTN.

In the case of Example 6-10, stripping the 2 digit before sending the call to the PBX system is not a desired behavior. Thus, the **no digit-strip** command prevents this automatic digit-stripping process.

> **Note:** The automatic digit-stripping function is specific to POTS dial peers. VoIP dial peers (discussed in the following section) do not automatically strip digits.

Configuring VoIP Dial Peers

In the scenario shown in Figure 6-7, the POTS dial peers now provide connectivity to the legacy voice equipment. However, CME_A and ROUTER_B are divided by an IP WAN connection that the legacy voice equipment must cross to achieve end-to-end communication.

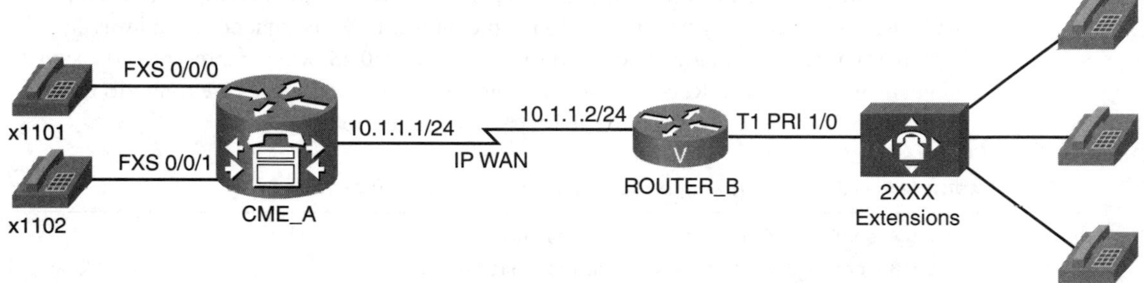

Figure 6-7 *Dial Peer Configuration Scenario*

To accomplish this connectivity, you must use VoIP dial peers, because the call is crossing an IP-based network. Example 6-11 configures the necessary VoIP dial peers on the CME_A and ROUTER_B devices.

Example 6-11 *Configuring VoIP Dial Peers*

```
CME_A(config)# dial-peer voice 2000 voip
CME_A(config-dial-peer)# destination-pattern 2...
CME_A(config-dial-peer)# session target ?
  WORD  A string specifying the session target
CME_A(config-dial-peer)# session target ipv4:10.1.1.2
CME_A(config-dial-peer)# codec ?
  clear-channel  Clear Channel 64000 bps (No voice capabilities: data transport
    only)
  g711alaw       G.711 A Law 64000 bps
  g711ulaw       G.711 u Law 64000 bps
  g722-48        G722-48K 64000 bps - Only supported for H.320<->H.323 calls
```

```
  g722-56         G722-56K 64000 bps - Only supported for H.320<->H.323 calls
  g722-64         G722-64K 64000 bps - Only supported for H.320<->H.323 calls
  g723ar53        G.723.1 ANNEX-A 5300 bps (contains built-in vad that cannot be
    disabled)
  g723ar63        G.723.1 ANNEX-A 6300 bps (contains built-in vad that cannot be
    disabled)
  g723r53         G.723.1 5300 bps
  g723r63         G.723.1 6300 bps
  g726r16         G.726 16000 bps
  g726r24         G.726 24000 bps
  g726r32         G.726 32000 bps
  g728            G.728 16000 bps
  g729br8         G.729 ANNEX-B 8000 bps (contains built-in vad that cannot be
    disabled)
  g729r8          G.729 8000 bps
  ilbc            iLBC 13330 or 15200 bps
CME_A(config-dial-peer)# codec g711ulaw

ROUTER_B(config)# dial-peer voice 1100 voip
ROUTER_B(config-dial-peer)# destination-pattern 110.
ROUTER_B(config-dial-peer)# session target ipv4:10.1.1.1
ROUTER_B(config-dial-peer)# codec g711ulaw
```

The primary difference between the POTS and VoIP dial peer configuration is the use of the **session target** command rather than the **port** command. When you use the context-sensitive help after the **session target** command, the router simply replies with WORD. This means that whatever you enter after the command is somewhat freeform. Most of the time, you will use the syntax **ipv4:**<*ip address*> to enter a remote IP address, as shown in Example 6-11. This command also allows you to direct calls to DNS names (by using **dns:**<*name*> syntax) or to a variety of call-management servers, such as H.323 gatekeepers or Session Initiation Protocol (SIP) proxy servers.

After you set the session target destination, you can optionally use the **codec** command to select the codec the router should use when placing a call to this destination.

Note: If the codec values do not match between the two routers, the call fails and returns a reorder tone (fast busy signal). This is commonly called a codec mismatch. The default codec value for VoIP dial peers is G.729.

Finally, notice that the dial peer 1100 on ROUTER_B uses the command destination-pattern 110. to direct all calls starting with the digits 110 to the CME_A router. Without this wildcard, you would need to create two VoIP dial peers on ROUTER_B: one for x1101 and one for x1102.

Note: Notice that dial-peer tag 2000 is used on the CME_A router for a VoIP dial peer and used on the ROUTER_B router for a POTS dial peer. This combination works just fine. The only restriction to keep in mind is that you cannot use the same dial peer tag value for different functions on the same router.

Using Dial Peer Wildcards

As you saw in the previous few sections, configuring dial peers (and destination patterns) without using wildcards would be extremely time consuming. By far, the most commonly used wildcard is the dot (.), which represents any dialed digit. You will find a few other wildcards useful in your configurations. Table 6-3 describes these wildcards.

Table 6-3 *Wildcards You Can Use with the destination-pattern Command*

Wildcard	Description
Period (.)	Matches any dialed digit from 0–9 or the * key on the telephone keypad. For example, 20.. matches any number from 2000 through 2099.
Plus (+)	Matches one or more instances of the preceding digit. For example, 5+23 matches 5523, 55523, 555523, and so on. This trend continues up to 32 digits, which is the maximum length of a dialable number.
Brackets ([])	Matches a range of digits. For example, [1-3]22 matches 122, 222, and 322. You can include a caret (^) before the entered numbers to designate a "does not match" range. For example, [^1-3]22 matches 022, 422, 522, 622, 722, 822, 922, and *22.
T	Matches any number of dialed digits (from 0–32 digits).
Comma (,)	Inserts a 1-second pause between dialed digits.

Note: The pound symbol (#) on a telephone keypad is not a wildcard symbol. This key immediately processes a dialed number when it is entered without waiting for additional digits.

Tip: If you plan to create a dial peer using only the T wildcard as the destination pattern, Cisco recommends that you create the destination as .T. This requires a user to dial at least one digit to match the destination pattern. Otherwise, a phone left off-hook for too long without a dialed digit will match the destination pattern.

Typically, the brackets wildcard is the most difficult to understand, primarily because it is the most flexible. Table 6-4 shows a few examples of how it can be used.

Table 6-4 *Destination-Pattern Brackets Wildcard Examples*

Pattern	Description
555[1-3]...	Matches dialed numbers beginning with 555, having 1, 2, or 3 as the fourth digit, and ending in any three digits.
[14-6]555	Matches dialed numbers where the first digit is 1, 4, 5, or 6 and the last three digits are 555.
55[59]12	Matches dialed numbers where the first two digits are 55, the third digit is 5 or 9, and the last two digits are 12.
[^1-7]..[135]	Matches dialed numbers where the first digit is not 1–7, the second and third digits are any number, and the last digit is 1, 3, or 5.

These wildcards are most often used when creating dial-plans for PSTN access. Initially, the most logical destination pattern choice for the PSTN may seem to be 9T (9 for an outside line followed by any number of digits). The problem with this is that Cisco designed the T wildcard to match variable-length strings from 0–32 digits. When a user dials an outside number, such as 14805551212, the router configured with the T wildcard will sit silently and wait for the user to dial more digits. By default, the router will wait for additional dialed digits for 10 seconds, which is the interdigit timeout (also called the T302 timer). Although you can force the router to process the call immediately after dialing the number by pressing the pound key (#), this is not something you would want to train all of your users to do.

Creating a PSTN dialing plan using wildcards other than T is not extremely difficult, as long as you think through the reachable PSTN numbers. Table 6-5 provides a sample PSTN dial-plan that you could use in the United States.

Table 6-5 *Sample PSTN Destination Patterns for North America*

Pattern	Description
[2-9]......	Used for 7-digit dialing areas
[2-9]..[2-9]......	Used for 10-digit dialing areas
1[2-9]..[2-9]......	Used for 11-digit long-distance dialing
[469]11	Used for service numbers, such as 411, 611, and 911
011T	Used for international dialing

Note: Although you can manually create an international dial-plan without using the T symbol, doing so can become tedious.

Example 6-12 illustrates the configuration of a North American PSTN dial-plan on a router. In this example, the T1 CAS voice port 1/0:1 is connected to the PSTN, and internal users must dial 9 for outside PSTN access.

Example 6-12 *Configuring a North American PSTN Dial-Plan*

```
VOICE_RTR(config)# dial-peer voice 90 pots
VOICE_RTR(config-dial-peer)# description Service Dialing
VOICE_RTR(config-dial-peer)# destination-pattern 9[469]11
VOICE_RTR(config-dial-peer)# forward-digits 3
VOICE_RTR(config-dial-peer)# port 1/0:1
VOICE_RTR(config-dial-peer)# exit
VOICE_RTR(config)# dial-peer voice 91 pots
VOICE_RTR(config-dial-peer)# description 10-Digit Dialing
VOICE_RTR(config-dial-peer)# destination-pattern 9[2-9]..[2-9]......
VOICE_RTR(config-dial-peer)# port 1/0:1
VOICE_RTR(config-dial-peer)# exit
VOICE_RTR(config)# dial-peer voice 92 pots
VOICE_RTR(config-dial-peer)# description 11-Digit Dialing
VOICE_RTR(config-dial-peer)# destination-pattern 91[2-9]..[2-9]......
VOICE_RTR(config-dial-peer)# forward-digits 11
VOICE_RTR(config-dial-peer)# port 1/0:1
VOICE_RTR(config-dial-peer)# exit
VOICE_RTR(config)# dial-peer voice 93 pots
VOICE_RTR(config-dial-peer)# description International Dialing
VOICE_RTR(config-dial-peer)# destination-pattern 9011T
VOICE_RTR(config-dial-peer)# prefix 011
VOICE_RTR(config-dial-peer)# port 1/0:1
VOICE_RTR(config-dial-peer)# exit
```

Two commands in this syntax deal with the automatic digit-stripping feature of POTS dial peers: **forward-digits** *<number>* and **prefix** *<number>*. The **forward-digits** *<number>* command allows you to specify the number of right-justified digits you want to forward. Notice the first dial peer 90 in Example 6-12. With a destination pattern of 9[469]11, the router would automatically strip the 9 and the two 1s from the pattern before sending the call. By entering the command **forward-digits 3**, the router forwards the right-justified three digits (411, 611, or 911) and only strips the 9.

The **prefix** *<number>* command adds any specified digits to the front of the dialed number before routing the call. This is useful for dial peer 93 in Example 6-12. Because international numbers can be a variable length, it is impossible to tell what value to enter for the **forward-digits** command. By using the **prefix 011** command, the automatic digit-stripping feature of POTS, dial peers strip the explicitly defined 9011 digits from the pattern, and the **prefix** command then adds the 011 back in its place.

Private Line Automatic Ringdown

Although not directly related to dial peer configuration, Private Line Automatic Ringdown (PLAR) configurations rely heavily on existing dial peers to complete a call. Ports configured with PLAR capabilities automatically dial a number as soon as the port detects an off-hook signal. The most obvious use for PLAR configurations is emergency phones in locations such as company elevators or parking garages. Example 6-13 designates x1101 (shown in Figure 6-8) as a PLAR extension that immediately dials x1102 as soon as a user lifts the receiver.

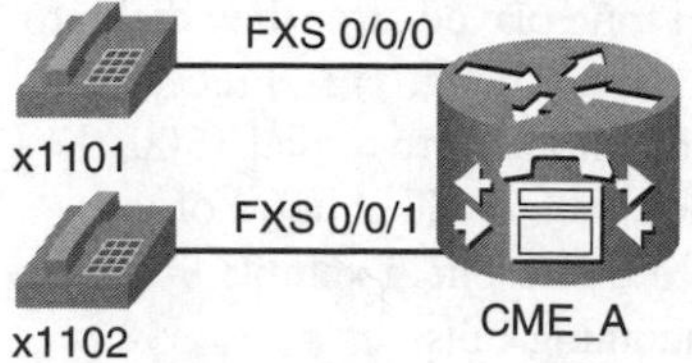

Figure 6-8 *PLAR Configuration*

Example 6-13 *FXS PLAR Configuration*

Key Topic

```
CME_A(config)# voice-port 0/0/0
CME_A(config-voiceport)# connection ?
  plar      Private Line Auto Ringdown
  tie-line  A tie line
  trunk     A Straight Tie Line
CME_A(config-voiceport)# connection plar ?
  WORD  A string of digits including wild cards
  tied  dedicated tie to this number
CME_A(config-voiceport)# connection plar 1102
```

The FXS voice port 0/0/0 is now hard-coded to dial the number 1102 as soon as a user lifts the handset.

PLAR can also be useful in a variety of other circumstances. One common scenario is using FXO connections to the PSTN, as shown in Figure 6-9.

Figure 6-9 *FXO PSTN Connections*

Although the **destination-pattern** command from dial peer configuration mode is useful for dictating what can go out the PSTN FXO ports, it is not too useful for handling what comes in the FXO ports. When the CME_A router shown in Figure 6-9 receives an incoming call from the PSTN, the call information sent from the PSTN carrier does not include dialed number information. (This is known as Dialed Number Identification Service [DNIS].) It includes caller ID information (known as Automatic Number Identification [ANI]), but this does not help the router to know where to send the call when it is received.

As a result, calls into the CME_A router hear a second dial tone played after they dial into the CME_A router from the PSTN. This is essentially the router saying, "Yes, I received your call; please tell me what to do now." If the caller on the phone were to dial 1500, the CME_A router would forward them to the receptionist. However, the likelihood of a PSTN caller doing this is slim. This is where PLAR comes to the rescue. Example 6-14 configures two analog FXO ports as PLAR connections for incoming calls.

Example 6-14 *FXO PLAR Configuration*

```
CME_A(config)# voice-port 2/0/0
CME_A(config-voiceport)# connection plar 1500
CME_A(config-voiceport)# exit
CME_A(config)# voice-port 2/0/1
CME_A(config-voiceport)# connection plar 1500
CME_A(config-voiceport)# exit
```

By entering the connection **plar 1500** command under both FXO ports, the router receives incoming calls from the PSTN and immediately forwards them to the receptionist phone rather than playing a second dial tone.

Note: Configuring PLAR connections for incoming calls is something you only need to do for analog FXO trunks. Digital PSTN connections (such as T1 or E1) receive DNIS information for incoming calls, which the router can use for Direct Inward Dial (DID) services.

Understanding Router Call Processing and Digit Manipulation

Understanding how the router processes dialed digits is critical to accurately implementing dial peers. There are two primary rules to guide you in your dial peer strategy:

- The most specific destination pattern always wins.
- When a match is found, the router immediately processes the call.

This section presents examples of these rules in action. Example 6-15 shows the dial peers for a router.

Example 6-15 *Sample Dial Peer Configuration 1*

```
dial-peer voice 1 voip
 destination-pattern 555[1-3]...
 session target ipv4:10.1.1.1
dial-peer voice 2 voip
 destination-pattern 5551...
 session target ipv4:10.1.1.2
```

If a user dials the number 5551234, both dial peers match, but the router chooses to use dial peer 2 because it is a more specific match (5551... matches 1000 numbers while 555[1-3]... matches 3000 numbers). Now, Example 6-16 shows what happens if you add a third dial peer to this configuration.

Example 6-16 *Sample Dial Peer Configuration 2*

```
dial-peer voice 1 voip
 destination-pattern 555[1-3]...
 session target ipv4:10.1.1.1
dial-peer voice 2 voip
 destination-pattern 5551...
 session target ipv4:10.1.1.2
dial-peer voice 3 voip
 destination-pattern 5551
 session target ipv4:10.1.1.3
```

If the user again dials 5551234, the router uses dial peer 3 to route the call. Likewise, the router processes only the 5551 digits and drops the 234 digits. This can be useful for emergency patterns such as 911 or 9911 (in North America) because the call is immediately routed when a user dials this specific pattern.

Tip: If you ever have a question of which dial peer will match a specific string, Cisco routers include a handy testing feature. From privileged mode, enter the command **show dialplan number** <*number*>, where *number* is the number you want to test. The router displays all the matching dial peers in the order in which the router will use them. The router lists more specific matches first.

Tip: Because the router immediately routes the call after it makes a specific match, it is best to avoid overlapping dial-plans if possible.

Avoiding overlapping dial-plans may be impossible at times. In these cases, you need to get creative with your dial peers to accomplish your objectives. For example, if you are required to have a dial peer matching the destination pattern 5551 while a second dial peer has the destination pattern 5551..., you could use a configuration like this as a solution:

```
dial-peer voice 2 voip
 destination-pattern 5551...
 session target ipv4:10.1.1.2
dial-peer voice 3 voip
 destination-pattern 5551T
 session target ipv4:10.1.1.3
```

Notice the T wildcard after 5551, which matches 0–32 digits. Users dialing extension 5551 now have to press the pound key (#) after they finish dialing or wait the 10-second interdigit timeout period. You could also accomplish this objective by using some fancy digit-manipulation techniques, which you learn about in an upcoming section.

Matching Inbound and Outbound Dial Peers

When a router receives a voice call, it must always match a dial peer in some way for the router to process the call. Although this might seem like a simplistic statement, there is actually a lot of strategy that must be in place to accomplish this in both the inbound and outbound direction. Take the scenario presented in Figure 6-10, which expands on the call leg scenario that opened this section on dial peers.

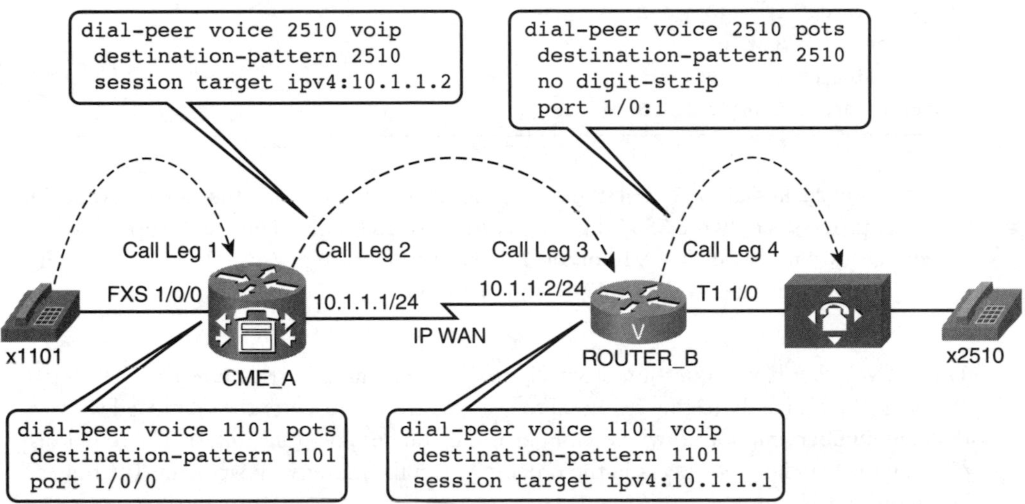

Figure 6-10 *Inbound and Outbound Dial Peers*

In addition to the call legs, Figure 6-10 displays the dial-peer configurations necessary to complete end-to-end calls from x1101 to x2510 and vice versa. Now, matching the outbound dial peers is easy: Take the dialed digits and compare them to the destination patterns under the dial peers you configured on the router to find the most specific match. For example, if x1101 dials x2510, the CME_A router looks at its dial peers and realizes there is a VoIP dial peer match directing the call to the IP address 10.1.1.2. When ROUTER_B receives the call, it realizes the dialed digits are an exact match to the POTS dial peer 2510, which causes the router to send the call out the T1 interface to the attached PBX system. This process explains how the router matches the outbound dial peers (shown in Figure 6-10 as call leg 2 and call leg 4), but how does the route match the inbound dial peers? A router matches inbound dial peers through the following five methods:

1. Match the dialed number (DNIS) using the **incoming called-number** dial peer configuration command.
2. Match the caller ID information (ANI) using the **answer-address** dial peer configuration command.

3. Match the caller ID information (ANI) using the **destination-pattern** dial peer configuration command.
4. Match an incoming POTS dial peer by using the **port dial-peer** configuration command.
5. If no match has been found using the previous four methods, use dial peer 0.

Look at Figure 6-10. Call legs 2 and 4 are accounted for as outbound dial peers matched by using the dialed number (DNIS) information against the **destination-pattern** command under the dial peers. Here's how the router uses the previous list of five rules to match the inbound dial peers.

For call leg 1:

1. (NO MATCH) 2510 (the dialed number) does not match an **incoming called-number** dial peer configuration command on the CME_A router, because this command does not exist in the configuration.
2. (NO MATCH) x1101 caller ID information (ANI) does not match an **answer-address** dial peer configuration command on the CME_A router, because this command does not exist in the configuration.
3. (NO MATCH) x1101 caller ID information (ANI) does not match the **destination-pattern** dial peer configuration command on the CME_A router, because x1101 does not have any caller ID information. That is, the phone itself does not provide caller ID information to the router, because an analog phone does not know its own phone number.
4. (MATCH) x1101 comes in FXS port 1/0/0, which matches an incoming POTS dial peer on the CME_A router by using the **port** dial peer configuration command (port 1/0/0).

Using the five-step matching process, the CME_A router is able to match an inbound dial peer using the incoming port value of the attached analog phone. The CME_A router then processes the outbound dial peer (call leg 2), and the call arrives at ROUTER_B. Once again, ROUTER_B works through the five-step process to match an inbound dial peer.

For call leg 3:

1. (NO MATCH) 2510 (the dialed number) does not match an **incoming called-number** dial peer configuration command on ROUTER_B, because this command does not exist in the configuration.
2. (NO MATCH) x1101 caller ID information (ANI) does not match an **answer-address** dial peer configuration command on ROUTER_B, because this command does not exist in the configuration.
3. (MATCH) x1101 caller ID information (ANI) does match the **destination-pattern** dial peer configuration command for the VoIP dial peer 1101 on ROUTER_B.

In this case, the VoIP dial peer 1101 on ROUTER_B doubles as both the outgoing dial peer for calls placed to x1101, and as an incoming dial peer for calls coming from x1101.

Now, to see the inbound matching process in its entirety, imagine that there is no VoIP dial peer 1101 on ROUTER_B, as shown in Figure 6-11.

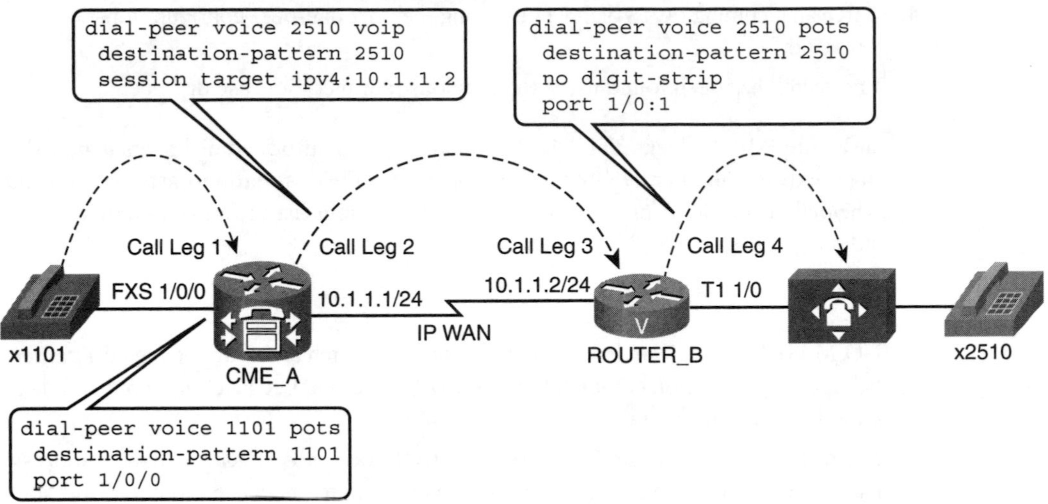

Figure 6-11 *Matching Inbound Dial Peers Using Dial Peer 0*

The first result is that you could not place calls to x1101 from ROUTER_B (or the PBX system attached to ROUTER_B). However, what if x1101 called x2510? The CME_A and ROUTER_B routers have enough information to match the outbound call legs. ROUTER_B is just missing the information for the inbound dial peer (call leg 3). Here's how the decision process would flow:

1. (NO MATCH) 2510 (the dialed number) does not match an **incoming called-number** dial peer configuration command on ROUTER_B, because this command does not exist in the configuration.
2. (NO MATCH) x1101 caller ID information (ANI) does not match an **answer-address** dial peer configuration command on ROUTER_B, because this command does not exist in the configuration.
3. (NO MATCH) x1101 caller ID information (ANI) does not match the **destination-pattern** dial peer configuration command, because the VoIP dial peer 1101 was removed on ROUTER_B.
4. (NO MATCH) x1101 did not come in a POTS interface (FXS, FXO, E&M, Voice BRI/T1/E1 digital interface) that could be matched using the **port** command; rather, x1101 came across a VoIP connection.
5. (MATCH) Because ROUTER_B could not find a match using the previous four methods, it uses dial peer 0.

So, this now begs the question, "What is dial peer 0?" Dial peer 0 is like a default gateway dial peer that appears when there is no dial peer match (this applies only for inbound dial peers, not for outbound dial peers). Although this allows the call to complete, you have no control over dial peer 0. You cannot configure it nor change any of its default settings. Dial peer 0 uses the following, unchangeable settings:

Key Topic

- **Any voice codec:** Dial peer 0 handles any incoming voice codec; it is not hard-coded to any specific codec.
- **No DTMF relay:** DTMF relay sends dialed digits outside of the audio stream. This is useful because compressed codecs often distort dialed tones on the call.
- **IP Precedence 0:** This is probably the most painful default of dial peer 0. Setting the traffic to IP Precedence (IPP) to 0 strips all QoS markings. The router now treats the voice traffic the same as the data traffic.
- **Voice Activity Detection (VAD) enabled:** VAD allows you to save bandwidth by eliminating voice traffic during periods of silence on the call.
- **No Resource Reservation Protocol (RSVP) support:** The lack of RSVP goes right along with the lack of any QoS for the voice calls. The router does not reserve any bandwidth specifically for dial peer 0 calls.
- **Fax-rate voice:** The router limits the bandwidth available to fax signals to the maximum allowed by the VoIP codec. This could devastate fax calls if you are using a low-bandwidth compressed codec.
- **No application support:** Dial peer 0 cannot refer calls to outside applications, such as an Interactive Voice Response (IVR) system.
- **No DID support:** Dial peer 0 cannot use the DID feature to automatically forward calls from an outside PSTN carrier to internal devices.

In light of this list of dial peer 0 features, it is best to always match an inbound dial peer where you can control the configuration.

Using Digit Manipulation

Digit manipulation is the process of adding or removing digits from a dialed number to help a call reach an intended destination. You have already seen a few of the digit manipulation commands during the discussion of the automatic digit-stripping feature of POTS dial peers (such as the no digit-strip and forward-digit commands). Before we look at some practical examples, Table 6-6 shows a list of common digit-manipulation commands you can use on a Cisco router.

Table 6-6 *Common Digit-Manipulation Methods on Cisco Routers*

Command	Mode	Description
prefix digits	POTS dial peer	Allows you to specify digits for the router to add before the dialed digits. Example: **prefix** 011 adds the numbers 011 to the front of the originally dialed number.
forward-digits number	POTS dial peer	Allows you to specify the number of right-justified digits to forward. Example: **forward-digits 4** forwards only the rightmost four digits from the dialed number.

Table 6-6 *Common Digit-Manipulation Methods on Cisco Routers*

Command	Mode	Description
[no] digit-strip	POTS dial peer	Enables or disables the default digit-stripping behavior of POTS dial peers. Example: **no digit-strip** turns off the automatic digit-stripping behavior under a POTS dial peer.
num-exp match digits set digits	Global	Transforms any dialed number matching the match string into the digits specified in the set string. Example: **num-exp 4... 5...** matches any four-digit dialed number beginning with 4 into a four-digit number beginning with 5 (4123 becomes 5123). Example: **num-exp 0 5000** matches the dialed digit 0 and changes it to 5000.
voice transla-tion-profile	Global and POTS or VoIP dial peer	Allows you to configure a translation profile consisting of up to 15 rules to transform numbers however you want. The translation profile is created globally and then applied to any number of dial peers (similar to an access list).

Following are four practical scenarios in which these digit-manipulation commands can prove to be useful.

Practical Scenario 1: PSTN Failover Using the prefix Command

One of the benefits of using VoIP communication over traditional telephony is the ability to have more than one path to a destination. Figure 6-12 shows a commonly encountered scenario encountered.

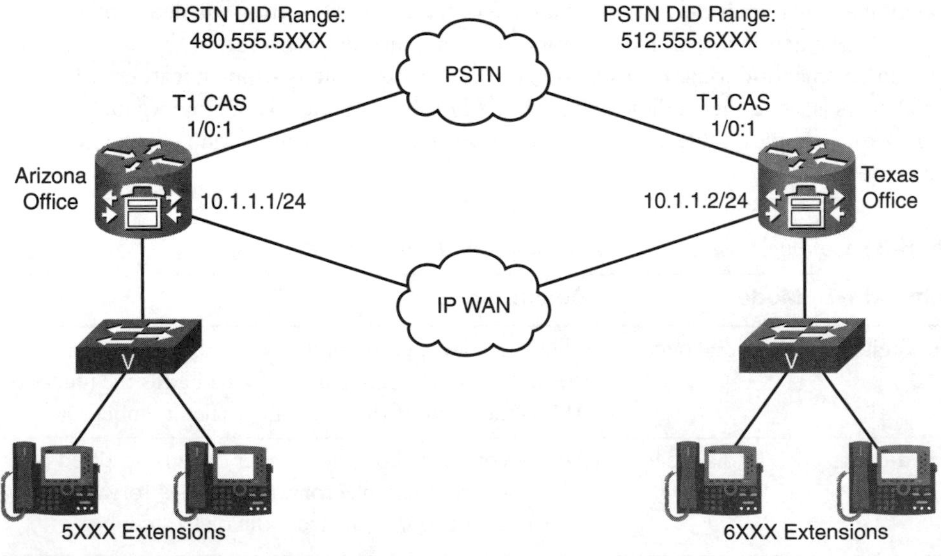

Figure 6-12 *Multiple Voice Paths*

The organization shown in Figure 6-12 prefers to use the IP WAN as its primary communication path between Arizona and Texas. However, if the IP WAN should fail, calls between the offices should use the PSTN as their communication path.

One of the benefits of using VoIP is the merging of voice networks into one, seamless communication path. Because calls are traveling over the IP WAN, users in the Arizona office can dial the users in the Texas office using their four-digit (6XXX) extension. Likewise, users in the Texas office can dial the users in the Arizona office using their four-digit (5XXX) extension. It would be inconvenient to require all the users in the Arizona office to dial the Texas office using the PSTN DID range rather than the four-digit extension (and vice versa).

Using a combination of the preference and prefix commands, you can allow this failover transformation to occur dynamically, as shown in Example 6-17.

Example 6-17 *Dynamic WAN to PSTN Failover Implementation*

Key Topic

```
Arizona(config)# dial-peer voice 10 voip
Arizona(config-dial-peer)# destination-pattern 6...
Arizona(config-dial-peer)# session target ipv4:10.1.1.2
Arizona(config-dial-peer)# preference 0
Arizona(config-dial-peer)# exit
Arizona(config)# dial-peer voice 11 pots
Arizona(config-dial-peer)# destination-pattern 6...
Arizona(config-dial-peer)# port 1/0:1
Arizona(config-dial-peer)# preference 1
Arizona(config-dial-peer)# no digit-strip
Arizona(config-dial-peer)# prefix 1512555
Texas(config)# dial-peer voice 10 voip
Texas(config-dial-peer)# destination-pattern 5...
Texas(config-dial-peer)# session target ipv4:10.1.1.1
Texas(config-dial-peer)# preference 0
Texas(config-dial-peer)# exit
Texas(config)# dial-peer voice 11 pots
Texas(config-dial-peer)# destination-pattern 5...
Texas(config-dial-peer)# port 1/0:1
Texas(config-dial-peer)# preference 1
Texas(config-dial-peer)# no digit-strip
Texas(config-dial-peer)# prefix 1480555
```

The **preference** command allows the router to determine which dial peer it should use in the case where the destination patterns are identical. It might seem counterintuitive, but the router considers lower preferences to be better than higher preferences (the preference value can be any number from 0–10). The default preference for a dial peer is 0. Thus, the **preference 0** command in Example 6-17 under dial peer 10 on both routers is redundant.

Tip: If you create multiple dial peers with exactly equal destination patterns and preferences, the router will randomly choose a dial peer to use.

Looking at the Arizona router in Example 6-17, you can see that dial peer 10 is the more preferred path to the Texas router. Because the connection uses VoIP dial peers, no automatic digit stripping occurs and no digit-manipulation commands are required. (Keep in mind that the no digit-strip, forward-digits, and prefix commands are only valid under POTS dial peers anyhow.)

If the IP connection between Arizona and Texas fails, the Arizona router begins using the next most preferred dial peer, which is dial peer 11. To overcome the automatic digit-stripping feature of POTS dial peers, the **no digit-strip** command is used. (Otherwise, the router would strip the 6 digit from the dialed number.) Because a four-digit number is invalid on the PSTN, the prefix 1512555 command adds the necessary prefix information to get the call across the PSTN.

Note: If the IP WAN fails, all the active calls established during the WAN failure will disconnect and be required to redial. There is no "dynamic failover" mechanism for calls already established.

Practical Scenario 2: Directing Operator Calls to the Receptionist

This practical scenario is fairly simple. The organization shown in Figure 6-13 wants to direct all calls to the operator number 0 to the receptionist at extension 5000.

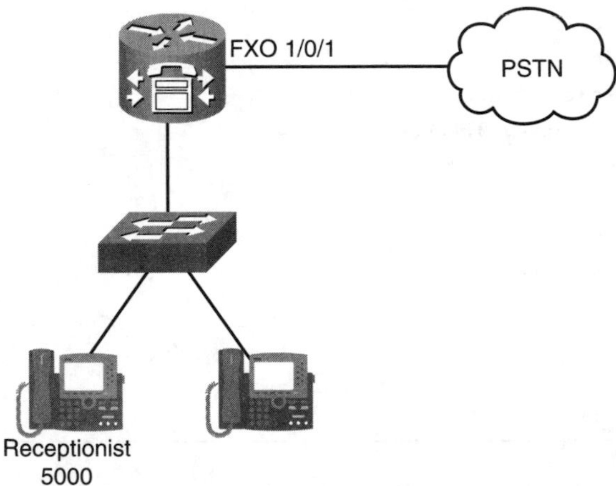

Figure 6-13 *Redirecting Operator Calls*

Because this is a "universal" transformation (you always want to change the dialed number 0 to 5000), you can accomplish this objective using the **num-exp** global configuration command, which is shown in Example 6-18.

Example 6-18 *Transforming Dialed Numbers Using num-exp*

```
Voice_RTR(config)# voice-port 1/0/1
Voice_RTR(config-voiceport)# connection plar 0
Voice_RTR(config-voiceport)# exit
Voice_RTR(config)# num-exp 0 5000
```

Now, anytime the number 0 is dialed from anywhere in the organization (could be an IP phone, FXS port, and so on), the voice router automatically transforms it to 5000 and then searches for a dial peer allowing it to reach the number 5000.

Note: The router applies the **num-exp** command the instant it receives a dialed number, even before it attempts to match an inbound dial peer.

Practical Scenario 3: Specific POTS Lines for Emergency Calls

As organizations move more to VoIP connections, they are finding cost-saving benefits by eliminating traditional telephony connections at remote offices in favor of centralizing all PSTN calls (and toll charges) at a central site. Figure 6-14 illustrates this type of network design.

This type of call routing allows an organization to get higher call volume from a single location, which typically allows the organization to negotiate cheaper long-distance rates with its PSTN carrier.

Note: Some countries restrict businesses from forwarding PSTN calls over the IP WAN. You should always check with the local government regulations before you do this. Thankfully, the United States is not one of those countries.

Although the centralization of PSTN calls offers significant cost savings, the remote sites need to keep at least one local PSTN connection for emergency calling. This is because PSTN carriers provide location information to emergency service providers based on the POTS connection. If emergency calls from the remote offices were to traverse the IP WAN and leave the PSTN connection at the central site, the emergency service provider would receive location information for the central site.

Depending on the size of the remote office, you can typically dedicate one or two analog FXO ports for emergency calls. The configuration in Example 6-19 configures the necessary dial peers for dual FXO ports connected to the PSTN. This example assumes the FXO ports are 1/0/0 and 1/0/1.

Example 6-19 *Dynamic WAN to PSTN Failover Implementation*

```
REMOTE_RTR(config)# dial-peer voice 10 pots
REMOTE_RTR(config-dial-peer)# destination-pattern 911
REMOTE_RTR(config-dial-peer)# port 1/0/0
REMOTE_RTR(config-dial-peer)# no digit-strip
REMOTE_RTR(config-dial-peer)# exit
```

```
REMOTE_RTR(config)# dial-peer voice 11 pots
REMOTE_RTR(config-dial-peer)# destination-pattern 9911
REMOTE_RTR(config-dial-peer)# port 1/0/0
REMOTE_RTR(config-dial-peer)# forward-digits 3
REMOTE_RTR(config-dial-peer)# exit
REMOTE_RTR(config)# dial-peer voice 12 pots
REMOTE_RTR(config-dial-peer)# destination-pattern 911
REMOTE_RTR(config-dial-peer)# port 1/0/1
REMOTE_RTR(config-dial-peer)# no digit-strip
REMOTE_RTR(config-dial-peer)# exit
REMOTE_RTR(config)# dial-peer voice 13 pots
REMOTE_RTR(config-dial-peer)# destination-pattern 9911
REMOTE_RTR(config-dial-peer)# port 1/0/1
REMOTE_RTR(config-dial-peer)# forward-digits 3
```

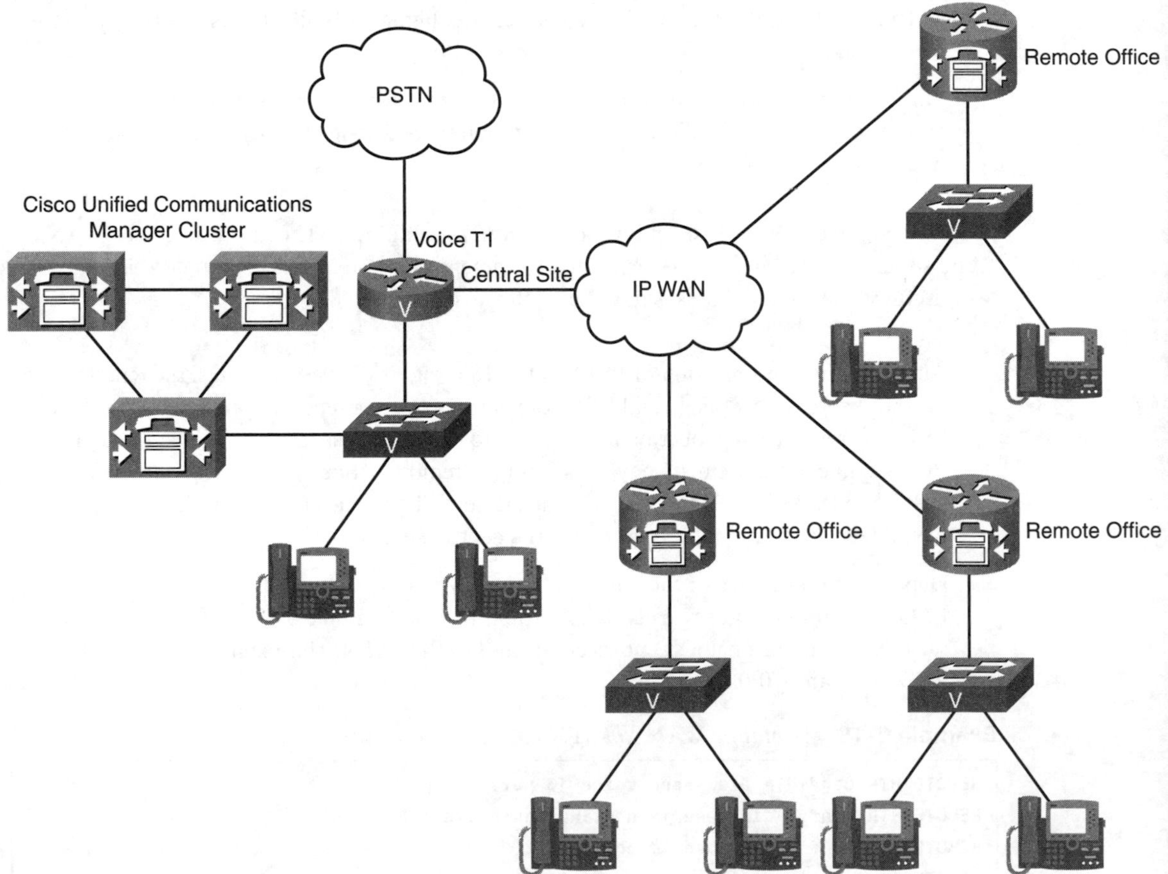

Figure 6-14 *Centralizing PSTN Access*

This configuration creates two identical destination patterns for the two FXO ports. Because the **preference** command is not used to indicate a more preferred dial peer, the router will randomly choose one of the FXO ports as an exit point anytime a user dials 911 or 9911 (the additional 9 may be entered if users are accustomed to dialing 9 for an outside line). The dial peers created for the 911 destination pattern (dial peers 10 and 12) are also assigned the **no digit-strip** command. Otherwise, the automatic digit-stripping rule of POTS dial peers would strip any explicitly defined digits (which are all of them in this case; the router would not send any digits to the PSTN). The dial peers created for the 9911 destination pattern (dial peers 11 and 13) are assigned the **forward-digits 3** command to send the right-justified three digits (911, in this case) to the PSTN and allow the automatic digit-stripping rule to remove the initial 9 access code.

Practical Scenario 4: Using Translation Profiles

The digit manipulation commands discussed thus far allow you to perform "minor translations" to a number. For example, you can add some digits using the **prefix** command or ensure digits do or do not get stripped with the **forward-digits** command. The **num-exp** command allows you to make the biggest changes of all, but these changes are applied globally to the router, which might not give you the flexibility all situations require. Translation profiles are useful to address these needs. If you find yourself saying, "I want to change this dialed number to that dialed number, but only when it goes out this port," you need a translation profile.

Working with translation profiles is definitely not as easy as working with the "simple" digit manipulation methods discussed earlier. Implementation of translation profiles requires a three-step process:

Step 1. Define the rules that dictate how the router will transform the number.

Step 2. Associate the rules into a translation profile.

Step 3. Assign the translation profile to a dial peer.

In a way, this is similar to access-list configuration on a router.

To demonstrate the configuration of translation profiles, consider the scenario illustrated in Figure 6-15.

The headquarters of this organization uses the DID range from a PSTN provider of 602.555.6XXX. This allows PSTN callers to dial directly into the organization without being redirected by a receptionist. Typically, when you lease a block of DID numbers, the PSTN carrier will strip the numbers down to a four-digit extension. In this case, the DID block assigned to the organization (6XXX) does not match its internal extension range (5XXX). The administrator of this network would like to translate all 6XXX dialed numbers to 5XXX, but only if these dialed numbers come in from the T1 PSTN interface, so as to not interfere with the numbering scheme of the remote office. To accomplish this, he cannot use the **num-exp 6... 5...** global configuration command because it will interfere with dialing the 6XXX extensions at the remote office. This situation is ideal for translation profiles.

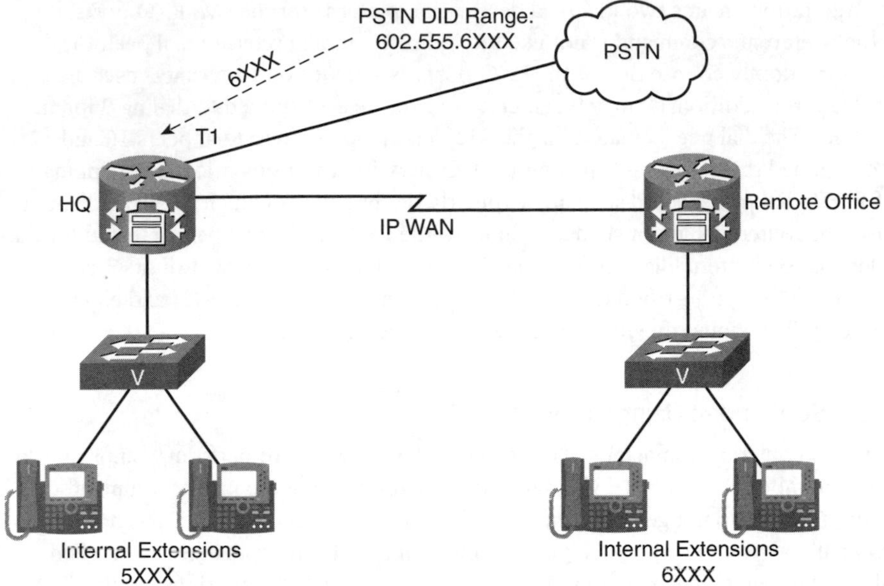

Figure 6-15 *Translating DID Ranges to Internal Extensions*

The first step to configure translation profiles is to create the translation rules. These use the general syntax shown in Example 6-20.

Example 6-20 *Translation Rule General Syntax*

```
Router(config)# voice translation-rule rule number
Router(cfg-translation-rule)# rule 1 /match/ /set/
Router(cfg-translation-rule)# rule 2 /match/ /set/
Router(cfg-translation-rule)# rule 3 /match/ /set/ ...and so on
```

Example 6-21 configures the necessary translation rule for the scenario in Figure 6-15.

Example 6-21 *Configuring Translation Rules*

```
HQ_RTR(config)# voice translation-rule 1
HQ_RTR(cfg-translation-rule)# rule 1 ?
  /WORD/  Matching pattern
  reject  Call block rule
HQ_RTR(cfg-translation-rule)# rule 1 /6/ ?
  /WORD/  Replacement pattern
HQ_RTR(cfg-translation-rule)# rule 1 /6/ /5/
```

The syntax in the rule 1 command may look a little cryptic. The first entry between the set of forward slashes (/) is the **match** statement. This tells the router, "Look for the number 6." The entry between the second set of forward slashes is the **set** statement. This tells the router, "Replace the 6 you found from the match statement with a 5." In this case, the router changes the first 6 that is found to a 5.

Thankfully, Cisco does not leave you "hoping" that the translation rule will work properly after it is applied to the interface. You can use the **test voice translation-rule** command from privileged mode to test the rules you create before you apply them, as shown in Example 6-22.

Example 6-22 *Testing Translation Rules*

```
HQ_RTR# test voice translation-rule 1 6546
Matched with rule 1
Original number: 6546   Translated number: 5546
Original number type: none      Translated number type: none
Original number plan: none      Translated number plan: none
HQ_RTR# test voice translation-rule 1 6677
Matched with rule 1
Original number: 6677   Translated number: 5677
Original number type: none      Translated number type: none
Original number plan: none      Translated number plan: none
```

Example 6-22's output indicates the translation rule tests successfully. 6546 is translated to 5546 and 6677 is translated to 5677.

Next, you need to take the voice translation rule and assign it to a translation profile. The translation profile designates whether the translation rule will change the calling (caller ID or ANI) or called (dialed number or DNIS) information. Example 6-23 assigns translation rule 1 to a translation profile called CHANGE_DID.

Example 6-23 *Assigning Translation Rules to a Translation Profile*

```
HQ_RTR(config)# voice translation-profile ?
  WORD  Translation profile name
HQ_RTR(config)# voice translation-profile CHANGE_DID
HQ_RTR(cfg-translation-profile)# translate ?
  called           Translation rule for the called-number
  calling          Translation rule for the calling-number
  redirect-called  Translation rule for the redirect-number
  redirect-target  Translation rule for the redirect-target
HQ_RTR(cfg-translation-profile)# translate called ?
  <1-2147483647>  Translation rule tag
HQ_RTR(cfg-translation-profile)# translate called 1
```

Example 6-23 assigns translation rule 1 as a called (dialed number) translation. Because the scenario requires you to change the DID information, this is the proper assignment. Assigning the translation rule as a calling translation would change the caller ID of a person calling into the organization.

The last step is to assign the translation profile. The following example assumes that the router is using POTS dial peer 100 as the inbound dial peer for calls coming from the PSTN:

```
HQ_RTR(config)# dial-peer voice 100 pots
HQ_RTR(config-dial-peer)# translation-profile incoming CHANGE_DID
```

Notice that the example applies the translation profile in the incoming direction. This causes it to affect calls coming in from the PSTN rather than outgoing calls. The translation profile is now in effect, accomplishing the objective of the scenario.

Note: You can do far more with translation profiles (and far more complex patterns that you can match with translation rules). This is covered more in the CVOICE exam of the CCNP Voice certification track.

Tip: If you are interested, I (Jeremy) recorded a free video that explains translation rules in depth. You can watch this video at www.cbtnuggets.com/series?id=440.

With all these various methods of digit manipulation, two questions quickly arise: Which method gets applied first? Will the router remove added prefix digits because of the automatic digit-stripping rule? Figure 6-16 answers these questions by displaying the order of operations for outgoing POTS dial peers. The order remains the same for VoIP dial peers; however, most digit-manipulation commands apply only to POTS dial peers.

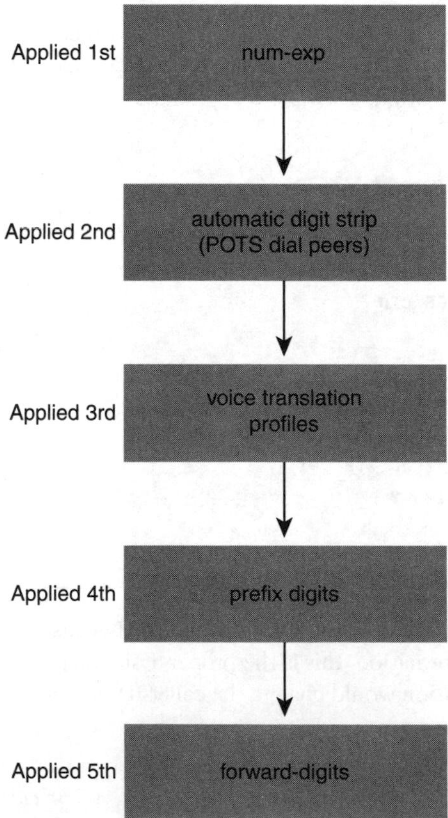

Figure 6-16 *Digit Manipulation Order of Operation for POTS Dial Peers*

Using CCP to Configure a CME Dial-Plan

If you prefer, you can also use the Cisco Configuration Professional (CCP) to modify the CME dial-plan. Cisco neatly organized all dial-plan configurations in a Dial-Plans folder, shown in Figure 6-17.

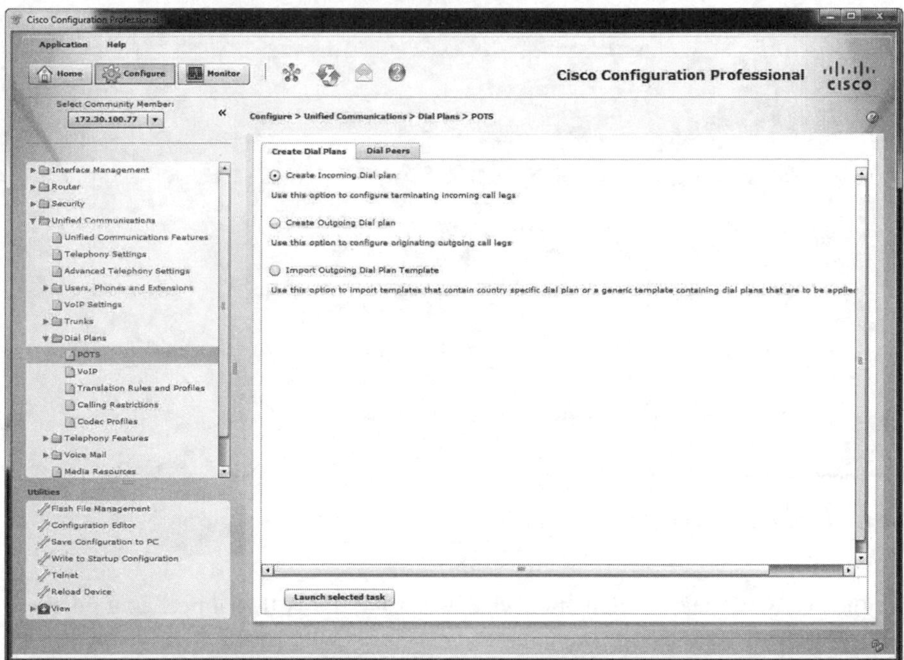

Figure 6-17 *Configuring a Dial-Plan Using CCP*

Notice that the first two items under the Dial-Plans folder of CCP is POTS and VoIP. These correlate to the two categories of dial peers you most frequently configure in CME. Selecting the POTS configuration item brings up the Create Dial-Plan menu. In an attempt to ease the configuration, Cisco provides three wizard-like configuration items:

- **Create Incoming Dial-Plan:** Allows you to configure inbound dial-peers using PSTN trunks.
- **Create Outbound Dial-Plan:** Allows you to configure outbound dial-peers using PSTN trunks. The wizard also provides you the ability to specify access digits (such as dialing 9 for an outside line) and caller ID information.
- **Import Outgoing Dial-Plan Template:** Allows you to import a dial-plan from a CSV file template.

After you get more comfortable with the CME configuration, you will likely access the Dial Peers tab (next to the Create Dial-Plans tab shown in Figure 6-17). In this configuration pane, you can create manually configured dial peers through a GUI configuration. Figure 6-18 shows the creation of a ten-digit dialing PSTN dial peer.

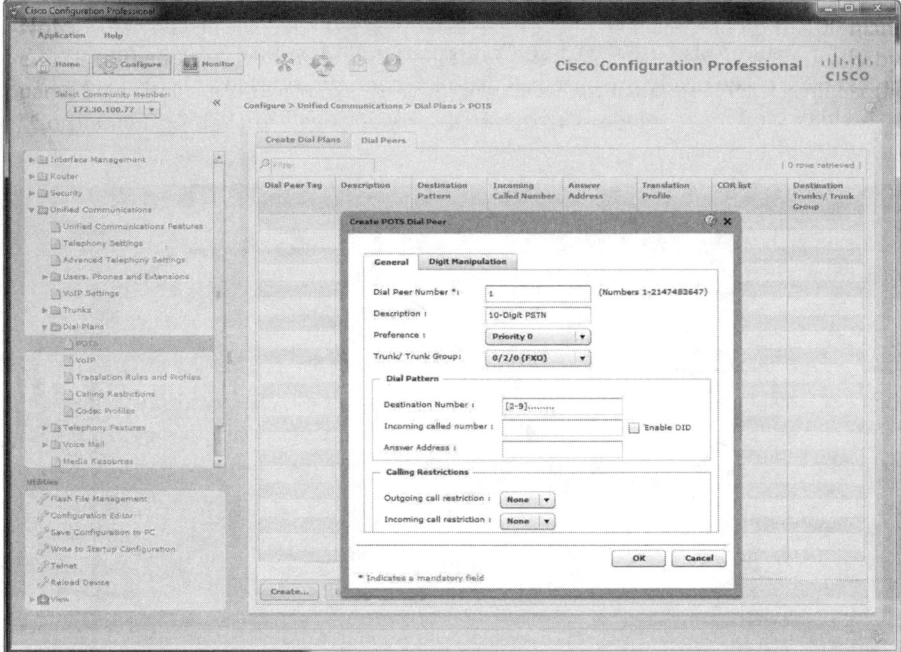

Figure 6-18 *Configuring Dial Peers Using CCP*

Unlike the POTS dial peer configuration window in CCP, the VoIP dial peer configuration does not have the wizard-based configuration items; it allows only the manual dial peer creation.

Understanding and Implementing CME Class of Restriction

If you implement what you've seen so far, you can create a powerful VoIP system that supports both internal (ephone) and external (dial peer) dialing. However, there might be times when you want to prevent some users from calling certain numbers, such as the following examples:

- Prevent standard employees from making international calls, but allow management to place international calls without restriction
- Block certain high-cost numbers (such as 1-900 numbers in the United States)
- Prevent certain internal phones from reaching executive office directory numbers

This list goes on and on, but you get the idea: Sometimes, it's necessary to place calling restrictions on users of the VoIP network. If you were implementing this feature in the full Cisco Unified Communications Manager (CUCM) platform, you would use a feature called Partitions and Calling Search Spaces (CSS). In the CME environment, the equivalent feature is called incoming and outgoing Class of Restriction (COR) lists.

Without a doubt, COR lists take some practice before the concept sinks in fully. At a high-level view, here's how the process works:

1. A user picks up his IP phone and is immediately associated with an incoming COR list (which lists the "tags" he can access).
2. The user dials, causing CME to match an outgoing dial-peer.
3. If the outgoing dial-peer requires a COR "tag," CME checks to see if that tag is listed in the user's incoming COR list.
4. If the "tag" is listed in the user's incoming COR list, CME permits the call.
5. If the "tag" is not listed in the user's incoming COR list, CME denies the call.

Note: I know the word "tag" is vague, but when you see the configuration of COR lists, you will understand why I chose this word.

Similar to the famous quote in *The Matrix*, "Unfortunately, no one can be told what COR lists are. You have to see them for yourself." Let's walk through a practical configuration example (see Figure 6-19).

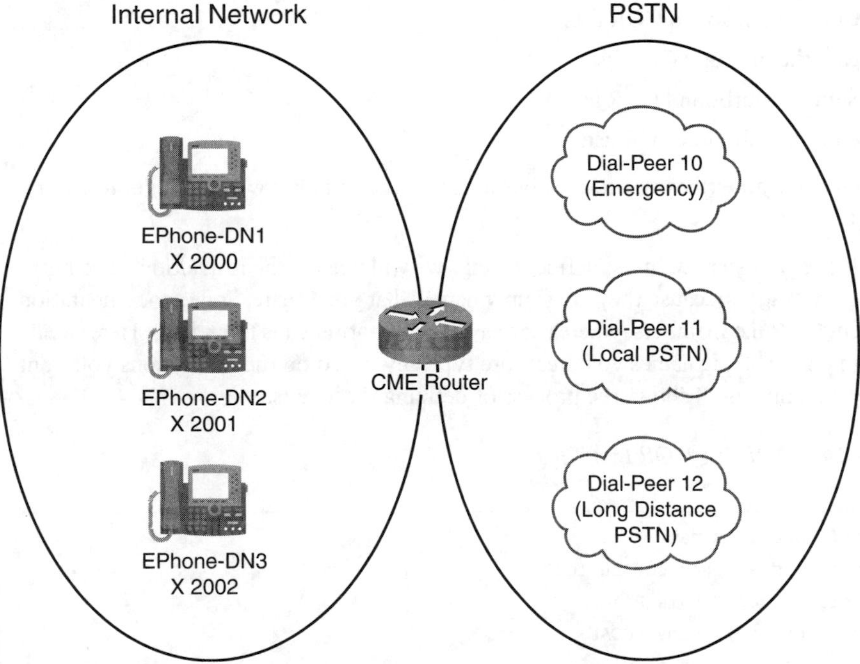

Figure 6-19 *Practical COR List Example*

As shown in Figure 6-19, three phones are on the internal network:

- Ephone-dn 1 (x2000)
- Ephone-dn 2 (x2001)
- Ephone-dn 3 (x2002)

In addition, there are three POTS dial-peers connecting to the PSTN:

- Dial-peer 10 (emergency calls, destination-pattern 911)
- Dial-peer 11 (local PSTN calls, destination-pattern [2-9]......)
- Dial-peer 12 (long-distance PSTN calls, destination-pattern 1..........)

A corporation wants to implement the following restrictions in CME:

- Ephone-dn 1 (x2000) represents a lobby phone that should only be able to dial internal extensions and place emergency calls.
- Ephone-dn 2 (x2001) represents an employee phone that should be able to dial internal extensions and place emergency and local PSTN calls.
- Ephone-dn 3 (x2002) represents a manager phone that should have no calling restrictions.

Based on these requirements, we can begin the COR list implementation on the CME router. It breaks down into the following steps:

1. Define the COR tags we will use for the restrictions.
2. Create the outbound COR lists.
3. Create the inbound COR lists.
4. Assign the outbound COR lists.
5. Assign the inbound COR lists.

This seems like an extensive process, but it goes rather quickly when you get to enter the syntax.

To tackle the first step, we must define the tags we will use for the restrictions. I've been calling them "tags" because they are simply names that you create. Some documentation calls them COR list members, whereas other documentation calls them keys. Here, we'll call them tags. The tag names you create are typically based on the restrictions you want to apply. Example 6-24 shows the process of defining these tags.

Example 6-24 *Defining COR List Tags*

```
Router# configure terminal
Router(config)# dial-peer cor custom
Router(config-dp-cor)# name 911
Router(config-dp-cor)# name LOCAL
Router(config-dp-cor)# name LD
```

At this point, we defined the tags 911, LOCAL, and LD, which represent the various outgoing restrictions we can place. At this point, these tags are doing absolutely nothing, but because we defined the names, the CME router allows us to use them to create COR lists. We'll first create the outgoing COR lists that we will apply to the PSTN dial peers, as shown in Example 6-25.

Example 6-25 *Creating Outgoing COR Lists*

```
Router(config)#dial-peer cor list 911-CALL
Router(config-dp-corlist)#member 911
Router(config-dp-corlist)#exit
Router(config)#dial-peer cor list LOCAL-CALL
Router(config-dp-corlist)#member LOCAL
Router(config-dp-corlist)#exit
Router(config)#dial-peer cor list LD-CALL
Router(config-dp-corlist)#member LD
Router(config-dp-corlist)#exit
```

We will eventually apply the COR lists shown in Example 6-25 as outgoing COR lists (to the PSTN dial peers). If you were to read these COR lists in plain English, the 911-CALL COR list would say, "For this COR list to allow the call, the caller must be assigned the 911 tag." The LOCAL-CALL COR list would say, "For this COR list to allow the call, the caller must be assigned the LOCAL tag." Hopefully, you get the idea, but keep in mind that the COR lists are not doing anything, because we have not yet applied them to the dial peers. Next, we create the incoming COR lists, as shown in Example 6-26.

Example 6-26 *Creating Incoming COR Lists*

```
Router(config)#dial-peer cor list 911-ONLY
Router(config-dp-corlist)#member 911
Router(config-dp-corlist)#exit
Router(config)#dial-peer cor list 911-LOCAL
Router(config-dp-corlist)#member 911
Router(config-dp-corlist)#member LOCAL
Router(config-dp-corlist)#exit
Router(config)#dial-peer cor list 911-LOCAL-LD
Router(config-dp-corlist)#member 911
Router(config-dp-corlist)#member LOCAL
Router(config-dp-corlist)#member LD
Router(config-dp-corlist)#exit
```

We will eventually apply the COR lists shown in Example 6-26 to the ephone-dns to grant calling privileges. If you were to read these COR lists in plain English, the 911-ONLY COR list would say, "Anyone assigned to this COR list can call dial peers requiring the 911 tag." The 911-LOCAL COR list would say, "Anyone assigned to this COR list can call dial peers requiring the 911 or LOCAL tags."

> **Note:** You might wonder, "How do these COR lists say different things when they're created the same way?" Great question! This is likely what makes COR lists so confusing. Both the inbound and outbound COR lists are created the same way. The effect they have is completely based on how you apply them. If you apply a COR list in the outbound direction, it says, "I will require a caller to have the defined tags to complete a call." If you

apply a COR list in the inbound direction, it says, "I will assign these tags to a caller, which grant the ability to place a call."

Now, we can move to the final two steps: assigning the inbound and outbound COR lists. Example 6-27 records this monumental event.

Example 6-27 *Assigning Outbound and Inbound COR Lists*

```
Router(config)#dial-peer voice 10 pots
Router(config-dial-peer)#corlist outgoing 911-CALL
Router(config-dial-peer)#exit
Router(config)#dial-peer voice 11 pots
Router(config-dial-peer)#corlist outgoing LOCAL-CALL
Router(config-dial-peer)#exit
Router(config)#dial-peer voice 12 pots
Router(config-dial-peer)#corlist outgoing LD-CALL
Router(config-dial-peer)#exit
Router(config)#ephone-dn 1
Router(config-ephone-dn)#corlist incoming 911-ONLY
Router(config-ephone-dn)#exit
Router(config)#ephone-dn 2
Router(config-ephone-dn)#corlist incoming 911-LOCAL
Router(config-ephone-dn)#exit
Router(config)#ephone-dn 3
Router(config-ephone-dn)#corlist incoming 911-LOCAL-LD
Router(config-ephone-dn)#exit
```

Note: Example 6-27 assumes dial peers 10, 11, and 12 have been previously configured with the necessary destination-pattern and port values to correctly route the call to the PSTN.

Example 6-27 puts it all together. You might need to flip back through the last couple pages to see the building examples to fill the pieces in your own mind. To see this in action, we follow a call originated from the lobby phone (ephone-dn 1):

1. Someone picks up the lobby phone and calls 4805551212.
2. CME immediately assigns the lobby phone the 911-ONLY COR list, which assigns the tag 911.
3. CME matches the outbound dial peer 11, which is assigned the outgoing COR list LOCAL-CALL.
4. The outgoing LOCAL-CALL COR list requires the LOCAL tag.
5. Because the lobby phone was only assigned the 911 tag (and not LOCAL), the call fails with a reorder tone.

Okay, we've now seen how COR lists deny a call. Let's now follow a call that succeeds from the employee phone (ephone-dn 2):

1. An employee picks up her phone and calls 4805551212.
2. CME immediately assigns the employee phone the 911-LOCAL COR list, which assigns the tags 911 and LOCAL.
3. CME matches the outbound dial peer 11, which is assigned the outgoing COR list LOCAL-CALL.
4. The outgoing LOCAL-CALL COR list requires the LOCAL tag.
5. Because the employee phone is assigned the 911 and LOCAL tags, the call completes successfully.

Note: In the previous COR list scenarios, the incoming COR list are applied to ephone-DNs and the outgoing COR list are applied to dial peers. Although this is the most common configuration, you can apply incoming and outgoing COR lists to any combination of calling or called entities. For example, you might apply an incoming COR list to a PSTN dial peer to restrict which internal extensions a PSTN caller can reach.

That's the foundation of COR lists! Now, with that understanding, there are a couple important rules of COR lists that you should know:

Rule 1: If there is no outgoing COR list applied, the call is always routed.

Rule 2: If there is no incoming COR list applied, the call is always routed.

Typically, Rule 1 usually makes perfect sense. For example, if you created a PSTN dial-peer that you'd like all phones to access, simply do *not* apply an outgoing COR list. After that is done, the PSTN dial peer does not require any COR list tags to pass the call through. Now, regardless of the incoming COR list applied to an ephone-dn (or a complete lack of COR list applied to the ephone-dn), the call to the PSTN dial peer completes successfully.

On the other hand, Rule 2 seems to make no sense at all. If a calling entity (like an ephone-dn) does *not* have an incoming COR list, it is able to call any other entity regardless of the outgoing COR list assigned. Nonetheless, this is the way the rule works. Think of it this way: If a calling entity does not have an incoming COR list, no calling restrictions are applied to the device. By using this approach, Cisco has lessened the potential of causing outages when applying COR to your CME router. COR always requires an inbound COR list meeting an outbound COR list. If either of those entities is missing, CME permits the call.

With these two rules in mind, we could make our previous configuration example more efficient. You might remember that the scenario required the manager phone (ephone-dn 3) to be able to call any of the listed dial-peers. To accomplish this, we created a COR list called "911-LOCAL-LD" that listed all three COR tags. However, in light of Rule 2, we could have simply *not* assigned ephone-dn 3 an incoming COR list, and we would have accomplished the same objective.

We could have applied a similar efficiency for emergency calling. Every ephone-dn in our scenario had the ability to make emergency calls. This was granted to them because every ephone-dn had the 911 tag assigned to their incoming COR list. However, in light of Rule 1, we could have *not* assigned an outgoing COR list to dial-peer 10 (which was used for emergency calls). At that point, any of the ephone-dns would be able to use the dial-peer regardless of their incoming COR list.

Tip: Even though the previous two rules can make your configuration more efficient, it might help other administrators who not as familiar with CME to add the extra configuration shown in Examples 6-24 through 6-27. Assigning all calling/called entities a COR list makes it less likely that one would "slip through the cracks."

Quality of Service

Quality of Service (QoS) is a topic that has been hinted at in nearly every chapter of this book. For a VoIP network to operate successfully, the voice traffic must have priority over the data traffic as it traverses its way from one end of the network to the other. The Cisco definition of QoS is as follows:

> Quality of service is the ability of the network to provide better or special service to a set of users and applications at the expense of other users and applications.

That sounds exactly like what the voice traffic needs as it crosses the network: better or "special" service than the typical data traffic, such as web browsing, FTP transfers, e-mail traffic, and so on. The voice traffic needs this not so much because of bandwidth requirements (VoIP uses very little bandwidth compared to most data applications), but rather delay requirements. Unlike data, the time it takes a voice packet to get from one end of the network to the other matters. If a data packet crossing the network experiences delay, a file transfer bar might take a couple more seconds to complete or a web page might take a half second longer to load. From a user's perspective, this is not a big deal. However, if voice traffic crossing the network experiences delay, conversations begin to overlap (a person begins speaking at the same time as another person), the conversation breaks up, and, in some extreme cases, the voice call drops.

To combat these issues, you need to ensure not only that there is bandwidth available for VoIP traffic, but that the VoIP traffic gets the first bandwidth available. This means if a bottleneck is in the network where a router queues traffic before it is sent, the router will move the waiting voice traffic ahead of the data traffic to be sent at the first available interval. Accomplishing this is the job of QoS. QoS is not a tool in itself, but rather, a category of many tools aimed at giving you complete control over the traffic crossing your network. There might be times when you just use a single QoS tool aimed at decreasing the delay of traffic. Other times, you might employ multiple QoS tools to control delay, reserve bandwidth, and compress data that is heading over the WAN. How and when you use each of the QoS tools depends on the network requirements of your traffic and the characteristics (such as bandwidth, delay, and so on) of the network supporting the traffic.

Understanding the Enemy

Before you can deploy QoS successfully, you need to know what you are fighting against. The following are the three enemies of your VoIP traffic:

- **Lack of bandwidth:** Multiple streams of voice and data traffic competing for a limited amount of bandwidth.
- **Delay:** The time it takes a packet to move from the original starting point to the final destination. Delay comes in three forms:
  - **Fixed delay:** Delay values that you cannot change. For example, it takes a certain amount of time for a packet to travel specific geographical distances. This value is considered fixed. QoS cannot impact fixed delay issues.
  - **Variable delay:** Delay values that you can change. For example, queuing delay (how long a packet waits in a router's interface queue) is variable because it depends on how many packets are currently in the queue. You can impact queuing delay by selectively moving voice packets ahead of data packets.
  - **Jitter (delay variations):** Describes packets that have different amounts of delay between them. For example, the first voice packet of a conversation might take 100 ms to reach a destination, whereas the second voice packet might take 110 ms. There is 10 ms of delay variation (jitter) between these packets.
- **Packet loss:** Packets lost because of a congested or unreliable network connection.

These enemies plague every network environment; however, the stakes are much higher when you add VoIP traffic to an existing data network. Users are accustomed to a PBX-style environment that has a separate network and dedicated bandwidth assigned just for voice traffic. The tolerance for crackling, echoing, or dropped calls from a voice network is very low.

QoS is designed to keep voice traffic running smoothly during temporary moments of congestion on the network. It is not a "magic bullet" that can solve any network scenario. For example, if there is a network environment in which the WAN link is constantly lacking bandwidth, adding voice to the link and expecting QoS to take care of the situation is similar to rearranging the deck chairs on the sinking Titanic. QoS can only do so much; either your data applications will perform so slowly they are no longer functional or your voice traffic will experience quality issues. This also goes the other way; if you have a network environment where fiber-optic cable is the norm and gigabit speeds abound, you might never experience network congestion. These environments will get little to no gain by using QoS because most QoS tools only engage during times of network congestion.

Your goal with QoS is to provide consistent bandwidth to voice traffic in such a way that there is low, steady delay from one end of the network to the other. To accomplish this, you need to have QoS in some form at each point of the network where congestion exists. This means doing an end-to-end audit of your network to determine the traffic types that exist and the service levels required for those traffic types.

Requirements for Voice, Video, and Data Traffic

The different traffic types that cross your network every day each have their own QoS requirements. Some of these requirements might be very loose; the network would essentially need to fail for the application to stop working. Other requirements might be very tight, requiring high-speed connectivity with low delay for the application to work successfully. This section describes general goals for voice, video, and data.

> **Tip:** A plethora of QoS software utilities are available that will analyze your network traffic and report the bandwidth and delay each traffic type receives from the network. A popular tool in the Cisco realm is NetQoS (www.netqos.com).

Network Requirements for Voice and Video

Unlike data traffic, voice traffic is predictable. Whereas data traffic can jump considerably if a large web download or file transfer is started, voice traffic remains a consistent value for each call entering and leaving the network. The actual amount of bandwidth required for voice is heavily dependent on the codec you are using.

In addition to bandwidth requirements, voice traffic has the following additional one-way requirements:

Key Topic

- **End-to-end delay:** 150 ms or less
- **Jitter:** 30 ms or less
- **Packet loss:** 1% or less

Video traffic has identical delay requirements as voice, but consumes a lot more bandwidth. In addition, the bandwidth can vary depending on how much movement is in the video (lots of movement increases the bandwidth required for video considerably).

Network Requirements for Data

It is impossible to give one sweeping guideline for all data applications, because every data application that exists has its own QoS requirement. When designing QoS for the data applications on your network, divide your applications into no more than four or five broad categories. For example:

- **Mission-critical applications:** These applications are critical to your organization and require dedicated bandwidth amounts.
- **Transactional applications:** These applications are typically interactive with users and require rapid response times. For example, a technical support employee might use a database application to retrieve caller information based on previous case ID values.
- **Best-effort applications:** These applications are noncritical or uncategorized. For example, web browsing, e-mail, and FTP file transfers fall into this category.
- **Scavenger applications:** These nonproductive applications typically have no business need, but consume excessive amounts of bandwidth. For example, peer-to-peer file-sharing applications, such as Kazaa, BitTorrent, and LimeWire, fall into this category.

You can assign each of these data application categories a specific level of QoS. You can then map the actual applications to these categories using a variety of methods (such as incoming interface, exit interface, access lists, and so on).

Key Topic

QoS Mechanisms

With the applications requiring different levels of QoS, multiple models and mechanisms emerged to address the needs. Today, the following models are available for you to deploy QoS:

- **Best Effort:** Best Effort makes the list simply because this is the model every network uses by default. On the positive side, the Best-Effort model requires absolutely no effort at all on your end to implement. No QoS mechanisms are used, and all traffic is treated on a first come, first served basis. Of course, this does not address the QoS requirements of most network environments today.
- **Integrated Services (IntServ):** The IntServ model works through a method of reservations. For example, if a user wanted to make an 80Kbps VoIP call over the data network, the network designed purely to the IntServ model would reserve 80Kbps on every network device between the two VoIP endpoints using the Resource Reservation Protocol (RSVP). For the duration of the call, 80Kbps of bandwidth would not be available for any other use other than the VoIP call. Although the IntServ model is the only model that provides *guaranteed* bandwidth, it also has scalability issues. If enough reservations are made, the network simply runs out of bandwidth.
- **Differentiated Services (DiffServ):** The DiffServ model is the most popular and flexible model to use for implementing QoS. In this model, you can configure every device to respond with a variety of QoS methods based on different traffic classes. You can specify what network traffic goes into each class and how each class is treated. Unlike the IntServ model, the traffic is not absolutely guaranteed (since the network devices do not completely reserve the bandwidth). However, DiffServ gets so close to guaranteed bandwidth (some Cisco documentation refers to it as "almost guaranteed" bandwidth), while at the same time addressing the scalability concerns of IntServ, that it has become the standard QoS model used by most organizations around the world.

The QoS model you use is primarily a strategy or mindset of how you design and implement QoS throughout your network. The QoS mechanisms themselves are a series of tools that combine together to deliver the levels of service your network traffic needs to survive. Each of these tools fit into one of the following categories:

- **Classification and Marking:** These tools allow you to identify and mark a packet so network devices can easily identify it as it crosses the network. Typically the first device that receives the packet identifies it using tools such as access-lists, incoming interfaces, or deep packet inspection (which looks at the application data itself). These tools can be processor intense and add delay to the packet, so after the packet is initially identified, it is then marked. The marking can be in the layer 2 (data link) header (allowing switches to read it) and/or the layer 3 (network) header so routers can read it. Then, as the packet crosses the rest of the network, the network devices simply look at the marking to classify it rather than digging deep in the packet.

- **Congestion Management:** All of the QoS queuing strategies fall under this umbrella, which are typically the primary tools you will use to implement QoS network-wide. The queuing strategies define the rules the router should apply when congestion occurs. For example, if a T1 WAN interface was completely saturated with traffic, the router would begin holding packets in memory (queuing) to send them when bandwidth is available. All the queuing strategies aim to answer that one golden question: When there is bandwidth available, what packet goes first?
- **Congestion Avoidance:** Most QoS mechanisms engage only when congestion occurs on the network. The aim of congestion avoidance tools are to drop enough packets of non-essential (or not-as-essential) traffic to the network to avoid heavy congestion occurring in the first place.
- **Policing and Shaping:** You can think of policing as one of the few "anti-QoS" mechanisms available (yes, you could argue this, but please don't). Rather than guaranteeing a certain amount of bandwidth, policing limits the amount of bandwidth certain network traffic can use. This is useful for many of the typical "bandwidth hogs" on the network: peer-to-peer applications, web surfing, FTP, and so on. You can also use shaping to limit the amount of bandwidth certain network traffic can use. It is designed for networks where the actual speed allowed is slower than the physical speed of the interface. The difference between the two mechanisms is shaping queues excess traffic (and tries to send it later), whereas policing typically drops excess traffic.
- **Link Efficiency:** As the name implies, this final group of tools focus on delivering the traffic in the most efficient way. For example, some low-speed links might work better if you take the time to compress your network traffic before it is sent (compression is one of the link efficiency tools).

Understanding QoS completely is a fairly massive undertaking, which is why Cisco has dedicated an entire certification exam to the topic in the CCNP Voice track. At the CCNA Voice level, Cisco has chosen to highlight two of the key QoS categories: link efficiency mechanisms and queuing algorithms.

Link Efficiency Mechanisms

As network technology progresses and spreads around the world, links slower than T1 speed (1.544Mbps) are becoming increasingly rare. However, there are still many of these slower (aka "speed challenged") links in existence. There are typically two challenges facing these connections:

- Lack of bandwidth makes it difficult to send the amount of data required in a timely fashion.
- Slower link speeds can have a significant impact on end-to-end delay due to the serialization process (the amount of time it takes the router to put the packet from its memory buffers onto the wire). On these slow links, the larger the packet, the longer the serialization delay. For example, sending a 1500-byte packet on a 56Kbps link adds 214ms just in serialization delay.

To address these challenges, the following link efficiency mechanisms have been introduced:

- **Payload Compression:** Compresses application data being sent over the network so the router sends less data across the slow WAN link.
- **Header Compression:** Some traffic (such as VoIP) may have a small amount of application data (RTP audio) in each packet but send many packets overall. In this case, the amount of header information becomes a significant factor and often consumes more bandwidth than the data itself. Header compression addresses this issue directly by eliminating many of the redundant fields in the header of the packet. Amazingly, RTP header compression (also called Compressed Real-time Transport Protocol (cRTP) reduces a 40-byte header down to 2–4 bytes!
- **Link Fragmentation and Interleaving (LFI):** LFI addresses the issue of serialization delay by chopping large packets into smaller pieces before they are sent. This allows the router to move critical VoIP traffic in between the now-fragmented pieces of the data traffic (which is called "interleaving" the voice). You can use LFI on PPP connections (by using multilink PPP) or on Frame Relay connections (using FRF.12 or FRF.11 Annex C).

Tip: One major thing to understand: Link efficiency mechanisms are not a magic way to get more bandwidth. Each of them has their own drawback: Compression adds delay, and processor load and link fragmentation increases the amount of actual data being sent on the line (because all the fragmented packets now need their own header information). Cisco does not recommend using these methods on links faster than T1 speed.

Queuing Algorithms

Queuing define the rules the router should apply when congestion occurs. The majority of network interfaces use basic First-in, First-out (FIFO) queuing by default. In this method, whatever packet arrives first is sent first. Although this seems fair, not all network traffic is created equal. The primary goal of queuing is to ensure that the network traffic servicing your critical or time-sensitive business applications gets sent before non-essential network traffic. Beyond FIFO queuing, there are three primary queuing algorithms in use today:

- **Weighted Fair Queuing (WFQ):** WFQ tries to balance available bandwidth among all senders evenly (thus the "fair" queuing). By using this method, a high-bandwidth sender gets less priority than a low bandwidth sender. On Cisco routers, WFQ is often the default method applied to serial interfaces.
- **Class-Based Weighted Fair Queuing (CBWFQ):** This queuing method allows you to specify guaranteed amounts of bandwidth for your various classes of traffic. For example, you could specify that web traffic gets 20 percent of the bandwidth, whereas Citrix traffic gets 50 percent of the bandwidth (you can specify values as a percent or a specific bandwidth amount). WFQ is then used for all the unspecified traffic (the remaining 30 percent, in the previous example).
- **Low Latency Queuing (LLQ):** LLQ is often referred to as PQ-CBWFQ because it is exactly the same thing as CBWFQ, but adds a priority queuing (PQ) component.

When you specify that certain network traffic should go into the priority queue, the router then not only guarantees that traffic bandwidth, it guarantees it the first bandwidth. For example, using pure CBWFQ, Citrix traffic might be guaranteed 50 percent of the bandwidth, but it may get that bandwidth after the router has fulfilled some other traffic guarantees. When using LLQ, the priority traffic always gets sent before any other guarantees are fulfilled. As you might guess, this works very well for VoIP, making LLQ the preferred queuing algorithm for voice.

Although there are many other queuing algorithms available, these three encompass the methods used by most modern networks.

Applying QoS

By nature, you can apply most of the QoS mechanisms discussed as the network traffic leaves a router (because you cannot control the order the router receives traffic; it simply arrives). Table 6-7 summarizes the QoS methods discussed and the direction you can apply them on your router.

Table 6-7 *Applying QoS to Input and Output Interfaces of a Router*

QoS Methods Applied as Traffic Enters the Router (Input)	QoS Methods Applied as Traffic Leaves the Router (Output)
Classification	Congestion Management
Marking	Marking
Policing	Congestion Avoidance
	Shaping
	Policing
	Compression
	Fragmentation and Interleaving

As you can see, you can apply some QoS methods (such as policing) in either direction.

Using Cisco AutoQoS

Deploying QoS can be complex (which is why there's a difficult CCNP Voice exam dedicated just to the topic). To help ease the learning curve for QoS, Cisco created a mechanism called AutoQoS, which allows you to enable a variety of QoS mechanisms with little QoS knowledge. AutoQoS works so well out of the box that many network administrators who have full knowledge of the QoS capabilities and configuration on Cisco devices use it. AutoQoS moved to this acclaimed status because it deploys a template QoS configuration in line with Cisco QoS best practices based on the bandwidth and encapsulation you configured under each of your router or switch interfaces. This template-based QoS deployment offers multiple advantages to manual QoS configuration:

- **Reduces the time of deployment:** Entering a single command on a device is much less time consuming than the potentially complex QoS configurations.

- **Provides configuration consistency:** Using a single-command QoS template on each device ensures that all the devices use a similar QoS configuration that is not as prone to forgotten commands or mistypes.
- **Reduces deployment cost:** It takes some time and training to get fully up to speed on everything QoS has to offer.
- **Allows manual tuning:** You can manually adjust and tune the template-based configuration deployed by AutoQoS to fit your specific network QoS requirements.

Before you can deploy AutoQoS on your network, you must first establish the trust boundary for your voice traffic. However, to understand the concept of a trust boundary, you must first have a basic understanding of QoS markings. As a device sends traffic, that traffic might or might not have QoS markings attached to it. These markings might or might not be trustworthy. For example, a Cisco IP phone marks all of its traffic with an extremely high priority. In this case, the markings are trustworthy because the audio traffic from the phone does indeed need high-priority service. However, a technology-savvy user might configure a computer to mark traffic from it with the same high-priority marking as the voice traffic. In this case, the marking is not trustworthy.

Now, we can jump back to the concept of a trust boundary. The trust boundary is the point of the network where you begin trusting that the network traffic is accurately identified with the correct QoS marking. Depending on the capabilities of the devices on your network, you can you can begin applying QoS markings close to the user devices, as shown in Figure 6-20.

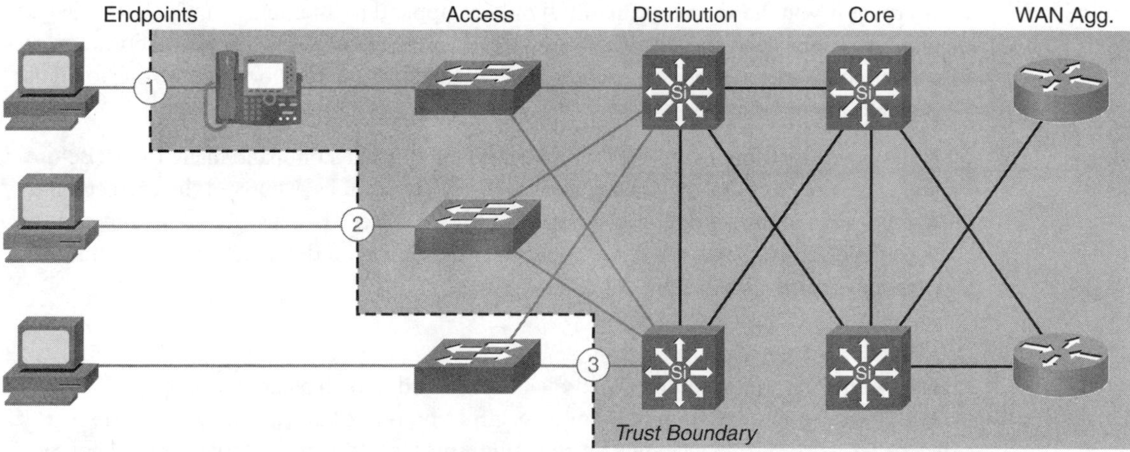

Figure 6-20 *Possible QoS Trust Boundaries*

Cisco IP phones have the ability to mark their own traffic as high priority and strip any high-priority markings from traffic sent by the attached PC. If you are using the Cisco IP phone to mark traffic, you have extended the trust boundary to point 1 shown in Figure 6-20. This is the ideal trust point because it distributes the QoS marking process to many

Cisco IP phones rather than forcing the switches to apply QoS markings to a higher volume of traffic.

If you have PCs attached to the network and you have access layer switches with QoS capabilities, you can begin marking at these devices (point 2 in Figure 6-20). If your access layer switches do not have QoS capabilities, then the first possible place you can apply QoS markings is at the distribution layer switches (point 3 in Figure 6-20). This will work just fine; however, it adds an extra load to the distribution layer switches. Likewise, you will have network traffic passing through access layer switches without any QoS treatment. Although this is usually a safe bet—because access layer switches typically have higher-speed connections, on which congestion is rare—it is always best to apply QoS in as many places as possible where there is a potential bottleneck.

Note: AutoQoS uses Cisco Discovery Protocol (CDP) to detect Cisco IP phones on Cisco switches and properly configure the QoS settings. This ensures that a user cannot disconnect their IP phone and attach another device to receive high-priority network treatment. Be sure you do not disable CDP on switches supporting Cisco IP phones.

Now, we have come to the point of configuring AutoQoS. Amazingly, by Cisco's design, enabling AutoQoS is accomplished through a single command applied under interface configuration mode. To enable AutoQoS in your network, you must first identify the interfaces to which applying AutoQoS makes sense. AutoQoS does not need to be applied under every switch and router interface in your network (although it probably won't hurt anything if you did this). It primarily should be applied to interfaces on which the devices or applications need special or preferred treatment over others. Figure 6-21 shows a typical network. The interfaces labeled A represent areas of the network where you would use AutoQoS.

As you can see from Figure 6-21, you'll be typing this one command many times. Before you enter the AutoQoS command, always ensure that you have entered the correct bandwidth statement under the serial interfaces of your routers, because a router cannot autodetect the actual speed of a WAN connection. A router can detect all other interfaces without requiring the bandwidth command.

Note: AutoQoS uses a sophisticated queuing method known as Low Latency Queuing (LLQ). This queuing method provisions a specific amount of bandwidth for the various types of network traffic, including voice. Using AutoQoS features with incorrectly configured bandwidth commands can cause substandard network service.

The AutoQoS command syntax might be slightly different depending on where you enter it. The syntax in Example 6-28 enables AutoQoS for the interfaces shown in Figure 6-21 that are connected to the Cisco IP phones.

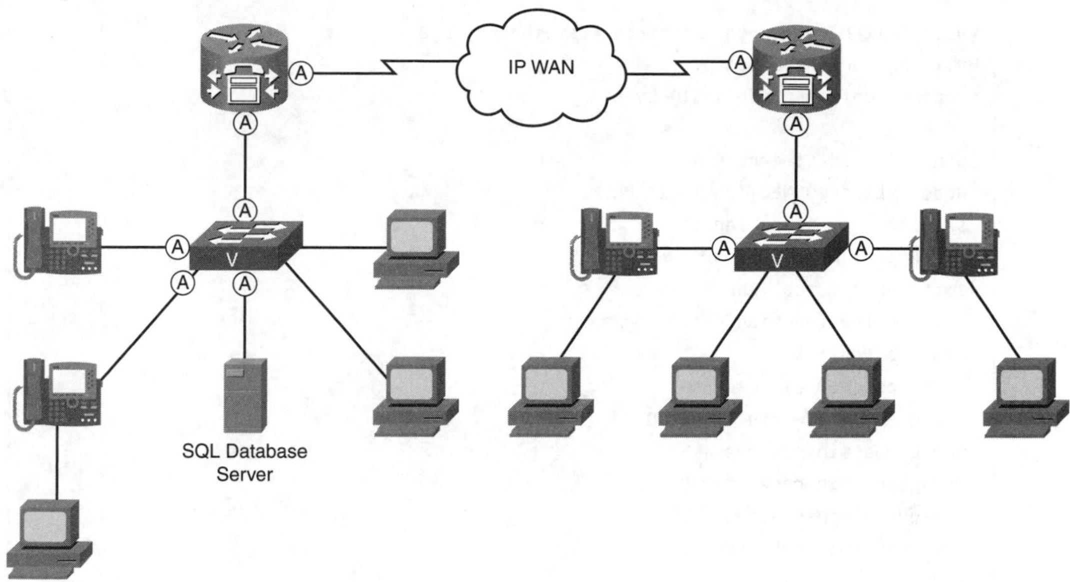

Figure 6-21 *AutoQoS Configuration Points*

Example 6-28 *Enabling AutoQoS on the Access Layer Switchports*

```
Voice_Switch# show run interface FastEthernet 0/3
Building configuration...
Current configuration : 169 bytes
!
interface FastEthernet0/3
 description CONNECTION TO IP PHONE
 switchport access vlan 10
 switchport mode access
 switchport voice vlan 5
 spanning-tree portfast
end
Voice_Switch# config term
Enter configuration commands, one per line.  End with CNTL/Z.
Voice_Switch(config)# interface fa0/3
Voice_Switch(config-if)# auto qos ?
  voip  Configure AutoQoS for VoIP
Voice_Switch(config-if)# auto qos voip ?
  cisco-phone       Trust the QoS marking of Cisco IP Phone
  cisco-softphone   Trust the QoS marking of Cisco IP SoftPhone
  trust             Trust the DSCP/CoS marking
Voice_Switch(config-if)# auto qos voip cisco-phone
Voice_Switch(config-if)# ^Z
```

```
Voice_Switch# show run interface FastEthernet 0/3
Building configuration...
Current configuration : 510 bytes
!
interface FastEthernet0/3
 description CONNECTION TO IP PHONE
 switchport access vlan 10
 switchport mode access
 switchport voice vlan 5
 mls qos trust device cisco-phone
 mls qos trust cos
 auto qos voip cisco-phone
 wrr-queue bandwidth 10 20 70 1
 wrr-queue min-reserve 1 5
 wrr-queue min-reserve 2 6
 wrr-queue min-reserve 3 7
 wrr-queue min-reserve 4 8
 wrr-queue cos-map 1 0 1
 wrr-queue cos-map 2 2 4
 wrr-queue cos-map 3 3 6 7
 wrr-queue cos-map 4 5
 priority-queue out
 spanning-tree portfast
end
```

Notice the options given by the context-sensitive help when the **auto qos voip ?** command was entered. Entering the command **auto qos voip cisco-phone** or **auto qos voip cisco-softphone** only enables the trust boundary if CDP detects a Cisco IP phone or Cisco IP Communicator (or equivalent Cisco IP SoftPhone device) attached to the port. If a user removes this device, the trust boundary is broken and is not restored until the device is reattached. If you enter the command **auto qos voip trust**, the switch trusts the markings from the attached device regardless of what it is. You need to use this command if you purchase non-Cisco IP phones. Keep in mind that if you use this command, the network susceptible to users removing the non-Cisco IP phone and attaching rogue devices.

Note: Before the **auto qos voip** command is entered under the FastEthernet 0/3 interface in Example 6-28, a **show run** command was performed so that you could see the current syntax entered under the interface. Notice how many commands are generated after entering the **auto qos voip** command. It is beneficial that the Cisco switch (and router) shows you all the individual commands so that you can optionally tune the settings to exactly fit your environment.

If the configuration generated by the **auto qos voip** command is not desired, you can remove this configuration simply by entering **no auto qos voip**.

Example 6-29 shows the AutoQoS syntax to use on the switch for the interface connecting to the router.

Example 6-29 *Enabling AutoQoS on the Switch-Router Uplink*

```
Voice_Switch# show run interface FastEthernet 0/1
Building configuration...
Current configuration : 169 bytes
!
interface FastEthernet0/1
 description CONNECTION TO ROUTER
 switchport access vlan 10
 switchport mode access
 spanning-tree portfast
end
Voice_Switch# config term
Enter configuration commands, one per line.  End with CNTL/Z.
Voice_Switch(config)# interface fa0/1
Voice_Switch(config-if)# auto qos voip trust
Voice_Switch(config-if)# ^Z
Voice_Switch# show run int fa0/1
Building configuration...
Current configuration : 369 bytes
!
interface FastEthernet0/1
 description CONNECTION TO ROUTER
 switchport access vlan 10
 switchport mode access
 mls qos trust cos
 auto qos voip trust
 wrr-queue bandwidth 10 20 70 1
 wrr-queue min-reserve 1 5
 wrr-queue min-reserve 2 6
 wrr-queue min-reserve 3 7
 wrr-queue min-reserve 4 8
 wrr-queue cos-map 1 0 1
 wrr-queue cos-map 2 2 4
 wrr-queue cos-map 3 3 6 7
 wrr-queue cos-map 4 5
 priority-queue out
end
```

You can configure the interface between the switch and router with the **auto qos voip trust** command, because you would consider the QoS markings from the router as trusted.

Finally, you can enable AutoQoS on the router's FastEthernet and Serial interfaces with the syntax shown in Example 6-30.

Example 6-30 *Enabling AutoQoS on Router Interfaces*

```
CME_Voice# show run int fa0/0
Building configuration...
!
interface FastEthernet0/0
 ip address 172.30.4.3 255.255.255.0
 ip nat inside
 ip virtual-reassembly
 duplex auto
 speed auto
end
CME_Voice# show run int s0/1/0
Building configuration...
!
interface Serial0/1/0
 bandwidth 512
 ip address 10.1.1.1 255.255.255.0
 encapsulation ppp
 no fair-queue
 clock rate 2000000
end
CME_Voice# configure terminal
Enter configuration commands, one per line.  End with CNTL/Z.
CME_Voice(config)# interface FastEthernet 0/0
CME_Voice(config-if)# auto ?
  discovery  Configure Auto Discovery
  qos        Configure AutoQoS
CME_Voice(config-if)# auto qos voip trust
CME_Voice(config-if)# exit
CME_Voice(config)# interface Serial 0/1/0
CME_Voice(config-if)# auto qos voip trust
CME_Voice(config-if)# ^Z
CME_Voice# show run int fa0/0
Building configuration...
!
interface FastEthernet0/0
 ip address 172.30.4.3 255.255.255.0
 ip nat inside
 ip virtual-reassembly
 duplex auto
 speed auto
```

```
 auto qos voip trust
 service-policy output AutoQoS-Policy-Trust
end
CME_Voice# show run int s0/1/0
Building configuration...
!
interface Serial0/1/0
 bandwidth 512
 no ip address
 encapsulation ppp
 auto qos voip trust
 no fair-queue
 clock rate 2000000
 ppp multilink
 ppp multilink group 2001100116
end
```

The changes to the router interfaces look relatively tame compared to the amount of syntax entered under the switch interfaces; however, what you do not see are the billion other commands (or perhaps slightly less than a billion) that were entered in other configuration modes of the router to create class maps, policy maps, multilink interfaces, and so on. The full explanation of those commands is saved for the CCNP Voice QoS material.

Note: In Example 6-30, after entering the **auto ?** command under the FastEthernet interface, notice that one of the options you are given is **auto discovery**. This enables a newer, ultra-incredible version of AutoQoS that allows the router to monitor your network for an extended time to discover known types of data, voice, and video traffic that are considered higher priority based on common high-priority application types. After the router captures enough traffic, it generates QoS policy recommendations that you can choose to apply or ignore.

Table 6-8 summarizes the different variations of AutoQoS commands you can enter on Cisco switch and router platforms.

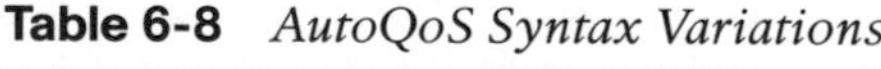
Table 6-8 *AutoQoS Syntax Variations*

Command	Platform	Description
auto qos voip	Router or Layer 3 switch	Enables AutoQoS without trusting any existing markings on packets. The router re-marks all traffic types using access lists or Network-Based Application Recognition (NBAR) to identify traffic (higher processor-utilization tasks).

Table 6-8 *AutoQoS Syntax Variations*

Command	Platform	Description
auto qos voip trust	Router or switch	This configuration explicitly trusts QoS markings set by the attached device and does not rely on CDP to verify a Cisco IP Phone is attached.
auto qos voip cisco-phone	Switch	Enables AutoQoS, trusting any existing QoS markings that enter the interface only if the switch detects a Cisco IP phone attached through CDP.
auto qos voip cisco-softphone	Switch	Enables AutoQoS, trusting any existing QoS markings that enter the interface only if the switch detects a Cisco IP SoftPhone (such as Cisco IP Communicator) attached through CDP.

Note: QoS engineers identify what have been called QoS markings in the previous section as Class of Service (CoS) and Type of Service (ToS) markings. CoS is a marking that exists in the Layer 2 header of a frame, which a switch can identify. ToS is a marking that exists in the Layer 3 header of a packet, which a router can identify. This topic is explored in depth in the material for the QoS CCNP Voice certification exam.

Exam Preparation Tasks

Review All the Key Topics

Review the most important topics in the chapter, noted with the key topics icon in the outer margin of the page. Table 6-9 describes these key topics and identifies the page number on which each is found.

Table 6-9 *Key Topics for Chapter 6*

Key Topic Element	Description	Page Number
Figure 6-1	Illustrates the use of analog FXS ports	108
Figure 6-2	Illustrates the use of analog FXO ports	111
Note	Signaling channel information for T1 and E1 interfaces	117
List	Description of POTS and VoIP dial peers	118
Figure 6-5	Illustrates the use of call legs to design dial peer configurations	119
Example 6-7	Basic POTS dial peer configuration	120
Text	Highlights the automatic digit-stripping rule of POTS dial peers	123
Example 6-11	Basic VoIP dial peer configuration	124
Table 6-3	Summarizes dial peer wildcards	126
Table 6-4	Provides examples of using the dial peer bracket wildcard	127
Table 6-5	Provides a sample PSTN dialing plan for North America	127
Example 6-13	Basic PLAR configuration using FXS interfaces	129
List	Highlights the rules Cisco routers use to handle overlapping dial peers	130
List	The method a router uses to match inbound dial peers	132
List	Characteristics of dial peer 0	135
Table 6-6	Summarizes digit-manipulation commands	135
Example 6-17	Implementing WAN to PSTN failover using preference and prefix commands	137
Tip	Tip on how the router handles identical dial peers	138
List	Two key rules of COR lists	151
List	Three areas of concern when deploying QoS	153

Table 6-9 *Key Topics for Chapter 6*

Key Topic Element	Description	Page Number
List	Key delay requirements for voice and video traffic	154
List	Three common QoS models	155
List	Categories of QoS mechanisms	155
List	Specific link efficiency mechanisms	157
List	Specific queuing algorithms	157
Table 6-7	Directions you can apply various QoS mechanisms	158

Definitions of Key Terms

Define the following key terms from this chapter, and check your answers in the Glossary:

Dialed Number Identification Service (DNIS), Automatic Number Identification (ANI), dial-peer, Foreign Exchange Station (FXS) ports, Foreign Exchange Office (FXO) ports, Private Line Automatic Ringdown (PLAR), Direct Inward Dial (DID), Class of Restriction (COR)

- **Configuring a Voice Network Directory:** This section walks through the creation of a local directory of CME devices, which gives your users an easier method to find and dial local DNs.
- **Configuring Call Forwarding:** This section discusses the concepts and configuration of the call-forwarding features in the CME environment.
- **Configuring Call Transfer:** This section discusses the concepts and configuration of the call-transfer features in the CME environment.
- **Configuring Call Park:** This section discusses the concepts and configuration of the call park features in the CME environment.
- **Configuring Call Pickup:** This section discusses the concepts and configuration of the call pickup features in the CME environment.
- **Configuring Intercom:** This section discusses the concepts and configuration of the intercom features in the CME environment.
- **Configuring Paging:** This section discusses the concepts and configuration of the paging features in the CME environment.
- **Configuring After-Hours Call Blocking:** This section discusses the methods you can use to allow or deny specific dialing patterns in the after-hours time frame for all or specific IP phones.
- **Configuring CDRs and Call Accounting:** This section discusses the configuration of CDRs and call-accounting features.
- **Configuring Music on Hold:** This section discusses the configuration of Music on Hold (MoH) with CME.
- **Configuring Single Number Reach:** This section discusses the configuration of Single Number Reach using CME.
- **Enabling the Flash-Based CME GUI:** This section walks through the installation and configuration of this utility.

CHAPTER 7

Configuring Cisco Unified CME Voice Productivity Features

After implementing the ephone, ephone-dn, and dial-peer concepts, you now have an IP Telephony network that is able to make and place internal and external calls. If only voice networks stayed this simple! Organizations expect modern telephony systems to support a whole host of features, such as call transfer, Music on Hold, conference calling, and so on. This chapter is dedicated to adding these types of features to the voice network.

"Do I Know This Already?" Quiz

The "Do I Know This Already?" quiz allows you to assess whether you should read this entire chapter or simply jump to the "Exam Preparation Tasks" section for review. If you are in doubt, read the entire chapter. Table 7-1 outlines the major headings in this chapter and the corresponding "Do I Know This Already?" quiz questions. You can find the answers in Appendix A, "Answers Appendix."

Table 7-1 *"Do I Know This Already?" Foundation Topics Section-to-Question Mapping*

Foundation Topics Section	Questions Covered in This Section
Configuring a Voice Network Directory	1
Configuring Call Forwarding	2–3
Configuring Call Transfer	4–5
Configuring Call Park	6–7
Configuring Call Pickup	8
Configuring Intercom	9
Configuring Paging	10
Configuring After-Hours Call Blocking	11
Configuring CDRs and Call Accounting	12
Configuring Single Number Reach	13

1. What process must you follow to build the local phone **directory** for the CME environment?

   a. Assign directory entries under each ephone-dn using the **directory** command.
   b. Allow CME to automatically build the directory when you assign caller-ID information using the name command.
   c. Assign directory entries under each ephone using the **directory** command.
   d. Enter the directory configuration mode and begin associating ephone-dn values with directory entry values.

2. You enter the command **call-forward max-length 0** from telephony service configuration mode. How does this affect the voice network?

   a. Users can forward their phone for an unlimited amount of time.
   b. Users can forward their phones to any destination that is reachable from their IP phone.
   c. The CFwdAll softkey on users' IP phones dims and becomes unavailable.
   d. All IP phones that are currently forwarding calls will transfer calls directly to voicemail.

3. Which of the following categories of standards prevents calls from hairpinning on the network when they are forwarded or transferred?

   a. H.450
   b. H.225
   c. H.323
   d. H.240

4. Which of the following transfer modes does a Cisco router support by default?

   a. Blind
   b. Consult
   c. Full-blind
   d. Full-consult

5. When you enter the command **transfer-pattern** 95... from telephony service configuration mode, what is the result?

   a. Call transfers are restricted to only numbers matching the 95... pattern.
   b. Transferred calls have 95 added to the front of the dialed number information.
   c. Transferred calls have 95 added to the end of the dialed number information.
   d. Users are now able to transfer calls to numbers matching the 95... pattern.

6. You entered the following configuration on the CME router:

```
CME_Voice(config)#ephone-dn 51
CME_Voice(config-ephone-dn)#number 3002
CME_Voice(config-ephone-dn)#park-slot timeout 60 limit 10 recall
```

What effect does this have on the voice environment?

   a. This creates a call park slot reserved for ephone-dn 3002 that can have up to ten parked calls for 60 seconds each.
   b. This creates a call park slot numbered 3002 that can have up to ten parked calls for 60 seconds each.
   c. This creates a call park slot numbered 3002 that can have a call parked for 60 seconds, at which point the original phone is recalled. If this occurs ten times, the parked call is disconnected.
   d. This creates a call park slot numbered 3002 that can have up to ten calls parked for 60 seconds each, at which point the original phone is recalled.

7. By default, what does pressing the PickUp softkey allow you to do in a Cisco Unified CME environment?
   a. Pick up a ringing phone in your group
   b. Pick up a ringing phone in another group
   c. Answer your own ringing phone
   d. Pick up a specific ringing extension (which you must specify)

8. You are watching an administrator configure an intercom line. After creating a new ephone-dn, she enters the command number A100 and presses Enter. What is the purpose of this command?
   a. To designate extension 100 as an intercom line
   b. To prevent the number from being dialed from an IP phone
   c. To match the number with number **B100** on the other side of the intercom connection
   d. To list the line as the first intercom in the configuration

9. What is the maximum number of paging groups to which a Cisco IP Phone can belong?
   a. 1.
   b. 5.
   c. 25.
   d. There is no practical limit.

10. What process can a user use to exempt his IP phone from an after-hours call block configured with the 24/7 keyword?

a. An after-hours 24/7 call block cannot be exempted.

b. Enter the correct PIN number.

c. You must configure the user's phone as exempt.

d. Enter the correct accounting code.

11. What are two destinations to which the Cisco Unified CME can send CDRs? (Choose two.)

a. TFTP server

b. Syslog server

c. Logging buffer

d. HTTP server

12. What two areas does Cisco Unified CME not allow you to modify from the CME GUI, by default? (Choose two.)

a. Ephone-dn

b. Ephone

c. Time on the router

d. System message

13. A user is on an active call at the office using his desk phone. Midway through the call, he presses the Mobility softkey on the screen of his IP phone. What process occurs?

a. The call transfers to his preconfigured Single Number Reach number.

b. CME places the call on hold and allows retrieval from a remote phone.

c. CME places the call on hold and allows retrieval from a predefined call park number.

d. The user logs out of the phone and then logs into a new phone where he retrieves the call.

Foundation Topics

Configuring a Voice Network Directory

When most people think of a corporate phone directory, visions of Microsoft Excel spreadsheets e-mailed out monthly come to mind. Isn't there a better way? Sure there is! Cisco IP Phones support a local directory that you can update from the Cisco Unified Communication Manager Express (CME) router as you are configuring devices.

You can enter names under ephone-dn configuration mode either as you are configuring new lines for the organization or separately, after you configure the lines. These names are used both for building the internal corporate phone directory (often called the local directory) and for caller ID information. Example 7-1 shows names being added to individual ephone-dns currently in use in the organization.

Example 7-1 *Configuring Local Directory and Internal Caller ID Information*

```
CME_Voice(config)# ephone-dn 20
CME_Voice(config-ephone-dn)# name Joshua Bellman
CME_Voice(config-ephone-dn)# exit
CME_Voice(config)# ephone-dn 21
CME_Voice(config-ephone-dn)# name Ruth Hopper
CME_Voice(config-ephone-dn)# exit
CME_Voice(config)# ephone-dn 22
CME_Voice(config-ephone-dn)# name Esther Billford
CME_Voice(config-ephone-dn)# exit
CME_Voice(config)# ephone-dn 23
CME_Voice(config-ephone-dn)# name Job Smith
CME_Voice(config-ephone-dn)# exit
CME_Voice(config)# ephone-dn 24
CME_Voice(config-ephone-dn)# name Samuel Oldham
CME_Voice(config-ephone-dn)# exit
```

After you enter these names in ephone-dn configuration mode, they immediately take effect. If ephone-dn 24 were to call ephone-dn 21, Ruth would see "Samuel Oldham" appear on her caller ID information, as shown in Figure 7-1.

In addition, all modern Cisco IP Phone models allow you to browse the corporate directory by pressing the Directory button on the phone itself. Some low-end IP phones may not have a dedicated Directory button, but instead have a menu-driven process to get there. After you press the Directory button, you are able to browse categories including Missed Calls, Received Calls, and so on. Move down to the option showing the Local Directory, as shown in Figure 7-2.

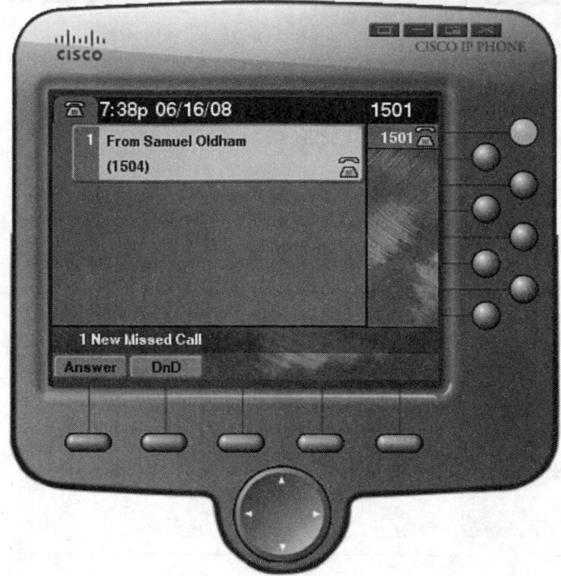

Figure 7-1 *Name Command Affects Directory and Caller ID Information*

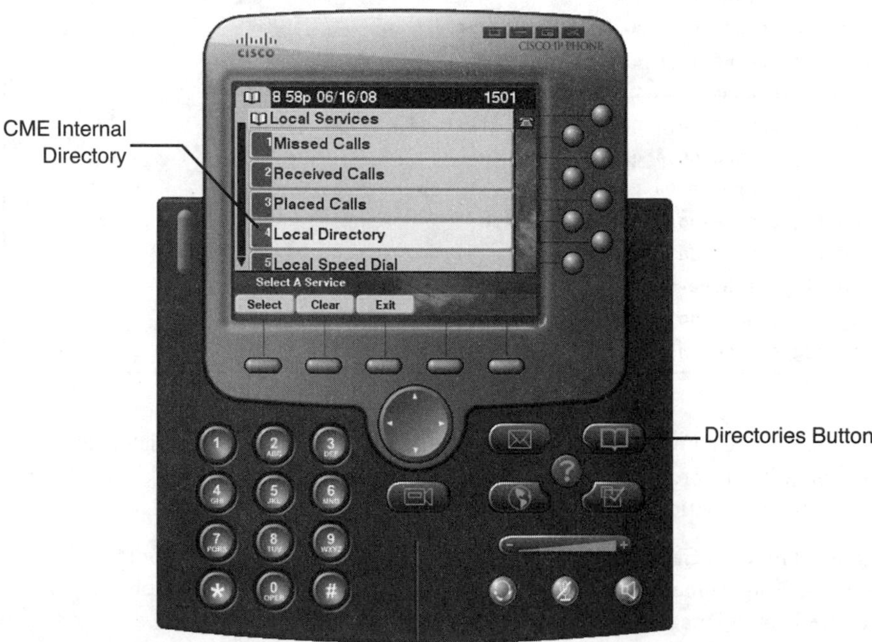

Figure 7-2 *Browsing Phone Directories*

After you select the Local Directory, the IP phone gives you the option to search by first or last name by typing in a user's name as a string on the IP phone. You can enter as many characters as you like to filter down the number of results or simply press the Select softkey to see the entire corporate directory, as shown in Figure 7-3.

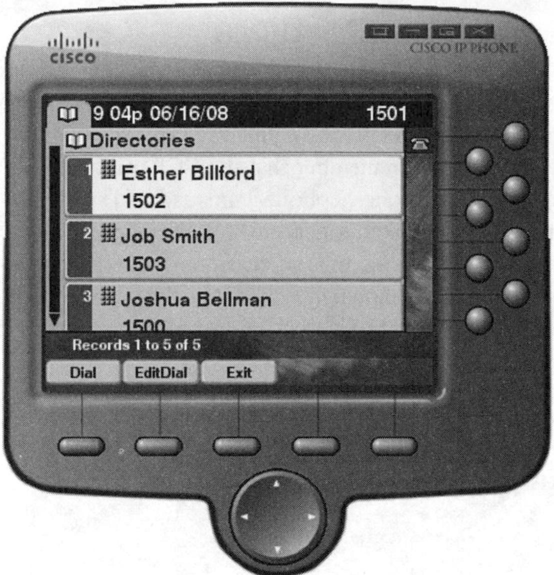

Figure 7-3 *Local CME Directory*

By default, Cisco Unified CME organizes the local directory alphabetically by first name. You can change this setting by using the **directory** command from telephony service configuration mode. In addition, you can also add manual entries to the directory by using the **directory entry** command. This is useful for devices in the company that do not have an explicit ephone-dn configuration. Example 7-2 demonstrates these two commands in action.

Example 7-2 *Configuring Manual Local Directory Entries*

```
CME_Voice(config-telephony)# directory ?
  entry              Define new directory entry
  first-name-first  first name is first in ephone-dn name field
  last-name-first   last name is first in ephone-dn name field
CME_Voice(config-telephony)# directory last-name-first
CME_Voice(config-telephony)# directory entry ?
  <1-100>  Directory entry tag
  clear    clear all directory entries
CME_Voice(config-telephony)# directory entry 1 ?
  WORD  A sequence of digits representing dir. number
CME_Voice(config-telephony)# directory entry 1 1599 ?
  name  Define directory name
CME_Voice(config-telephony)# directory entry 1 1599 name ?
  LINE  A string - representing directory name (max length: 24 chars)
CME_Voice(config-telephony)# directory entry 1 1599 name Corporate Fax
```

Note: As you can see from the context-sensitive help, you can add up to 100 manual entries to the local CME directory. Also, keep in mind that sorting alphabetically by last name flips all the information in the directory to list last name first. CME will list the "Corporate Fax" directory entry just added as "Fax Corporate."

If you are using the Cisco Configuration Professional (CCP) to manage caller ID and Local Directory configurations, the GUI performs the Caller-ID assignment for you when you associate a user with a Phone/Extension. You might remember that the CCP utility does not associate extensions (ephone-dns) directly to phones (ephones). Instead, after you create the necessary extensions and phones, they are linked together through the user account. Once you add a First Name and Last Name to the user account, the name is applied to the extension associated to that user account. As shown in Figure 7-4, the user account Peter Rock is associated with extension 1501 (using the Phones/Extensions tab, not pictured in Figure 7-4).

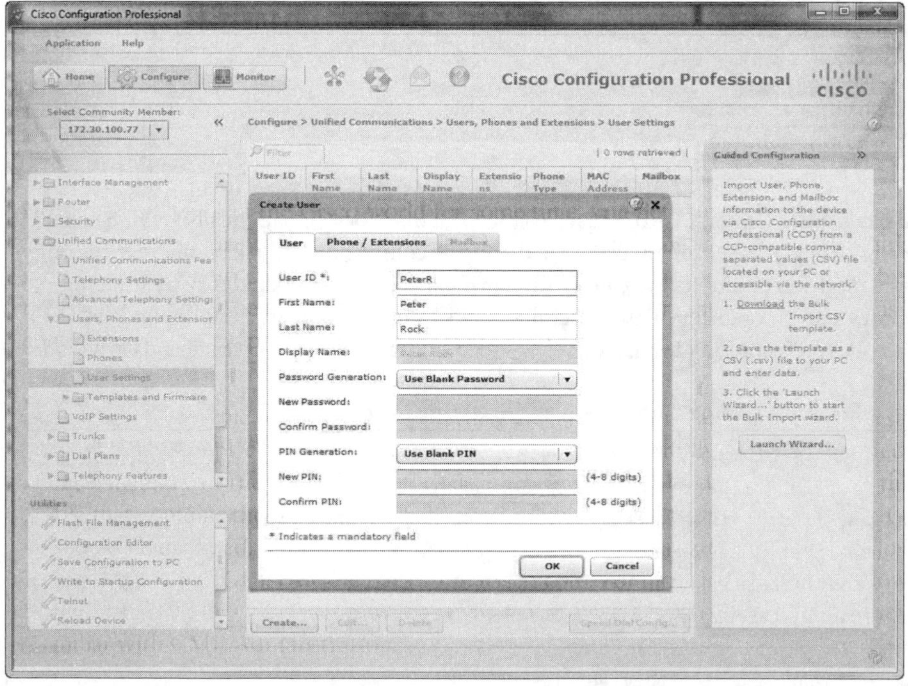

Figure 7-4 *Assigning Directory Information Using CCP*

Likewise, you can find the directory sorting option under **Unified Communications > Advanced Telephony Settings > System Config** (shown in Figure 7-5).

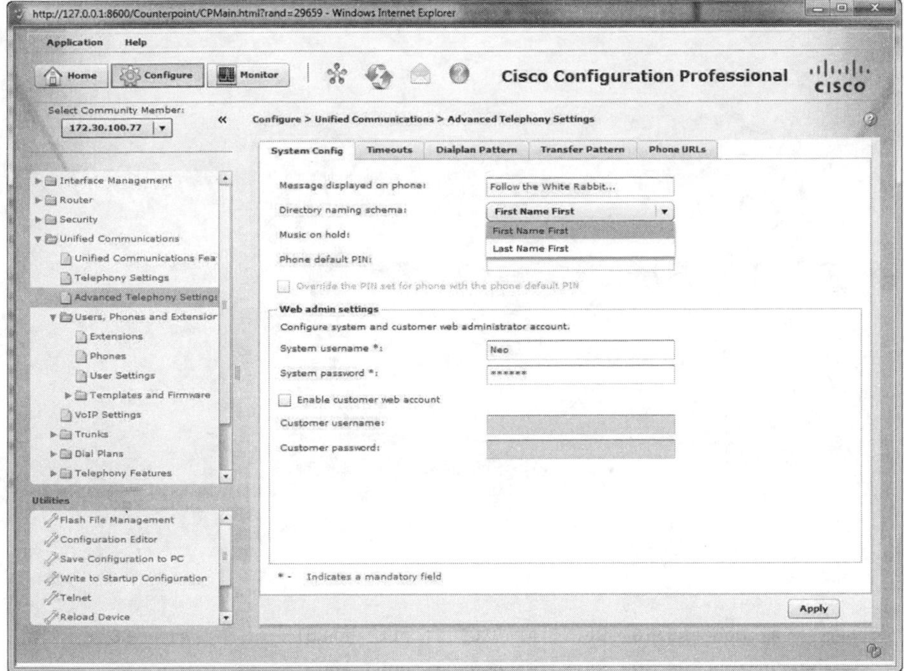

Figure 7-5 *CCP Directory Sorting Options*

Finally, you can create manual directory entries by navigating to **Unified Communications > Telephony Features > Directory Services** (shown in Figure 7-6).

Configuring Call Forwarding

There are two methods used to forward calls to a different destination: from the IP phone (the user's method) and from the Cisco IOS CLI (the administrator's method). This section describes both methods and also provides an overview of the call-forward pattern command.

Forwarding Calls from the IP Phone

To forward calls from the IP phone, simply press the CFwdAll softkey button, as shown in Figure 7-7. The IP phone beeps twice and allows you to enter a number. Enter the number to which all calls on the IP phone will forward, and then press the pound key (#) on the phone so that it knows you are done entering the number. To cancel call forwarding, press the CFwdAll button a second time.

Tip: If you want to forward all calls directly to voicemail, press the CFwdAll button followed by the messages button on the IP phone.

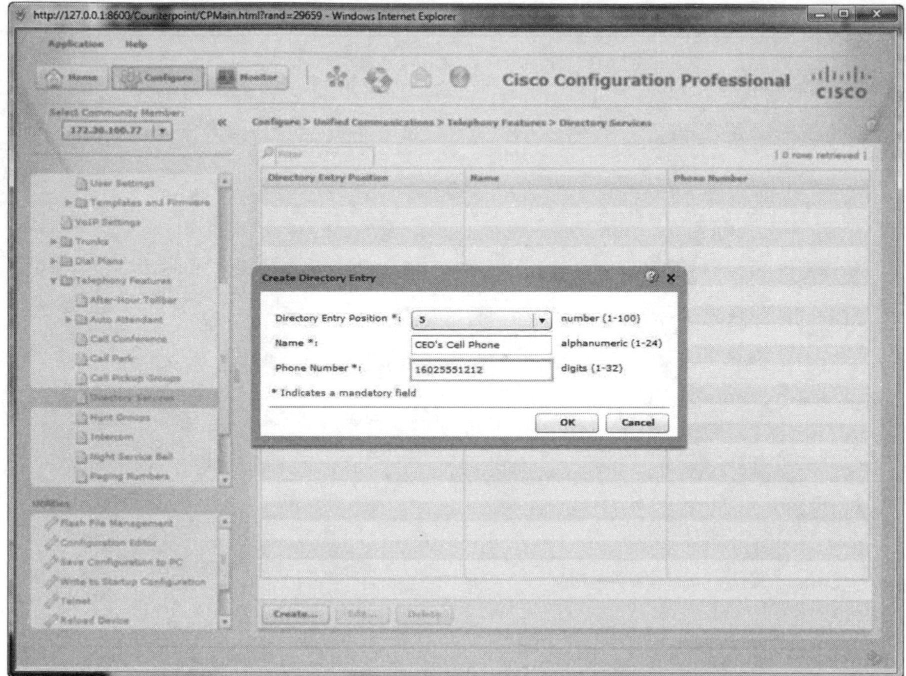

Figure 7-6 *Adding Manual Directory Entries with CCP*

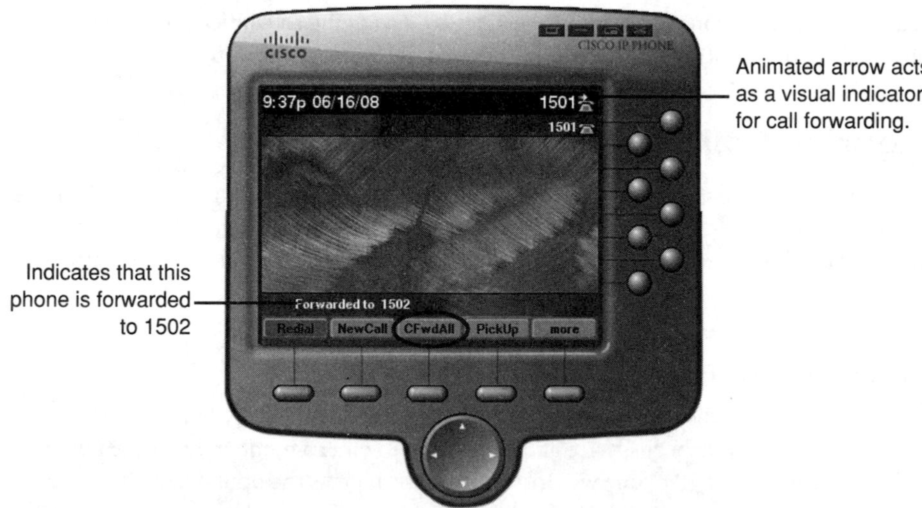

Figure 7-7 *Forwarding Calls from the Cisco IP Phone*

Forwarding Calls from the CLI

Forwarding calls from the command line gives you more options than does forwarding calls from the IP phone, as shown in Example 7-3.

Example 7-3 *Forwarding Calls from the Cisco IOS CLI*

Key Topic

```
CME_Voice(config)# ephone-dn 21
CME_Voice(config-ephone-dn)# call-forward ?
  all            forward all calls
  busy           forward call on busy
  max-length     max number of digits allowed for CFwdAll from IP phone
  night-service  forward call on activated night-service
  noan           forward call on no-answer
CME_Voice(config-ephone-dn)# call-forward busy 1599
CME_Voice(config-ephone-dn)# call-forward noan 1599 ?
  timeout  Ringing no answer timeout duration
CME_Voice(config-ephone-dn)# call-forward noan 1599 timeout ?
  <3-60000>  Ringing no answer timeout duration in seconds
CME_Voice(config-ephone-dn)# call-forward noan 1599 timeout 25
```

These options allow you to forward calls that are busy or not answering (noan) to a different extension. Although this is typically a voicemail number (which 1599 represents in Example 7-3), this could also be another IP phone if this DN was a member of a hunt group.

Tip: In the United States, the phone rings for 2 seconds followed by 4 seconds of silence. Knowing this can be useful in calculating a good no answer (noan) timeout value.

Also notice that you can specify a max-length value after the **call-forward** command. Using this, you can restrict the IP phone from forwarding to external destinations. If you enter the command **call-forward** max-length 0, CME makes the IP phone call forwarding feature unavailable to the Cisco IP Phone. The CFwdAll button will dim on the IP phone and become inaccessible.

Tip: At this point, you should have a good idea that plenty of configurations under each ephone-dn are similar to all the others. Make an ephone-dn (and ephone) template in Notepad (or some other text editor) in which you list all the common configuration commands you'll be applying in your environment. That way, if you ever need to add new ephone-dns, you will already have a template listing the common commands you need to enter.

Using the **call-forward pattern** Command to Support H.450.3

There is one additional command to discuss here, which is available from telephony service configuration mode: **call-forward pattern.** This command allows you to enter a pattern for numbers that will support the H.450.3 call forwarding standard.

To understand the benefits of H.450.3, you must first understand what happens with typical VoIP forwarding. When a call enters the network and hits a forwarded device, that device takes responsibility for the call and becomes a tandem hop in the call flow. That means that the voice traffic now forwards through the IP phone that forwarded the call. This can cause quality problems if the device that forwarded the call is a large geographical distance away from the phone receiving the forwarding call. The H.450.3 standard represents a method that allows the CME router to redirect the call directly to the final destination instead of acting as a tandem hop. Figure 7-8 illustrates this concept.

Key Topic

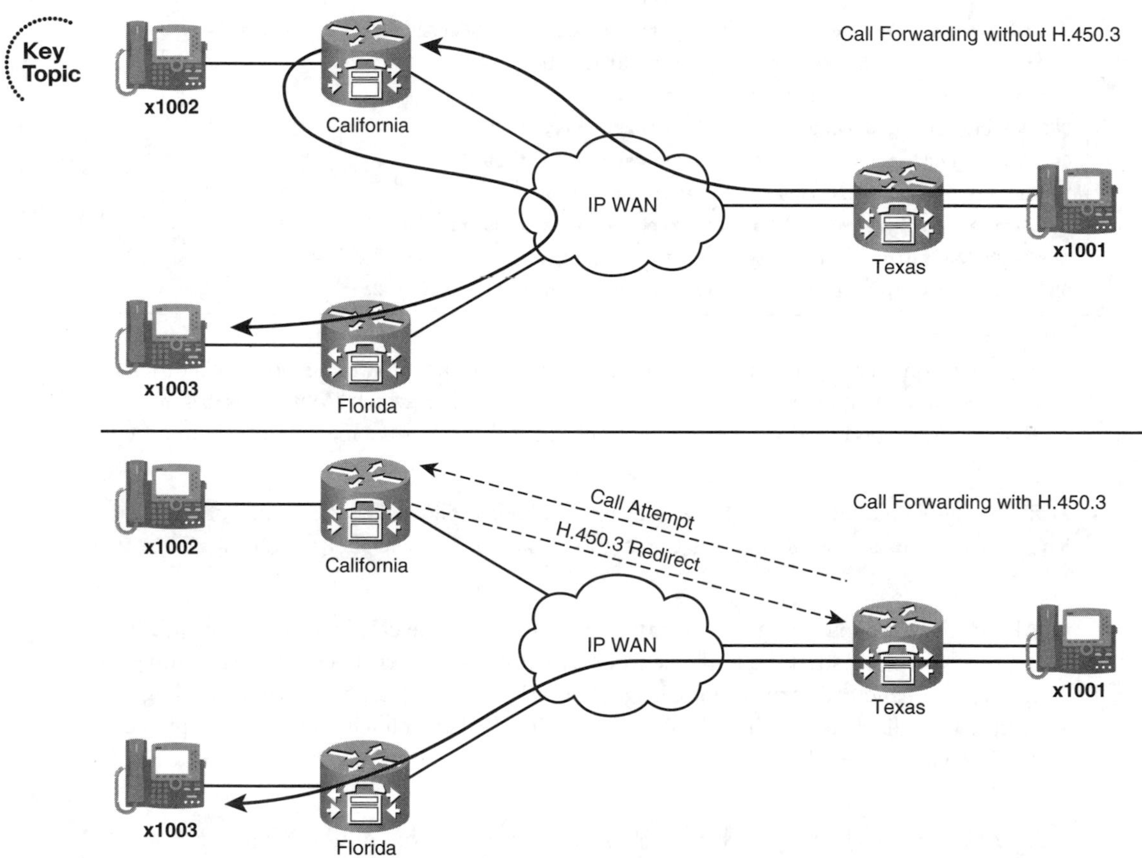

Figure 7-8 *Forwarding Calls With and Without H.450.3 Standards*

In Figure 7-8, the IP phone with x1002 is forwarded to the IP phone with x1003. The top part shows the VoIP call flow without H.450.3 when x1001 places a call to x1002. Notice that the VoIP traffic must pass through the California CME router to reach Florida. This can cause intense quality of service (QoS) problems with the call, such as audio clipping, distortion, and even call drops. This symptom is commonly called *hairpinning* the call.

The bottom part shows the call with H.450.3 support enabled. When the call reaches California, CME sends an H.450.3-based redirect message, instead of accepting the call and forwarding it on to Florida. The VoIP traffic then travels directly from x1001 in Texas to Florida rather than passing through California to get there.

Entering **call-forward pattern** *<pattern>* from telephony service configuration mode tells CME which numbers should support the H.450.3 standard. Entering the pattern 15.. tells CME, "I want all four-digit numbers that begin with 15 to support H.450.3." Thus, all calls to 15XX extensions would support H.450.3 call forwarding.

> **Note:** There is much more to be said about the H.450.3 standard. There is also more configuration that should be in place to fully support H.450.3. This is intended to be a "sneak peak" of the standard, which the CCNP Voice certification track fully explores.

If you prefer to use CCP to make forwarding modifications, you can find the settings under the Advanced tab of the Extensions configuration window (**Unified Communication > Users, Phones, and Extensions > Extensions**), as shown in Figure 7-9.

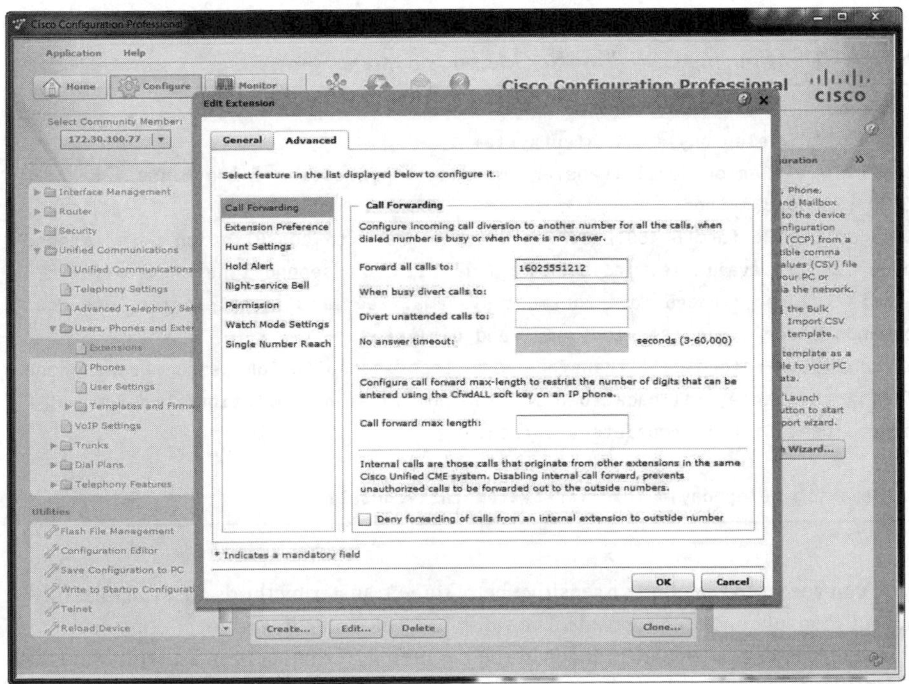

Figure 7-9 *Configuring Call Forwarding Using CCP*

Configuring Call Transfer

Transferring calls represents another common function in voice networks. To transfer a call, hit the Trnsfer softkey while on an active call. (Note that this is not a typo: Trnsfer without the *a* is correct.) When you do, you hear another dial tone, at which point you can dial the phone number to which you want to transfer your active call. What happens from there depends on the transfer method configured on the CME router. Two transfer methods are available:

- **Consult:** Consult transfer allows you to speak with the other party before transferring the call. After you dial the number to which you want to transfer the call, you can wait for the other party to answer and speak with them before transferring the call. Pressing the Trnsfer softkey a second time transfers the call, dropping you out of the conversation. Consult transfers require a second line (or dual-line configuration). This is the default transfer mode in CME. However, the specific method of transfer used by default in CME is Full Consult.
- **Blind:** Blind transfer immediately transfers the call after you dial the number (you do not hit the Trnsfer softkey a second time). Blind transfers can work in a single-line configuration.

To configure the transfer method used, see Example 7-4.

Example 7-4 *Configuring CME Transfer Methods System-Wide*

```
CME_Voice(config)# telephony-service
CME_Voice(config-telephony)# transfer-system ?
  full-blind      Perform call transfers without consultation using H.450.2 or SIP
    REFER standard methods
  full-consult    Perform H.450.2/SIP call transfers with consultation using second
    phone line if available, fallback to full-blind if second line unavailable.
    This is the recommended mode for most systems. See also 'supplementary-service'
    commands under 'voice service voip' and dial-peer.
  local-consult  Perform call transfers with local consultation using second phone
    line if available, fallback to blind for non-local consultation/transfer
    target. Uses Cisco  proprietary method.

CME_Voice(config-telephony)# transfer-system full-consult
```

As you can see from the context-sensitive help, three transfer methods are available: full-blind, full-consult, and local-consult. The full-blind, full-consult, and local-consult describe the transfer methods introduced at the beginning of this section. The full-blind and full-consult methods use the industry-standard H.450.2 method of transferring. Just like call forwarding, you don't want to hairpin the call and cause potential QoS issues each time you transfer. By using the H.450.2 standard when transferring a call, the CME router completely drops the call from the transferring phone and starts a new call at the phone to which the call was transferred.

The local-consult method uses a Cisco proprietary transfer method that performs a consult transfer if multiple lines or dual-line configurations are available, but will revert to blind transfers if only a single line is available. Cisco proprietary transfers work similar to the H.450 standard. The only problem is this transfer method results in hairpinned calls if you have non-Cisco IP telephony systems on your network.

> **Note:** You can also configure transfer modes individually for each ephone-dn by using the transfer-mode <blind/consult> syntax from ephone-dn configuration mode. Configuring the transfer mode this way uses H.450 standards and overrules the system-wide setting.

By default, the Cisco router restricts transfers to devices that are not locally managed. This is usually a good policy, because transferring outside of the company can result in toll fraud. For example, a user could transfer an outside caller to an international number, causing the toll charges to be billed to the organization rather than the outside caller. If you would like to allow transfers outside of the locally managed devices, you can use the **transfer-pattern** *<pattern>* command from telephony service mode, where pattern represents numbers to which you would like to allow transfers. Example 7-5 configures the Cisco Unified CME router to allow transfers to 5XXX extensions and local ten-digit PSTN numbers.

Example 7-5 *Configuring CME Transfer Patterns to Allow Outside Transfers*

```
CME_Voice(config)# telephony-service
CME_Voice(config-telephony)# transfer-pattern ?
  WORD  digit string pattern for permitted non-local call transfers

CME_Voice(config-telephony)# transfer-pattern 5...
CME_Voice(config-telephony)# transfer-pattern 9..........
```

Cisco CCP also allows the configuration of transfer patterns. These are found under **Unified Communications > Advanced Telephony Settings**. The simple configuration window shown in Figure 7-10 allows you to simply click the **Add** button and add manually configured transfer patterns directly into the CCP interface.

Configuring Call Park

Typically, when you place a call on hold, you can retrieve the call only from the original phone where you placed the call on hold. Shared-line systems bend the rules by allowing you to retrieve the call from any phone with the same shared line assignment. The call park feature takes this one step further by allowing you to retrieve the call from any phone in the organization. Call park "parks" the caller on hold at an extension rather than on a specific line. Any IP phone that is able to dial the park extension number can retrieve the call.

The call park system works by finding free ephone-dns in the Cisco Unified CME configuration that you have not assigned to an IP phone and have specifically designated as a call park slot. You can either allow CME to park calls randomly at the first available ephone-dn or allow users to choose the extension where the call is parked. Each of these scenarios fits different environments. Calls being parked at random extensions might work well for a warehouse environment with a voice-paging system. When an employee has a call, the receptionist could announce, "Larry, you have a call on 5913," over the loudspeaker, at which point Larry could go to a phone and dial the extension to pick up the call on hold.

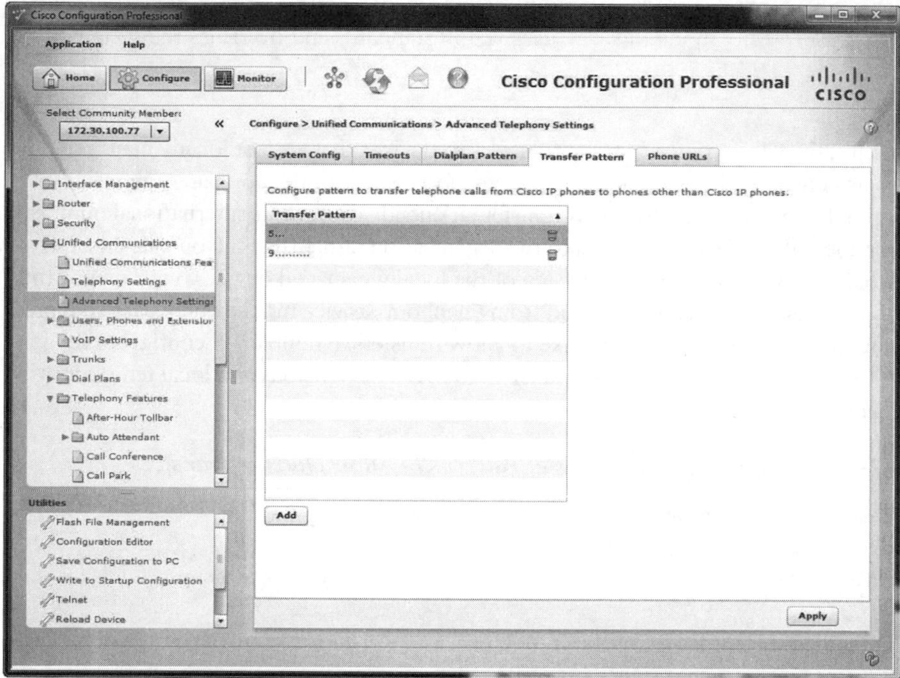

Figure 7-10 *Using CCP to Configure Transfer Patterns*

Choosing extensions would work well for an electronics superstore in which each department responded to a known extension number. For example, software could be extension 301, cameras could be extension 302, and so on. The receptionist can then park multiple calls on a single call park number (this requires multiple ephone-dns assigned the same extension). As the specific department retrieves the calls, CME distributes them in the order in which they were parked. The call parked longest is answered first.

You can configure call park simply by adding an ephone-dn designated for call park purposes. Example 7-6 creates two ephone-dns designated for call park.

Example 7-6 *Configuring Call Park Ephone-DNs*

```
CME_Voice(config)# ephone-dn 50
CME_Voice(config-ephone-dn)# number 3001
CME_Voice(config-ephone-dn)# name Maintenance
CME_Voice(config-ephone-dn)# park-slot
```

```
CME_Voice(config-ephone-dn)# exit
CME_Voice(config)# ephone-dn 51
CME_Voice(config-ephone-dn)# number 3002
CME_Voice(config-ephone-dn)# name Sales
CME_Voice(config-ephone-dn)# park-slot ?
  reserved-for  Reserve this park slot for the exclusive use of the phone with the
  extension  indicated by the transfer target extension number
  timeout       Set call park timeout
  <cr>
CME_Voice(config-ephone-dn)# park-slot timeout ?
  <0-65535>  Specify the park timeout (seconds) before the call is returned to the
    number it was  parked from
CME_Voice(config-ephone-dn)# park-slot timeout 60 ?
  limit  Set call park timeout count limit
CME_Voice(config-ephone-dn)# park-slot timeout 60 limit ?
  <1-65535>  Specify the number of park timeout cycles before the call is
    disconnected
CME_Voice(config-ephone-dn)# park-slot timeout 60 limit 10 ?
  notify    Define additional extension number to notify for park timeout
  recall    recall transfer back to originator phone after timeout
  transfer  Transfer to originator or specified destination after timeout limit
    exceeded
  <cr>
CME_Voice(config-ephone-dn)# park-slot timeout 60 limit 10 recall ?
  alternate  Transfer to alternate target if original target is busy
  retry      Set recall/transfer retry interval if target is in use
  <cr>
CME_Voice(config-ephone-dn)# park-slot timeout 60 limit 10 recall
```

Look at the configuration of ephone-dn 50 in Example 7-6. Designating a call park extension is as simple as entering the **park-slot** command under ephone-dn configuration mode.

> **Note:** When planning to configure call park, keep in mind that each parked call consumes an ephone-dn slot (regardless of single- or dual-line configurations). You may need to increase the number of ephone-dns (max-dn) that your CME deployment supports.

Example 7-6 also shows that you have many options when you designate call park–specific ephone-dns. Table 7-2 explains where you can use these options.

Table 7-2 *Options for Use with the park slot Command*

Command	Function
reserved-for *<dn>*	Allows you to reserve the call park slot for the directory number (DN) you enter. Other phones are not able to use the call park slot.

Table 7-2 *Options for Use with the park slot Command*

Command	Function
timeout <*seconds*>	Specifies the number of seconds CME should wait before notifying the phone that parked the call that the call is still parked. To notify, CME rings that phone for one second and displays a message on the LCD display.
limit <*count*>	Limits the number of timeout intervals a parked call can reach. After this limit is reached, the parked call is disconnected. As a side note, setting this value high is recommended. Customers tend to get bothered when they are on hold for an extended period and then are disconnected.
notify <*dn*>	Notifies a different DN, in addition to the phone that parked the call, when the parked call reaches timeout period.
only	Used with the prior notify syntax; instructs CME to only ring the DN specified with the notify command rather than ring the original phone.
recall	Causes the call to return (transfer back) to the original phone that parked the call after the parked call reaches the timeout period.
transfer <*dn*>	Causes the call to transfer to a specified DN after the parked call reaches the timeout period.
alternate <*dn*>	Allows you to specify an alternate transfer destination should the destination DN specified in the transfer command be on the phone.
retry <*seconds*>	Sets the amount of time before CME attempts to transfer a parked call again.

There's plenty of flexibility in configuring your call park options. After you have at least one ephone-dn designated for call park (by using the **park-slot** command), the Park softkey appears on the IP phones on an active call.

Note: You must restart or reset the IP phones after you configure the initial ephone-dn designated call park before the Park softkey will appear on active calls. You can accomplish this by using the **restart** or **reset** command from telephony service configuration mode.

To park a call, simply press the Park softkey while on an active call. CME finds a parking slot for the call and send a message back to the phone that parked the call, as shown in Figure 7-11.

When the user parks the call, CME allocates the first available park slot. Sometimes, you might want to designate which parking slot the call gets, in cases such as those in which each department of the company is assigned a unique call park number. In this case, you can transfer the call (using the Trnsfer softkey) directly into the parking slot you want.

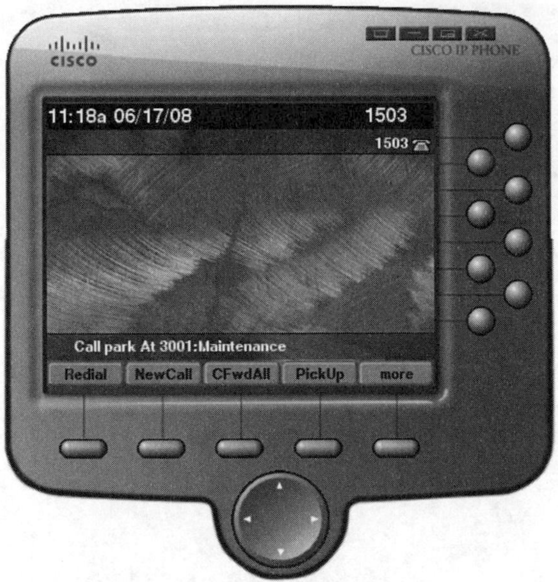

Figure 7-11 *IP Phone After Parking a Call*

> **Note:** If you want to use a Call Park system in which each department has its own Call Park slot, it may be beneficial to configure multiple ephone-dns assigned to each department designated for Call Park. Otherwise, you will be able to park only one call for each department.

You can answer parked calls in one of three ways:

- Dial directly into the call park slot. For example, lifting a phone handset and dialing 3001 answers whatever call is parked at 3001.
- Press the PickUp softkey and dial the call park number that you want to answer.
- From the phone at which the call was parked, press the PickUp softkey followed by an asterisk (*) to recall the most recently parked call back to the phone.

Using CCP to configure call park features automates the process quite a bit. First, navigate to the call park configuration window (**Unified Communications > Telephony Features > Call Park**). Once you arrive there, you can click the **Create** button to bring up the Create Call Park Entry configuration window. Entering a name creates a description label in the IOS for the call park entry. CCP then gives you the option to select the number of slots (call park numbers) as well as the starting number for slots. For example, applying the configuration shown in Figure 7-12 creates ten park slots numbered from 1300 through 1309.

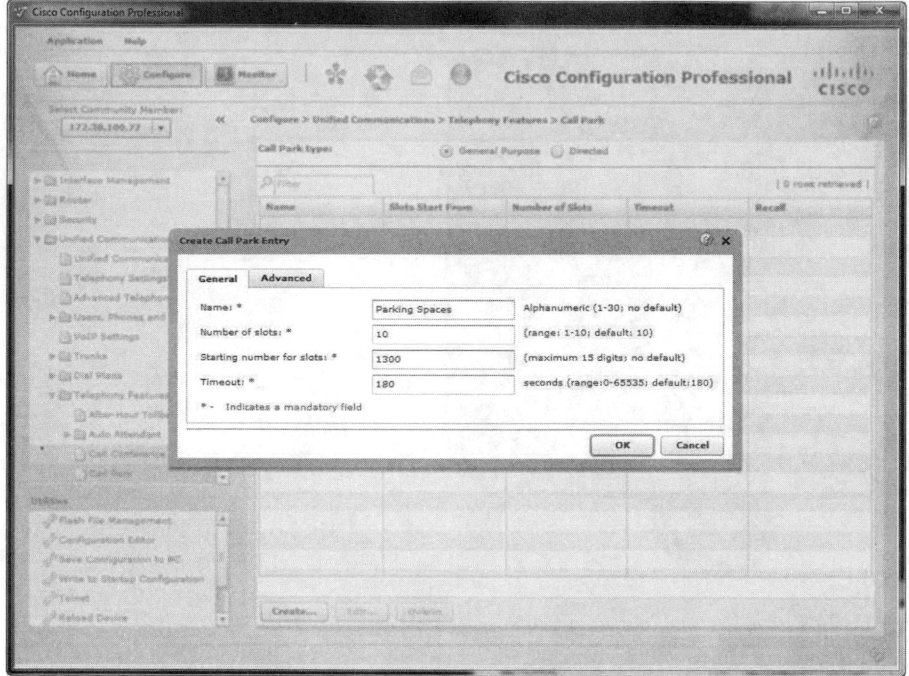

Figure 7-12 *Configuring Call Park Using CCP*

In addition, the Advanced tab of the Call Park Entry configuration window (shown in Figure 7-13) gives you numerous options (discussed in Table 7-2) to modify call park features.

Configuring Call Pickup

Michael works in the sales department at Widget Things, Inc. Being the newest member to the group, he works the late shift, covering calls from 10:30 a.m. to 7:30 p.m. Around 6:00 p.m., the last coworker leaves, and Michael handles all the incoming calls alone. Unfortunately, many of Widget Things' customers have the direct contact information for other sales employees, so a typical evening for Michael consists of running around answering phone calls coming in on the IP phones of the five other sales reps. This is where call pickup features can help.

Call pickup allows you to answer another ringing phone in the organization from your local phone. This is accomplished by pushing the PickUp softkey on the IP phone while another phone is ringing. The call automatically transfers to the local phone, where you can answer it. Of course, the organization is large, and there could be many ringing phones at the same time, so call pickup gives you the opportunity to divide the phones into groups. You assign each of these groups a number in the CME configuration, as shown in Figure 7-14.

Based on the softkey used, the users can answer other ringing phones in their own group or enter other group numbers to answer the ringing phones in that group.

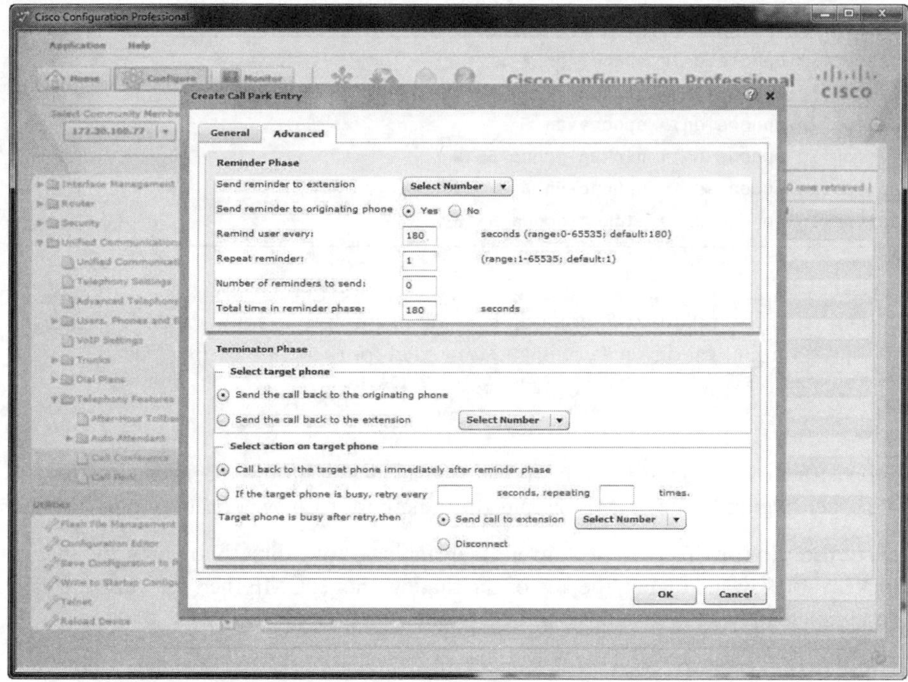

Figure 7-13 *Advanced Call Park Features Available from CCP*

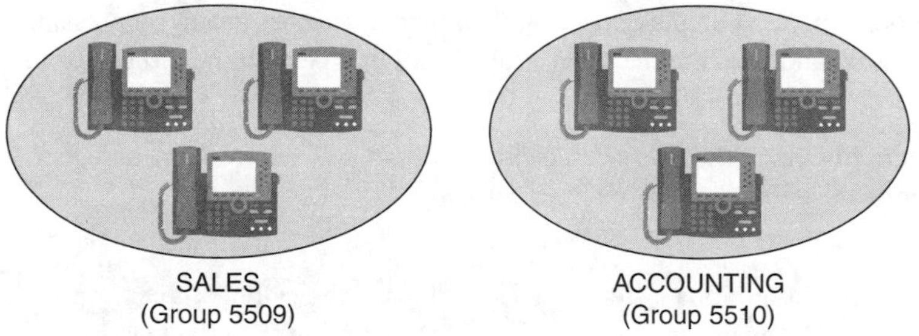

Figure 7-14 *Designing Call Pickup Groups*

The configuration of call pickup is incredibly simple: Just design your groups of phones and assign the ephone-dns to the groups. Example 7-7 assigns ephone-dns 1, 2, and 3 to the SALES group and ephone-dns 4, 5, and 6 to the ACCOUNTING group, as shown in Figure 7-14.

Example 7-7 *Configuring Call Pickup*

Key Topic

```
CME_Voice(config)# ephone-dn 1
CME_Voice(config-ephone-dn)# pickup-group 5509
CME_Voice(config-ephone-dn)# ephone-dn 2
CME_Voice(config-ephone-dn)# pickup-group 5509
CME_Voice(config-ephone-dn)# ephone-dn 3
```

```
CME_Voice(config-ephone-dn)# pickup-group 5509
CME_Voice(config-ephone-dn)# ephone-dn 4
CME_Voice(config-ephone-dn)# pickup-group 5510
CME_Voice(config-ephone-dn)# ephone-dn 5
CME_Voice(config-ephone-dn)# pickup-group 5510
CME_Voice(config-ephone-dn)# ephone-dn 6
CME_Voice(config-ephone-dn)# pickup-group 5510
```

Note: When you assign the first ephone-dn to a call pickup group number, CME creates the call pickup group. No additional command is needed for the call pickup group creation.

After you assign the ephone-dns to the respective call pickup groups, users can begin answering other ringing phones. CME permits three methods to answer other ringing phones:

- **Directed pickup:** You can pick up another ringing phone directly by pressing the PickUp softkey and dialing the DN of the ringing phone. CME then transfers the call and immediately answers it at your local phone.
- **Local group pickup:** You can pick up another ringing phone in the same call pickup group as your phone by pressing the GPickUp button and entering an asterisk (*) when you hear the second dial tone.
- **Other group pickup:** You can pick up a ringing phone in another group by pressing the GPickUp button and entering the other group number when you hear the second dial tone.

If multiple phones are ringing in the user's call pickup group, CME answers the oldest ringing phone when the user invokes call pickup.

Note: The GPickUp softkey functions differently depending on the call pickup configuration in CME. If there is only one group configured in CME, pressing the GPickUp button automatically answers the call from your own group number. You will not hear a second dial tone, and you do not need to dial an asterisk to signify your own group, because only one group is defined. After you configure multiple groups in CME, you hear a second dial tone after pressing the GPickUp softkey; at this point, you can dial either an asterisk for the local group or another group number.

Tip: By default, users can pick up other ringing phones managed by CME by using the directed pickup method, described previously, regardless of the destination device being assigned to a Call Pickup group. To disable this feature, enter the command no **service directed-pickup** from telephony service configuration mode. After you enter this command, the PickUp softkey on the IP phones operates as a local group pickup button. Pressing the softkey then immediately answers ringing calls in your own local pickup group.

To create pickup groups in CCP, navigate to the Call Pickup Groups configuration window (**Unified Communications > Telephony Features > Call Pickup Groups**) and click the **Create** button. The user-friendly configuration window allows you to define the pickup group number and allocate which extensions you want to include in the pickup group, as shown in Figure 7-15.

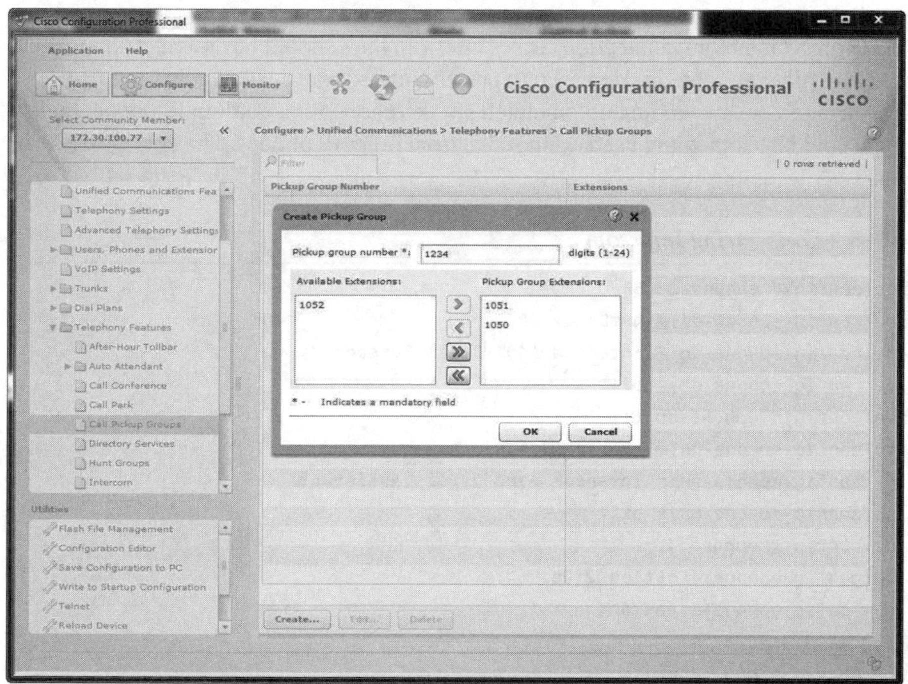

Figure 7-15 *Configuring Call Pickup Groups in CCP*

If you were to click the Deliver button with the CCP configuration shown in Figure 7-15, the CCP delivers the following commands to the Cisco router:

```
ephone-dn 2 dual-line
 pickup-group 1234
 exit
ephone-dn 1 dual-line
 pickup-group 1234
 exit
```

Configuring Intercom

Intercom configurations are common in traditional phone systems. This feature allows an administrative assistant and executive to work closely together by having a speakerphone "tether" between them.

Technically, the way intercom deployments work is through a speed-dial and auto-answer speed-dial configuration. If the administrative assistant presses the button configured as an intercom, it speed dials the executive's phone, which auto-answers the call on muted speakerphone. To establish two-way communication, the executive deactivates mute (by pressing the Mute button). Understanding this helps make the intercom configuration much clearer.

To configure intercom functionality, you must configure two new ephone-dns, one for each side of the intercom connection. These intercom lines should be assigned a number, just like any other ephone-dn. However, to prevent others from accidentally (or purposely) dialing the intercom and ending up on muted speakerphone for a random IP phone, the number should be something users cannot dial from other IP phones. The configuration in Example 7-8 accomplishes this objective.

Key Topic

Example 7-8 *Configuring Intercom*

```
CME_Voice(config)# ephone-dn 60
CME_Voice(config-ephone-dn)# number A100
CME_Voice(config-ephone-dn)# intercom A101 label "Manager"
CME_Voice(config-ephone-dn)# exit
CME_Voice(config)# ephone-dn 61
CME_Voice(config-ephone-dn)# number A101
CME_Voice(config-ephone-dn)# intercom A100 label "Assistant"
CME_Voice(config-ephone-dn)# exit
CME_Voice(config)# ephone 1
CME_Voice(config-ephone)# button 2:60
CME_Voice(config-ephone)# restart
restarting 0014.1C48.E71A
CME_Voice(config-ephone)# exit
CME_Voice(config)# ephone 2
CME_Voice(config-ephone)# button 2:61
CME_Voice(config-ephone)# restart
restarting 0019.D122.DCF3
```

Notice the number assigned to ephone-dn 60 is A100. You cannot dial this number from a Cisco IP Phone keypad, but you can assign it to a speed-dial button. The **intercom** command acts like a speed-dial button on the ephone-dn. In the case of ephone-dn 60, the command **intercom A101** dials the number A101, which is assigned to ephone-dn 61. Because ephone-dn 61 is also configured with the **intercom** command, it auto-answers the incoming call on muted speakerphone. The label syntax allows you to assign a logical name to the speed-dial; otherwise, the A101 or A100 label will show up next to the line button on the phone. There are three other arguments you can use with the **intercom** command to tune the functionality:

- **barge-in:** Automatically places an existing call on hold and causes the intercom to immediately answer.
- **no-auto-answer:** Causes the phone to ring rather than auto-answer on speakerphone.

- **no-mute:** Causes the intercom to answer with unmuted speakerphone rather than muted. Although this is beneficial to allow immediate two-way conversation, you run the risk of one side barging into existing conversations or background noise.

CCP can also configure Intercom functionality. To do this, navigate to **Unified Communications > Telephony Features > Intercom** and click the **Create** button. The configuration window shown in Figure 7-16 allows you to select the user and speed-dial button you want to assign Intercom functionality.

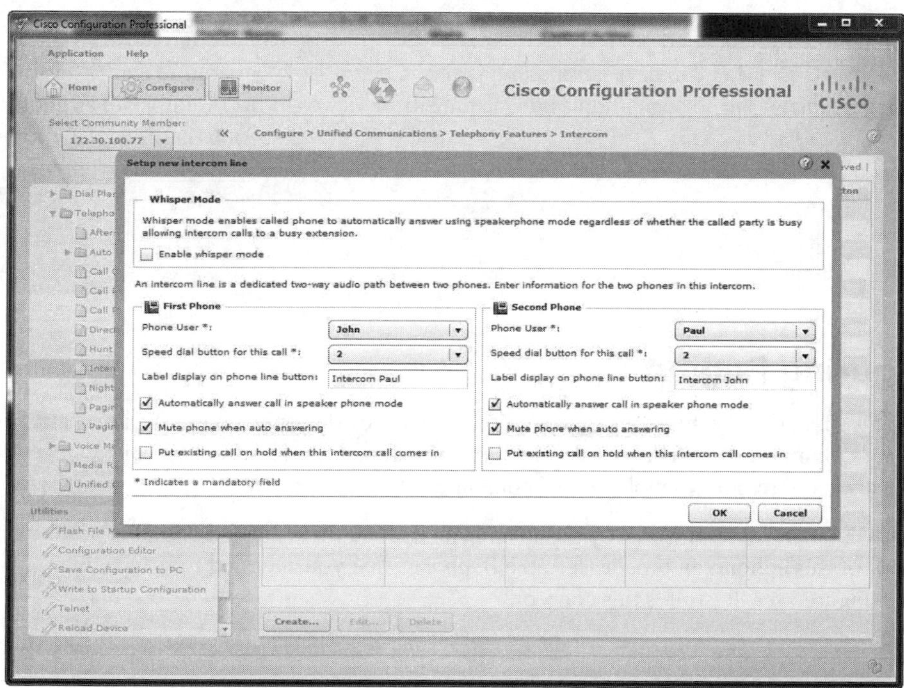

Figure 7-16 *Configuring Intercom in CCP*

After you click the **Deliver** button, CCP applies the following syntax to the router. Notice that CCP also uses alpha-numeric speed dials to prevent other users from accessing the intercom functionality inadvertently:

```
ephone-dn 5
 number A105002
 description Intercom
 intercom A105102 label "Intercom Paul"
 exit

ephone-dn 4
 number A105102
 description Intercom
```

```
 intercom A105002 label "Intercom John"
 exit
!
ephone-dn 5
 name ""
 exit
ephone 2
 button  2:5
 restart
 exit

ephone-dn 4
 name ""
 exit
ephone 3
 button  2:4
 restart
 exit
```

Configuring Paging

Paging is similar to the intercom concept; however, it provides only a one-way automatic path for communication. This is useful to allow broadcast messages, such as emergency notifications or to notify employees of holding calls.

The CME paging system works by designating an ephone-dn as a paging number. Calls to the DN of this ephone-dn broadcast to the IP phones that you assigned to this paging group. Figure 7-17 illustrates this concept.

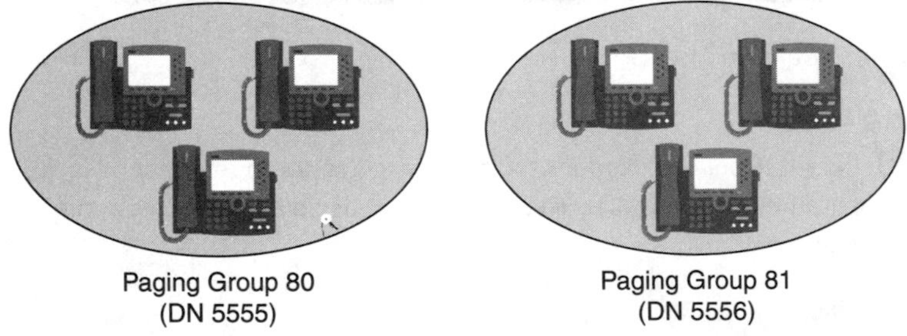

Figure 7-17 *Call Paging Functionality*

As shown in Figure 7-17, calls to DN 5555 page the three phones assigned to that paging group. Calls to 5556 do the same for the paging group 81.

Note: You can assign an IP phone to only one paging group. However, CME allows you to create paging numbers that page multiple paging groups, thus providing directed and company-wide paging functionality.

CME supports paging in unicast and multicast configurations. Paging in unicast configuration causes the CME router to send individual messages to each one of the IP phones in the group. So, if six IP phones were assigned to paging group 80, a page to the group would cause the CME router to stream six individual audio signals to the devices. Because of the overhead this causes, CME limits unicast paging groups to a maximum of ten IP phones.

Multicast configuration allows the CME router to send one audio stream, which only the IP phones assigned to the paging group will receive. This allows a virtually limitless number of IP phones in each paging group. Sounds like the winning option, right? The catch is this: To support multicast paging, you must configure the foundation network environment to support multicast traffic. Some of these configurations can get complex and are covered in the CCNP certification track.

The three paging configurations are unicast paging, multicast paging, and multiple-group paging. Example 7-9 shows unicast, single-group paging.

Example 7-9 *Configuring Unicast, Single-Group Paging*

```
CME_Voice(config)# ephone-dn 80
CME_Voice(config-ephone-dn)# number 5555
CME_Voice(config-ephone-dn)# paging
CME_Voice(config-ephone-dn)# exit
CME_Voice(config)# ephone 1
CME_Voice(config-ephone)# paging-dn 80
CME_Voice(config-ephone)# exit
CME_Voice(config)# ephone 2
CME_Voice(config-ephone)# paging-dn 80
```

Calls to the paging number 5555 now page both ephones 1 and 2 using unicast paging. To convert the configuration in Example 7-9 to multicast paging, you could modify the paging command with the following syntax:

```
CME_Voice(config)# ephone-dn 80
CME_Voice(config-ephone-dn)# paging ip 239.1.1.100 port 2000
```

The IP address that follows the paging command is a multicast address. Think of this as a "radio frequency" that the IP phones tune to each time a page occurs. Just like a car radio tuning to a specific FM frequency to hear a radio station, the IP phones tune into the IP address 239.1.1.100 and hear the audio stream for the paging system. As previously mentioned, you must configure your network to properly support multicast traffic. Otherwise, your switches treat this multicast traffic just like it treats broadcasts, flooding your network on all ports each time a page occurs.

The paging configuration in Example 7-10 demonstrates the configuration of a multiple-group paging system. Ephones 1 and 2 continue to use ephone-dn 80 as their dedicated paging group. Ephones 3 and 4 use ephone-dn 81. This time, a third paging group enables you to page both paging groups at once. This gives an organization the flexibility to page specific departments or the company as a whole.

Example 7-10 *Configuring Multiple-Group Paging*

```
CME_Voice(config)# ephone-dn 80
CME_Voice(config-ephone-dn)# number 5555
CME_Voice(config-ephone-dn)# paging
CME_Voice(config-ephone-dn)# exit
CME_Voice(config)# ephone-dn 81
CME_Voice(config-ephone-dn)# number 5556
CME_Voice(config-ephone-dn)# paging
CME_Voice(config-ephone-dn)# exit
CME_Voice(config)# ephone-dn 82
CME_Voice(config-ephone-dn)# number 5557
CME_Voice(config-ephone-dn)# paging group 80,81
CME_Voice(config-ephone-dn)# exit
CME_Voice(config)# ephone 1
CME_Voice(config-ephone)# paging-dn 80
CME_Voice(config-ephone)# exit
CME_Voice(config)# ephone 2
CME_Voice(config-ephone)# paging-dn 80
CME_Voice(config-ephone)# exit
CME_Voice(config)# ephone 3
CME_Voice(config-ephone)# paging-dn 81
CME_Voice(config-ephone)# exit
CME_Voice(config)# ephone 4
CME_Voice(config-ephone)# paging-dn 81
```

With the configuration shown in Example 7-10, a call to DN 5555 pages ephones 1 and 2, a call to DN 5556 pages ephones 3 and 4, and a call to DN 5557 pages all ephones. You do not need to assign any ephones to paging-dn 82, because this ephone-dn represents a group of both paging-dns 80 and 81.

Note: CME allows you to list up to ten paging numbers using the **paging group** command.

If you choose to use CCP to configure paging, navigate to **Unified Communications > Telephony Features > Paging Numbers.** Clicking the **Create** button allows you to define numbers and associate phones, as shown in Figure 7-18. Think of this as the single-group paging configuration.

CCP also allows the configuration of multiple-group paging. By navigating to **Unified Communications > Telephony Features > Paging Groups**, you can create paging group numbers that associate with other paging groups rather than individual extensions, as shown in Figure 7-19.

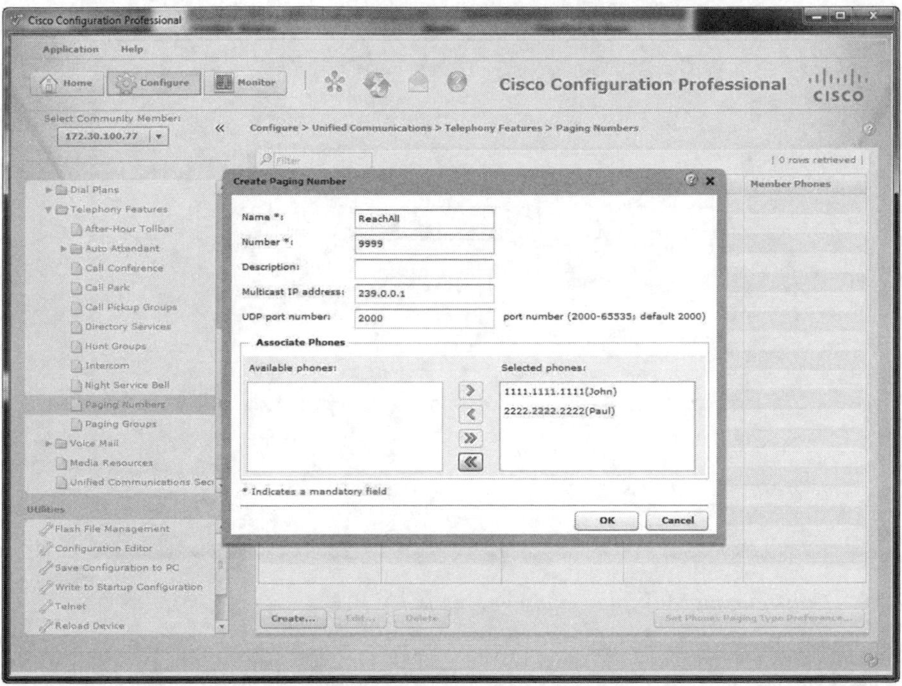

Figure 7-18 *Configuring Single-Group Paging in CCP*

Configuring After-Hours Call Blocking

In the traditional telephony realm, there have been many recorded incidents of unauthorized phone calls being placed after-hours, when most, if not all, staff has left for the evening. To prevent this, you can implement after-hours call blocking on CME.

After-hours call blocking allows you to define ranges of times specified as after-hours intervals. You can then list number patterns that are disallowed during those intervals. If a user places a call during the after-hours time range that matches one of the defined patterns, CME will play a reorder tone and disconnect the call.

Of course, there are exceptions to every rule. You may want to have some phones completely exempt from the policy, or give users a "back door" around the restrictions if they are working late and need to make business-related calls that CME would typically restrict. The after-hours call blocking configuration on CME provides for both of these scenarios. You have the option to completely exempt certain IP phones from the after-hours restrictions or provide users with a PIN they can enter into the IP phone. If they enter the PIN correctly, CME exempts the IP phone from the after-hours policy for a configurable amount of time.

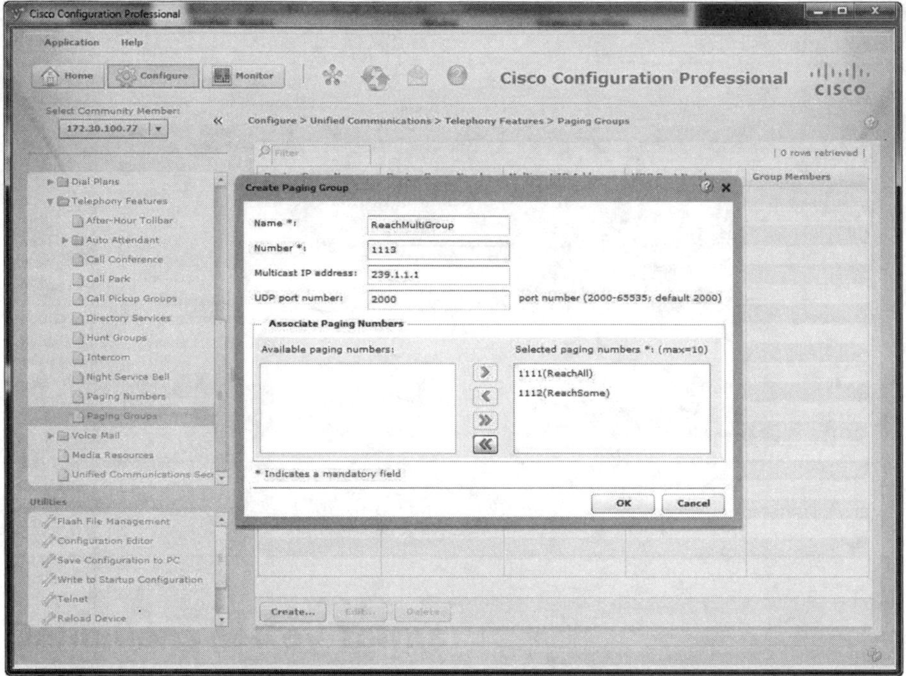

Figure 7-19 *Configuring Multiple-Group Paging in CCP*

Note: There are some patterns that may be beneficial to block all the time. For example, 1-900 numbers in the United States often represent high-cost, less-than-reputable businesses that are typically banned from all corporate environments. CME also allows you to create a 24/7, nonexemptible pattern that is disallowed at all times, using the after-hours call blocking system.

After-hours call blocking has three major steps of configuration:

Step 1. Define days and/or hours of the day that your company considers off-hours.

Step 2. Specify patterns that you want to block during the times specified in Step 1.

Step 3. Create exemptions to the policy, if needed.

You will perform most of the configuration for the after-hours call blocking restrictions from telephony service configuration mode. Example 7-11 demonstrates the configuration of after-hours time intervals.

Example 7-11 *Configuring After-Hours Time Ranges and Dates*

```
CME_Voice(config)# telephony-service
CME_Voice(config-telephony)# after-hours ?
  block  define after-hours block pattern
  date   define month and day
```

```
  day    define day in week
CME_Voice(config-telephony)# after-hours day ?
  DAY  day of week (Mon, Tue, Wed, etc)
CME_Voice(config-telephony)# after-hours day mon ?
  hh:mm  Time to start (hh:mm)
CME_Voice(config-telephony)# after-hours day mon 17:00 ?
  hh:mm  Time to stop (hh:mm)
CME_Voice(config-telephony)# after-hours day mon 17:00 8:00
CME_Voice(config-telephony)# after-hours day tue 17:00 8:00
CME_Voice(config-telephony)# after-hours day wed 17:00 8:00
CME_Voice(config-telephony)# after-hours day thu 17:00 8:00
CME_Voice(config-telephony)# after-hours day fri 17:00 8:00
CME_Voice(config-telephony)# after-hours date ?
  MONTH  Month (Jan, Feb, Mar, etc)
CME_Voice(config-telephony)# after-hours date dec ?
  <1-31>  day of month in date
CME_Voice(config-telephony)# after-hours date dec 25 ?
  hh:mm  Time to start (hh:mm)
CME_Voice(config-telephony)# after-hours date dec 25 00:00 ?
  hh:mm  Time to stop (hh:mm)
CME_Voice(config-telephony)# after-hours date dec 25 00:00 00:00
CME_Voice(config-telephony)# after-hours date jan 1 00:00 00:00
```

The configuration in Example 7-11 defines weekdays, from 5:00 p.m. to 8:00 a.m. the next day, as after-hours, along with the entire day on December 25 (Christmas) and January 1 (New Year's Day).

In the next step of the after-hours configuration, you define the patterns that CME should block during the after-hours time slots you have configured (see Example 7-12).

> **Note:** Chapters 8 and 9 discuss the patterns you can use for matching fully. For now, examples in this chapter use the "." wildcard, which matches any digit dialed, and the "T" wildcard, which matches any number of digits.

Example 7-12 *Configuring After-Hours Block Patterns*

Key Topic

```
CME_Voice(config)# telephony-service
CME_Voice(config-telephony)# after-hours block ?
  pattern  block pattern
CME_Voice(config-telephony)# after-hours block pattern ?
  <1-32>  index of patterns
CME_Voice(config-telephony)# after-hours block pattern 1 ?
  WORD  digits string for after hour block pattern
CME_Voice(config-telephony)# after-hours block pattern 1 91..........
CME_Voice(config-telephony)# after-hours block pattern 2 9011T
CME_Voice(config-telephony)# after-hours block pattern 3 91900....... ?
```

```
  7-24  block pattern works for 7 * 24
  <cr>
CME_Voice(config-telephony)# after-hours block pattern 3 91900....... 7-24
```

You might have noticed based on the context-sensitive help output in Example 7-12 that the CME router allows you to configure up to 32 indexes of block patterns. The initial block pattern 1 matches and blocks long distance numbers; block pattern 2 matches and blocks international numbers; block pattern 3 matches and blocks 1-900 toll calls. Notice that block pattern 3 is followed by the 7-24 keyword. This additional syntax tells the CME router to block calls to this pattern at all times. If you enter block patterns with this keyword, phones exempted from other after-hours blocked numbers are not exempt from these patterns.

Note: If you need more flexibility than after-hours blocking provides, you can also use Class of Restriction (COR) features with CME. *Cisco Voice over IP (CVOICE) Authorized Self-Study Guide,* Third Edition, by Kevin Wallace (Cisco Press, 2008), has more information on the configuration of COR on Cisco routers.
For more information on the web, go to http://tinyurl.com/64256f.

The final step in the configuration of after-hours blocking is to allow any necessary exemptions to the policy, as shown in Example 7-13. You can add exemptions on a per-IP phone basis or by using one or more PIN numbers to allow on-demand access to block patterns (with the exception of the patterns defined with the 7-24 keyword) from any IP phone. Example 7-13 configures ephone 1 to be exempt from the after-hours call-blocking policy. Ephones 2 and 3 are configured with PIN numbers. In order to become exempt from the after-hours call blocking policy, the user using the phone must enter the necessary PIN number.

Example 7-13 *Configuring After-Hours Exemptions*

```
CME_Voice(config)# ephone 1
CME_Voice(config-ephone)# after-hour exempt
CME_Voice(config-ephone)# exit
CME_Voice(config)# ephone 2
CME_Voice(config-ephone)# pin ?
  WORD  A sequence of digits - representing personal identification number
CME_Voice(config-ephone)# pin 1234
CME_Voice(config-ephone)# exit
CME_Voice(config)# ephone 3
CME_Voice(config-ephone)# pin 4321
CME_Voice(config-ephone)# exit
CME_Voice(config)# telephony-service
CME_Voice(config-telephony)# login timeout 120 clear 23:00
```

Note: The PIN number can be any number between four and eight digits.

The last line in Example 7-13 is a key to allowing the PIN numbers to function properly. By default, all the CME-supported Cisco IP Phones have a Login softkey on the LCD display. This softkey is dimmed and unusable until you enter the **login** command from telephony service configuration mode. The timeout value that follows this command represents the amount of idle time before the phone automatically revokes the last PIN number entered. The clear value is an absolute time at which the last-entered PIN number becomes invalid. In the case of Example 7-13, the PIN will clear at 11:00 p.m., regardless of the last time it was entered. This does not prevent users from logging back in by entering their PIN number a second time after 11:00 p.m.

Note: The default timeout for the login command is 60 minutes. Also, you need to restart or reset all phones before the **login** command takes effect.

CCP provides a fairly spectacular interface for configuring after-hours call blocking. When navigating to **Unified Communications > Telephony Features > After-Hour Toolbar**, CCP greets you with a window allowing you to define the specific patterns you want to block (shown in Figure 7-20). You can create schedules and PIN number overrides by clicking the Weekly Schedule, Holiday Schedule, or Override (Softkey Login) window panes at the bottom of the After-Hour Toolbar configuration window.

Configuring CDRs and Call Accounting

"Who made that call?" That question could arise for many reasons. Perhaps the entire police and fire departments arrive at the front door of your company because of an emergency call originating from your business. Perhaps management is reviewing the recent long-distance bill and came across an international call to Aruba that was four hours in length. Whatever the reason, you can find the answer by looking through the archived Call Detail Records (CDR), as long as you have configured the CME router to support them.

CDRs contain valuable information about the calls coming into, going out of, and between the IP phones on your network. These records contain all the information you need to find who called whom and how long they were talking. The CME router can log CDRs to the buffered memory (RAM) of the router, to a syslog server, or to both. Storing the CDRs in the RAM of the router is better than nothing, but not very effective. If the CME router ever loses power, all the CDRs will be lost. Likewise, the RAM of the router has limited storage and is not an effective solution. Viewing CDRs from the log file on the CME router is very cryptic and tedious to understand. Example 7-14 demonstrates the syntax you can use to enable logging of CDRs to the buffered memory of the router.

Example 7-14 *Configuring CDR Logging to Buffered Memory*

```
CME_Voice(config)# logging buffered 512000
CME_Voice(config)# dial-control-mib ?
  max-size        Specify the maximum size of the dial control history table
```

```
  retain-timer  Specify timer for entries in dial control history table
CME_Voice(config)# dial-control-mib retain-timer ?
  <0-35791>  Time (in minutes) for removing an entry
CME_Voice(config)# dial-control-mib retain-timer 10080
CME_Voice(config)# dial-control-mib max-size ?
  <0-1200>  Number of entries in the dial control history table
CME_Voice(config)# dial-control-mib max-size 700
```

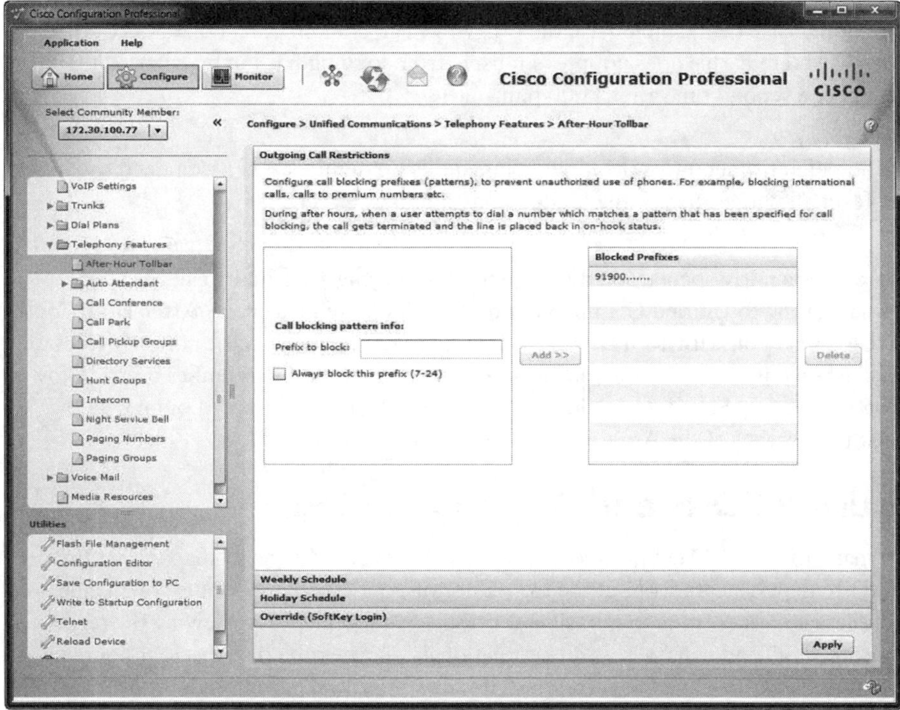

Figure 7-20 *Configuring After-Hours Restrictions Using CCP*

Example 7-14 specifies the following parameters for CDR buffered logging:

- 512,000 bytes of memory dedicated to the logging functions of the router.
- CDRs are kept for 10,080 minutes (7 days).
- The CME router keeps a maximum of 700 CDRs in memory.

To view the CDRs recorded by CME, use the **show logging** command, as shown in Example 7-15.

Example 7-15 *show logging Command Output*

```
CME_Voice# show logging
Syslog logging: enabled (12 messages dropped, 1 messages rate-limited,
                0 flushes, 0 overruns, xml disabled, filtering disabled)
```

```
    Console logging: level debugging, 168 messages logged, xml disabled,
                     filtering disabled
<...output omitted>
Log Buffer (512000 bytes):

*Jun 18 01:57:08.987: %SYS-5-CONFIG_I: Configured from console by Jeremy on vty0
  (172.30.3.28)
*Jun 18 01:57:48.640: %VOIPAAA-5-VOIP_CALL_HISTORY: CallLegType 1, ConnectionId
  B71427FB3C1011DD80EEB6A01B061E9, SetupTime *18:57:17.970 ARIZONA Tue Jun 17
  2008, PeerAddress 1503, PeerSubAddress , DisconnectCause 1    , DisconnectText
  unassigned number (1), ConnectTime *18:57:48.640 ARIZONA Tue Jun 17 2008,
  DisconnectTime *18:57:48.640 ARIZONA Tue Jun 17 2008, CallOrigin 2, ChargedUnits
  0, InfoType 2, TransmitPackets 0, TransmitBytes 0, ReceivePackets 0,
ReceiveBytes
  0
*Jun 18 01:57:48.640: %VOIPAAA-5-VOIP_FEAT_HISTORY: FEAT_VSA=fn:CFBY,ft:06/17/2008
  18:57:18.623,frs:0,fid:129,fcid:B77841E83C1011DD80F3B6A01B061E9,legID:0,frson:2,
  fdcnt:1,fwder:1501,fwdee:1503,fwdto:1599,frm:1501,bguid:B71427FB3C1011DD80EEB6A
  001B061E9
*Jun 18 01:57:48.640: %VOIPAAA-5-VOIP_FEAT_HISTORY: FEAT_VSA=fn:TWC,ft:06/17/2008
  18:57:17.967,cgn:1503,cdn:,frs:0,fid:126,fcid:B71427FB3C1011DD80EEB6A01B061E9,
  legID:4B,bguid:B71427FB3C1011DD80EEB6A001B061E9
<...output omitted>
```

What you see here are three CDR entries that record a call from x1503 to x1501. If you are able to decode most of what is displayed in that log, you are definitely ahead of the game.

Sending messages to a syslog server is better than sending them to the RAM of the CME router. A syslog server is a PC or server running a dedicated application that receives and stores logging messages from one or more devices. There are many syslog server platforms available for download on the Internet.

> **Note:** The Kiwi Syslog Daemon (www.kiwisyslog.com) is by far my (Jeremy) favorite syslog platform available. The fact that it is free helps, but it is also easy to install and manage.

After you set up a syslog server, you can direct the CME router to send CDR records to it by using the following syntax:

```
CME_Voice(config)# gw-accounting syslog
CME_Voice(config)# logging 172.30.100.101
```

The initial command in this syntax directs the CDR records to the syslog server. The second command tells the CME router where the syslog server is located; in this case, 172.30.100.101. Figure 7-21 shows the CDR records being received by the Kiwi Syslog application.

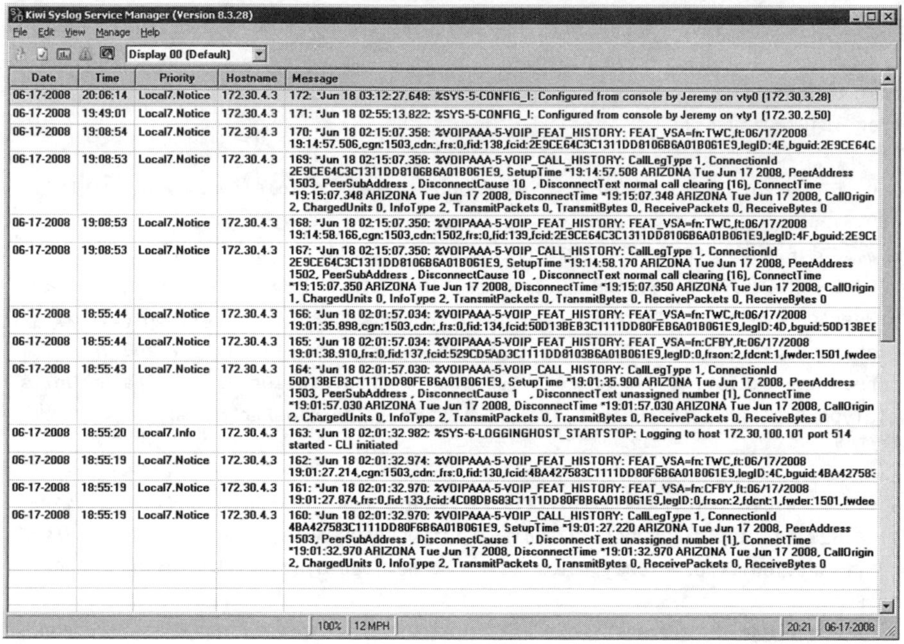

Kiwi Syslog Service Manager (Version 8.3.28)

File Edit View Manage Help

Display 00 (Default)

Date	Time	Priority	Hostname	Message
06-17-2008	20:06:14	Local7.Notice	172.30.4.3	172: *Jun 18 03:12:27.648: %SYS-5-CONFIG_I: Configured from console by Jeremy on vty0 (172.30.3.28)
06-17-2008	19:49:01	Local7.Notice	172.30.4.3	171: *Jun 18 02:55:13.822: %SYS-5-CONFIG_I: Configured from console by Jeremy on vty1 (172.30.2.50)
06-17-2008	19:08:54	Local7.Notice	172.30.4.3	170: *Jun 18 02:15:07.358: %VOIPAAA-5-VOIP_FEAT_HISTORY: FEAT_VSA=fn:TWC,ft:06/17/2008 19:14:57.506,cgn:1503,cdn:,frs:0,fid:138,fcid:2E9CE64C3C1311DD8106B6A01B061E9,legID:4E,bguid:2E9CE64C
06-17-2008	19:08:53	Local7.Notice	172.30.4.3	169: *Jun 18 02:15:07.358: %VOIPAAA-5-VOIP_CALL_HISTORY: CallLegType 1, ConnectionId 2E9CE64C3C1311DD8106B6A01B061E9, SetupTime *19:14:57.508 ARIZONA Tue Jun 17 2008, PeerAddress 1503, PeerSubAddress , DisconnectCause 10 , DisconnectText normal call clearing (16), ConnectTime *19:15:07.348 ARIZONA Tue Jun 17 2008, DisconnectTime *19:15:07.348 ARIZONA Tue Jun 17 2008, CallOrigin 2, ChargedUnits 0, InfoType 2, TransmitPackets 0, TransmitBytes 0, ReceivePackets 0, ReceiveBytes 0
06-17-2008	19:08:53	Local7.Notice	172.30.4.3	168: *Jun 18 02:15:07.350: %VOIPAAA-5-VOIP_FEAT_HISTORY: FEAT_VSA=fn:TWC,ft:06/17/2008 19:14:58.166,cgn:1503,cdn:1502,frs:0,fid:139,fcid:2E9CE64C3C1311DD8106B6A01B061E9,legID:4F,bguid:2E9CE
06-17-2008	19:08:53	Local7.Notice	172.30.4.3	167: *Jun 18 02:15:07.350: %VOIPAAA-5-VOIP_CALL_HISTORY: CallLegType 1, ConnectionId 2E9CE64C3C1311DD8106B6A01B061E9, SetupTime *19:14:58.170 ARIZONA Tue Jun 17 2008, PeerAddress 1502, PeerSubAddress , DisconnectCause 10 , DisconnectText normal call clearing (16), ConnectTime *19:15:07.350 ARIZONA Tue Jun 17 2008, DisconnectTime *19:15:07.350 ARIZONA Tue Jun 17 2008, CallOrigin 1, ChargedUnits 0, InfoType 2, TransmitPackets 0, TransmitBytes 0, ReceivePackets 0, ReceiveBytes 0
06-17-2008	18:55:44	Local7.Notice	172.30.4.3	166: *Jun 18 02:01:57.034: %VOIPAAA-5-VOIP_FEAT_HISTORY: FEAT_VSA=fn:TWC,ft:06/17/2008 19:01:35.898,cgn:1503,cdn:,frs:0,fid:134,fcid:50D13BEB3C1111DD80FEB6A01B061E9,legID:4D,bguid:50D13BEE
06-17-2008	18:55:44	Local7.Notice	172.30.4.3	165: *Jun 18 02:01:57.034: %VOIPAAA-5-VOIP_FEAT_HISTORY: FEAT_VSA=fn:CFBY,ft:06/17/2008 19:01:38.910,frs:0,fid:137,fcid:529CD5AD3C1111DD8103B6A01B061E9,legID:0,frson:2,fdcnt:1,fwder:1501,fwdee
06-17-2008	18:55:43	Local7.Notice	172.30.4.3	164: *Jun 18 02:01:57.030: %VOIPAAA-5-VOIP_CALL_HISTORY: CallLegType 1, ConnectionId 50D13BEB3C1111DD80FEB6A01B061E9, SetupTime *19:01:35.900 ARIZONA Tue Jun 17 2008, PeerAddress 1503, PeerSubAddress , DisconnectCause 1 , DisconnectText unassigned number (1), ConnectTime *19:01:57.030 ARIZONA Tue Jun 17 2008, DisconnectTime *19:01:57.030 ARIZONA Tue Jun 17 2008, CallOrigin 2, ChargedUnits 0, InfoType 2, TransmitPackets 0, TransmitBytes 0, ReceivePackets 0, ReceiveBytes 0
06-17-2008	18:55:20	Local7.Info	172.30.4.3	163: *Jun 18 02:01:32.982: %SYS-6-LOGGINGHOST_STARTSTOP: Logging to host 172.30.100.101 port 514 started - CLI initiated
06-17-2008	18:55:19	Local7.Notice	172.30.4.3	162: *Jun 18 02:01:32.974: %VOIPAAA-5-VOIP_FEAT_HISTORY: FEAT_VSA=fn:TWC,ft:06/17/2008 19:01:27.214,cgn:1503,cdn:,frs:0,fid:130,fcid:4BA427583C1111DD80F6B6A01B061E9,legID:4C,bguid:4BA42758
06-17-2008	18:55:19	Local7.Notice	172.30.4.3	161: *Jun 18 02:01:32.970: %VOIPAAA-5-VOIP_FEAT_HISTORY: FEAT_VSA=fn:CFBY,ft:06/17/2008 19:01:27.874,frs:0,fid:133,fcid:4C08DB683C1111DD80FBB6A01B061E9,legID:0,frson:2,fdcnt:1,fwder:1501,fwdee
06-17-2008	18:55:19	Local7.Notice	172.30.4.3	160: *Jun 18 02:01:32.970: %VOIPAAA-5-VOIP_CALL_HISTORY: CallLegType 1, ConnectionId 4BA427583C1111DD80F6B6A01B061E9, SetupTime *19:01:27.220 ARIZONA Tue Jun 17 2008, PeerAddress 1503, PeerSubAddress , DisconnectCause 1 , DisconnectText unassigned number (1), ConnectTime *19:01:32.970 ARIZONA Tue Jun 17 2008, DisconnectTime *19:01:32.970 ARIZONA Tue Jun 17 2008, CallOrigin 2, ChargedUnits 0, InfoType 2, TransmitPackets 0, TransmitBytes 0, ReceivePackets 0, ReceiveBytes 0

100% 12 MPH 20:21 06-17-2008

Figure 7-21 *CDR Records Logged to a Kiwi Syslog Server*

The output shown on the syslog server is the same messages received in the buffered logging. Although it is easier to read than scrolling through wrapped terminal output, the messages are just as cryptic. For this reason, many third-party vendors created CDR interpreters that format the syslog data into easy-to-understand spreadsheets and HTML pages.

Tip: Cisco offers a web-based utility that can show many third-party software applications geared around CDR interpretation. You can find the partner search application at http://tinyurl.com/5okclk.

It is common for an organization to use these CDRs for billing purposes. Businesses track the long-distance and international calls to the department level to assist in budget accounting. Although it is possible to keep track of the extension numbers that are in each department and the calls they make, the call data is easier to manage if CME can flag the CDR with an account code.

Businesses can distribute account codes to each department in the organization. For example, the East Coast sales group might get account code 1850, the West Coast sales group 1851, management 1852, and so on. You could then train the users in each department to enter this account code each time they make a long distance or international call by pressing the Acct softkey on the phone. This softkey appears when the IP phone is in the ring out or connected state, as shown in Figure 7-22.

After the user presses the Acct softkey, an Acct prompt appears at the bottom of the phone, where the user can enter their department account number followed by the pound key (#). Entering this number during the ring out or connected call state does not interrupt the call in any way. After the user enters the account number, CME flags the CDR records with the account number dialed. This allows for easy filtering and accurate billing to each department.

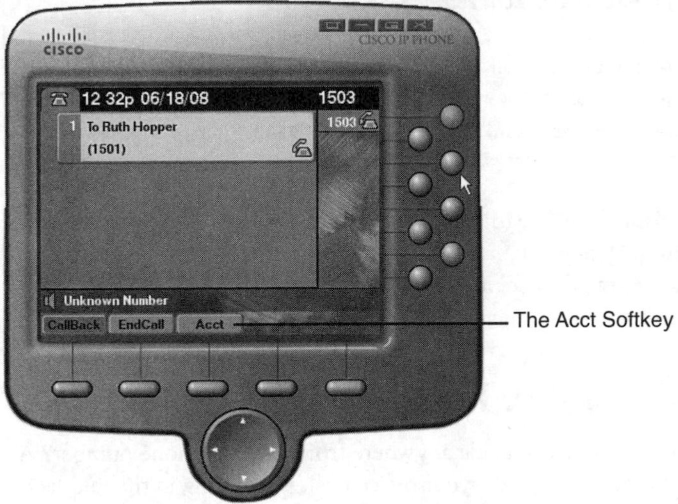

Figure 7-22 *Acct Softkey in the Ring Out State*

Configuring Music on Hold

What voice network would be complete without the sound of music coming through handsets on hold everywhere? CME has the ability to stream Music on Hold (MoH) from specified WAV or AU audio files that you copy to the flash memory of the router. CME can stream this audio either in multiple unicast streams (which is more resource intensive) or in a single multicast stream (which is less resource intensive, but requires a multicast network configuration). In addition, CME can stream the MoH using G.711 or G.729 codecs.

> **Note:** Because the G.729 audio codec is designed for human voice, the quality of MoH streamed using G.729 is significantly lower than MoH streamed using G.711. In addition, CME uses transcoding DSP resources to convert the MoH to the G.729 codec. With all these factors, using G.711 for your MoH, if at all possible, is highly recommended.

Example 7-16 configures a CME router to support multicast MoH, streaming music from a file in flash called bonjovi.wav.

Example 7-16 *Configuring MoH Support*

```
CME_Voice(config)# telephony-service
CME_Voice(config-telephony)# moh ?
  WORD  music-on-hold filename containing G.711 A-law or u-law 8KHz encoded audio
    file (.wav or .au format). The file must be loaded into the routers flash
```

```
    memory.
CME_Voice(config-telephony)# moh bonjovi.wav
CME_Voice(config-telephony)# multicast moh ?
  A.B.C.D  Define music-on-hold IP multicast address from flash
CME_Voice(config-telephony)# multicast moh 239.1.1.55 ?
  port  Define media port for multicast moh
CME_Voice(config-telephony)# multicast moh 239.1.1.55 port ?
  <2000-65535>  Specify the RTP port: 2000 - 65535
CME_Voice(config-telephony)# multicast moh 239.1.1.55 port 2123
```

Note: Because the U.S. government sees MoH as a type of broadcasting, you must purchase a license if you intend to play any songs covered by a copyright (such as music by Bon Jovi) over MoH. With that in mind, there are thousands of royalty-free songs available on the Internet that you can use for MoH.

Configuring Single Number Reach

Wouldn't it be wonderful if you were reachable anywhere from a single phone number? A call to your single phone number could make your office desk phone ring in the middle of the day, your cell phone ring in the middle of dinner, or your home phone ring in the middle of the night. Single Number Reach (SNR) allows you to link an additional device to a "parent" number. For example, you could link your mobile phone to your desk extension. When a call comes in for your DN, the office phone begins to ring. After a specified timeout interval, your mobile phone begins to ring along with your office phone. If neither phone answers within a specified timeout, CME transfers the call to the corporate voicemail server.

Tip: Single Number Reach in CME is a lightweight version of Mobile Connect, which is a CUCM feature allowing you (or the user) to assign multiple devices to ring simultaneously. The first one to answer receives the call.

In addition to a simultaneous-ring feature, Single Number Reach also allows a mid-call transfer. For example, you could be sitting at your desk speaking on a Cisco VoIP Phone with a valued customer when you suddenly realize that you're late for your daughter's birthday party. You can simply press the Mobility softkey on your office phone and CME transfers the call to your preconfigured Single Number Reach destination. CME can always transfer the call back by pressing the Resume softkey.

Note: Using Single Number Reach may cost you additional voice trunks to the PSTN. The mystical feature allowing you to move a call between your office phone and cell phone is only available because CME is maintaining the call at all times. For example, if a user received a call on his desk phone and then pressed the mobility button to send it to his cell phone, there will be two PSTN trunks active: one for the incoming call to the office phone and one for the outgoing call to the cell phone.

You can configure Single Number Reach from the command line or using CCP. If using CCP, navigate to the Extension configuration window (**Unified Communications > Users, Phones, and Extensions > Extensions**). You can then edit any extension where you want to enable Single Number Reach. After bringing up the Edit Extension window, click the Advanced tab and choose the Single Number Reach menu option (see Figure 7-23).

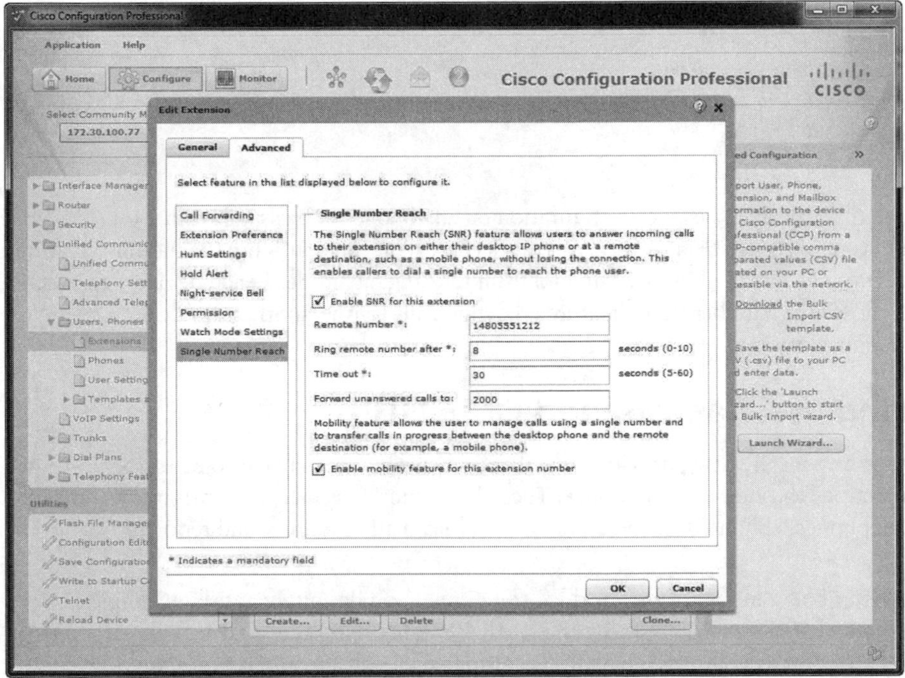

Figure 7-23 *Configuring SNR Using CCP*

From here, you can define the following options:

- **Enable SNR for this extension:** Enables the feature and allows you to configure the following fields.
- **Remote Number:** This required field defines the remote number CME should ring after a specified timeout. Remember to enter the number accordingly if your CME dial plan requires an access code (such as 9) for an outside line.
- **Ring remote number after:** How long (in seconds) CME should wait before ringing the remote number defined in the previous field.
- **Timeout:** The amount of time (in seconds) CME should wait before considering the call unanswered.
- **Forward unanswered calls to:** This optional field allows calls to forward to an additional number (such as an operator or hunt group) when the timeout value is reached.

Note: Keep in mind the amount of time the phone rings in your country when defining the "remote ring number after" and "timeout" values. For example, in the United States, phones ring for 2 seconds and then remain silent for 4 seconds. Defining a "ring remote number after" value of 8 allows two full rings and then immediately begins ringing the remote cell phone.

Selecting the options shown in Figure 7-22 generates the following command-line syntax:

```
ephone-dn 2 dual-line
 snr 14805551212 delay 8 timeout 30 cfwd-noan 2000
 mobility
 exit
```

Notice that CME applies the configuration on an ephone-dn basis (allowing you to enable this feature only for select users). Also, notice that you can configure the mobility feature (allowing transfers of active calls to and from the Single Number Reach device) separately from the Single Number Reach feature (allowing calls to ring another device after a specified amount of time).

Enabling the Flash-Based CME GUI

In addition to CCP, Cisco provides a GUI that allows you to manage some of the CME basic functions through a web interface. These basic functions include configuring and managing ephones, ephone-dns, some system and voicemail functions, and reports.

Note: After being in the Cisco world for some time, you definitely get the feeling that "real Cisco techs" use the CLI. Seeing the CME GUI only reinforces that feeling; although it does enable you to configure some basic settings, you can accomplish far more by using the command line. The GUI suffices for a phone administrator whose primary job is the configuration of new phones and phone lines.

Before you are able to access the GUI, there are a few preliminary configuration steps you need to have in place. First and foremost, you need to ensure that you have loaded into the flash memory of the CME router the files that power the GUI. If you extracted the TAR file that contains the full CME installation into the flash of the CME router, the GUI files should be included. If you installed the CME files individually, be sure to download and install the CME GUI TAR file pack from Cisco.com. For more information on downloading and installing CME files into the flash of the router, check out Chapter 4, "Getting Familiar with CME Administration."

Tip: You can always verify that you have installed the GUI files by performing a directory list of your router's flash. Different CME versions organize the file structure differently, but should have relatively the same files. Here's a directory listing of CME version 4.3:

```
CME_Voice# dir flash:
Directory of flash:/
```

```
     1  drw-           0  Jun 10 2008 14:57:20 -07:00  bacdprompts
    13  -rw-       22224  Jun 10 2008 14:57:30 -07:00  CME43-full-readme-
  v.2.0.txt
    14  drw-           0  Jun 10 2008 14:57:30 -07:00  Desktops
    27  drw-           0  Jun 10 2008 14:57:36 -07:00  gui
    45  -rw-      496521  May 12 2008 21:30:00 -07:00  music-on-hold.au
    46  drw-           0  May 12 2008 21:30:00 -07:00  phone
   127  drw-           0  May 12 2008 21:35:46 -07:00  ringtones
   161  -rw-    45460908  Jun 12 2008 15:17:22 -07:00  c2801-adventerprisek9_
  ivs-mz.124-15.T5.bin
129996800 bytes total (16123904 bytes free)
CME_Voice# dir flash:gui
Directory of flash:/gui/

    28  -rw-         953  Jun 10 2008 14:57:36 -07:00  Delete.gif
    29  -rw-        3845  Jun 10 2008 14:57:38 -07:00  admin_user.html
    30  -rw-      647358  Jun 10 2008 14:57:40 -07:00  admin_user.js
    31  -rw-        1029  Jun 10 2008 14:57:40 -07:00  CiscoLogo.gif
    32  -rw-         174  Jun 10 2008 14:57:40 -07:00  Tab.gif
    33  -rw-       16344  Jun 10 2008 14:57:42 -07:00  dom.js
    34  -rw-         864  Jun 10 2008 14:57:42 -07:00  downarrow.gif
    35  -rw-        6328  Jun 10 2008 14:57:42 -07:00  ephone_admin.html
    36  -rw-        4558  Jun 10 2008 14:57:42 -07:00  logohome.gif
    37  -rw-        3724  Jun 10 2008 14:57:42 -07:00  normal_user.html
    38  -rw-           0  Jun 10 2008 14:57:42 -07:00  normal_user.js
    39  -rw-         843  May 12 2008 21:29:56 -07:00  sxiconad.gif
    40  -rw-        1347  May 12 2008 21:29:56 -07:00  Plus.gif
    41  -rw-        2399  May 12 2008 21:29:56 -07:00  telephony_service.html
    42  -rw-         870  May 12 2008 21:29:56 -07:00  uparrow.gif
    43  -rw-        9968  May 12 2008 21:29:56 -07:00  xml-test.html
    44  -rw-        3412  May 12 2008 21:29:58 -07:00  xml.template
129996800 bytes total (16123904 bytes free)
```

Because you will be accessing the GUI through a web interface, you need to turn the CME router into a mini-web server to serve up the CME pages. The configuration in Example 7-17 accomplishes this feat.

Example 7-17 *Configuring the CME Router as a Web Server*

```
CME_Voice(config)# ip http server
CME_Voice(config)# ip http secure-server
% Generating 1024 bit RSA keys, keys will be non-exportable...[OK]

CME_Voice(config)# ip http path flash:/gui
CME_Voice(config)# ip http authentication local
```

Example 7-17 enables both the HTTP (**ip http server**) and HTTPS (**ip http secure-server**) services on the router. Of course, HTTPS is the preferred method of accessing the CME GUI because the CME router encrypts all communication. The **ip http path flash:/gui** command sets the HTTP server to use files from the GUI subdirectory of flash memory of the CME router for HTTP requests. Finally, the CME router uses its local user database to authenticate users attempting to access the web interface.

Note: You might need to change the argument of the **ip http path** command based on where the HTML files are located for the CME GUI. Earlier versions of CME placed all CME files directly into flash without any directory structure. In this case, the command should be entered as **ip http path flash:**.

The next step in enabling the CME GUI is to create a user account with permission to access and manage the CME router. Example 7-18 configures this.

Example 7-18 *Creating a CME Web Administrator Account and Adding Permissions*

```
CME_Voice(config)# telephony-service
CME_Voice(config-telephony)# web admin ?
  customer  customer admin
  system    system admin
CME_Voice(config-telephony)# web admin system ?
  name      admin username
  password  admin password
  secret    secret password
CME_Voice(config-telephony)# web admin system name ?
  WORD  username for admin
CME_Voice(config-telephony)# web admin system name NinjaAdmin ?
  password  admin password
  secret    secret password
CME_Voice(config-telephony)# web admin system name NinjaAdmin secret ?
  0  UNENCRYPTED password will follow
  5  ENCRYPTED password will follow
CME_Voice(config-telephony)# web admin system name NinjaAdmin secret 0 ?
  WORD  Admin password
CME_Voice(config-telephony)# web admin system name NinjaAdmin secret 0 cisco
CME_Voice(config-telephony)# dn-webedit
CME_Voice(config-telephony)# time-webedit
```

The CME router is now equipped with a user account called NinjaAdmin with the password cisco.

By default, the CME GUI is not able to add ephone-dns to the CME configuration or modify the time on the CME router. The **dn-webedit** and **time-webedit** commands unlock these functions.

Note: If you are synchronizing your router's clock via NTP, do not enter the **time-webedit** command, to ensure that the time remains set to the more accurate NTP server.

The CME router's web interface is now ready to go. The final step is to connect to the CME router using a supported web-browser platform. From the supported web browser, enter the URL http://<CME_IP_Address>/ccme.html to access the CME GUI. After authenticating with your web admin account, the CME management console is displayed, as shown in Figure 7-24.

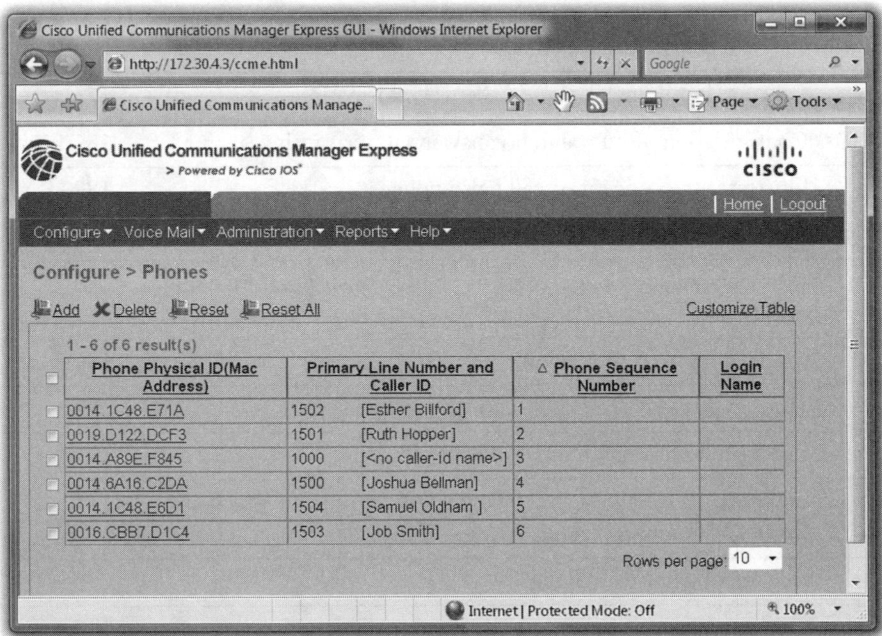

Figure 7-24 *CME Web-Based Management Interface*

Note: Figure 7-24 was captured after clicking Configure > Phones in the CME GUI. This gives you something more exciting to look at than the opening gray screen.

Note: Even though the built-in CME GUI is still available for download from the Cisco website, the CCP management utility has far surpassed its functionality and ease of use. The only benefit offered by the CME GUI is its ability to run from the flash of the router, which allows you to access it with no additional applications installed on your PC.

Exam Preparation Tasks

Review All the Key Topics

Review the most important topics in the chapter, noted with the key topics icon in the outer margin of the page. Table 7-3 lists a reference of these key topics and the page numbers on which each is found.

Table 7-3 *Key Topics for Chapter 7*

Key Topic Element	Description	Page Number
Example 7-1	Creating local directory and caller ID information	175
Example 7-3	Configuration of busy and no answer call forwarding	181
Figure 7-8	Illustrates the concept of call hairpinning	182
List	Differentiation between blind and consult call transfer methods	184
Example 7-5	Allowing outside call transfers using the transfer-pattern command	185
Example 7-7	Configuration of call pickup	191
List	Three types of call pickup methods	192
Example 7-8	Configuration of intercom	194
Note	Key note regarding the number of paging groups to which a Cisco IP Phone can belong	197
Example 7-9	Configuring unicast paging	197
Example 7-11	Configuring after-hours time designations	200
Example 7-12	Configuring after-hours block pattern	201
Example 7-17	Configuring the CME router as a web server	211
Example 7-18	Configuring the CME web administrator account	212

Definitions of Key Terms

Define the following key terms from this chapter, and check your answers in the Glossary:

local directory, H.450.3, H.450.2, hairpinning, call park, call pickup, directed pickup, local group pickup, other group pickup, Single Number Reach

This chapter includes the following topics:

- **Describe the CUCM GUI and CLI:** This section provides an overview of the graphical and command-line administrative interfaces for CUCM.
- **Describe the CUC GUI and CLI:** The section reviews the GUI and CLI administrative interfaces for CUC.
- **Describe the CUP GUI and CLI:** This section discusses the GUI and CLI administrative interfaces for CUP.

CHAPTER 8

Administrator and End-User Interfaces

Each Unified Communications application has several different administrative interfaces to provide configuration and maintenance functionality. End users also have an interface to allow them to customize their own environment. This chapter reviews the interfaces for the Cisco Unified Communications Manager (CUCM), Cisco Unity Connection (CUC), and Cisco Unified Presence (CUP) products.

"Do I Know This Already?" Quiz

The "Do I Know This Already?" quiz allows you to assess whether you should read this entire chapter or simply jump to the "Exam Preparation Tasks" section for review. If you are in doubt, read the entire chapter. Table 8-1 outlines the major headings in this chapter and the corresponding "Do I Know This Already?" quiz questions. You can find the answers in Appendix A, "Answers Appendix."

Table 8-1 *"Do I Know This Already?" Foundation Topics Section-to-Question Mapping*

Foundation Topics Section	Questions Covered in This Section
Describe the CUCM GUI and CLI	1–4
Describe the CUC GUI and CLI	5–7
Describe the CUPS GUI and CLI	8–10

1. How many administrative web interfaces does CUCM provide?

   a. 3

   b. 4

   c. 5

   d. 6

2. The account named CMMasterAdmin was created at install as the application Administration account. The account named CMPlatformAdmin was created as the Platform Administration account. Which two CUCM interfaces can CMMasterAdmin *not* log into?
   a. Unified CM Administration
   b. Unified Serviceability
   c. Unified OS Administration
   d. Disaster Recovery System
3. Which of the following best describes the interaction between CUCM Users, Groups, and Roles?
   a. User accounts can be assigned a Role, which defines the users' administrative role in the company. Users may be placed into Groups to allow simpler directory searches.
   b. User Groups are associated with one or more Roles that define a level of privilege to an application's resources. The user inherits those privileges when they are added to the Group.
   c. Of the 24 Roles defined by default, only 12 are active. Others must be activated via the Unified Serviceability interface.
   d. One Group and one Role are defined by default (the Standard CCM Super Users Group and the Standard CCM Admin Users Role). Other custom Groups and Roles can be defined by the administrator.
4. Bob needs to restart the TFTP service on a CUCM server. Where can he do this?
   a. Unified Serviceability > Tools > Service Activation
   b. Unified Serviceability > Tools > Control Center > Feature Services
   c. Unified Serviceability > Tools > Control Center > Network Services
   d. Only at the command line
5. Cisco Unity Connection provides five web-based administrative interfaces. Which one of the following is *not* one of them?
   a. Cisco Unified Serviceability
   b. Disaster Recovery System
   c. Cisco Unity Connection Serviceability
   d. Cisco Unified OS Administration
   e. Cisco Unity Connection Administration
   f. Cisco Unified Messaging Administration

6. Which of the following are valid call-handler types in CUC 8.x? (Choose three.)
   a. System Call Handlers
   b. Holiday Call Handlers
   c. Directory Call Handlers
   d. Interactive Voice Response Call Handlers
   e. Interview Call Handlers
7. Which of the following is true of the Cisco Unified Serviceability Application in Cisco Unity Connection?
   a. It is identical to the Cisco Unified Serviceability Application in CUCM.
   b. It is also known as Cisco Unity Connection Serviceability.
   c. It is reached via the same URL (with a different IP) as the CUCM Unified Serviceability.
   d. It is replaced by Cisco Unity Connection Serviceability in CUC 8.x.
8. Cisco Unified Presence Server has several administrative components in common with other Unified Communications applications. Which of the following are CUPS administrative interfaces? Choose all that apply.
   a. Cisco Unified Serviceability
   b. Cisco Unified OS Administration
   c. Disaster Recovery System
   d. Command-Line Interface
9. CUPS uses two different protocols for integration with Microsoft Office Communicator and third-party services, such as Google Voice. Under which administrative menu are they configured?
   a. CUPS Administration > System
   b. CUPS Administration > Messaging
   c. CUPS Administration > Presence
   d. CUPS Administration > Application
10. CUPS can be configured for regulatory compliance for IM retention. Under which administrative menu can this be done?
    a. CUPS Administration > System
    b. CUPS Administration > Messaging
    c. CUPS Administration > Presence
    d. CUPS Administration > Application

Foundation Topics

Describe the CUCM GUI and CLI

Administrative Web access to CUCM is possible only via HTTPS (or Secure Shell [SSH] for the command line). There are six separate interfaces:

- Cisco Unified Communications Manager Administration (https://<node_ip>/ccmadmin)
- Cisco Unified Serviceability (https://<node_ip>/ccmservice)
- Disaster Recovery System (https://<node_ip>/drf)
- Cisco Unified Operating System Administration (https://<node-ip>/cmplatform)
- Cisco Unified Reporting (https://<node-ip>/cucreports)
- Command-Line Interface (CLI)

Each of these (with the exception of the CLI) is accessible via its own URL or by using the navigation pull-down at the top right of the page and clicking the Go button next to it.

The account and password defined at install for platform administration is used to access to Disaster Recovery System and the Operating System Administration pages. Likewise, the application administration account and password defined during install is used to access the CM Administration, Serviceability, and Unified Reporting interfaces. Additional accounts can be created and assigned administrative privileges so that they may also access these interfaces.

During install, an additional password is defined as the security password. This password is needed to connect to the Publisher database (for Subscriber servers or other Unified Communications applications that use the Publisher database).

Cisco Unified Communications Manager Administration Interface

The CM Administration interface (as shown in Figure 8-1) includes nine menus for the following (a brief description is given for the tasks that each menu can perform; this is by no means an exhaustive list, and there are many tasks not listed):

- **System menu:** Includes tasks for the configuration of CM groups, Presence groups, Device Mobility Groups, Device Pools, Regions, Locations, Enterprise and Service Parameters, Survivable Remote Site Telephony (SRST), and others.
- **Call Routing menu:** Includes tasks to define the call routing system, Call Hunting, Class of Control, Intercom, and features such as Call Park, Call Pickup, and many more.
- **Media Resources menu:** Under this menu, resources such as Music On Hold (MOH), annunciator, media termination points, and transcoders can be defined and hold-music files managed.
- **Advanced Features menu:** Under this menu, VoiceMail integrations, Inter-company Media Engine Configuration, Extension Mobility Cross-Cluster, and VPN features are configured.

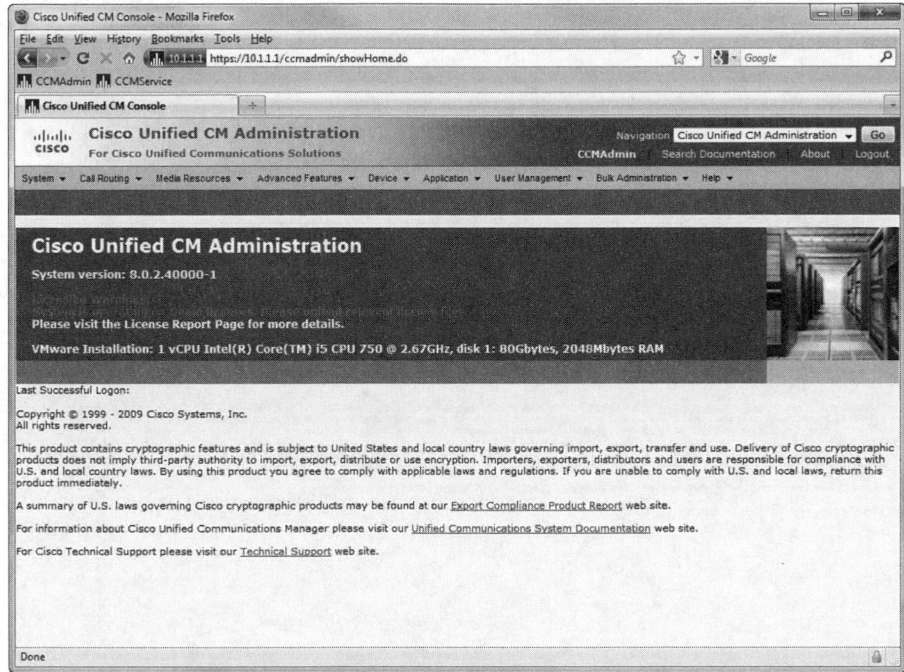

Figure 8-1 *CUCM Administration Home Page*

- **Device menu:** Provides configuration pages for gateways, gatekeepers, trunks, IP phones, and Remote Destinations, plus many device settings, including phone button and softkey templates.
- **Application menu:** Accesses the CUCM Assistant Configuration Wizard and the Plug-ins menu.
- **User Management menu:** Accesses the Application User, End User, Group, and Role configuration pages.
- **Bulk Administration menu:** Provides many options to perform repetitive configuration tasks (such as adding many users or phones) in an automated way. There are many additional and powerful capabilities of the BAT tool not listed here.
- **Help menu:** Provides access to the local searchable help files, the This Page help, and the About information page.

Cisco Unified Serviceability Administration Interface

The Cisco Unified Serviceability interface (shown in Figure 8-2) provides five menus, each with submenus, as summarized in the following:

- **Alarm menu:** Provides Configuration and Definition options for alarms to monitor system performance and health.
- **Trace menu:** Provides the Configuration and Troubleshooting Trace Settings submenus, which monitor the system and troubleshoot.

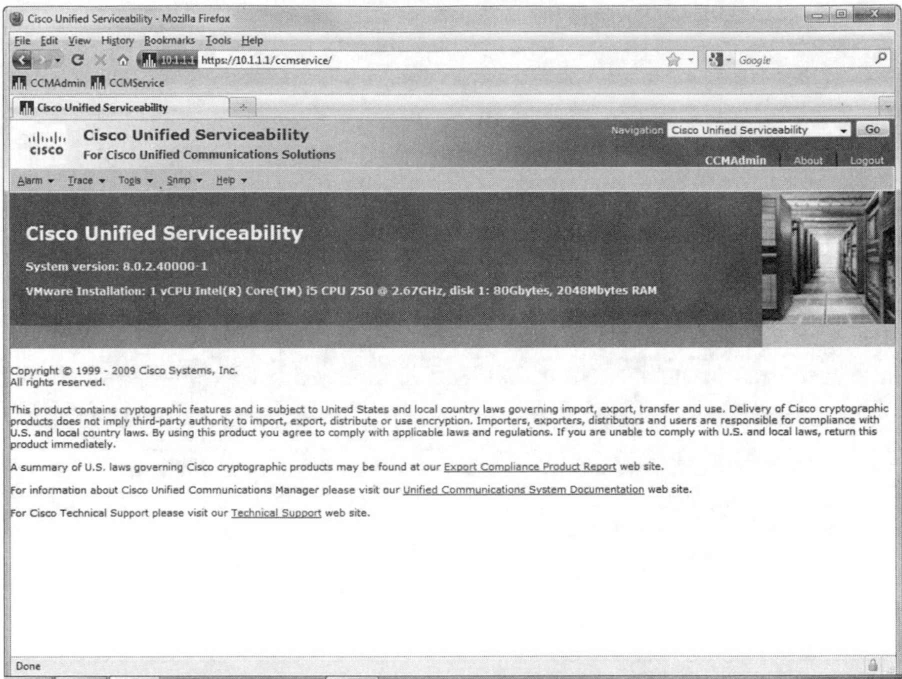

Figure 8-2 *Cisco Unified Serviceability Interface for CUCM*

- **Tools menu:** Under this menu, the CDR Analysis and Reporting submenu provides an interface to gather call logs and report on calls made using the system. The Service Activation screen provides the interface to activate installed services for the first time (or deactivate them later). There are two Service Control Centers: Network and Feature (see the following Note). Using this interface, administrators can stop, start, or restart activated services. The Serviceability Reports Archive provides access to the reporting interface for system and trend analysis. The CDR Management interface allows administrators to configure and check Call Detail Record (CDR) storage disk utilization. The Audit Log Configuration page provides settings for what will be included in audit logs.
- **SNMP menu:** The submenus (V1/V2c, V3, and SystemGroup) control Simple Network Management Protocol (SNMP) connectivity and authentication to network management applications.
- **Help menu:** Provides access to the searchable Contents help, This Page help, and About information.

Note: What is the difference between Feature services and Network services? Network services are automatically activated and required for server operation (such as Cisco CallManager Admin Service, DB Replicator, and CDP). Network services cannot be deactivated, but can be started, stopped, and restarted.

Feature services are optional services that can be activated using the Service Activation page. These services might or might not be activated on a particular server in a large cluster where the design calls for assigning a particular role to a server. For example, the Cisco CallManager, TFTP, or IP Voice Media Streaming App services might be active or inactive depending on the server's job(s) in the cluster.

Cisco Unified Operating System Administration Interface

The Unified Operating System interface (shown in Figure 8-3) allows an administrator to monitor and interact with the Linux-based operating system platform. Administrative tasks that can be performed here include the following:

- Monitor hardware-resource utilization (CPU, disk space)
- Check and upgrade software versions
- Verify and change IP address information
- Manage Network Time Protocol (NTP) server IP addresses
- Manage server security including IPSec and digital certificates
- Create a TAC remote assistance account
- Ping other IP devices

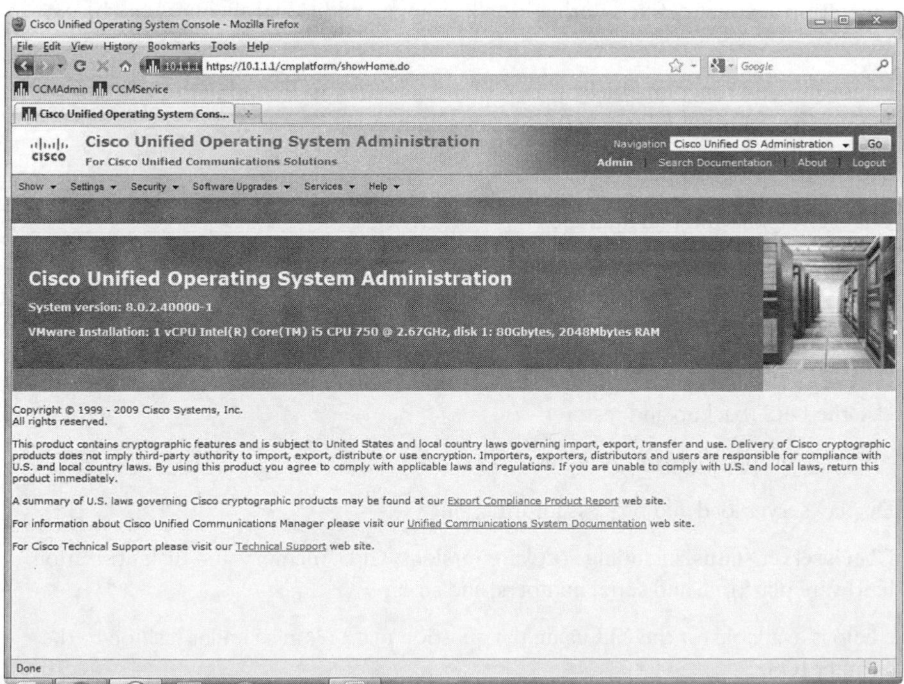

Figure 8-3 *Cisco Unified OS Administration Interface for CUCM*

Disaster Recovery System Interface

The Disaster Recovery System (DRS) provides a backup (with scheduler) and restore capability. Access to this interface uses the Platform Administration account defined at install (as does the OS Admin interface). Additional accounts can be created for access by other individuals.

Backups can be written to a local DLT tape drive or to an SFTP server. A scheduler is provided for automated backups, or an immediate start to the backup can be selected. Individual server or full cluster backups may be performed.

Cisco Unified Reporting Interface

The Cisco Unified Reporting interface provides a simplified method to access system reports. These reports gather information from existing logs and format the data into simple, useful, one-click reports. Data is collected from logs across the cluster (Publisher and Subscribers) to provide summarized information and highlight issues or irregularities that might impact operation of the cluster. The interface also warns if running a particular report could adversely impact server operation and affect performance or take excessive time.

CLI

The CLI is typically accessed using SSH, although it is possible to directly connect a keyboard and monitor. Initially, the only account that can log on using the CLI is the Platform Administration account defined during install, although additional accounts can be created to allow access.

The commands and functionalities of the CLI include all those found in the OS Administration interface, plus the following (this is not a comprehensive list):

- Shut down or restart the system
- Change versions after an upgrade
- Start, stop, and restart services
- Modify network settings (IP Address, mask, gateway, and so on)
- Use network tools such as ping, traceroute, and packet capture
- Use the DRS (backup and restore)
- Add and modify Platform Administration accounts
- Display server load and process information
- Check server status, including software versions, CPU, memory and disk utilization, hardware platform, and serial numbers, and so on

Inline help is available for the CLI using the question mark (?) in a similar fashion to the Cisco router IOS.

Caution: The CLI is a powerful interface. The Enter key commits the command immediately, and you are very rarely asked to confirm your intent. This effectively means the CLI will do exactly what you just told it to do, including the possibility of negatively impacting the operation of the server. Review the CLI Administration Guide and be certain of the commands you are entering before using the CLI.

User Management in CUCM: Roles and Groups

The CUCM application defines a consistent and simple method of assigning (and limiting) administrative privilege to users. Users are assigned to groups; Groups are assigned Roles; Roles define privileges to applications. The following sections provide some detail on this structure.

Roles

Roles define a set of privileges to the resources in an application. Resources may be a CUCM administration web page, a report tool, or a feature section within a CUCM web page.

The privilege assigned for each resource can be one of the following:

- **No Access:** The role denies access to the resource; for example, a web page will not load, and an error appears instead.
- **Read:** The role allows the resource to be displayed but not edited: for example, a web page may load, but none of the fields or settings can be modified. Buttons such as Add, Insert, Delete, or Update do not appear.
- **Update:** The role allows full access to the resource, including editing and modification (and deletion where applicable).

Other applications (for example, Cisco Unified Serviceability or Cisco Extension Mobility) define access privileges that are relevant and specific to the application.

There are 39 Standard roles defined by default in CUCM 8.0; in most cases, one of these roles provide the administrative functions required. If a Standard role is not exactly correct for a particular scenario, custom roles can be defined that exactly define the administrative privilege required in any context. Custom roles can be defined for any of the following applications:

- CUCM Administration
- CUCM Serviceability
- Cisco Computer Telephone Interface (CTI)
- CUCM Administrative XML (AXL) database
- Cisco Extension Mobility
- CUCM End User

Figure 8-4 shows the Find and List Roles page.

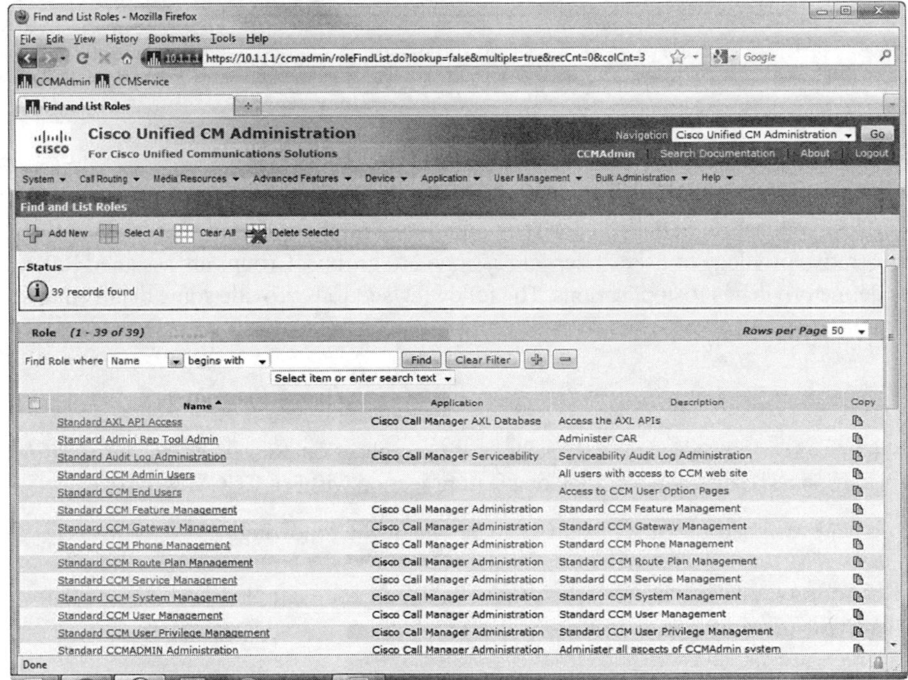

Figure 8-4 *Roles List in CUCM*

Groups

The CUCM application defines 24 Standard Groups by default. These Groups are each associated with one or more Standard Role(s) that provide various levels of privilege to the various applications accessible via CUCM Administration pages. Most Groups have no members by default; when end users are created or imported in CUCM, these accounts may be added to one or more Groups. The privileges defined by the Role are inherited by the user account by way of their membership in the Group.

As with Roles, it is likely that the administrative requirements in most scenarios will be met by using the Standard Groups and Role associations. If this is not the case, custom Groups can be created and associated with Standard or Custom Roles to exactly meet administrative requirements.

Because a user account can be a member of multiple Groups and therefore inherit multiple levels of privilege, it is important to understand the effect of conflicting Role privileges. Consider this scenario:

> Bob is a member of two User Groups, called A and B. Group A is associated with Role X, which provides Update privilege to an application resource. Group B is associated with Role Z, which provides Read privilege to the same resource. The question is: What are Bob's effective privileges for the resource?

The answer is determined by the Enterprise Parameter **Effective Access Privileges for Overlapping User Groups and Roles**. The default setting is **Maximum**, meaning that, by default, Bob has Update privileges. The parameter can be changed to **Minimum**, changing Bob's effective privilege to Read.

> **Note:** Changing the enterprise parameter as described affects all groups except the Standard CCM Super Users group (the maximum-privilege group, the only default member of which is the CUCM Application Administrator defined at install [for example, CCMAdministrator]).
>
> Interestingly, the default setting of Maximum privilege is the opposite of the default security settings of most network applications and operating systems. This fact is not an issue as long as the administrator is aware of the impact and significance of the setting, regardless of what setting is chosen.

Figure 8-5 shows the Find and List User Groups page in CUCM.

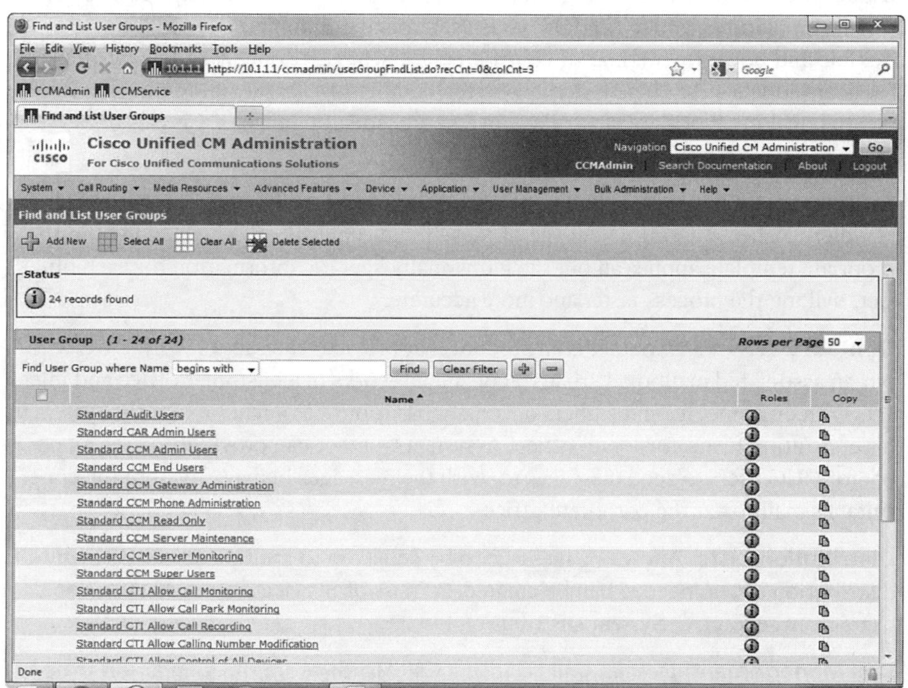

Figure 8-5 *User Groups List in CUCM*

Describe the CUC GUI and CLI

CUC provides the following six administrative interfaces:

- Cisco Unity Connection Administration (https://<ip-address>/cuadmin)

- Cisco Unified Serviceability (https://<ip-address>/ccmservice)
- Cisco Unity Connection Serviceability (https://<ip-address>/cuservice)
- Cisco Unified Operating System Administration (https://<ip-address>/cmplatform)
- Disaster Recovery System (https://<ip-address>/drf)
- Command-Line Interface

Cisco Unity Connection Administration

The CUC Administration interface uses a tree structure similar to the Windows Explorer, making finding and navigating to the desired page easy. The main menu items are described in the following sections; more detailed information is presented in Chapter 13, "Voicemail Integration with Cisco Unity Connection."

- **Users:** Provides access to the local user database to create or edit users. User import and synchronization menu options are also listed here.
- **Class of Service:** Defines the features available to the user, including Licensed and Advanced features, and may apply other user interaction limits. Class of Service is a flexible and powerful method of controlling users' interaction with CUC, particularly for licensed features. There is no limit to the number of Classes of Service that can be created, making it possible to build a diverse and specific range of Classes of Service to cover any combination of requirements.
- **Templates:** Provide a way to define common settings for Users, Contacts, or Call Handlers. When creating a new one of these three types of object, selecting the appropriate template applies all but the individually specific information for each object, making the process faster and more accurate.
- **Contacts:** A System Contact is an account that provides interaction with CUC without an associated mailbox. Perhaps a consultant works in a customer office and interacts with customer team members on a daily basis, but has a mailbox on the system at his consulting firm's corporate office. A System Contact can provide a directory entry for CUC users to send messages to, with the messages being relayed to the consultant's mailbox at the corporate office.
- **Distribution Lists:** Allow one message to be delivered to multiple users. As many distribution lists as needed can be created. A Class of Service setting can prevent users from sending to System Distribution Lists.
- **Call Management:** Defines Call Handlers. Call Handlers are the foundation structures of CUC and can be configured to answer calls, play greetings, route calls, and take messages. Specialized call handlers include *Directory Handlers*, which allow a caller to search the directory and either call or leave a message for the selected user, and *Interview Handlers*, which interact with caller by playing a series of recorded questions, then collecting the answers in a single message.
- **Message Storage:** Allows mailbox quotas to be set, enforcing limits on mailbox size to prevent running out of disk space.

- **Networking:** Configures multiple CUC systems in either a Digital Networking or VPIM environment.
- **Dial-Plan:** Provides for the creation of additional Partitions and Search Spaces to control visibility and access to messaging components. It is useful to "hide" part of the CUC system from certain users or functions.
- **System Settings:** The System Settings submenus provide global configuration settings. Some of the submenus are summarized here:
  - **Licenses:** Tracks and displays the licenses available to the system. Licenses are tied to the MAC address of the NIC on the server.
  - **Holiday Schedules:** The three system schedules (**All Hours**, **Weekdays**, and **Voice Recognition Update Schedule**) can be modified but not deleted; new schedules can be added to customize the system.
  - **External Services:** CUC can be configured to access user calendar and contact information held by Microsoft Exchange. This information can then be incorporated into Personal Call Routing Rules. Likewise, if Cisco Unified MeetingPlace is available in the network, CUC can pull conference information from the server so users can view and join meetings from their Personal Communications Assistant or from the phone.
  - **LDAP:** The submenus allow the configuration of LDAP Synchronization (User import) and LDAP Authentication (redirection of password authentication to LDAP).
  - **SMTP:** CUC can notify users of new messages via e-mail; the submenus configure details for the integral SMTP server.
  - **Advanced:** Several configuration submenus are found under the Advanced submenu, including the SMPP entry, in which CUC can be configured to send Short Message Peer to Peer/Short Message Service text message notifications to mobile phones.
- **Telephony Integrations:** Lists and configures the phone systems with which CUC is integrated, port groups and ports.
- **Tools:** Includes the Bulk Administration Interface and the Task Management (automated maintenance and troubleshooting) system.

Figure 8-6 shows the CUC home page.

Cisco Unity Connection Serviceability

The CUC Unified Serviceability application is similar to the app of the same name in CUCM. CUC provides an additional application called Cisco Unity Connection Serviceability, which despite the similar name provides a very different functionality. CUC Serviceability is primarily a troubleshooting tool, with tools to define alarms, traces and logs, plus service controls (Activate/Deactivate, Start/Stop/Restart) for CUC-specific feature services. If an Active-Active redundant cluster pair is in use, the tools to manage the cluster are provided here. A reports interface is also available.

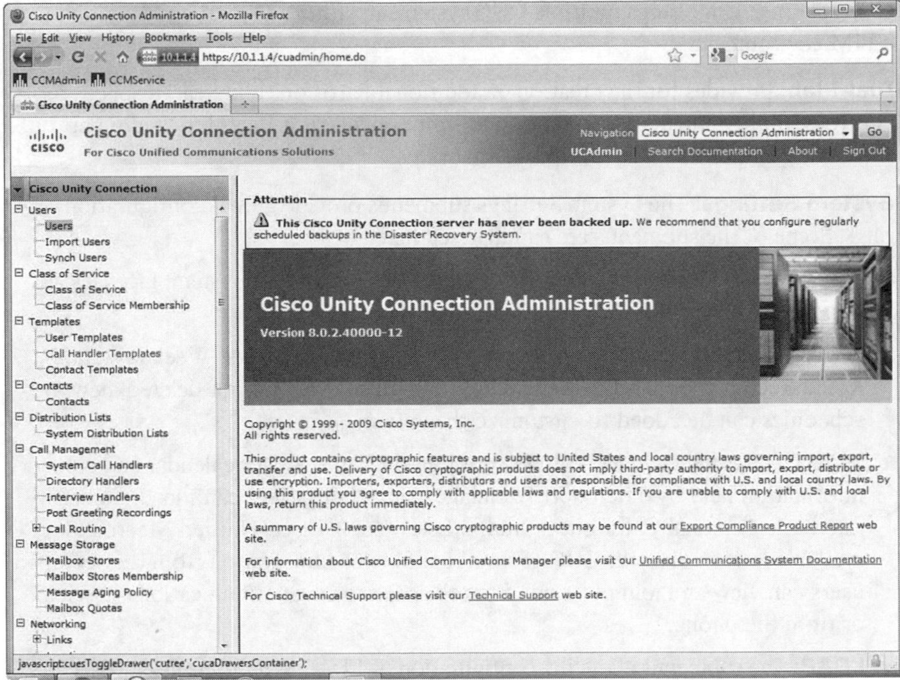

Figure 8-6 *CUC Administration Home Page*

Describe the Cisco Unified Presence Server GUI and CLI

Cisco Unified Presence Server (CUPS) provides five administrative interfaces:

- Cisco Unified Presence Administration (https://<ip-address>/cupadmin)
- Cisco Unified Serviceability (https://<ip-address>/ccmservice)
- Cisco Unified Operating System Administration (https://<ip-address>/cmplatform)
- Disaster Recovery System (https://<ip-address>/drf)
- Command-Line Interface

As with other CUCM and CUC, the account defined at install as the Platform Administrator is, by default, the only one that can log into the OS Administration, DRS, and CLI interfaces. The Application Administration account is the only one, by default, that can log into the CUPS Administration and Unified Serviceability interfaces. Additional Application and Platform Administration accounts can be created as required.

Cisco Unified Presence Administration Interface

The CUPS Administration menus include the following:

- **System:** Provides integration configuration for CUCM, definition of inbound and outbound Access Control Lists, and Licensing status and management.
- **Presence:** Provides definition of gateways providing Presence information from CUCM or calendar integration with Microsoft Outlook. Interdomain federation across different presence domains using Session Initiation Protocol (SIP) or Extensible Messaging and Presence Protocol (XMPP) can be configured. SIP is commonly used with Microsoft Office Communications Server (OCS), while XMPP is typically used with Google Talk.
- **Messaging:** CUPS can be used with external databases (PostgreSQL compliant) or third-party servers to enable IM retention regulatory compliance (Persistent messaging).
- **Application:** CUPS applications, such as Desk Phone Control and IP Phone Messenger, are configured here.
- **User Management:** Follows a Groups-and-Roles structure similar to CUCM. User accounts can be imported and synchronized from CUCM.
- **Bulk Administration:** A similar interface to CUCM provides bulk configuration of repetitive tasks, with scheduler capability.
- **Diagnostics:** Accesses system status and troubleshooting tools, plus a System Dashboard for quick review of system configurations.

Cisco Unified Presence Serviceability

The Cisco Unified Presence Serviceability interface provides similar functionality to the CUC Serviceability interface, including the following:

- Alarms and events monitoring for troubleshooting purposes
- Access to CUPS service trace logs
- Monitor real-time CUPS component behavior via CUCM RTMT
- Feature Service activation, deactivation, and control; Network Service control
- Reports archive
- SNMP configuration
- Disk usage monitoring for log partition on local and cluster servers

Exam Preparation Tasks

Review All the Key Topics

Review the most important topics in the chapter, noted with the key topics icon in the outer margin of the page. Table 8-2 describes these key topics and identifies the page number on which each is found.

Table 8-2 *Key Topics for Chapter 8*

Key Topic Element	Description	Page Number
Paragraph	Default accounts' access to administrative interfaces	220
Section	Roles in CUCM	225
Section	User Groups in CUCM	226

Definitions of Key Terms

Define the following key terms from this chapter, and check your answers in the Glossary:

role (CUCM), user group (CUCM), application (CUCM), resource

This chapter includes the following topics:

- **Implementing IP Phones in CUCM:** This section reviews the required network services and systems configurations to support IP Phones, details the startup and registration processes, and reviews manual, automatic, and bulk administration tasks for adding phones.
- **Describe End Users in CUCM:** This section describes the characteristics of End User configuration in CUCM.
- **Implementing End Users in CUCM:** This section reviews the methods by which End Users may be added to CUCM, including manual addition, Bulk Administration, and LDAP Synchronization and Authentication.

CHAPTER 9

Managing Endpoints and End Users in CUCM

IP Phones and End Users are important parts of a Unified Communications deployment; after all, without phones or people to use them, what is the point of having the system? This chapter reviews the configuration of endpoints and users in CUCM, including setting up basic network services, registering phones, configuration, and bulk administration.

"Do I Know This Already?" Quiz

The "Do I Know This Already?" quiz allows you to assess whether you should read this entire chapter or simply jump to the "Exam Preparation Tasks" section for review. If you are in doubt, read the entire chapter. Table 9-1 outlines the major headings in this chapter and the corresponding "Do I Know This Already?" quiz questions. You can find the answers in Appendix A, "Answers Appendix."

Table 9-1 *"Do I Know This Already?" Foundation Topics Section-to-Question Mapping*

Foundation Topics Section	Questions Covered in This Section
Implementing IP Phones in CUCM	1–6
Describe End Users in CUCM	7
Implementing End Users in CUCM	8–10

1. Which of the following protocols is critical for IP Phone operation?
   - **a.** DNS
   - **b.** DHCP
   - **c.** NTP
   - **d.** TFTP
2. What file does an IP Phone first request from TFTP during its startup and registration process?
   - **a.** SEP<mac_address>.cnf.xml
   - **b.** None. The phone receives all information via SCCP signaling.
   - **c.** SEP<mac_address>.xml
   - **d.** XMLDefault.cnf.xml

3. Which of the following statements is true?

- **a.** SCCP phone configuration files contain all settings, including date/time and softkey assignments.
- **b.** SIP phone configuration files are larger than SCCP phone configuration files.
- **c.** SCCP phone configuration files are exactly the same as SIP phone configuration files.
- **d.** SIP phone configuration files are much smaller than SCCP configurations files because of the limited feature set of SIP phones.

4. Which of the following is true of DHCP in CUCM?

- **a.** The DHCP server capability is no longer available as of CUCM v8.x.
- **b.** The DHCP service is a basic capability intended for supporting up to 1000 IP Phones.
- **c.** DHCP is mandatory for IP Phones.
- **d.** CUCM supports a proprietary IP address assignment protocol called LLDP.

5. Which of the following is not a Device Pool setting?

- **a.** Cisco Unified Communications Manager Group
- **b.** Local Route Group
- **c.** Region
- **d.** Common Phone Profile

6. Bob asks you to provide a third DN button and a BLF Speed Dial for the Auto Parts Desk's 12 7965 IP Phones. Which of the following is the best choice?

- **a.** Modify the Standard User Softkey Template.
- **b.** Copy the Standard User Softkey Template, name it PartsDesk, and add the requested features.
- **c.** Copy the Standard 7965 SCCP Phone Button Template, name it PartsDesk, and add the requested features.
- **d.** It is not possible to add a third DN and a BLF Speed Dial to a 7965 IP Phone.

7. Pete recently learned that he can add his own speed dials, subscribe to phone services, and do other useful things via his User Web Page. He comes to you complaining that he can't do any of these things. Why can't Pete modify his own phone?

- **a.** The Active Directory GPO is limiting Pete's permissions.
- **b.** Pete's account needs to be associated with his phone in the Device Associations settings in his User Configuration Page.
- **c.** Additional licensing is required to support User Web Page functionality.
- **d.** Pete must be part of the CCM Super Users Group to make these changes.

8. Angie changes her Windows domain login password, but notices that her password for her User Web Page in CUCM has not changed. Which of the following is true?
   a. LDAP Synchronization has not been configured.
   b. Cisco Unified Services for Windows Domains has not been configured.
   c. Angie must wait 24 hours for the password to synchronize.
   d. LDAP Authentication has not been configured.
9. Which of the following is not true of LDAP Synchronization?
   a. Application User accounts are configured only in LDAP and are replicated to CUCM.
   b. End User accounts and passwords are replicated to CUCM.
   c. LDAP checks the user accounts in CUCM and replicates only those that also exist in LDAP.
   d. End User accounts that exist in LDAP are replicated to CUCM, unless the LDAP **sn** attribute is blank.
10. Which is true of LDAP Synchronization Agreements?
   a. The User Search Base defines the point in the tree where CUCM begins searching. CUCM can search all branches below that point.
   b. The User Search Base defines the point in the tree where CUCM begins searching. CUCM can search all branches above and below that point.
   c. The User Search Base must specify the root of the domain; LDAP Custom Filters must be used to limit the search returns.
   d. All Synchronization Agreements must run on a regular scheduled basis.
   e. Only one Synchronization Agreement can be made with a single LDAP system.

Foundation Topics

Implementing IP Phones in CUCM

The implementation of IP Phones is remarkably simple, considering the myriad of services, protocols, and processes going on in the background to make the system work well. This section reviews these "hidden" processes and details some of the administrative tasks required to easily and reliably run IP Phones in CUCM.

Special Functions and Services Used by IP Phones

A variety of standards-based and proprietary protocols and services support IP Phones in CUCM. In no particular order, they include the following:

- Network Time Protocol (NTP)
- Cisco Discovery Protocol (CDP)
- Dynamic Host Configuration Protocol (DHCP)
- Power over Ethernet (PoE)
- Trivial File Transfer Protocol (TFTP)
- Domain Name System (DNS)

The next section describes each of these services, how IP Phones use them, and how to configure them in CUCM (or other systems as appropriate).

NTP

NTP is an IP standard that provides network-based time synchronization. There are many good reasons to use NTP beyond the convenience and consistency of having the same time on all devices. Call Detail Records (CDR) and Call Management Records (CMR) are time stamped, as are log files. Comparing sequential events across multiple platforms is much simpler and easier to understand if the relative time is exactly the same on all those devices. Some functions and features can also be time (calendar) based, too, so time synchronization is important for those functions to operate properly.

In a typical NTP implementation, a corporate router synchronizes its clock with an Internet time server (such as a NIST atomic clock or a GPS satellite clock). Other devices in the corporate network then sync to the router.

The Cisco Unified Communications Manager (CUCM) Publisher is one such device; during installation, CUCM asks for the IP address of an NTP server. (Alternatively, it can use its internal clock, which is not recommended because of its inaccuracy compared to NTP.) The Subscriber servers then sync their clocks to the Publisher, and the IP Phones get their time from their Subscribers via Skinny Client Control Protocol (SCCP) messages. Session Initiation Protocol (SIP) phones need an NTP reference (detailed later), but in the absence of one, they can get the time from the time stamp in the SIP OK response from the Subscriber server.

CDP

CDP is a Cisco-proprietary Layer 2 protocol that provides network mapping information to directly connected Cisco devices. (You learned about CDP in your CCNA studies, so we do not detail it here.) Cisco IP Phones generate CDP messages and use CDP to learn the voice VLAN ID from the Cisco switch to which they are connected. The IP Phone then tags the voice frames it is transmitting with that VLAN ID in the 802.1Q/P frame header.

DHCP

DHCP is a widely used IP standard that can provide the following information to IP Phones:

- IP Address
- Subnet Mask
- Default Gateway
- DNS Server(s)
- TFTP Server(s)

Although it is possible to statically configure IP Phones with all that information, it would be time-consuming and error-prone. DHCP is faster, easier, more scalable, and a widely accepted practice. DHCP can be provided by an existing DHCP server (because most deployments already have one), a local router, or even by CUCM itself, although this is not generally recommended, particularly for larger deployments. Later sections review the configuration of DHCP services in CUCM and router IOS.

PoE

PoE is a standards-based feature that delivers DC power supply over Ethernet cabling. IP Phones can use this feature, and doing so means less cabling to clutter the desk, no power supplies to buy for the phones, and potential cost savings. PoE is generally assumed to be provided by the switch that the phones connect to, but it may also be provided by a powered patch panel or inline power injector.

TFTP

TFTP is a critical service for IP Phones. The phones use TFTP to download their config files, firmware, and other data. Without TFTP, the phones simply do not function properly. When you make a configuration change to a device, CUCM creates or modifies a config file for the device and uploads it to the TFTP server(s). TFTP services must therefore be provided by one (or more in large deployments) of the CUCM servers in the cluster; a generic TFTP server will not have the integrated capability that a CUCM TFTP server does and will not correctly fulfill the role.

DNS

DNS provides hostname-to-IP address resolution. DNS services are not critical to IP Phones. (In fact, in most deployments, it is recommended to eliminate DNS reliance from the IP Phones [see Chapter 10, "Understanding CUCM Dial-Plan Elements and Interactions"].) But in some circumstances, it is desirable. A DNS server must be external to the CUCM cluster; DNS is not a service that CUCM can offer.

IP Phone Registration Process

The steps that each phone goes through as it registers and becomes operational are more complex than you might think. The following section reviews these steps:

1. The phone obtains power (PoE or AC adapter).
2. The phone loads its locally stored firmware image.
3. The phone learns the Voice VLAN ID via CDP from the switch.
4. The phone uses DHCP to learn its IP address, subnet mask, default gateway, and TFTP server address. (Other items may be learned also.)
5. The phone contacts the TFTP server and requests its configuration file. (Each phone has a customized configuration file named SEP<mac_address>.cnf.xml created by CUCM and uploaded to TFTP when the administrator creates or modifies the phone.)
6. The phone registers with the primary CUCM server listed in its configuration file. CUCM then sends the softkey template to the phone using SCCP messages.

> **Note:** What is in that SEP<mac_address.cnf.xml file?
> The file contains a list of CUCM server, in order, that the phone should register with. It lists the TCP ports it should use for SCCP communication. It also lists the firmware version for each device model and the service URLs that each device should be using.
>
> The CUCM server sends other configurations, such as DNs, softkeys, and speed dials, via SCCP messages in the last phase of the registration process. The configuration files for SIP phones are generally larger than the equivalent files for SCCP phones. This is because SIP phones have no equivalent mechanism for configuring items that are set by SCCP messages on SCCP phones; these items must be included in the configuration file downloaded from TFTP.

SIP Phone Registration Process

SIP phones use a different set of steps to achieve the same goal. Steps 1 to 4 are the same as SCCP phones. The following are the rest of the steps:

1. The phone contacts the TFTP server and requests the Certificate Trust List file (only if the cluster is secured).
2. The phone contacts the TFTP server and requests its SEP<mac_address>.cnf.xml configuration file.
3. The phone downloads the SIP Dial Rules (if any) configured for that phone.
4. The phone registers with the primary CUCM server listed in its configuration file.
5. The phone downloads the appropriate localization files from TFTP.
6. The phone downloads softkey configurations from TFTP.
7. The phone downloads custom ringtones (if any) from TFTP.

Preparing CUCM to Support Phones

Before we add phones, a certain amount of work should be done on the CUCM servers. Doing this setup work makes adding phones easier, more consistent, and more scalable, assuming that we follow our design plan.

The tasks we review in this section are as follows:

- **Configure and Verify Network Services:** Set up NTP, DHCP, and TFTP.
- **Configure Enterprise Parameters:** Modify and verify cluster-wide default settings.
- **Configure Service Parameters:** Tune application settings and behavior.

Service Activation

Many required services are deactivated by default on CUCM. Using the Unified Serviceability admin page, you must activate the one you need. For our purposes, we activate the Cisco CallManager, Cisco TFTP, and CISCO DHCP Monitor services. Figure 9-1 shows the Unified Serviceability Service Activation page with those services activated.

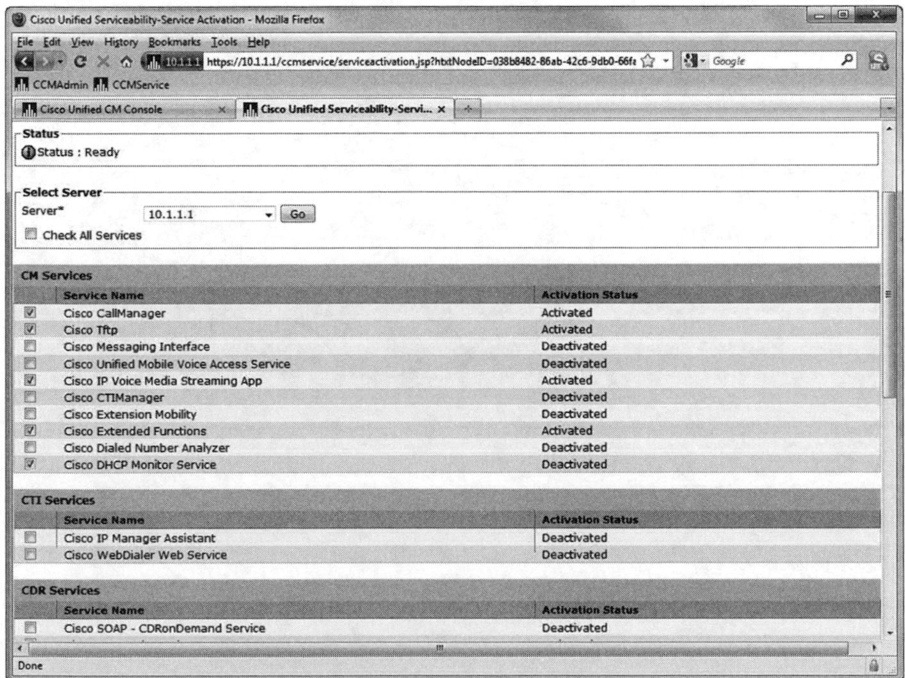

Figure 9-1 *Activating Required Services*

DHCP Server Configuration

CUCM includes a basic DHCP server capability. It is intended to support only IP Phones, and not very many of them: only up to 1,000 phones (this is the maximum recommended due to heavy CPU load). There is no native capability for DHCP server redundancy and only one DHCP server is supported per cluster, typically running on the Publisher. Multiple subnets (scopes) can be configured on the server.

Setting up the DHCP service is straightforward. We already activated the DHCP Monitor Service, so now we follow these basic steps:

1. Navigate to **System > DHCP > DHCP Server.**
2. Click **Add New.**
3. Select the server running the DHCP Monitor Service from the pull-down.
4. Configure the desired settings.

The settings that can be configured on the Server page include the following (among others):

- Primary DNS Server IPv4 Address
- Primary TFTP Server IPv4 Address
- IP Address Lease Time

Any settings you configure on the Server page are inherited by the Subnet configuration (shown next); however, any setting you change on the Subnet page overrides the Server setting. Figure 9-2 shows the DHCP Server configuration page.

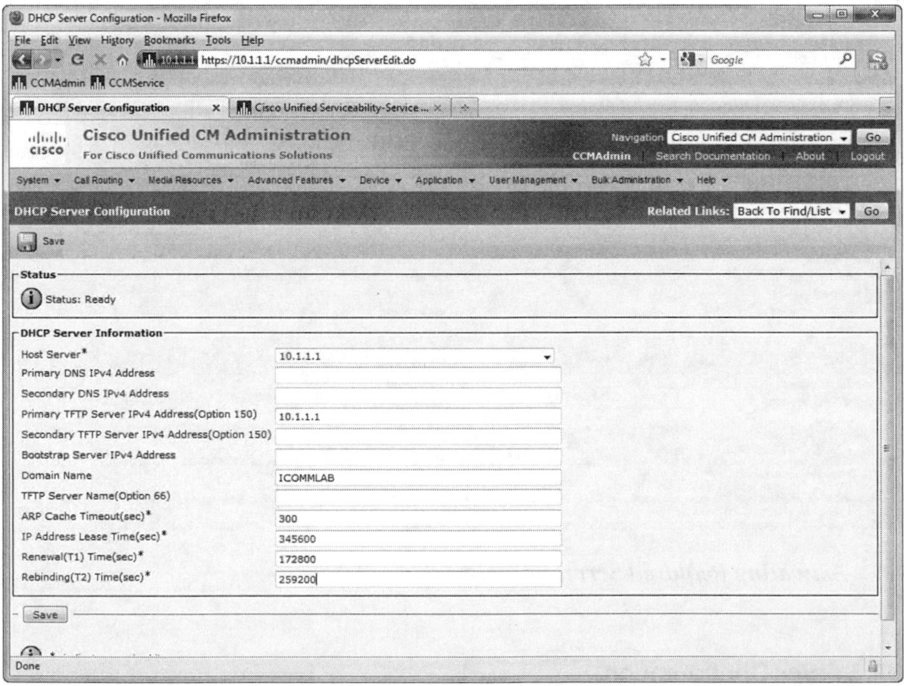

Figure 9-2 *DHCP Server Configuration*

Configuring DHCP Subnets requires some understanding of IP subnetting and assumes that you have an IP addressing plan in place. Because these topics are covered in the

CCNA prerequisite, we assume you have a grasp of these fundamentals. To configure DHCP subnets, navigate to System > DHCP > DHCP Subnet. Click **Add New** to create subnets; you can create multiple subnets as needed for your environment design. On the Subnet Configuration page, select the server from the DHCP Server drop-down list. You can then configure the following (some other settings are not listed):

- Subnet address
- Primary range start IP
- Primary range end IP
- Primary router IP address (default gateway)
- Subnet mask
- Primary DNS server IP address
- TFTP server IP address

Remember that settings in the subnet configuration override the same settings in the server configuration. Figure 9-3 shows the DHCP Subnet Configuration page.

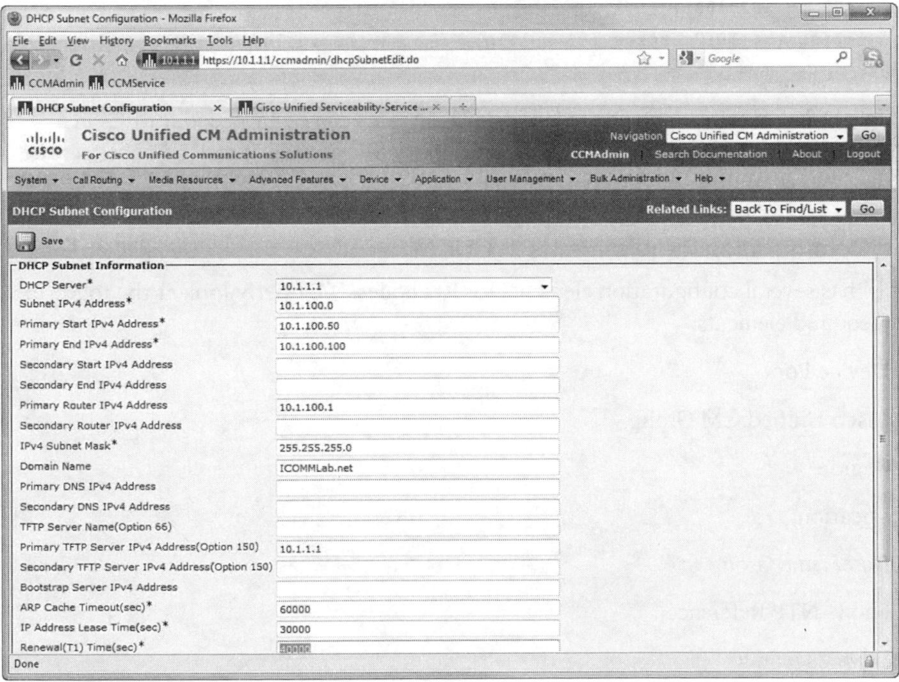

Figure 9-3 *DHCP Subnet Configuration*

Configuring DHCP in Router IOS

Cisco routers support basic DHCP server functionality, commonly used in small office environments where a dedicated DHCP server is not needed or available. The following commands show a typical DHCP configuration, with commands annotated for reference:

```
service dhcp
! Enables the DHCP service
!
ip dhcp excluded-address 10.1.1.1 10.1.1.10
! Specifies a start / end range of addresses that DHCP will NOT assign
ip dhcp pool name IP_PHONES
! Creates a pool of addresses (case-sensitive name) and enters DHCP configura-
tion mode
!
network 10.1.1.0 255.255.255.0
! Defines the subnet address for the pool
default-router address 10.1.1.1
! Defines the default gateway
dns-server address 192.168.1.10 192.168.1.11
! Identifies the DNS server IP address(es) - up to 8 IPs
!
option 150 ip 192.168.1.2
! Identifies the TFTP server IP. Multiple IPs may be included, separated by
spaces.
```

Multiple DHCP pools can be created, so DHCP services can be provided for PCs in a small office by the same router. For some third-party SIP phones, it may be necessary to specify Option 66 (the TFTP server DNS name).

IP Phone Configuration Requirements in CUCM

CUCM has several configuration elements for IP Phones. We briefly look at the following basic required elements:

- Device Pool
- Cisco Unified CM Group
- Region
- Location
- Date/Time Group
- Phone NTP Reference
- Device Defaults
- Softkey Template
- Phone Button Template
- SIP Profile

- Phone Security Profile
- Common Phone Profile

Device Pool

Device Pools provide a set of common configurations to a group of devices; think of a Device Pool as a template to apply several different settings all at once, quickly and accurately. You can create as many Device Pools as you need, typically one per location, but they can also be applied per function. (For example, all the phones in the call center may use a different Device Pool from the rest of the phones in the administration offices, although they are all at the same location.) There are several settings within the Device Pool; some of the ones relevant to us are described next.

Cisco Unified CM Group A CM Group defines a top-down ordered list of redundant subscriber servers to which the phones can register. The list can include a maximum of three subscribers (plus an optional Survivable Remote Site Telephony [SRST] reference). The first server in the list is the primary subscriber; the second is the backup, and the third is the tertiary. In normal operation, phones send primary registration messages to the primary, backup registration messages to the backup, and nothing to the tertiary. If the primary server fails or otherwise becomes unavailable, the phone sends a primary registration message to the backup server (and registers with it) and begins sending backup registration messages to the tertiary.

The number of CM Groups created depends on the number of subscribers in the cluster; the goal is to provide server redundancy to the phones while distributing phone registrations evenly as planned in the system design. A subscriber may be listed in more than one CM Group to provide an overlapping depth of coverage, as long as its performance capacity will not be exceeded in any foreseeable failure circumstance. This is simply another requirement of a good design.

Region A region is a virtual assignment that allows the system designer to control the bit rate for calls. For example, if we define two regions, called Vancouver_HQ_REG and Ottawa_BR_REG, we can set the bit rate for calls within the Vancouver region to 256 kbps, within the Ottawa region to 64 kbps, and between the two regions to 16 kbps.

We are actually selecting the codec to be used for these calls; the codec in turn generates a known bit rate, which in turn uses a predictable amount of bandwidth and provides a predictable voice quality. In general, it is assumed that WAN bandwidth is limited; selecting a lower bit rate reduces the amount of bandwidth per call at the expense of call quality.

Location As we just saw, we can select the appropriate bit rate for calls and, therefore, the bandwidth used by each call. Given that WAN bandwidth is usually limited, we need to be able to limit the amount of bandwidth used by calls to a particular location. Location defines a maximum amount of bandwidth used by calls to a particular location; each call is tracked, and the bandwidth it uses is deducted from the total for that location. When the bandwidth remaining is not enough to support another call at a given bit rate, that call is dropped by default (but may be rerouted over the PSTN if AAR is correctly configured). This is one mechanism for Call Admission Control, which is described later in this book.

Date/Time Group As discussed earlier, it is recommended to use NTP for time synchronization of all devices. The problem is that NTP references Greenwich Mean Time, which makes the time displayed on devices "wrong" if they are not in the GMT time zone. Date/Time Groups allow us to offset the correct time learned via NTP to match the local time zone of the device. Date/Time Groups also allow us to display the time and date in the desired format, which can vary from place to place.

Phone NTP Reference SIP phones need an NTP server address from which they can obtain the time using NTP. (This is not required for SCCP phones, which are configured to the correct time using SCCP signaling.) It is preferred that the NTP reference be local to the phones that need it.

It is common to have groups of phones with similar configurations. Using a Device Pool for each group simplifies and speeds up administrative tasks, while making them less error-prone in the bargain. Figure 9-4 shows part of a Device Pool Configuration page.

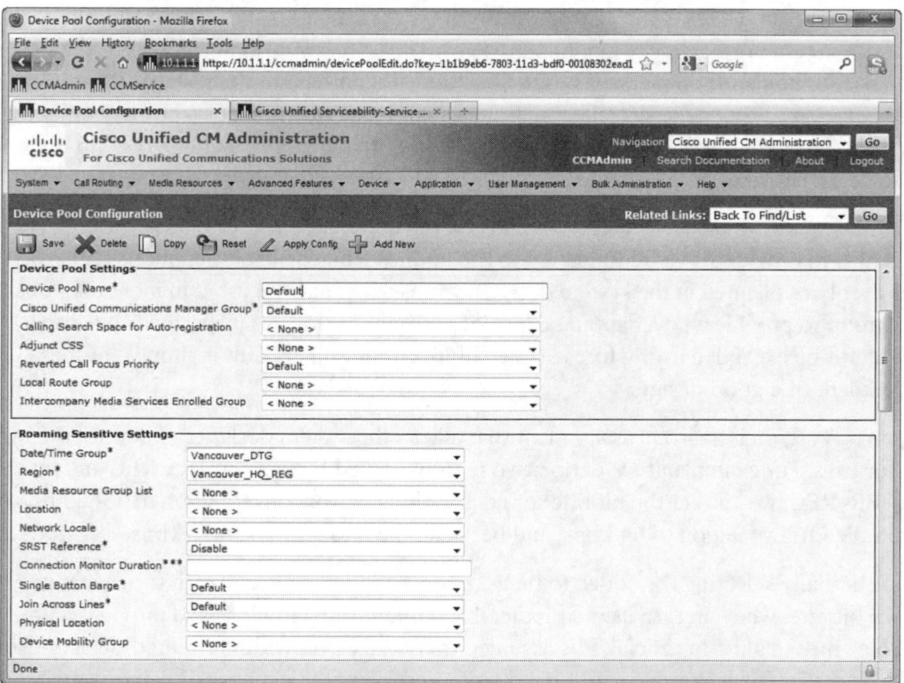

Figure 9-4 *Device Pool*

Device Defaults

The Device Defaults page lists all the supported endpoints (with separate entries for SCCP and SIP as necessary), and the firmware load, Device Pool, and Phone Button Template each endpoint uses by default. This allows an administrator to set useful system-wide defaults for any newly registered device of each type.

Softkey Template and Phone Button Template

The Softkey Template controls what softkey button functions are available to the user; these are typically used for feature access (Conference, Transfer, Park, Extension Mobility, and so on). Seven softkey templates are available by default, and you can create as many more as your design requires.

The Phone Button Template defines the behavior of the buttons to the right of the phone screen (for most models). Eighty (or more) are defined by default, because there are unique templates for each supported phone type—and for most phones, a separate template for SCCP and SIP. The default templates typically provide two lines and as many speed dials as there are remaining buttons on a particular phone model; you can add and customize the templates to assign each button one of many different functions.

Profiles

Profiles allow for a one-time configuration of repetitive tasks; several types of profiles exist, and you can create many versions of each type to be applied to phones as needed.

Phone Security Profile

A default Phone Security Profile exists for each type of phone/protocol. These default profiles have security disabled; you can choose to configure the device as secured, set encrypted TFTP configuration files, and modify Certificate Authority Proxy settings.

Common Phone Profile

The Common Phone Profile includes settings that control the behavior of the phone, including the following:

- DND settings
- Phone personalization capabilities
- VPN settings
- USB port behavior
- Video capabilities
- Power-save options

Adding Phones in CUCM

Phones can be added to CUCM in four ways:

- **Manual Configuration:** The administrator creates a new phone, configuring all settings in real time on the Phone Configuration page.
- **Autoregistration:** The administrator configures CUCM to dynamically configure and add to the database any new IP Phone that connects to the network.

- **Bulk Administration Tool (BAT):** Using templates provided by CUCM, the administrator creates .csv files that contain all the required information to create multiple phones in one operation.
- **Auto Register Phone Tool (TAPS):** An Interactive Voice Response (IVR) server enhances the Autoregister and BAT functionality, providing an automated method of adding potentially thousands of phones at a time.

The following sections provide more detail on each of these operations.

Manual Configuration of IP Phones

The basic steps for manually adding an IP Phone are as follows:

1. Navigate to Device > Phone, and then click Add New.
2. Choose the IP Phone Model from the drop-down list.
3. Choose the device protocol (either SCCP or SIP; some phones will support only one protocol, and this step will be skipped).
4. Select, or enter, the required specific information for the phone. The four required fields that do not have default values (must be manually configured) include the following:
   - **MAC Address:** The MAC address is the unique identifier that links the IP Phone hardware to the software configuration in CUCM. If you are building a phone for Bob, you must obtain the MAC address of the phone that will end up on Bob's desk, or else Bob will not see the correct settings, DN, and so forth.
   - **Device Pool:** The Device Pool (as described earlier in this chapter) applies many common settings to the phone that are relevant to its physical location and desired behavior.
   - **Phone Button Template:** The Phone Button Template (also detailed earlier in this chapter) defines what functions are assigned to the buttons on the phone (DNs, Speed Dials, Services, and so on).
   - **Device Security Profile:** Applies a set of security-related configurations, as described previously in this chapter.
5. Click Save.

When the page reloads, a new pane labeled **Association Information** appears on the left, in which you can configure the phone buttons functions. The base functionality (Line, Speed Dial, Intercom, Service, and so on) is defined by the Phone Button Template specified previously; here is where you specify what the DN number on the line(s) will be, what Service is accessed, or which Intercom DN is dialed. Figure 9-5 shows the Phone Configuration page, including the Association Information pane.

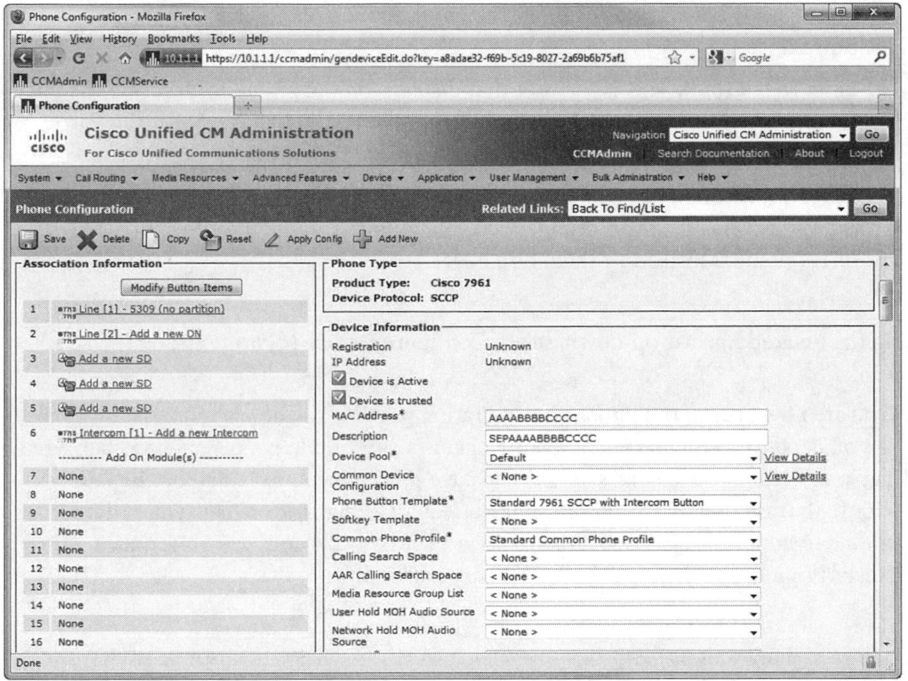

Figure 9-5 *The Phone Configuration Page*

In the Association Information pane, continue the basic phone configuration steps, as follows:

6. Click **Line [1] - Add New DN.** The Directory Number Information page opens, in which you must enter a Directory Number, and optionally set the Partition and other optional configurations. The following points highlight a few of the settings found on the Directory Number Configuration page:

   - **Route Partition:** As discussed in Chapter 10, "Understanding CUCM Dial-Plan Elements and Interactions," the Partition is part of the calling privileges system or Class of Control.
   - **Alerting Name:** This is the name to display on the caller's phone when this phone is ringing. Some PSTN connections might not support this functionality.
   - **Call Forward and Call Pickup Settings:** This is where the administrator can determine how to forward a call if the DN is busy or does not answer, or for Call Forward All. The user can set Call Forward All at the phone itself using the CallFwdAll softkey, or on their User Web Page; other Call Forward settings (such as Busy and No Answer) are available to the user only on the user's User Web Page.
   - **Display:** The text entered in the Display field serves as an internal Caller ID. When this DN calls another IP Phone, the Display text replaces the calling DN number. In other words, if Bob's DN is 5309 and the Display field is blank, when Bob calls Ethan, Ethan's phone shows that 5309 is calling. If the Display field on

Bob's phone has "Bob Loblaw" as the entry, Ethan's phone displays the caller as Bob Loblaw.

- **Line Text Label:** This is the text that displays on the phone to describe the line; for example, if the second button on the phone is the shared DN for the Parts Desk, the Line Text Label for Line 2 might read "Parts Line."
- **External Phone Number Mask:** If this phone makes an off-net call (typically to the PSTN), this field can change the Calling Line ID (CLID) to present a full PSTN number instead of the internal DN.

7. Click **Save**.
8. In the **Related Links** drop-down, select **Configure Device (<*Phone*>)**, and then click **Go**.
9. You are now back at the Phone Configuration page for the new phone. At this point, if you need to continue making config changes you can do so, or you can click **Save** again to commit the changes so far. The page prompts you to "Click on the Apply Config button to have the changes take effect." This happens because in order for the phone to adopt the changes, it has to reload with its new config. This requires either a Restart or a Reset, depending on what was changed.

Note: There is a great deal of confusion about **Restart**, **Reset**, and **Apply Config**. The differences are explained in the following points:

- A **Reset** reboots both the firmware and the configuration of the phone. Some information such as firmware version, locale changes, SRST, or Communications Manager Group changes require a full reset so that the phone will pull a new file from the TFTP server. A Reset can be triggered from the Administration web page, or from the phone itself by entering **Settings** > **#** (using the keypad).
- A **Restart** unregisters the phone, and then the phone comes right back and registers again. Because Communications Manager reads the database for this device when it registers, it is a good way to refresh information that is not passed through the configuration file. Button changes, names, and forwarding would only require a Restart. A Restart is faster than a Reset, because the firmware is not rebooted as well.
- The confusion between Restart and Reset was such that in CUCM 8.X, a new function labeled **"Apply Config"** was introduced. This button intelligently triggers either a Reset or a Restart as appropriate, depending on what changes were made to the device. In all cases, the phone has to be registered for the Reset or Restart to be sent to the phone.

It is common, especially if advanced features such as Extension Mobility or Cisco Unified Personal Communicator are in use, to associate a user with a particular Device (IP Phone). It is *required* to associate the user with the Device if you want users to be able to use the User Web Pages to customize their phone. The End User is associated with the Device (IP Phone), and the Device is associated with one or more DNs. This allows the user not only to access the User Web Pages to configure this phone, but for other applications and processes to interact with the user through the phone system.

So, what happens if you delete an End User who is associated with a device that is associated with a DN? Nothing. Although the association exists and is important and useful, the three database entities of User, Device, and DN are independent of each other. The Device and the DN do not go away if the User is deleted, and the same result applies if the Device or DN are deleted (although a phone without a DN, or a DN without a phone, can't make calls).

Autoregistration of IP Phones

CUCM includes the Autoregistration feature, which dynamically adds new phones to the database and allows them to register, including issuing each new phone a DN so it can place and receive calls. Autoregistration is supported by all Cisco IP Phones.

To enable Autoregistration, perform the following steps:

1. Verify your Autoregistration Phone Protocol. Access this setting under System > Enterprise Parameters; choose either SCCP (default) or SIP. Phones that do not support the chosen protocol will still autoregister using their native protocol.
2. Verify that at least one CM Group has Autoregistration enabled (by selecting the checkbox for **Auto-registration Cisco Unified Communications Manager Group**).
3. Enable and configure autoregistration on one or more CUCM servers within the CM Group enabled for autoregistration:
   - Enable Autoregistration by deselecting the **Auto-registration Disabled on this Cisco Unified Communications Manager** checkbox; it is disabled by default, so unchecking the box enables it.
   - Configure the range of DNs that will be dynamically and sequentially issued to auto-registering phones. The default **Starting Directory Number** is 1000; if you change the **Ending Directory Number** to anything higher than 1000, Autoregistration is automatically enabled. If you set the Starting and Ending DNs to the same value, Autoregistration is automatically disabled. (Autoregistration is disabled by default, because both the Starting and Ending Directory Numbers are set to 1000.) You will want to choose a range of DNs that fits in well with your dial plan to avoid overlap and confusion.
   - Set the Partition that will be assigned to the autoregistered DNs. This is optional, but it is one good way to limit and control autoregistered phones.
   - Verify that the **Auto-registration Disabled on this Cisco Unified Communications Manager** checkbox is unchecked, and then click **Save**.

A simple way to test Autoregistration is to plug in a new phone; if it receives a DN in the range you specified (or a DN in the range of 1000–1999 if you left it at the defaults), Autoregistration is working.

Some administrators see Autoregistration as a security weakness, because any IP Phone will be dynamically added to the database and potentially begin making calls, perhaps even to the PSTN if it is not restricted. It is common to enable Autoregistration only when it is needed to prevent the occurrence of "rogue phones."

Figure 9-6 shows the Auto-Registration Information section of the Unified CM Configuration page.

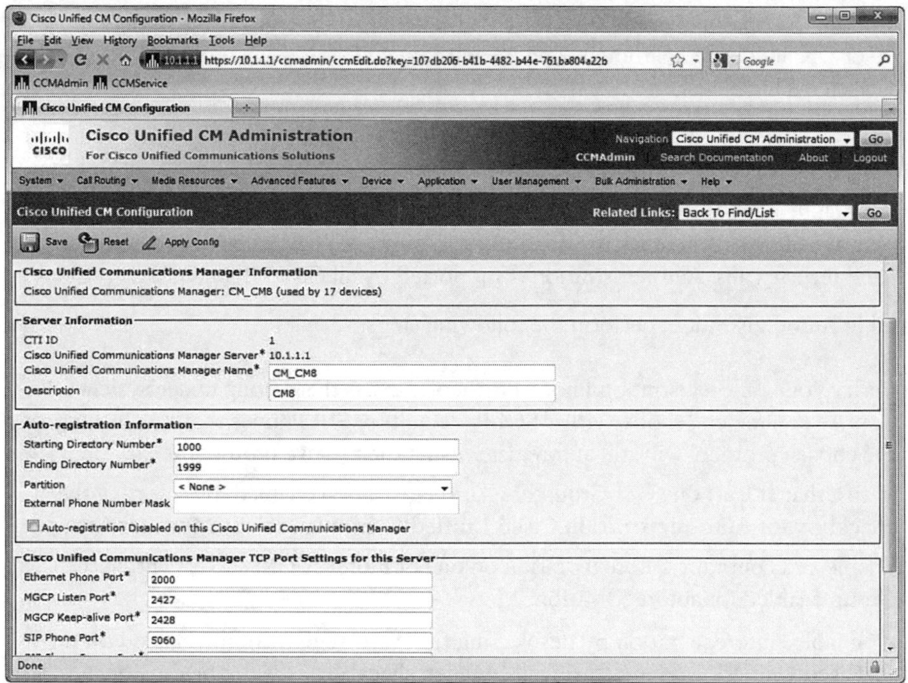

Figure 9-6 *Autoregistration Configuration*

Bulk Administration Tool

The Bulk Administration Tool (BAT) allows administrators to perform database inserts, modifications, or deletions in bulk. This makes it feasible to add a great many phones, users, or other elements more quickly and with fewer errors; it also allows the administrator to schedule the operation to happen automatically and unattended.

The BAT Export feature allows the administrator to pull selected records from the database and export them. The administrator can then modify the records and re-import them into the database, making bulk changes faster and more accurate.

BAT can be used to add, modify, or delete almost any component in CUCM, including Phones, Users, Forced Authorization Codes and Client Matter Codes, User Device Profiles, the Region Matrix, Gateway devices, and many others.

The components of BAT include an Excel template that provides the required fields and formatting for the new unique data, server-side templates that configure the common data, and a set of web page interfaces for preparing and executing the many operations that BAT supports.

The Excel template is downloaded from the CUCM server. The administrator then customizes the templates for the needs of this BAT operation, populates the required fields with the correct data, and uploads the resulting .csv file to the server.

Using the BAT interface appropriate for the operation (Insert Phones, Insert Users, create Call Routing components, and so on), the administrator may need to create a server-side BAT Template for adding new devices, or in some cases simply select the uploaded .csv file for processing. If templates are required (as they would be if adding phones, for example), the template specifies all the settings that all of the phones have in common, whereas the .csv file specifies all of the unique settings for each phone, such as DN, Line Text Label, and so forth.

The only trick to adding phones with the BAT tool is that the MAC address of each phone must be specified. Using a barcode scanner to scan the MAC barcode label on the phone into the .csv file makes things faster and more accurate, but there is another challenge waiting for you: You create a detailed config for the phone, including DNs and other user-specific settings, and you specify the MAC address of the new phone. Now you must make sure that the physical phone with that MAC gets to the user it was built for; this is no easy task if several hundred phones are being deployed at once.

A couple of alternate strategies are available to make BAT deployments easier. One is to use Autoregistration to get all the phones working, and then use the BAT tool to modify the phones' configurations after the fact. This approach still has some weaknesses, notably that you must still be positive of the MAC address of the physical phone that sits on the desk and match it to the database entry that BAT changes.

Auto Register Phone Tool

The most sophisticated strategy involves the use of the Auto Register Phone Tool (formerly known as the Tool for Auto Registered Phone Support, but which is still known as TAPS because it's a better acronym than ARPT). TAPS goes one step further in the automation of new IP Phone deployments, as summarized in the following steps:

1. An IP-IVR server is built and configured to support TAPS, and the CUCM server is integrated with the IP-IVR server. The IP-IVR functionality is supported by several Cisco applications, including Unified Contact Center Express.
2. The administrator prepares a BAT job, specifying a Device Template for all the common phone settings and a detailed .csv file with all the unique phone settings. The administrator runs the BAT job, substituting fake "dummy" MAC addresses for the as-yet-unknown real ones (a simple checkbox in the BAT interface does this substitution automatically).
3. The new phones are autoregistered and receive a DN. They can now place calls.
4. Using Bob's phone as an example: Bob (or perhaps an administrator if Bob feels uncomfortable doing so) picks up his new, autoregistered phone that currently has DN 1024 (from the default autoregistration range) and dials the specially-configured IP-IVR pilot number.
5. The IP-IVR may prompt Bob to authenticate (this is an optional but more secure approach). When Bob has authenticated successfully, the IP-IVR prompts Bob to enter

the extension his phone should have; in a new deployment, this may be provided to Bob on an information sheet, or it may simply be the same extension (let's assume 5309 in this case) that he had on the old phone system that is being migrated to CUCM.

6. When Bob enters the extension, the IP-IVR records his input of "5309" and captures the MAC address of the phone Bob is using. The IP-IVR sends all this information to CUCM.
7. CUCM looks up the extension of 5309 in the database, and finds it in the record for one of the newly-added BAT job phones—the one that will become Bob's phone. CUCM replaces the dummy MAC address in the BAT record with the real MAC captured and forwarded by the IP-IVR. The database record is now complete and accurate, including the real MAC address of the phone that sits on Bob's desk.
8. CUCM restarts Bob's phone, and when it comes back online, it is fully configured with all of the specific details from the BAT record for Bob's phone.

This is an exceptionally powerful way to deploy thousands of IP Phones. With some minor tweaks and some training of the users, it requires minimal administrator involvement in the phone deployment. The downside is that it requires the IP-IVR hardware and software and a capable administrator to configure it, and still involves either training users to set up their own phones or using administrators to perform repetitive simple tasks, which are not cost-effective uses of their time.

Describe End Users in CUCM

It is technically true that a phone system doesn't need End Users. A person parks himself in front of a phone and starts using it; it doesn't really matter who the person is as long as the phone does what that person needs it to do. But a Unified Communications system provides far more than just phone functionality; it has a massive array of features that can be provided to and customized by individual users. Converged networks are increasingly complex, and End Users expect an increasing simplicity of use. The configuration of End Users is an integral part of a full-featured system.

End Users Versus Application Users

CUCM makes a clear distinction between End Users and Application Users. The distinction is simple: End Users are typically people who type in a username and password into a login screen (usually a Web page) to access features or controls. An Application User is typically an application that sends authentication information inline with a request to read or write information to a system (perhaps a third-party billing application accessing the CDR/CAR database, for example). Table 9-2 lists some of the characteristics and limitations of End Users versus Application Users.

Table 9-2 *End Users Versus Application Users*

End Users	Application Users
Associated with an actual person	Associated with an application
For individual use in interactive logins	For noninteractive logins
Used to assign user features and administrative rights	Used for application authorization
Included in the user directory	Not included in the user directory
Can be provisioned and authenticated using LDAP	Must be provisioned locally (no LDAP)

Credential Policy

The Credential Policy defines preset passwords, end-user PINs, and application-user passwords. The default Credential Policy applies the application password specified at install to all Application Users.

Administrators can define additional policies that can specify the allowed number of failed login attempts, minimum password length, minimum time between password changes, number of previous passwords stored, and the lifetime of the password. The policy can also check for weak passwords. A strong password:

- Contains three of the four characteristics: uppercase, lowercase, numbers, and symbols
- Cannot use the same number or character more than three times consecutively
- Cannot include the alias, username, or extension
- Cannot include consecutive numbers or characters

Similar rules exist for phone PINs:

- Cannot use any number more than two times consecutively
- Cannot include the user mailbox or extension, nor the reverse of them
- Must contain at least three different numbers (for example, 121212 is invalid)
- Cannot be the dial-by-name version of the user name (such as Mike = 6453)
- Cannot contain repeated digit patterns, nor any patterns that are dialed in a straight line on the phone keypad (for example, 2580 or 357)

Features Interacting with User Accounts

The following features use the End-User-Account login process, with either the username/password or PIN as the authentication:

- Unified CM Administration web pages
- User web pages

- Serviceability
- OS Administration
- Disaster Recovery System
- Cisco Extension Mobility
- Cisco Unified Communications Manager Assistant
- Directories
  - IP Phone Services
  - Data Associated with User Accounts

User account information is divided into three categories, with fields for specific data in each category:

1. Personal and Organizational Settings:
   - UserID
   - First, Middle, Last Name
   - Manager UserID
   - Department
   - Phone Number, Mail ID
2. Password Information:
   - Password
3. CUCM Configuration Settings:
   - PIN
   - SIP Digest Credentials
   - User Groups and Roles
   - Associated PCs, controlled devices, and DNs
   - Application and feature parameters (Extension Mobility, Presence Group, CAPF)

Application User accounts use a subset of the previous attributes.

User Locale

User locales allow different languages to be displayed on the IP Phone and the User Web Pages. Additional locales are installed on the CUCM server; then specific locale files are downloaded to the phone via TFTP. This allows for the customization of the primary interfaces for users in a wide range of available locales/languages.

Device Association

For users to be able to control their own devices (setting up their own speed dials, services, and ring preferences, for example), the End User account must be associated with the device. IN CUCM, End Users can be associated with IP Phones, Cisco IP Communicator (CIPC), and Cisco Extension Mobility profiles.

Because the End User account must have a unique User Attribute Name in the CUCM database, it is possible to dial a user by name. Cisco Unified Presence Server (CUPS) tracks the availability status of a user and his communication capabilities (such as voice, video, and chat).

Implementing End Users in CUCM

End Users can be added to the CUCM database via three main methods:

- Manual, one-at-a-time entry
- Bulk import using the Bulk Administration Tool
- LDAP synchronization (and optional authentication)

This section reviews each of these methods.

Manual Entry

The CUCM database includes fields for comprehensive user information. Only some of these fields are required, including the following:

- User ID
- Last Name
- Presence Group (defaults to Shared Presence Group)
- Remote Destination Limit (defaults to 4)

Given that the last two required fields are populated by default, it is clear that CUCM does not require much information to create a new user. The User ID must be unique, which implies that you should have a naming convention that accommodates many users with similar names.

There are many optional fields on the End User Configuration page, including Password, PIN, First Name, Telephone Number, and Device Association. The more users you have, the more likely it is that these optional fields will be populated to implement features, improve searching and reporting, or improve security. Figure 9-7 shows part of the End User Configuration page.

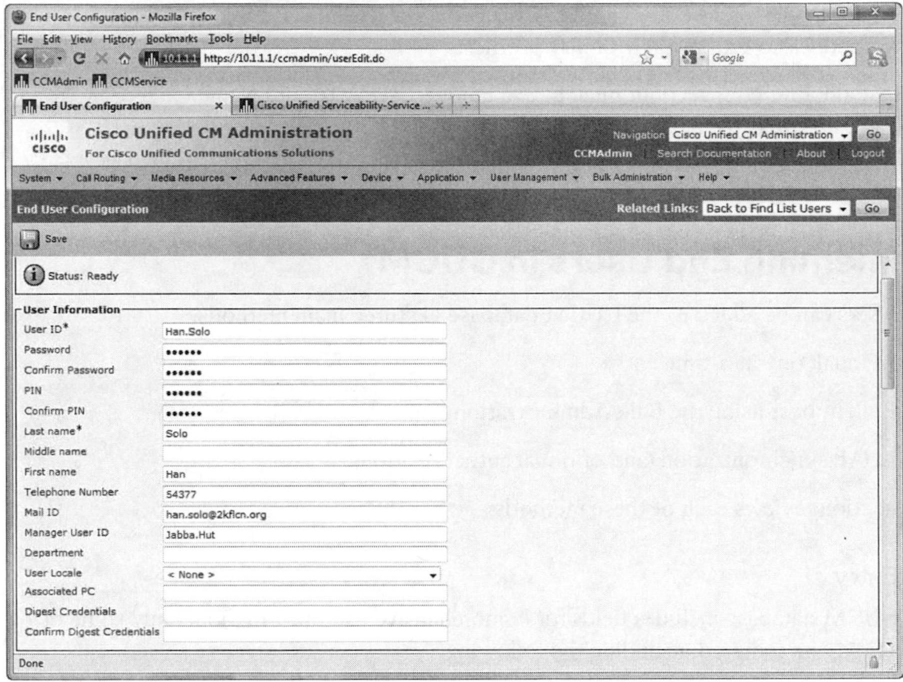

Figure 9-7 *End User Configuration Page*

Bulk Import Using BAT

Instead of adding potentially hundreds or thousands of users one at a time, the administrator can add users in bulk using the Bulk Administration Tool (BAT). BAT allows the administrator to create and upload a CSV file with all the users' information populated and insert the data into the database in an automated way. BAT is a fast way to add, remove, or modify database entries for almost every part of the CUCM database.

LDAP Integration

CUCM supports integration with Lightweight Directory Access Protocol (LDAP). LDAP is a standards-based (with some significant vendor-specific exceptions) system that allows an organization to create a single, centralized directory information store. LDAP holds information about user accounts, passwords, and user privileges. The information centralized in LDAP is available to other applications, so that separate directories do not need to be maintained for each application. Using LDAP simplifies user administration, and makes using systems slightly easier for users because they only need to maintain their information and passwords in one place.

> **Note:** Only End Users are replicated by LDAP Sync. Application users are always and only maintained in the CUCM database.

CUCM supports LDAP integration with several widely used LDAP systems, including the following:

- Microsoft Active Directory (2000, 2003, 2008)
- Microsoft Active Directory Application Mode 2003
- Microsoft Lightweight Directory Services 2008
- iPlanet Directory Server 5.1
- Sun ONE Directory Server (5.2, 6.x)
- Open LDAP (2.3.39, 2.4)

CUCM can interact with LDAP in two ways: **LDAP Synchronization** populates the CUCM database with user attributes from LDAP, and **LDAP Authentication** redirects password authentication to the LDAP system. Typically, Synchronization and Authentication are enabled together. In either case, some information that now comes from LDAP is no longer configurable in CUCM—the fields actually become read-only in CUCM, because the information can only be edited in LDAP. The following sections review LDAP Synchronization and Authentication in more detail.

LDAP Synchronization

Implementing LDAP Synchronization (LDAP Sync) means that some user data (but not all) is maintained in LDAP and replicated to the CUCM database. When LDAP Sync is enabled, user accounts must be created and maintained in LDAP and cannot be created or deleted in CUCM; the user attributes that LDAP holds become read-only in CUCM. However, some user attributes are not held in LDAP and are still configured in CUCM because those attributes exist only in the CUCM database.

Perhaps most important to understand is that with LDAP Sync, the user passwords are still managed in the CUCM database. This means that, although a user account in LDAP is replicated to the CUCM database, the user password must be maintained in both the LDAP system and in CUCM; this is likely to confuse and annoy the user.

CUCM uses the DirSync service to perform LDAP Sync. The synchronization can be configured to run just once, on demand, or on a regular schedule. The choice depends on the system environment and the frequency of changes to LDAP content; the need for up-to-date information must be balanced against the load on the servers and network if the sync is frequent or takes place during busy times.

Note: If LDAP Authentication is enabled and LDAP fails or is inaccessible, the only End-User account that will be able to log on to the CUCM system is the Application Administrator account defined during install. This may cause drastic unified communications service interruption, depending on how users normally interact with the system. Of course, if LDAP has failed, it is likely to be a serious issue already, causing many applications to cease functioning.

LDAP Authentication

LDAP Authentication redirects password authentication requests from CUCM to the LDAP system. End-User account passwords are maintained in the LDAP system and are not configured, stored, or replicated to CUCM. Because one of the benefits (particularly to the End User) of LDAP is a centralized password system (making single sign-on possible), it is typical and desirable to implement LDAP Authentication with LDAP Sync.

LDAP Integration Considerations

A common misconception regarding CUCM LDAP Integration is that all user data resides in LDAP. This is absolutely false. With LDAP Sync, certain LDAP user attributes are held in the LDAP directory and are replicated to the CUCM database as read-only attributes. The balance of the user attributes in the CUCM database (fields such as Associated Devices, PINs, Extension Mobility Profile, and so on) are still held and managed only in the CUCM database.

There is a similar misconception with LDAP Authentication: Remember that the LDAP password is *not* replicated to the CUCM database; rather, the entire authentication process is redirected to the LDAP system.

The interaction of CUCM with LDAP varies with the type of LDAP implementation. The primary concern is how much data is replicated with each synchronization event. For example, Microsoft Active Directory 2000/2003/2008 performs a full sync of all records every time; this can mean a very large amount of data is being synchronized, potentially causing network congestion and server performance issues. For this reason, sync intervals and scheduling should be carefully considered to minimize the performance impact.

Synchronization with all other supported LDAP systems is incremental (for example, only the new or changed information is replicated), which typically greatly reduces the amount of data being replicated, thereby reducing the impact on the network and servers.

LDAP Attribute Mapping The user attribute field names that LDAP uses are most likely different form the equivalent attribute field names in the CUCM database. Therefore, the various LDAP attributes must be mapped to the appropriate CUCM database attribute. Creating an LDAP Sync agreement involves identifying the one LDAP user attribute that will map to the CUCM **User ID** attribute. In a Microsoft Active Directory integration, for example, the LDAP attribute that will become the CUCM User ID can be any one of the following:

- sAMAccountName
- uid
- mail
- TelephoneNumber

It doesn't matter which one is chosen, but for consistency and ease of use, the attribute that the users are already using to log in to other applications should be used.

After the initial User ID mapping is selected, some other LDAP attributes should be manually mapped to CUCM database fields. Table 9-3 lists the fields in the CUCM database that map to the possible equivalent attribute in each type of supported LDAP database.

Table 9-3 *LDAP User Attribute Mapping*

CUCM	Microsoft AD	Other Supported LDAP
User ID	sAMAccountName mail employeeNumber telephoneNumber UserPrincipalName	uId mail employeeNumber telephonePhone
First Name	givenName	Givenname
Middle Name	middleName initials	Initials
Last Name	**sn**	**sn**
Manager ID	manager	manager
Department	department	department
Phone Number	telephoneNumber ipPhone	telephonenumber
Mail ID	mail sAMAccountName	mail uId

LDAP Sync Requirements and Behavior Keep these points in mind when planning and implementing an LDAP Sync:

- The data in the LDAP attribute that is mapped to the CUCM User ID field must be unique in the LDAP (and therefore CUCM) database. Some LDAP fields allow duplicate entries, but the CUCM User ID must be unique, so it is necessary to verify that the LDAP data is unique before the Sync agreement is built.
- The **sn** attribute (surname/last name) in LDAP must be populated with data or the record will not be replicated to CUCM.
- If the LDAP attribute that maps to the CUCM **User ID** attribute contains the same data as an existing Application User in CUCM, that entry is skipped and not imported into the CUCM database.

LDAP Sync Agreements

An LDAP Sync agreement defines what part of the LDAP directory will be searched for user accounts. Many LDAP systems have a highly organized structure, with different containers for different functions, departments, locations, or privileges. The synchronization agreement specifies at which point in the tree the search for user accounts will begin. CUCM has access to the container specified in the agreement, and all levels below that in the tree; it cannot search higher up the tree than the start point, nor can it search across to other branches in the tree that must be accessed by going higher than the starting point then back down.

The agreement can specify the root of the domain, but although this is a simple agreement to create, it causes the entire LDAP structure to be searched, which may return unwanted accounts or simply too many accounts.

LDAP Sync Mechanism

The LDAP Sync agreement specifies when to begin synchronizing and when to repeat the synchronization (a schedule). It is possible to have a synchronization run only once, although this is somewhat unusual.

The first time the synchronization happens, the following events take place:

1. All existing end-user accounts in the CUCM database are deactivated (not deleted).
2. Accounts whose CUCM **User ID** exactly matches a user in LDAP are reactivated, and any settings from LDAP are updated or applied in the CUCM database.
3. Accounts that exist only in LDAP are created in the CUCM database.
4. Any accounts that remain deactivated (meaning they do not exist in LDAP) are deleted from the CUCM database after 24 hours.

LDAP Custom Filters

The default behavior of LDAP Sync is to import all user accounts from the start point in the tree on down. This may cause accounts to be imported that are not wanted. Using a Custom Filter allows an administrator to limit which accounts are imported; for example, a filter could specify that only user accounts in a particular Organizational Unit (OU) are imported. If the filter is changed, a full LDAP sync must be performed for the change to take effect.

Configure LDAP Sync

Setting up LDAP Sync is surprisingly simple. The main difficulty is typically gaining a full understanding of the target LDAP structure, knowing what containers hold the users to be imported, and knowing where to start the LDAP search.

The basics steps to set up LDAP Sync are as follows:

1. Activate the Cisco DirSync service.
2. Configure the LDAP system.
3. Configure the LDAP directory.
4. Configure LDAP Custom Filters.

For CUCM to be able to access and search LDAP, an account must be created in LDAP for CUCM. Configurations may vary between LDAP systems, but the account must essentially have read permissions on everything in the search base.

Activate DirSync

Using the Unified Serviceability application, navigate to **Tools > Service Activation.** From the Server drop-down list, choose the Publisher. Find the Cisco DirSync service, check the box next to it, and click **Save.** Figure 9-8 shows the DirSync service activated.

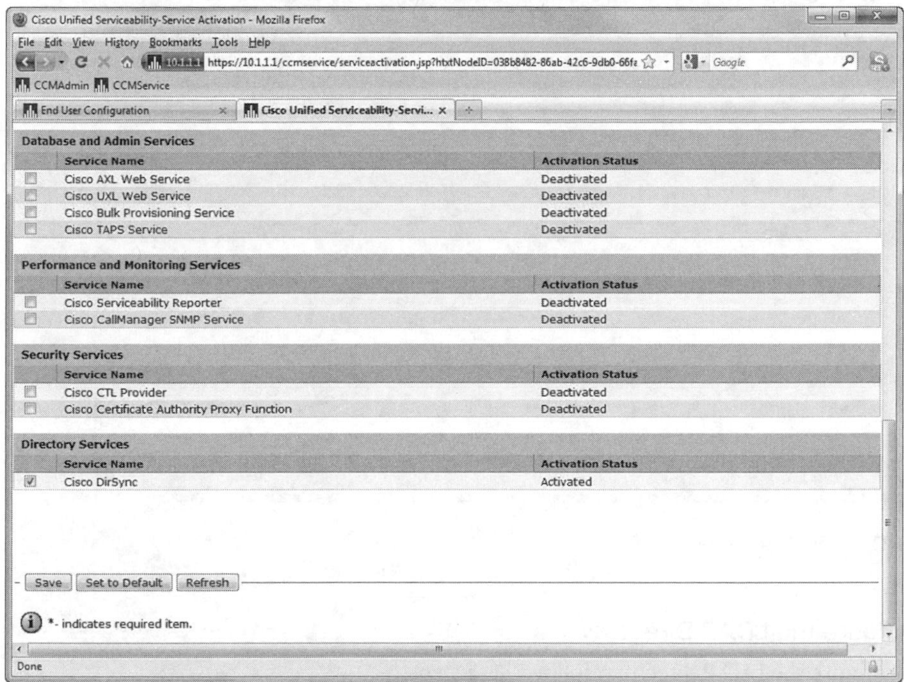

Figure 9-8 *Activating Cisco DirSync*

Configure the LDAP System

Follow these steps to enable LDAP Sync in CUCM:

1. Using the Unified CM Administration application, navigate to **System** > **LDAP** > **LDAP System.**
2. Check the **Enable Synchronizing from LDAP Server** box.
3. From the **LDAP Server Type** drop-down, choose the type of LDAP system with which CUCM will synchronize.
4. From the **LDAP Attribute for User ID** drop-down, select which LDAP attribute will map to the CUCM User ID attribute.
5. Click **Save.**

Figure 9-9 shows the LDAP System Configuration page.

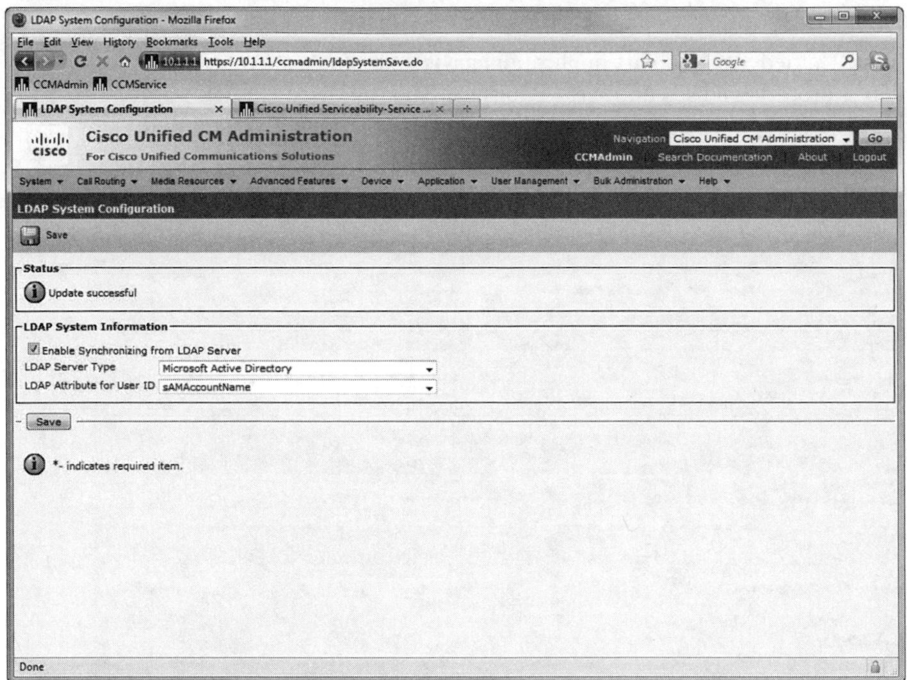

Figure 9-9 *LDAP System Configuration*

Configure the LDAP Directory

To configure the LDAP directory, follow these steps:

1. Using the Unified CM Administration application, navigate to **System > LDAP > LDAP Directory.**
2. Specify a name for this LDAP Sync agreement in the **LDAP Configuration Name** field.
3. Add the account name and password that CUCM will use to access LDAP.
4. Define the User Search Base. This will be the full LDAP path syntax (for example, ou=Users,dc=Pod1,dc=com).
5. Set the synchronization schedule.
6. Specify the LDAP user fields to be synchronized (mapping CUCM fields to LDAP fields).
7. Specify at least one (up to three for redundancy) LDAP server IP addresses. Specify SSL to secure the LDAP Sync process (requires similar configuration on the LDAP system).

Figure 9-10 shows the LDAP Directory configuration page.

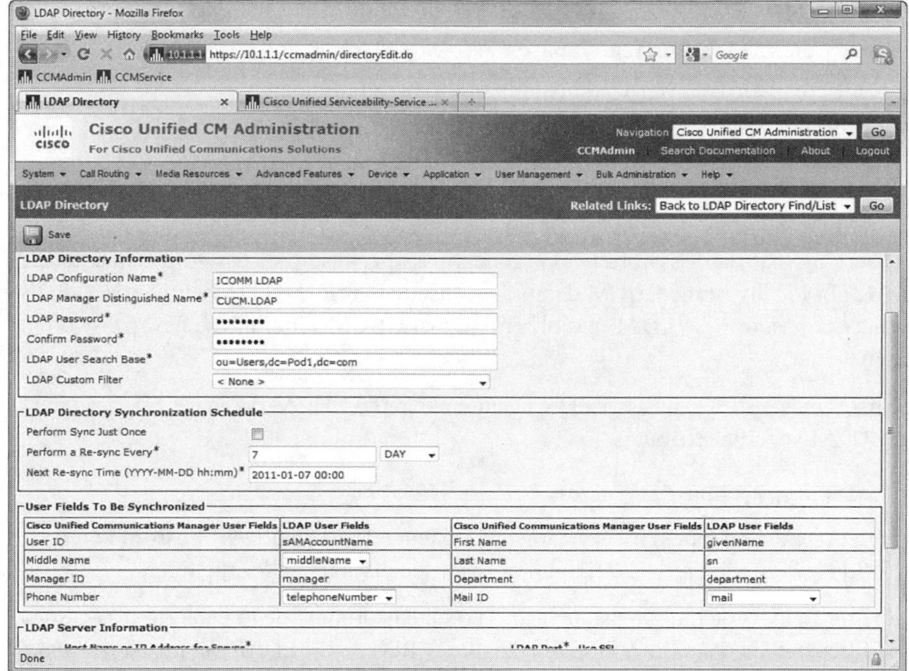

Figure 9-10 *LDAP Directory Configuration Page*

Verify LDAP Sync

The simplest way to verify that LDAP Sync is working is to do a quick search of the End Users on the CUCM. In the column under **LDAP Sync Status**, the users' status will be listed as Active or Inactive; there may also be new users added from the LDAP directory, depending on whether those users were previously configured in CUCM. Active users are being synced from LDAP; inactive users did not exist in LDAP and will be deleted in 24 hours unless they are created in LDAP and re-synced.

When you open the configuration page for an LDAP-synced user, you see that the **User ID**, **Last Name**, **Middle Name**, **Telephone Number**, **Mail ID**, **Manager User ID**, and **Department** fields are not editable; this is because they are synced with LDAP and can only be edited in the LDAP system.

Configuring LDAP Authentication

Configuring CUCM to redirect authentication to the LDAP system is normally done as part of an LDAP integration. It is not typical to sync all the users, but still make them maintain a separate password in CUCM.

To set up LDAP authentication, follow these steps:

1. Navigate to **System > LDAP > LDAP Authentication.**
2. Check the box next to **Use LDAP Authentication for End Users.**

3. Specify the account and password CUCM will use to access the LDAP system.
4. Specify the **LDAP User Search Base**.
5. Specify the LDAP server IP address (up to three for redundancy).
6. Click **Save**.

Verify LDAP Authentication

Verifying LDAP Authentication can be achieved by opening a user configuration page and observing that the Password field is gone; this is because the password is maintained in LDAP, not locally in the CUCM database. A user can test the LDAP authentication by changing her password in LDAP and observing that CUCM requires the new password to log in.

Note that the user PIN is always locally maintained in the CUCM database, as are all the other CUCM-specific attributes.

Create LDAP Custom Filters

Create LDAP Custom Filters by navigating to **System > LDAP > LDAP Custom Filter**. Click **Add New**. In the Filter Configuration page, specify a name for the filter.

In the **Filter** field, type the filter statement. The statement must be in parentheses ( ). Some sample filter statements follow; for more detail, see RFC 4515, *LDAP: String Representation of Search Filters*:

- (cn=Milton Macpherson)
- (!(cn=Milton Macpherson))
- (&(objectClass=Person)(|(sn=Macpherson)(cn=Milton M*)))
- (sn=M*)

Exam Preparation Tasks

Review All the Key Topics

Review the most important topics in the chapter, noted with the key topics icon in the outer margin of the page. Table 9-4 describes these key topics and identifies the page number on which each is found.

Table 9-4 *Key Topics for Chapter 9*

Key Topic Element	Description	Page Number
Topic	IP Phone registration process	240
Topic	IP Phone configuration requirements in CUCM	244
Topic	End Users versus Application Users	254
Topic	LDAP Integration	258

This chapter includes the following topics:

- **Call Flows in CUCM:** This section describes how call signaling and audio traffic flow in a CUCM system.
- **CAC and AAR:** This section explains CAC and the use of AAR.
- **Call-Routing Components:** This section discusses the sources, destinations, and interaction of the various call-routing components in CUCM.
- **Class of Control:** This section discusses the capabilities and configuration of Class of Control elements in CUCM.

CHAPTER 10

Understanding CUCM Dial-Plan Elements and Interactions

Chapter 9, "Managing Endpoints and End Users in CUCM," discussed the administration of IP Telephony endpoints and users in Cisco Unified Communications Manager (CUCM). This chapter reviews the components, behaviors, and interactions of the CUCM dial-plan.

"Do I Know This Already?" Quiz

The "Do I Know This Already?" quiz allows you to assess whether you should read this entire chapter or simply jump to the "Exam Preparation Tasks" section for review. If you are in doubt, read the entire chapter. Table 10-1 outlines the major headings in this chapter and the corresponding "Do I Know This Already?" quiz questions. You can find the answers in Appendix A, "Answers Appendix."

Table 10-1 *Understanding CUCM Dial-Plan Components*

Foundation Topics Section	Questions Covered in This Section
CUCM Call Flows	1–3, 6–9
CAC and AAR	4–5
Class of Control	10–12

1. Which two of the following are reasons to eliminate IP phone reliance on DNS?

a. Elimination of additional licensing costs for Cisco Unified DNS Server
b. Elimination of single point of failure
c. Reduce delay caused by name resolution lookups
d. Eliminate delays caused by ARP resolution

2. Which of the following is true regarding call flows in CUCM? (Choose two.)

a. Signaling traffic using RTP is sent directly to the CUCM from the IP phone.
b. Signaling traffic using SCCP/SIP is sent directly to the CUCM from the IP phone.
c. Voice bearer stream traffic flows through the CUCM server to maintain QoS policy.
d. Voice bearer stream traffic flows direct from phone to phone.

3. Which of the following is true of SRST?

- **a.** SRST allows IP phones in a branch to be controlled by the local router in the event that WAN failure causes a loss of connectivity to the CUCM.
- **b.** SRST performs dynamic gateway services, allowing keepalives and signaling to be sent over the backup PSTN link.
- **c.** The NM-SRST module is supported on the 2900 and 3900 series ISR platforms only.
- **d.** SRST is a legacy feature; the replacement feature set is called Service Advertisement Framework.

4. Call Admission Control serves what purpose?

- **a.** Limiting user access to toll calls (for example, preventing unauthorized long-distance calling).
- **b.** Throttling the number of concurrent call attempts to prevent Code Yellow events.
- **c.** Rerouting calls over the PSTN in the event of WAN failure.
- **d.** Tracing malicious calls received from the PSTN to verify their origin.
- **e.** Preventing oversubscription of IP WAN voice bandwidth by dropping calls that exceed the configured voice queue size.

5. Which of the following is the only event that will trigger AAR?

- **a.** WAN failure
- **b.** CFUR rejection
- **c.** SRST registration
- **d.** Local Route Group failure
- **e.** CAC call rejection
- **f.** International Talk Like a Pirate Day

6. Which of the following is the correct order of configuration of call-routing components?

- **a.** Device, Route Group, Route List, Route Pattern
- **b.** Route Pattern, Route List, Route Group, Device
- **c.** Route Pattern, Route Group, Route List, Device
- **d.** Route Group, Device, Route Pattern, Route List

7. Which of the following Route Patterns is the closest match for the dialed number 98675308?

- **a.** 9.1[2-9]XXXXXX
- **b.** 9.[2-9]XXXXXX
- **c.** 9.@
- **d.** 9.8[67][67]XXXX
- **e.** 9.8XXXXXX
- **f.** 9.86[^012345689]5308

8. Which two of the following signaling methods provide digit-by-digit analysis? (Choose two.)

- **a.** SIP en-bloc
- **b.** MGCP en-bloc
- **c.** SIP using station dial rules
- **d.** SIP using CUCM dial rules
- **e.** SCCP
- **f.** SIP using KPML

9. Which of the following are line group distribution algorithms? Choose all that apply.

- **a.** First-in, First-out
- **b.** Broadcast
- **c.** Directed Broadcast
- **d.** Top-down
- **e.** Longest Idle
- **f.** Circular

10. Which of the following is true?

- **a.** No Partitions or search spaces exist by default in a CUCM installation.
- **b.** One Partition exists by default; it is named "default."
- **c.** Only the default CSS has access to the default Partition.
- **d.** All CSSs have access to the default Partition.

11. Which statement is correct regarding Calling Search Spaces?

a. If a CSS is applied to the phone, it overrides the CSS on the line.

b. If a CSS is applied to the line, it invalidates the CSS on the phone.

c. If a CSS is applied to the line, it overrides the CSS applied to the phone.

d. If a CSS is applied to the phone, a CSS cannot be applied to the line.

12. A Partition is linked with a schedule that is in effect every day from 8:00 a.m. to 5:00 p.m. The Partition contains a translation pattern that causes the dialed number of 5555309 to ring extension 2112. The Partition is listed last in the CSS applied to a gateway. Another translation pattern is created that causes the dialed number 5555309 to ring the Auto-Attendant pilot. The translation pattern is not assigned to a Partition. What happens when a PSTN phone calls 5555309 at 7:00 p.m. on Saturday?

a. Extension 2112 rings.

b. The caller gets an error message.

c. The Auto-Attendant answers.

d. None of the options are correct.

Foundation Topics

CUCM Call Flows

Chapter 6, "Understanding the CME Dial-Plan," discussed how Cisco Unified Communications Manager Express (CUCME) selects call-routing targets. The dial-plan in Cisco Unified Communications Manager (CUCM) is more complex because it is a distributed system that uses remote components, such as gateways to route calls. CUCM is intended to scale to large-enterprise environments, and consequently has a greater capacity for call-routing complexity and redundancy. This chapter introduces and discusses call signaling and voice traffic flow in different scenarios, the components of the call-routing system, the call-routing decision process, component configuration, redundancy, and restriction.

Call Flow in CUCM if DNS Is Used

Generally speaking, Domain Name System (DNS) is not recommended for use with Cisco IP phones. If DNS is used, the IP phones must complete a DNS name resolution lookup to learn the IP address of the CUCM server before any signaling can occur. At best, doing so introduces delay; at worst, it allows the possibility of a misconfiguration or failure of the DNS system that could cause the phones to stop working.

When the DNS lookup has completed successfully, the call flow consists of signaling (using either SCCP or SIP) between the phone and the CUCM, and the voice bearer streams (using Real-Time Transport Protocol [RTP]) directly between the phones. Note that the phones do not signal each other directly, nor does any voice traffic usually flow through the CUCM. Figure 10-1 illustrates call flow when DNS is used by the phones.

Note: The exception to the last statement is if the CUCM is hosting a voice conference; in that case, the voice streams from all conference participants flow into the CUCM and the combined streams (minus the listener's own stream) flow out of the CUCM back to the participants.

Call Flow in CUCM if DNS Is Not Used

Eliminating IP phone reliance on DNS is recommended to eliminate unnecessary delay and potential points of failure. If DNS is not used, the call flow is similar, except that the initial DNS lookup is eliminated, there remains only the signaling flow between the phones and the CUCM, and the voice bearer streams directly between the phones. Figure 10-2 illustrates call flow without DNS in use by the phones.

The elimination of DNS reliance is simple to configure. The default installation of CUCM lists the host name in the database field used by the phone configuration file to identify the CUCM server(s) the phone should use for registration. To change the value in this field, follow these steps:

1. In CUCM Administration, navigate to **System** > **Server**.

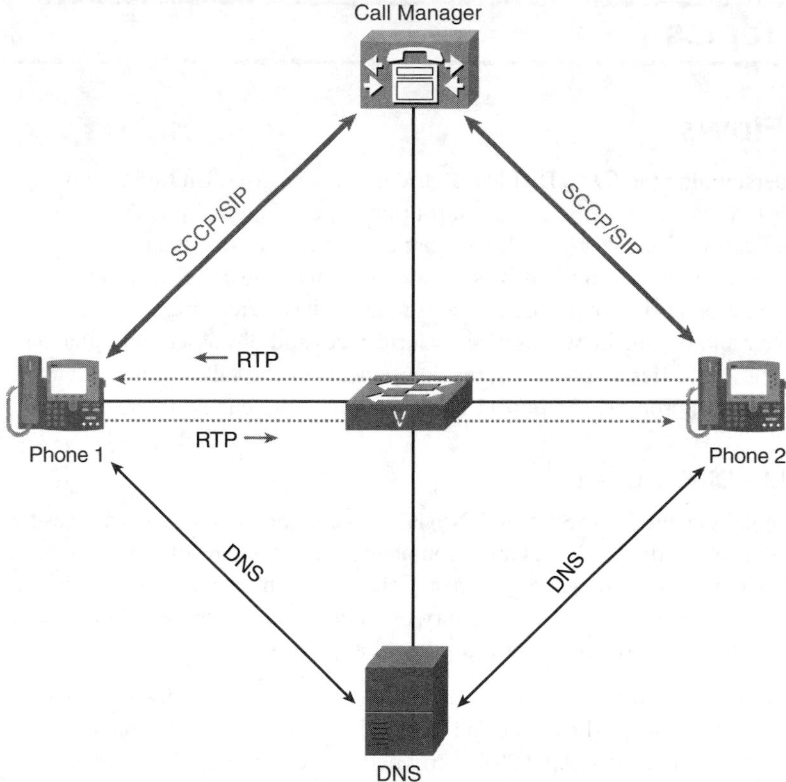

Figure 10-1 *Call Flow with DNS*

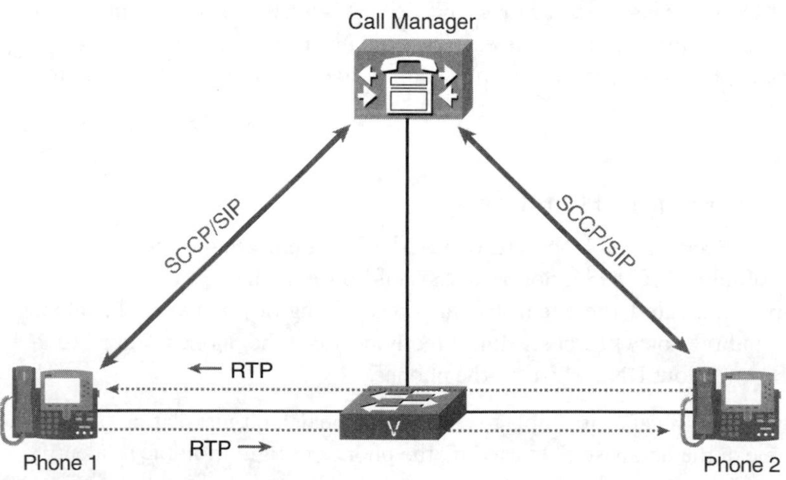

Figure 10-2 *CUCM Call Flow Without DNS*

2. Select a server and, in the Server Configuration page, change the value of the Hostname/IP Address field from the hostname to the host's IP address, as shown in Figure 10-3.

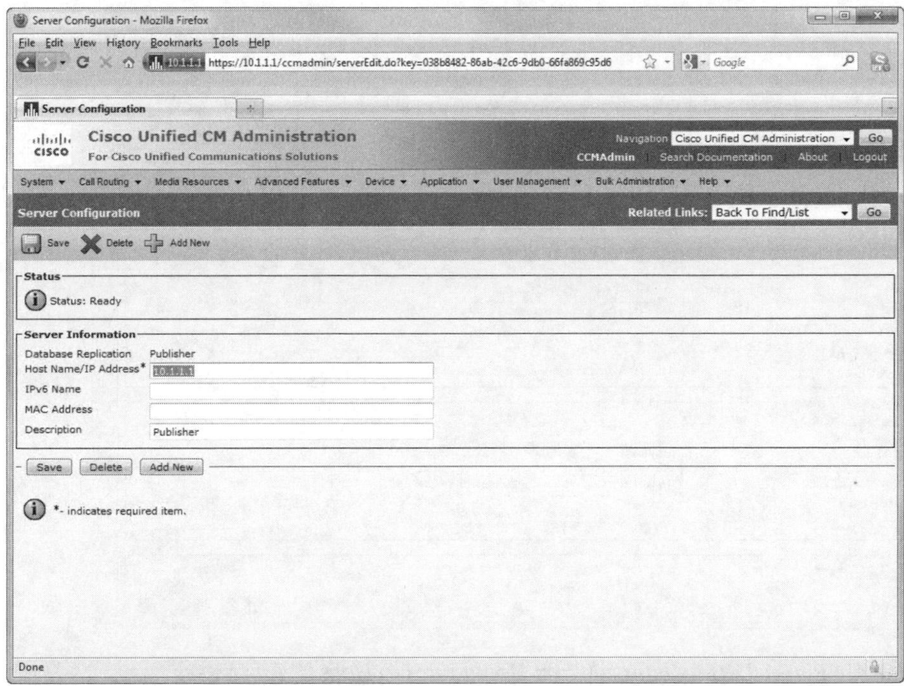

Figure 10-3 *Hostname Changed to IP Address in Server Configuration Page*

3. Click Save.
4. Repeat the previous steps for each server.

The next configuration task is similar, but carried out under System > Enterprise Parameters. Scroll to the section titled **Phone URL Parameters** (shown in Figure 10-4), and in each field, change the URL to use the host IP address instead of the hostname.

> **Note:** The elimination of DNS reliance is recommended for IP phones, but there are some circumstances where it is useful, including when changing IP address schemes (when the DNS name stays stable, but the IP address it represents changes). The integration of Cisco Unified Presence Server (CUPS) with CUCM requires DNS service, but this might not affect the hardware IP phones. If DNS is used, it must be correctly managed and maintained so that it does not cause failures.

Centralized Remote Branch Call Flow

In a centralized deployment, the CUCM servers are located at the main location, with one or more locations at remote branches, connected by an IP WAN. The branch-office IP phones use the IP WAN for both signaling and on-net voice traffic. Off-net calls from branch IP phones might be routed out the branch gateway or the main location gateway, depending on design.

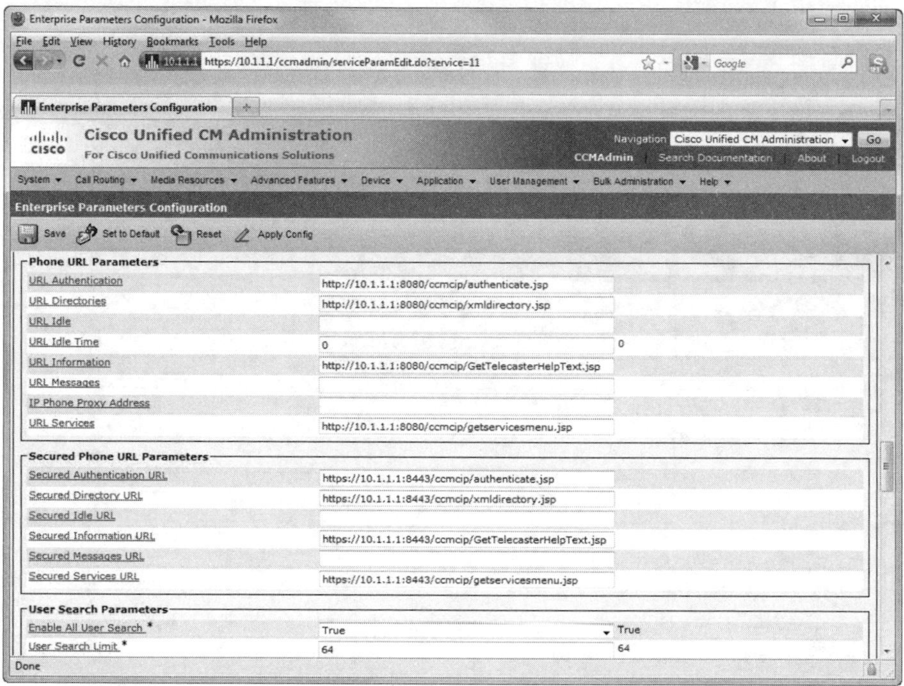

Figure 10-4 *Phone URLs Changed from Hostnames to Host IP Addresses*

The signaling and voice bearer stream paths are the same as in the single-site deployment; the difference is simply that the branch phones send all signaling traffic over the WAN to the CUCM server, along with on-net RTP voice streams to IP phones at other locations. However, if a branch phone calls another branch phone, both phones signal to the CUCM across the WAN, but the RTP voice streams stay local to the branch—direct from phone to phone, as always. Figure 10-5 illustrates Centralized Remote Branch call flow.

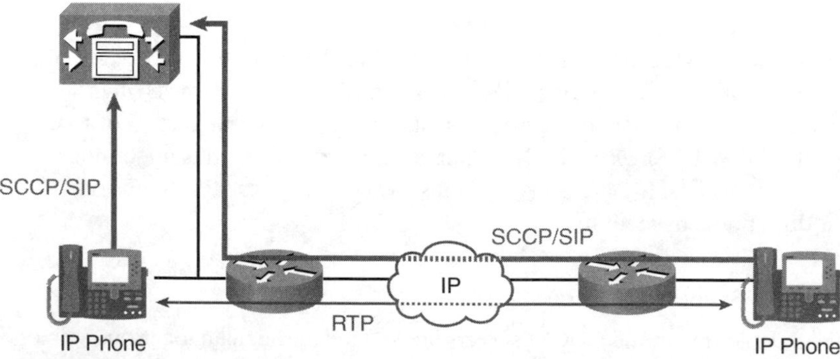

Figure 10-5 *Centralized Remote Branch Call Flow*

Centralized Deployment PSTN Backup Call Flow

Key Topic

In the event the IP WAN fails, all phones in the branch lose connectivity to the CUCM at the main location and no longer function. In this event, Survivable Remote Site Telephony (SRST) is recommended to provide registration and call control signaling local to the branch phones. SRST provides on-net IP-to-IP calling between branch phones, but if calls from branch phones to other locations are placed (whether on-net or off-net), they will fail. This is because of a lack of call-routing configuration on the branch gateway router. Typically, the branch gateway is configured with a subset of CUCME capabilities to provide public switched telephone network (PSTN) access to branch phones. If the SRST router dial-plan is configured properly, the branch users can still dial the same on-net extensions for calls to other locations, and the SRST router transparently modifies the dialed digits for PSTN routing. Most users will not be aware of the WAN failure.

From the main location CUCM perspective, the branch phones have de-registered and are considered unreachable. This is not entirely accurate (because they are reachable via the PSTN) and is an undesirable failure of the system. To provide continuity of service, call-routing options and settings are configured on the CUCM that provide an alternate path to the branch via the PSTN:

1. The call-routing table has a second option added to provide a PSTN gateway and appropriate digit manipulation to provide the required PSTN dialed digits.
2. The Call Forward UnRegistered (CFUR) option is configured on each branch phone to specify the full PSTN number needed to reach the branch phone.

These two configurations, in combination with the appropriate SRST configuration, ensure that branch phones can both reach and be reached by the main location in the event of WAN failure.

When the WAN recovers, the branch phones de-register from the SRST router, re-register with the CUCM, and the normal WAN-based signaling and calling patterns resume.

Centralized Deployment Considerations and Limitations

Deploying CUCM in a centralized call-processing environment is one of the most popular choices, providing full features and capabilities to branch offices while centralizing equipment and management efforts at the headquarters. The following points should be considered during the design and planning of a Centralized deployment:

- CUCM supports a maximum of 1000 locations and a maximum of 1100 H.323 or Media Gateway Control Protocol (MGCP) devices per cluster.

  Although there is no limit to the number of phones at a branch, the number of phones supported by the SRST router will be limited based on the router hardware model and the configuration applied to it. The maximum number of phones supported in SRST mode is 1500 on a 3945E branch router. The correct configuration to allocate resources to support that many phones must be applied, and the appropriate amount of RAM must be installed.

- The WAN must be properly provisioned and QoS enabled to allocate the priority queue bandwidth appropriate to the number of concurrent calls, and to limit delay for both RTP voice and signaling traffic.

- Call Admission Control (CAC), as explained in the next section, should be implemented, either CUCM location-based or Resource Reservation Protocol (RSVP)-based. Regardless of the CAC method used, the goal is to prevent more calls than the design specifies from being extended across the WAN, thereby protecting call quality for all IP calls.

PSTN Backup Using CAC

CAC prevents IP calls from being extended across a WAN link, if the additional WAN bandwidth required would exceed the Quality of Service (QoS)-allocated bandwidth for concurrent calls. For example, if the design specifies a maximum of ten calls using G.729, the QoS configuration would create a Priority Queue (LLQ) sized to serve those ten calls. (The QoS configurations are beyond the scope of the CCNA Voice exam.) If an 11th call were extended to the gateway, the additional bandwidth would overrun the input buffer for the LLQ, and packets from all 11 calls would start to drop, causing unacceptable packet loss and resultant deterioration of voice quality for all 11 calls.

With location-based CAC implemented, CUCM tracks how many calls are extended to a given location, subtracting the bandwidth used for each concurrent call from the bandwidth specified for that location. When the remaining bandwidth is less than the amount used by a single call (which varies based on the codec used for the call, which is in turn defined by the Region setting), the default behavior of CAC is to drop the call. Users get re-order tone or an annunciator message indicating that the call has failed.

In most cases, dropping the call is not desirable, and Automated Alternate Routing (AAR) is implemented to reroute the call across the PSTN. AAR is exclusively triggered by CAC; when CAC prevents the call from extending across the WAN, CUCM checks to see if an AAR Group is configured for the calling phone. If an AAR Group is configured, it specifies what digit manipulation is required to retry the call with a full PSTN Dialed Number Information Service (DNIS). The operation is transparent to the user (although it is possible that he might notice a difference in call quality using the PSTN instead of the WAN).

If the call is extended over the PSTN, the call-signaling path includes the PSTN gateways at the main and branch locations, and the RTP voice streams are converted to the appropriate PSTN transport.

Note: Be clear about the difference between PSTN rerouting due to WAN failure versus CAC and AAR being triggered by a lack of available WAN bandwidth for calls. If the WAN fails, calls may be rerouted to the PSTN using the hierarchical design of the dial plan and CFUR on the CUCM side, and SRST on the branch side. If the WAN has no available bandwidth, CAC triggers AAR, which will redial the call with a full PSTN number, but only if it is correctly configured to do so. Figure 10-6 summarizes the sequence of events if the WAN fails.

Distributed Deployment Call Flow

In a distributed deployment, one CUCM cluster signals another CUCM cluster across a WAN. The signaling flows from the calling phone to the local CUCM, then from the local CUCM across the WAN to the remote CUCM, then from the remote CUCM to the remote (called) phone. The RTP voice streams extend from each phone across the WAN to the other phone. Figure 10-7 illustrates call flow in a distributed deployment.

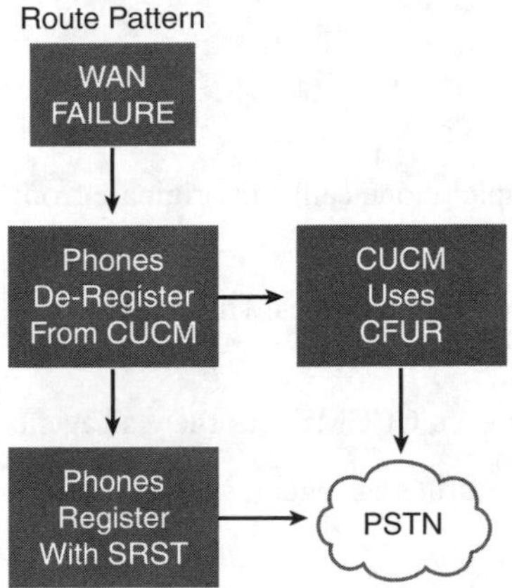

Figure 10-6 *WAN Failure to SRST/CFUR*

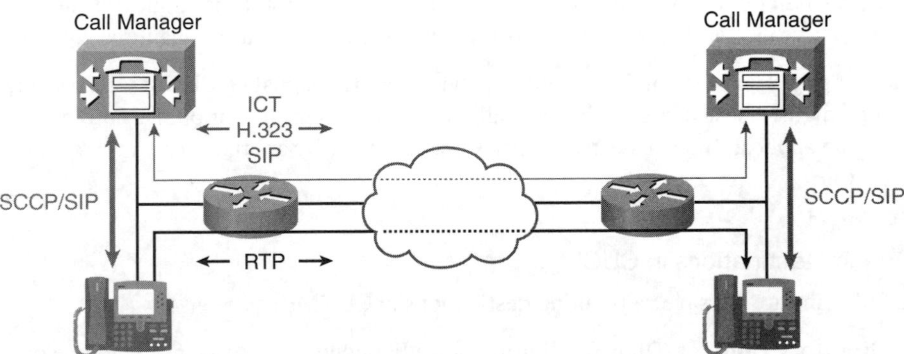

Figure 10-7 *Distributed Deployment Call Flow CUCM Call-Routing Components*

The signaling protocols available include Inter-Cluster Trunk (ICT), H.323, and Session Initiation Protocol (SIP). The CUCM clusters at each site control their local phones using Skinny Client Control Protocol (SCCP) or SIP. The use of gatekeepers (along with a well-designed dial-plan) allow scalability to thousands of sites, with either a full CUCM cluster for large sites or a CUCME for smaller sites (third-party solutions may also be integrated). PSTN backup for WAN failure is achieved using a hierarchical dial-plan, and gatekeeper CAC may be used when WAN bandwidth for calls is not available.

Note: Location-based CAC does not work in a distributed deployment because the clusters do not communicate their current bandwidth utilization to other clusters. Instead, Gatekeeper CAC provides a centralized service to track available bandwidth between clusters and trigger AAR when necessary. Gatekeeper is a router IOS feature set that may be configured on one or more gateway routers in the system.

The building blocks of the CUCM dial-plan are simple and consistent. The potential for complexity exists when multiple hierarchical paths are implemented, with different features and capabilities enabled. This section identifies the core elements and explains their interactions.

Call-Routing Sources in CUCM

A call-routing request (including but not limited to a simple phone call) can originate from any of the following:

- **IP phone:** Places a routing request using one of its configured lines. This may be a manually dialed number, a speed dial, feature button, or softkey.
- **Trunk:** Signals inbound calls from another CUCM cluster, CUCME, or other call agent.
- **Gateway:** Signals inbound calls from the PSTN or another call agent, such as a private branch exchange (PBX).
- **Translation pattern:** Matches the original called digits and immediately transforms them to a new dialed string. The new string is resubmitted to digit analysis for routing.
- **Voicemail port:** Can be the source of a call-routing request if the application attempts a call, transfer, or message notification on behalf of a user's mailbox.

A call-routing request is simply one of the previous entities signaling CUCM with a string of dialed digits. These digits can be manually dialed by a user on their phone, automatically by an application such as Cisco Unity Connection, or by another system via a trunk or gateway.

Call-Routing Destinations in CUCM

The following are possible call-routing destinations in CUCM:

- **Directory Number (DN):** Each button on an IP phone can be assigned a unique on-net extension.
- **Translation Pattern:** Matches a string of dialed digits and transforms them to a new dialed string, which is in turn resubmitted to digit analysis and routed to a different target.
- **Route Pattern:** Matches a set of dialed digits and triggers a call-routing process that can include one or more potential paths, providing a hierarchical set of call-routing options.
- **Hunt Pilot:** A specific pattern of digits that, when matched, triggers a customizable call-coverage system.
- **Call Park Number:** A pattern or range of patterns that CUCM can use to temporarily hold a call until a user dials the call park number to pick it up from any IP phone.
- **Meet-Me Number:** The conference call initiator dials into the Meet-Me number to begin the conference, and one or more other users dial into the same number to join the conference.

All the previous destinations are represented by strings of dialable digits. The dial-plan must allocate ranges of numbers for all these targets, and CUCM must be configured to route to them appropriately (including not routing to them, if necessary).

Call-Routing Configuration Elements

The primary components of the CUCM call-routing system are the following:

- Route Patterns
- Route Lists
- Route Groups
- Gateways/Trunks

Route Pattern

A route Pattern matches a string of dialed digits. The pattern may be specific, matching a single dialable number, or it may be general, matching hundreds of thousands of possible numbers. This variable precision is configured using wildcard digits in the pattern. Route Patterns allow the administrator to specify the target of any given string of dialed digits.

Route Patterns are necessary to provide PSTN dial access. They may also be used to integrate the CUCM dial-plan with an existing PBX dial-plan; in this instance, the Route Pattern would match all the DNs (extensions) controlled by the PBX. In fact, a Route Pattern may be customized to allow users to dial any number and reach any desired end station.

Route Patterns are associated with either a Route List or a Specific Gateway.

> **Note:** If a Route Pattern is directly associated with a gateway (as opposed to a Route List), the selected Gateway can no longer be referenced by a Route Group; the gateway is "locked in" to the Route Pattern. In small deployments, this may not be problematic, but in large deployments, doing so limits the flexibility of the system.

Route List

A Route List is an ordered list of Route Groups. The first entry in the list is the preferred call-routing path; if that path is unavailable (due to failure or no circuit/channel available), if a second choice is configured, it will be used instead. There may be several choices in the list; each new call uses the choices in top-down order.

The hierarchical order of the Route List entries allows the administrator to provide depth of coverage for calls while controlling which resources are used for each call. For example, if the Route Pattern that matches a national long-distance number is associated with a Route List that lists its first choice as a Route Group providing access to an inexpensive Inter-Exchange Carrier (IXC) PRI circuit, the call is routed to that IXC circuit. Subsequent calls matching the same Route Pattern are also routed to the IXC circuit until no channels are available.

At that point, the routing request to the IXC is rejected and the Route List's second choice is tried. If the second choice is a Local Exchange Carrier (LEC) PSTN PRI circuit, the call is dialed over the LEC PSTN link, as long as a channel is available. This call costs more than using the IXC circuit, but it works—the call does not fail, and business continuity is maintained.

In this example, cost is the design driver: The IXC is less expensive than the LEC circuit, so the IXC is the desired target for all long-distance calls. However, if the IXC is busy (or has failed), it is acceptable that long-distance calls be placed out the LEC circuit so that service is not interrupted.

Route Group

A Route Group is a list of devices (gateways or trunks) that are configured to support circuits to the PSTN or to remote CUCM clusters in distributed designs. Route Groups are commonly configured to contain devices with common signaling characteristics (for example, a set of PSTN PRI gateways in one group and a set of WAN IP trunks to a remote cluster in another).

The distribution algorithm of a Route Group is configurable; selecting Top-Down causes the devices in the group to be used in top-down order for each new call, while selecting Circular uses the devices in round-robin order. The specific context and requirements of the system determines which algorithm is appropriate.

Note: The Local Route Group feature allows the administrator to define a Route Group in the Device Pool, and reference that local Route Group in the Route List. Doing so effectively decouples the location of a PSTN gateway from the Route Patterns that target the gateway. This feature greatly reduces the complexity of dial-plan design in systems with many locations.

Gateways and Trunks

Gateways and trunks are the devices that physically terminate and support circuits to the PSTN, to digital or analog PBXs, and to IP WAN circuits to remote clusters or IP-TSP circuits to service providers.

CUCM supports various gateway devices and interfaces, controlling them with either peer-to-peer gateway protocols (H.323 and SIP) or gateway control protocols (MGCP and SCCP).

Figure 10-8 shows the call-routing elements previously described.

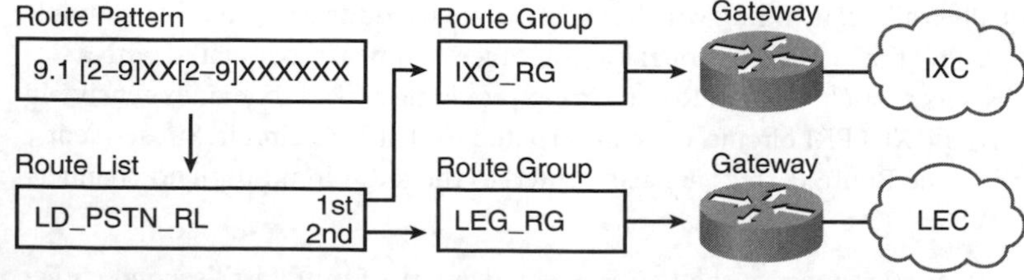

Figure 10-8 *Call-Routing Elements*

Key Point: The configuration order of the call routing elements is: Devices, Route Groups, Route Lists, Route Patterns. The call flow is the reverse: The dialed digits match a Route Pattern, which points to a Route List, which points to a Route Group, which references Devices.

Call-Routing Behavior

Digit analysis is the process by which CUCM matches dialed digits to possible targets for call routing. Different protocols and devices perform digit analysis in different ways: Dialed digits are collected digit-by-digit on SCCP phones and SIP phones that use Keypad Markup Language (KPML), and en-bloc (all at once as a set of digits) on basic SIP phones, trunks, and most gateways.

Digit Analysis

The digit analysis logic that CUCM uses to select the target for the call-routing request is based on the closest match. Consider the following set of Route Patterns:

- 1111
- 1211
- 1[23]XX
- 131
- 13[0-4]X
- 13!

With these Route Patterns in mind (these are not intended to be realistic; they just illustrate the pattern-matching logic), let's look at three examples of how CUCM will process different dialed strings:

Example 1 If User A dials 1111, there is an exact match with the pattern "1111." No other patterns match, and the call is extended to that target.

Example 2 If User B dials 1211, CUCM has two possible matches:

- **1211:** Matches 1 digit string.
- **1[23]XX:** Matches 200 digit strings.

CUCM selects the closest match target (the one that matches the fewest possible strings) (in this case, 1211).

Example 3 If User C dials 131, CUCM has four possible matches:

- **1[23]XX:** Matches 200 digit strings.
- **131:** Matches 1 digit string.
- **13[0-4]X:** Matches 50 digit strings.
- **13!:** Matches almost 10,000 *quintillion* digit strings. (For our purposes, that is practically an unlimited number.)

The pattern 131 matches exactly, but other patterns match, too; these other patterns are longer, so CUCM has to wait to see if User C dials another digit. If she does, 131 no longer matches and is discarded as a possible target. The wait time is set by the T.302 Inter-Digit Timeout value, which defaults to 15 seconds. If User C dials a fourth digit, the T.302 timer starts again, because there is still a longer pattern that might match ("13!"). If User C stops dialing after 4 digits, after 15 seconds (the T.302 timer wait), the call is extended to the target of the "13[0-4]X" pattern. If she dials another digit within the T.302 timer count, the "13[0-4]X" pattern is discarded because it no longer matches. When User C finishes dialing, the T.302 timer must exhaust before the call is routed.

Key Concept: Digit-by-digit analysis means that CUCM collects digits one at a time as they are dialed. As digits are collected, patterns that no longer match the string are discarded as possible routing targets.

The Closest Match logic will choose a pattern target according to the following criteria:

- The pattern matches the dialed string.
- Among all the potential matches, it matches the fewest strings other than the actual dialed string.

Hunt Groups

A *Hunt Group* is a set of IP phones (technically, the Directory Numbers [DN] on the phones) that are able to be reached by calling a common number. The classic example is the Helpdesk; users dial 7777, and all the DNs of the helpdesk staff ring in sequence until one of them picks up the call.

The components of a Hunt Group are

- **Line Groups:** Contain the DNs that will be rung sequentially. The Line Group settings allow the selection of the call-distribution algorithm: Top-Down, Circular, Longest Idle, or Broadcast. The settings also control when, or if, to proceed to the next available Line Group in the Hunt List.
- **Hunt Lists:** Contain a top-down ordered list of Line Groups. Each new call is routed to the first Line Group in the list; if that group cannot provide call coverage, the next Line Group in the Hunt List is tried until the list is exhausted.
- **Hunt Pilot:** This is a call-routing entry (much like a Route Pattern) that matches a dialed string and targets a Hunt List (which in turn targets a Line Group). Hunt Pilot numbers may be on-net, E.164, or any format as required.

Class of Control

Class of Control is defined as the ability to apply calling restrictions to devices. Typical examples might include the following:

- Preventing certain individuals from placing long-distance calls
- Routing the same called number to different targets at different times of day
- Routing the same called number to different targets at different locations

Class of Control is configured using Partitions and Calling Search Spaces (discussed in the following sections).

Partition

A Partition is a grouping of things with similar reachability characteristics. In general, you can think of a Partition as being assigned to things you can dial, such as the following:

- DNs
- Route Patterns
- Translation Patterns
- Voicemail Ports
- Meet-Me Conference Numbers

By default, one Partition exists; it is called the Null Partition, although it is listed in the CUCM web pages as <none>. Up to 75 additional Partitions can be created at once.

Calling Search Space

A Calling Search Space is a top-down ordered list of Partitions. A Calling Search Space can be applied to a device (such as an IP phone or gateway), or to a line on the IP phone. One CSS exists by default, and by default, it contains only the Null Partition. You can think of a CSS as being assigned to things that can place calls.

Interaction of Partitions and Calling Search Spaces

The essential thing to understand is this: If the target that is being dialed does not exist in one of the Partitions in the CSS of the caller, the call will fail. This behavior allows us to design specific calling-privilege schemes and apply them to different calling devices or lines. Consider the following example: A company wants to implement call restriction such that the lobby phones cannot place long-distance calls. A new Partition is created called "PSTN_Local_PTN." The Route Patterns that match 7-digit and 10-digit (toll free) PSTN calls are assigned to the PSTN_Local_PTN Partition.

A new CSS is created called Lobby_CSS. The PSTN_Local_PTN Partition is added to the Lobby_CSS.

The Lobby CSS is applied to the lobby phones configuration. As soon as that change is made, the lobby phones will get the reorder tone when they try to dial any number that does not match the 7- or 10-digit patterns in the Lobby_CSS. Thus, no one can abuse the lobby phones and cost the company toll fees by making long-distance calls.

A few related points:

- If a user tries to call 911 from the lobby phone, that call will fail too, unless the emergency call patterns are also added to the PSTN_Local_PTN Partition or to another Partition that is in turn added to the Lobby_CSS. It is generally a good idea to ensure that every phone can place emergency calls.

- When a Route Pattern is moved from the default Partition <none>, it is no longer accessible to the default CSS. This means that calls matching the Route Pattern will fail until a new, functional CSS/Partition structure is completed. For this reason, it is best to plan and implement Class of Service (CoS) configurations before phones are in use to avoid service interruption.
- Every CSS includes the default Partition <none>, at the end of the list of custom Partitions. This means that any target that is left in the default Partition <none> is reachable by every calling device.

Line-Device Configuration

So far, we have assumed that the CSS is applied only to the device (that is, the IP phone). It is also possible to apply a CSS to the line on the phone; the line CSS may be very different and include other Partitions, which in turn contain different Route Patterns, and so on.

If both a Device and a Line CSS are applied, the Partitions in both CSS are concatenated in sequential top-down order, with the line CSS Partitions listed first, and the device CSS Partitions second.

CUCM will analyze the list of Partitions in that order, looking for a match to the dialed digits among the patterns in the list. If a match is found, the call is routed to the target. If more than one identical match is found, the Partition containing the match that is first in the concatenated list will be the target. In other words, the Line CSS overrides the Device CSS.

There are several benefits to using the Line/Device method. In general, best practices suggest setting up the Device CSS to allow full calling privilege to all patterns, with the call routing appropriate to the device's location (for example, PSTN calls placed out the local gateway). The calling restrictions are then applied using the Line CSS, which can contain Route Patterns that match long-distance numbers, but are configured to block those calls. The result is that fewer total CSS need to be configured, which makes it simpler to manage and scale the dial-plan.

Exam Preparation Tasks

Review All the Key Topics

Review the most important topics in the chapter, noted with the key topics icon in the outer margin of the page. Table 10-2 lists and describes these key topics and identifies the page number on which each is found.

Table 10-2 *Key Topics for Chapter 10*

Key Topic Element	Description	Page Number
Section	Call-routing behavior and features in the event of WAN failure	277
Section	Call-routing behavior in the event of a lack of WAN bandwidth for voice	278
Section	Describes the configuration and use of Route Patterns, Route Lists, Route Groups, and Devices	281
Figure 10-8	Illustrates call-routing components	282
Section	Explains the logic CUCM uses in dialed digit analysis for call routing	284
Class of Control	Explains the structure and behavior of Partitions and Calling Search Spaces	284
Line/device Class of Control	Explains interaction of CSS applied to both Phone and Line	286

This chapter includes the following topics:

- **Describe Extension Mobility in CUCM:** This section describes the Extension Mobility feature, its advantages, disadvantages, and integration into the CUCM cluster.
- **Enable Extension Mobility in CUCM:** This section describes how to enable the Extension Mobility feature.
- **Describe Telephony Features in CUCM:** This section describes CUCM telephony features including Call Coverage, Intercom, and Presence.
- **Enable Telephony Features in CUCM:** This section describes how to enable the telephony features described above.

CHAPTER 11

Enabling Telephony Features with CUCM

Getting Cisco Unified Communications Manager (CUCM) to the point where it will ring phones is only part of the fun. Users of a contemporary business phone system expect it to have a comprehensive feature set. This chapter explores just a few of the features available in CUCM 8.x and the basics of implementing them.

"Do I Know This Already?" Quiz

The "Do I Know This Already?" quiz allows you to assess whether you should read this entire chapter or simply jump to the "Exam Preparation Tasks" section for review. If you are in doubt, read the entire chapter. Table 11-1 outlines the major headings in this chapter and the corresponding "Do I Know This Already?" quiz questions. You can find the answers in Appendix A, "Answers Appendix."

Table 11-1 *"Do I Know This Already?" Foundation Topics Section-to-Question Mapping*

Foundation Topics Section	Questions
Extension Mobility in CUCM	1–2
Enable Telephony Features in CUCM	3–10

1. Which of the following defines Cisco Extension Mobility?
   a. A user can log in to any IP Phone in the cluster; that phone is dynamically configured with the user's DN, speed dials, and other custom configurations.
   b. A user can move his phone anywhere within the cluster, and its calls will be routed out the local gateway.
   c. A user can define several remote destinations so that a call to his IP Phone rings his mobile and home destinations simultaneously.
   d. Users can log into any phone in the cluster and receive a DN from a predefined range.
   e. Users can choose one of several wireless IP Phone models for mobile IP communication throughout the WLAN coverage area.

2. Which of the following is not an administrative option when a user attempts to log in to multiple devices using Cisco Extension Mobility?
   a. Allow Multiple Logins
   b. Prompt User
   c. Deny Login
   d. Auto Logout
3. How can call-forwarding options be configured in CUCM? (Choose all that apply.)
   a. Administratively, using the CM Administration pages
   b. Automatically, using Device Defaults
   c. By the user from the IP Phone
   d. By the user from the User Web Pages
   e. By the user using Cisco Unified Call Forward Central
4. A user hears another phone ringing, presses a softkey, enters a number, and the call is extended to his phone. What feature did the user just invoke?
   a. Call Pickup
   b. Group Call Pickup
   c. Other Group Pickup
   d. Call Intercept
5. Which of the following is the correct order of call flow through a Call Hunting system?
   a. Hunt Pilot > Hunt Group > Hunt List > DN
   b. Hunt Pilot > Hunt List > Hunt Group > DN
   c. Hunt Group > Hunt List > Line Group > DN
   d. Hunt Pilot > Hunt List > Line Group > DN
6. Which of the following are distribution algorithm choices for Hunt Lists? (Choose all that apply.)
   a. Top-Down
   b. Round-Robin
   c. Simultaneous
   d. Broadcast
   e. Longest Idle
   f. Circular
   g. Multicast

7. True or False: You can only configure an Intercom button that speed dials the target; you cannot create an Intercom button that allows you to dial the target manually.
   a. True
   b. False

8. John is on the phone with Guy. Lesley uses the Whisper Intercom feature to speak to John. What happens?
   a. John and Guy hear Lesley.
   b. John hears Lesley and Guy does not.
   c. John hears Lesley, Guy does not, and Lesley hears John.
   d. John hears Lesley, Guy does not, and Lesley does not hear John.

9. CUCM includes a native Presence capability. What three IP Phone states is CUCM Native Presence able to monitor?
   a. On Hook
   b. Logged Out
   c. Off Hook
   d. Unregistered

10. Which of the following describes the interaction of Presence Groups and Subscribe Calling Search Space?
    a. BLF Speed Dials depend on both in order to function.
    b. The Subscribe CSS overrides the Presence Group subscription permission.
    c. The Presence Group subscription permission overrides the Subscribe CSS.
    d. Both the Subscribe CSS and the Presence Group subscription permission must both allow subscription in order to allow presence indications to work properly.

Foundation Topics

Describe Extension Mobility in CUCM

Cisco Extension Mobility (EM) allows a user to log in to any phone in the CUCM cluster (and as of version 8.x, cross-cluster). In environments where workers move from desk to desk, this allows their personal configurations, such as directory numbers (DN) and speed dials to be dynamically set up on the IP Phones they are currently using, making them reachable at the same extension number regardless of which phone they are using.

EM operates as an IP Phone service, applying user-specific Device Profiles to the phone the user logs in to. Once the phone(s) are subscribed to the Extension Mobility Service, the user selects that service on the phone and enters his User ID and PIN when prompted (using the phone keypad in alphanumeric mode, similar to texting on a cell phone). CUCM then applies the user Device Profile settings and resets the phone. Separate Device Profiles must be created for each phone model a user might log in to. For example, suppose the user's primary phone is a 7965 with six buttons set up for two DNs and four speed dials. When the user logs in to a 7942 with only two buttons, the Device Profile defines what to configure on the two buttons: one DN and a speed dial, two DNs, or whatever the user needs. The user must select the correct profile from a list if multiple profiles exist.

If a user logs in to multiple phones concurrently, administrators have three options to determine how the system behaves:

- **Allow multiple logins:** The user can be logged in to multiple phones at the same time. When this happens, the effect is that of a shared line: All the phones will ring when the DN configured on each of them is called.
- **Deny login:** The user can only be logged in to one device at a time. When he attempts to log in to a second device, he receives an error message. He must log out of the first device before the second login will succeed.
- **Auto-logout:** The user can only be logged into one device at a time. When he attempts to log in to the second device, the system automatically logs him out of the first device, and the second login succeeds.

When the device is logged out, either another Device Profile can be applied (typically a logout Device Profile, perhaps allowing on-net and emergency calls only), or the settings in the phone configuration page are applied.

The Device Profile settings include the following:

- User Music on Hold (MoH) audio source
- Phone button template
- Softkey template
- User locale
- Do not disturb (DND)

- Privacy setting
- Service subscriptions
- Dialing name

A Default Device Profile exists to allow users without a user Device Profile for a particular IP Phone type to log in using EM.

Enable EM in CUCM

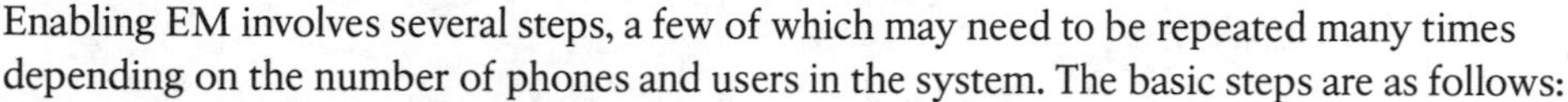

Enabling EM involves several steps, a few of which may need to be repeated many times depending on the number of phones and users in the system. The basic steps are as follows:

1. Activate the Cisco EM Service.
2. Configure EM service parameters.
3. Add the EM service.
4. Create a default Device Profile for each model of phone in use.
5. Create Device Profiles and subscribe them to the EM service.
6. Create end users and associate them with Device Profiles.
7. Enable EM for phones and subscribe phones to the EM service.

The steps are described in more detail in the following sections.

Step 1: Activate the EM Service

1.1 In the Serviceability web page, navigate to **Tools > Service Activation.**

1.2 Select **Cisco Extension Mobility**, and then click **Save**.

Step 2: Configure EM Service Parameters

2.1 In the CM Administration web page, navigate to **System > Service Parameters.** Select the server(s) you want to configure from the **Server** drop-down, and then select the **Cisco Extension Mobility Service** from the **Service** drop-down.

2.2 Scroll down to the Clusterwide Parameters section. Here, you can select whether to force EM logout after a maximum login time has expired, and how long that timer is. This is also where you can set the behavior for multiple logins (Multiple Logins Not Allowed, Multiple Logins Allowed, or Auto Logout), as described earlier. Additional settings here include enabling Alphanumeric User IDs (or using numeric only), choosing to remember and display on the phone (or not) the last User ID logged in to the phone, and whether or not to clear the call lists for the last logged in user on logout.

Step 3: Add the EM Service

3.1 Navigate to **Device > Device Settings > Phone Services.**

3.2 Click **Add New.**

3.3 Give the EM service a name and description, if desired.

3.4 Type (or copy and paste from an external source) the following URL into the Service URL field:
http://<IP_address_of_Publisher>:8080/emapp/EMAppServlet?device=#DEVICENAME#.

3.5 You may choose to add the Secure Service URL as well; if both are configured, and the IP Phone supports HTTPS, the secure URL will be used preferentially.

3.6 Make sure the **Enable** checkbox is selected.

3.7 You may choose to select the **Enterprise Subscription** checkbox as well. Doing so automatically subscribes all IP Phones that support service subscription to the EM service; the administrator does not need to add the service manually to each device.

Figure 11-1 shows the EM Services Configuration page.

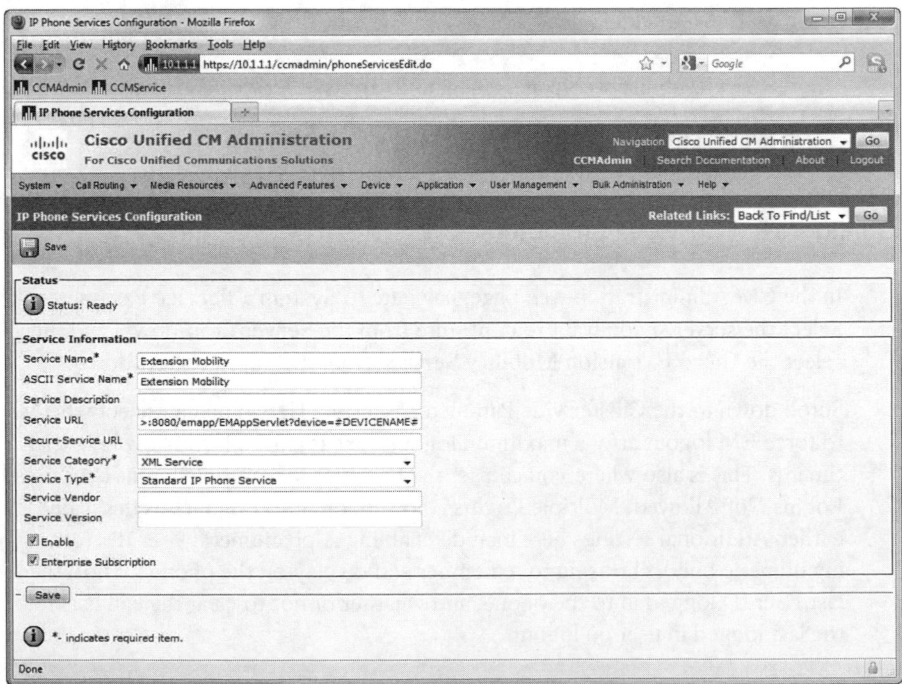

Figure 11-1 *Extension Mobility Service Configuration*

Step 4: Create Default Device Profiles

4.1 In the CM Administration web pages, navigate to **Device** > **Device Settings** > **Default Device Profile.**

4.2 Click **Add New.**

4.3 Select the **Product Type** (the phone model) and **Device Protocol.**

The available settings depend on the phone model and protocol chosen. You may select the Phone Button and Softkey Templates, but you are not able to configure DNs or other specific phone button settings. Figure 11-2 shows the Default Device Profile page.

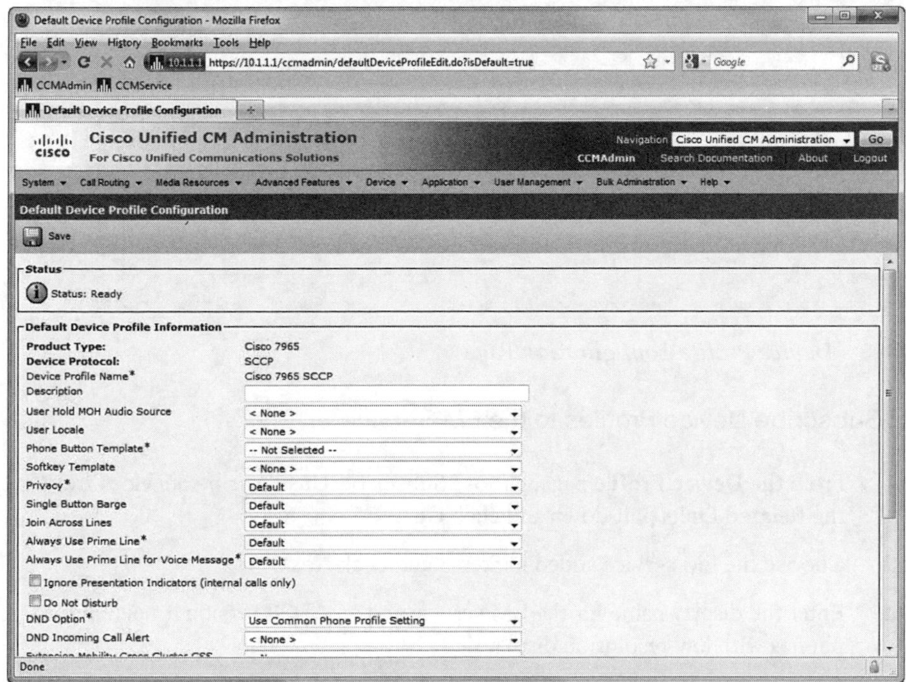

Figure 11-2 *Default Device Profile Configuration*

Step 5a: Create Device Profiles

5a.1 Navigate to **Device > Device Settings > Device Profile.**

5a.2 Click **Add New.**

5a.3 Select the phone model and protocol for a particular user's phone.

5a.4 Enter a name for the profile.

5a.5 Configure user-specific settings, including DN, button customizations, and other parameters.

Figure 11-3 shows a Device Profile Configuration page.

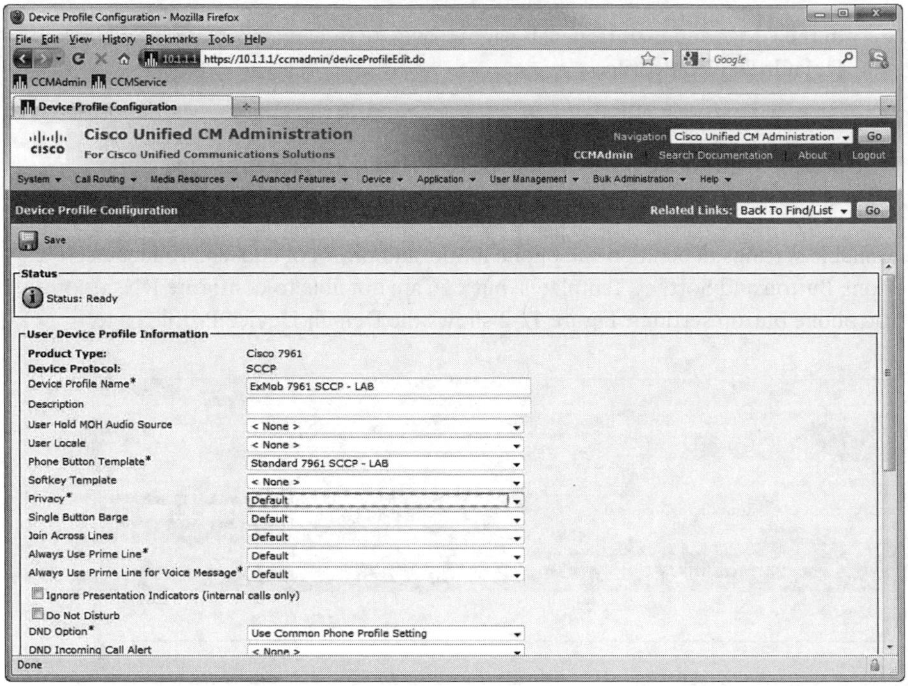

Figure 11-3 *Device Profile Configuration Page*

Step 5b: Subscribe Device Profiles to the EM Service

5b.1 From the **Device Profile** page, choose **Subscribe/Unsubscribe Services** from the **Related Links** pull-down and click **Go**.

5b.2 Choose the EM service added in Step 3 and click **Next**.

5b.3 Enter the display name for the EM service and an ASCII version if needed for phones with low-resolution displays.

5b.4 Click **Subscribe**, and then click **Save**.

Note: If the Enterprise subscription checkbox was selected in Step 3, Step 5b is not required.

Figure 11-4 illustrates this procedure.

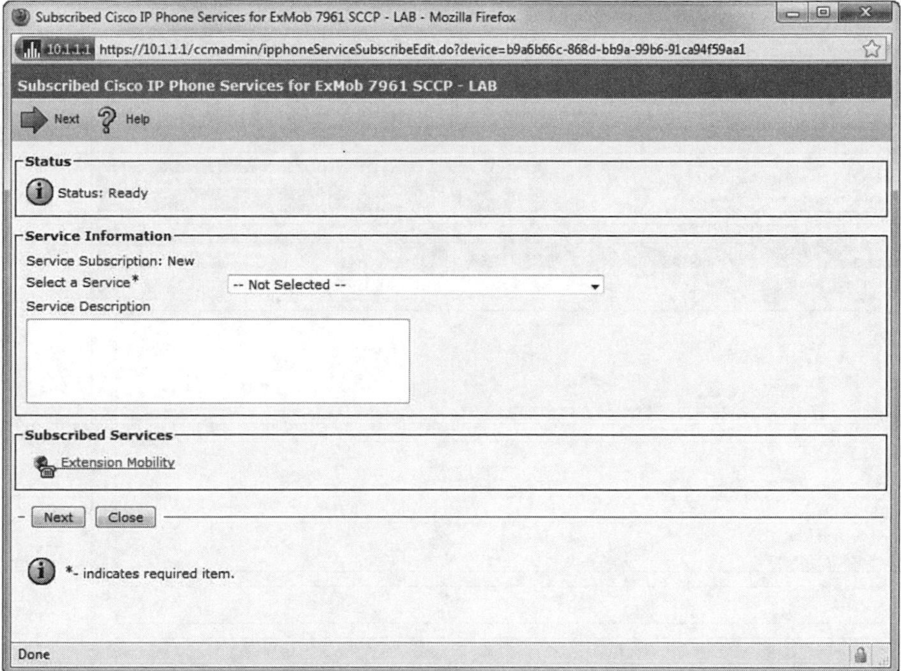

Figure 11-4 *Subscribe Device Profile to EM Service*

Note: You must subscribe both the Device Profiles and the IP Phones to the EM service. If you do not, the user won't have access to the EM phone service after she logs in and her Device Profile is applied—and she will not be able to log out!

Step 6: Associate Users with Device Profiles

6.1 Navigate to **User Management > End User**.

6.2 Select the user for whom you want to create a profile association, or, if necessary, create a new user (see Chapter 9, "Managing Endpoints and End Users in CUCM").

6.3 In the user configuration, choose the Device Profile(s) that should be associated with the user. If more than one is assigned, the user must select the one she wants to use after she logs into EM. The Default Profile option puts the selected profile at the top of the list of choices.

Figure 11-5 illustrates associating an end user with a Device Profile.

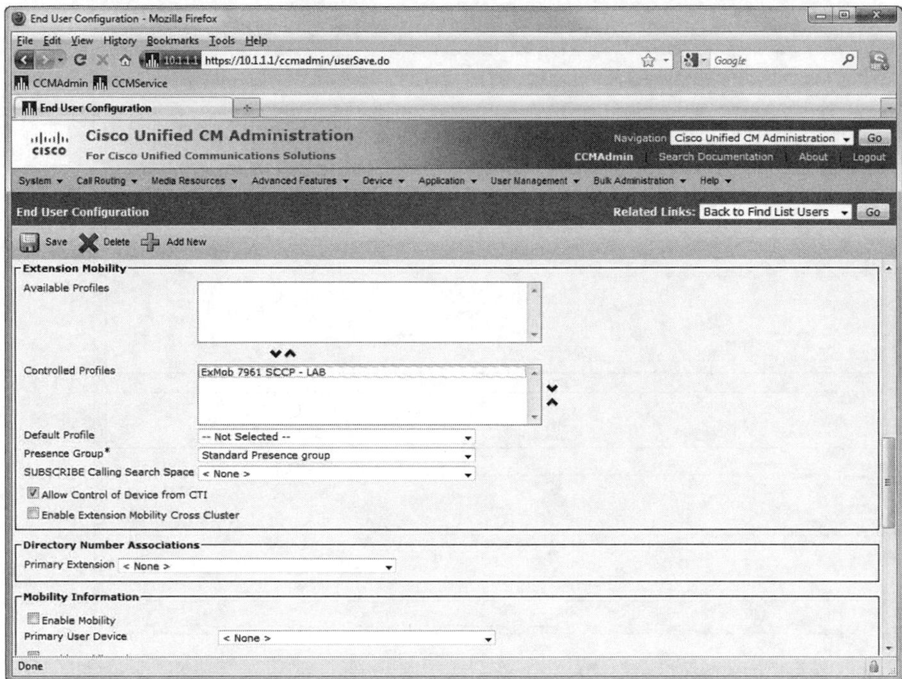

Figure 11-5 *Associating an End User to a Device Profile*

Step 7a: Enable EM for Phones

7a.1 Navigate to **Device > Phone** and select the phone you want to configure for EM.

7a.2 In the **Extension Mobility** section, check the **Enable Extension Mobility** box.

7a.3 Choose either a specific Device Profile or the currently configured device settings (recommended) in the **Log Out Profile** pull-down.

Note: The Log Out Profile is the configuration that is applied to the phone when no-one is logged into it. Often, this profile includes emergency, internal, and sometimes local calling capabilities.

Figure 11-6 shows the phone configuration for EM.

Step 7b: Subscribe Phones to EM Service

7b.1 In the phone configuration page, choose **Subscribe/Unsubscribe Services** from the Related Links pull-down to open the Subscribed Cisco IP Phone Services window. (This step is not necessary if the Enterprise Subscription checkbox is selected in the EM Service Parameters page.)

7b.2 Choose the EM service from the Select a Service pull-down.

7b.3 Enter the name of the service as you want it to appear on the IP Phone.

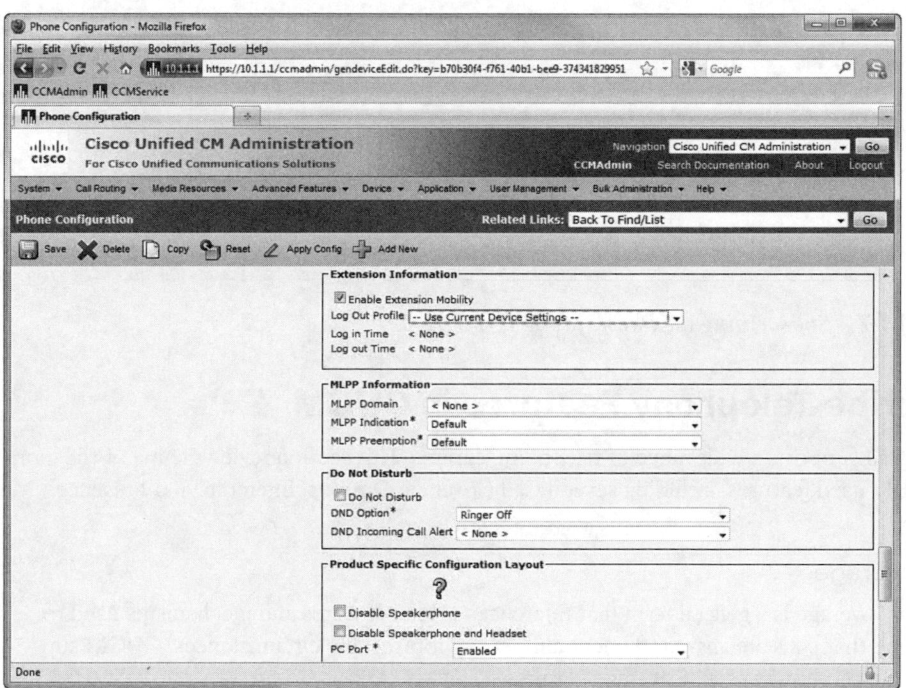

Figure 11-6 *Configuring the Phone for EM*

You should now be able to go any phone that you have subscribed to the EM service and log in as any user you have configured with a Device Profile for that type of phone.

Figure 11-7 illustrates adding the EM Service to a phone.

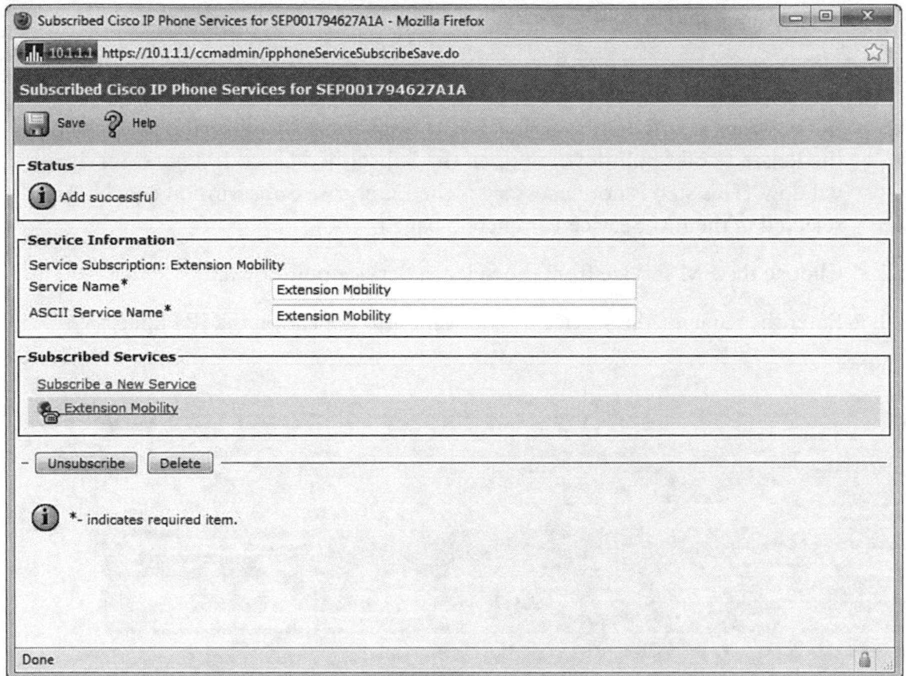

Figure 11-7 *Subscribing the Phone to the EM Service*

Describe Telephony Features in CUCM

CUCM supports a wide range of telephony features. This section reviews some of the more widely used features, including several Call Coverage Options, Intercom, and Presence.

Call Coverage

Call Coverage is a general term that references several features and mechanisms used to ensure that calls are answered under almost any foreseeable circumstances. CUCM supports the following Call Coverage features:

- Call Forward
- Shared Lines
- Call Pickup
- Call Hunting
- Call Park

Call Forward

There are several Call Forward options configurable in CUCM:

- **Call Forward All (CFA):** CFA causes all calls to be forwarded to the destination number specified by the user or administrator, either at the phone itself or on the CUCM user or administrative web pages, without ringing the original dialed number. When CFA is enabled, the Call Forward Search Space is used and line/device search spaces are ignored. For this reason, the Call Forward Search Spaces should be configured to avoid call failures in a system that uses custom Partitions and Search Spaces.

Note: If the Voice Mail checkbox is selected, CUCM ignores the destination in the forward setting and Calling Search Space fields and forwards the call to the Voice Mail Pilot number specified in the configured Voice Mail Profile.

Shared Lines

If two (or more) IP Phones have the same DN configured on one of their lines, calling the DN causes both phones to ring. The first phone to be picked up takes the call; the second phone cannot also pick up the call without invoking the Barge feature (if configured). If the first phone places the call on hold, the second phone can pick up the held call.

Barge and Privacy

If two phones have a shared line configured and one of the phones is using that line, the second phone can force a three-way conference with the first phone by using the Barge feature. The conference is hosted on the first phone's built-in conference bridge. (For IP Phone models not supporting a built-in bridge, an external conference bridge can be configured.) When the second phone barges in on the call, all parties hear a beep (by default). If the barge fails (typically because of a lack of conference resources), the barging phone displays an error message.

A Privacy softkey can be configured that, when enabled, prevents barging into a call in progress. Both the Barge and Privacy capabilities can be enabled and disabled both cluster-wide and at the individual phone configuration pages.

Call Pickup

A DN can be made a member of a Call Pickup Group, which is simply a numbered assignment. Three types of Call Pickup can be configured using these Pickup Group assignments:

- **Call Pickup:** If multiple DNs have the same group number and one of them is ringing, another phone with a DN in the same pickup group can invoke the Call Pickup softkey and the call is immediately extended to that phone instead.

- **Group Call Pickup:** If two phones have DNs in different Call Pickup Groups and one of them is ringing, the other phone can invoke the GPickup (Group Call Pickup) softkey, dial the group number of the ringing DN, and the call is immediately extended to that phone instead. A variant of this feature called *Directed Call Pickup* allows a user to enter a specific DN that is ringing to pick up that specific call, rather than the first call that started ringing in the group.

- **Other Group Pickup:** Introduced in CUCM v.7, Other Group Pickup allows the administrator to set up group associations between Call Pickup Groups. This allows a phone to pick up a call ringing in a different associated group without having to enter the other group's number. The OGroup softkey accesses this feature.

For all Call Pickup implementations, the administrator can configure audio/visual/both notifications on group member phones to indicate that a phone in one of their pickup groups is ringing. This is particularly useful if the phones are not within earshot of each other.

Call Hunting

A more advanced call coverage system can be built in CUCM using a Call Hunting structure. Call Hunting allows a single dialed DN (or PSTN number) to distribute calls to several phones in sequence. This is typically set up for helpdesk or customer-service groups that are not very large; large implementations would be better served by a dedicated call-center application. Call Hunting consists of the following components and configurations:

- **DNs and voicemail ports:** The ultimate targets of the Call Hunting system. These are assigned to Line Groups.
- **Line groups:** Assigned to Hunt Lists; one or more can be assigned to a single Hunt List. The line group configuration provides for different hunt algorithms (specifically, Top-Down, Circular, Longest Idle, and Broadcast) and other hunt options.
- **Hunt Lists:** A Hunt List is a top-down ordered list of line groups. Calls flowing through the Call Hunting system are sent to the first line group in the Hunt List. If no member of that line group can answer the call, it may be returned to the Hunt List, which then tries the second line group. This process may repeat until the call is answered, the list of line groups is exhausted, or the caller hangs up.
- **Hunt Pilots:** A Hunt Pilot is associated with a Hunt List. The Hunt Pilot may be a unique DN, a shared line, or a PSTN number.

During the hunting process, the Call Forwarding configuration of line group members is ignored; for example, if the DN is busy, the next DN in the line group would be chosen rather than using the CFB setting for that DN.

Call Park

Call Park allows a user to temporarily attach a call to a Call Park slot (effectively a DN). Any user can pick up the call by dialing the Call Park number. For example: If Mark is on the phone with a customer, and the customer asks about the HayBailer 9000 series product, Mark can say, "That's LuAnn's product line. Let me find LuAnn for you." Mark presses the Call Park softkey, and CUCM displays a message on the phone indicating the Call Park slot number at which it parked the call. Mark then has to contact LuAnn (perhaps using a paging system or just yelling across the showroom), tell her the Call Park number, and LuAnn simply dials that number to pick up the call.

A variation of this feature, called Directed Call Park, requires LuAnn to enter a prefix code (effectively a password) to retrieve the call.

Intercom

The Intercom feature allows a button to be configured that calls an intercom line on another phone. The recipient phone auto-answers in speakerphone mode, with the microphone muted. A one-way audio stream now exists from the caller to the recipient; the recipient can hear the caller, but the caller cannot hear the recipient. This is known as *Whisper Intercom*.

When the recipient presses their Intercom button, a second one-way audio stream is established back to the caller and they can each hear the other. For both these intercom calls, an auto-answer tone is heard when the recipient phone answers the call.

If the recipient is in a call when the Whisper Intercom call is made, the recipient hears the one-way audio from the intercom caller, but the other party does not.

Intercom lines are different from normal DNs. Intercom lines cannot call DNs, and DNs cannot call intercom lines. Intercom lines have their own dial plan and permissions (Intercom Partitions and CSS). An Intercom button can be a speed dial (with one preconfigured target intercom line), or you can configure an Intercom button that requires the user to dial the target intercom line.

CUCM Native Presence

Presence can be defined as "signaling one's capability and willingness to communicate." Presence indications can include Instant Messaging status indications, such as Online, Offline, Busy, Out to Lunch, In a Call, and so on, or in a telephone system, simply On Hook or Off Hook. CUCM supports a built-in capability to track the on or off hook status of a DN.

Presence status can be monitored using either a Busy Lamp Field (BLF) speed dial or Presence-enabled call and directory lists. A BLF Speed Dial is a speed dial button that lights up when the target of the speed dial is off hook. Presence-Enabled Call Lists display an icon that indicates that the entry in the list is one of the following:

- **Unknown:** The entry is not being watched, or the device displaying the list does not have permission to watch the target's presence status. The icon displayed is a blank phone keypad.
- **On Hook:** The entry is on hook. The icon displayed is a telephone over a blank keypad.
- **Off Hook:** The entry is off hook. The icon displayed is two handsets over a blank keypad.

Figure 11-8 shows an IP Phone displaying a Presence-Enabled Call List with each of the three icons showing the current status of the targets of recent placed calls. In the figure, 85122001 and 3001 display Unregistered status, 2002 is Off Hook, and 2001 is On Hook.

Presence Architecture

Key Topic

In some cases, it may not be desirable to have every phone watch every DN in the system. Presence visibility can be controlled in the following ways:

- BLF Speed Dials can be configured only by an administrator; users cannot create or edit their own.

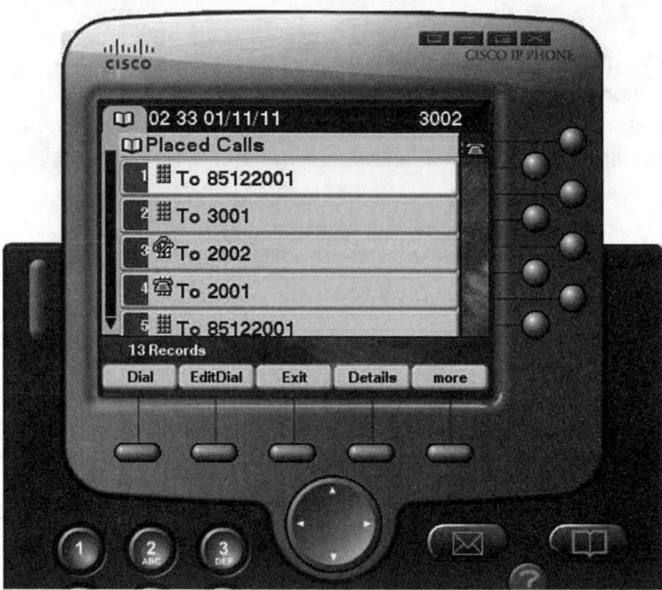

Figure 11-8 *Presence Indications*

- Visibility for Presence-enabled call and directory lists can be limited through the use of Partitions and subscribe Calling Search Spaces. A Subscribe CSS is specific to a Presence monitoring system: If the DN to be watched is in the watcher's Subscribe CSS, presence indications are visible; if it is not, the DNs appear as Unregistered status to the watcher.
- Presence Groups allow different sets of watchers to be assigned (or denied) permission to watch the Presence status of DNs in other Presence Groups. Phones, DNs, and users can be assigned to Presence Groups. All users are in the standard Presence Group by default, but can be assigned to custom Presence Groups as desired. The enterprise parameter configuration page **Inter-Presence Group Subscribe Policy** setting defines the default setting for whether Presence Groups have permission to watch each other's Presence status or not; this setting may be overridden in the Presence Group settings. Members of a Presence Group are always able to watch other members of the same group, unless the Subscribe CSS prevents the subscription. The Subscribe CSS and Presence Group settings may be used independently of one another or together. If both are used, both must allow a subscription in order for a watcher to monitor the Presence status of a DN.
- Inter-Presence Group settings apply only to BLF Call Lists and Directories and do not affect BLF Speed Dials.

Enable Telephony Features in CUCM

The following sections detail the necessary steps to configure the telephony features outlined earlier, including Call Coverage, Intercom, and Presence.

Enabling Call Coverage

This section describes the configuration steps to enable the following Call Coverage features:

- Shared Lines
- Barge and CBarge
- Call Pickup
- Call Park and Directed Call Park
- Call Hunting

Configuring Shared Lines

When two or more devices are configured with the same DN, it is called a *shared line*. The configuration is straightforward: Simply add the same DN to multiple phones, either from the DN Configuration page or the Phone Configuration page via the Phone Button configuration. Figure 11-9 shows the DN configuration page, with two devices associated to the same DN.

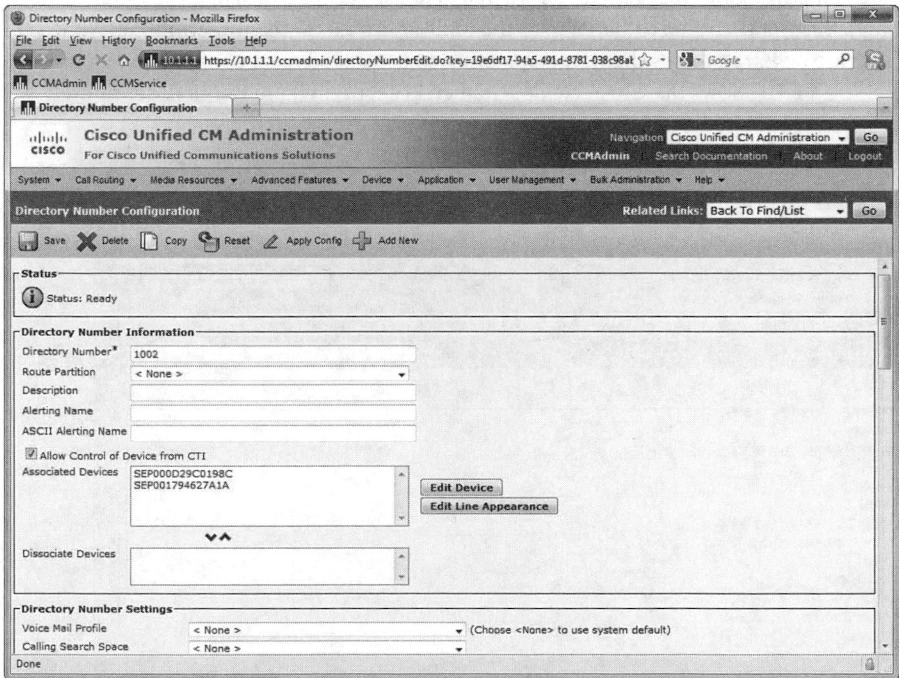

Figure 11-9 *DN Configuration for Shared Line*

Configuring Barge

The Barge feature (as described earlier) allows a user with a shared line to force a three-way conference with another user of that shared line. To enable the feature, the built-in conference bridge (available on most IP Phone models) must be activated. The Privacy feature removes the call information from all phones that share lines and blocks other shared lines from barging in on its calls. The following steps describe how to configure Barge and Privacy:

1. In CM Administration, navigate to **System > Service Parameters.** Select the server you want to configure from the **Server** drop-down.
2. Select the **Cisco CallManager** service from the **Service** drop-down.
3. Scroll down to the **Clusterwide Parameters (Device-Phone)** section. Set **Built-in** Bridge **Enable** to **On.**
4. Set the **Privacy** Setting to True (the default).

Note: The Built-in Bridge, Privacy, and Single Button Barge settings can be overridden at the Device Pool or the individual phone if desired.

5. Scroll down to **Clusterwide Parameters (Feature-Join Across Lines And Single Button Barge Feature Set)** and set the **Single Button Barge/CBarge Policy.** Your choices are **Off, Barge,** or **CBarge.** This setting allows pressing the shared line button to cause a Barge onto the shared line when it is in use (instead of using the Barge softkey). Setting the value to Barge uses the built-in bridge on the target phone, while setting it to CBarge forces the Barge operation to use an external conference resource. Figure 11-10 shows the Clusterwide Parameters section where Barge is configured.

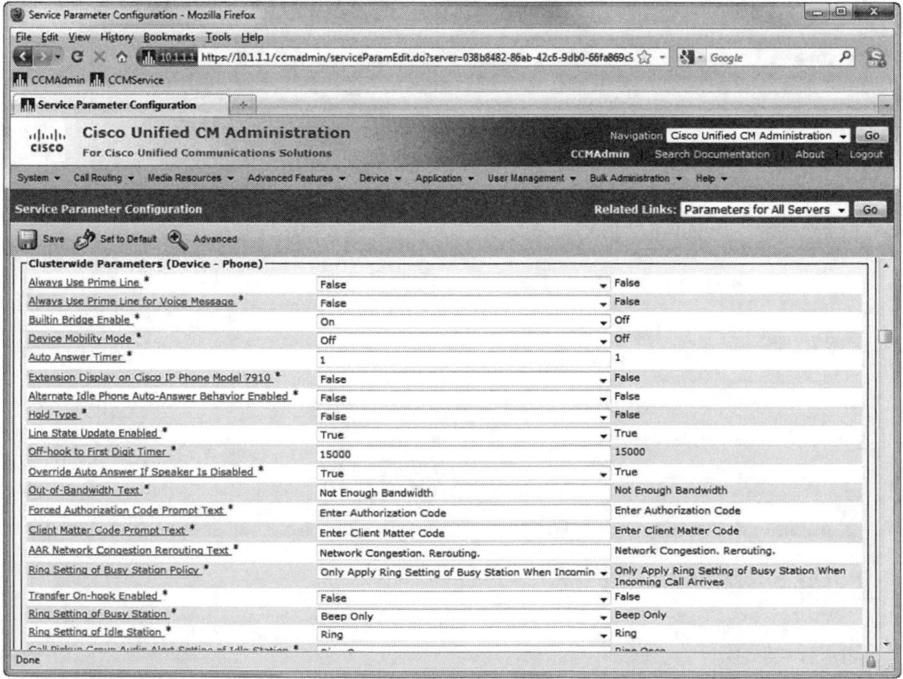

Figure 11-10 *Service Parameter Configuration for Barge*

Note: Using the Barge softkey method requires that you configure the softkey template to include the Barge softkey for the phones involved with the Barge operation. Using the Single Button Barge method does not.
If CBarge is required, an external conference resource must be configured and made available to the phones.

Configuring Call Pickup

To configure Call Pickup, you must create and apply Call Pickup Groups to DNs that must pick up each other's calls. Follow these steps:

1. In CM Administration, navigate to **Call Routing > Call Pickup Group.**
2. Click **Add New.**
3. Enter a name and a unique number. (The number cannot contain wildcards.)
4. Select a **Partition** (normally the same as the DN Partition; however, selecting a different Partition allows the administrator to restrict access to the pickup group by modifying the CSS of the phones if desired).

It is possible to preconfigure associated Call Pickup Groups so that the Other Group Pickup feature (described earlier) can be used if desired; this option is only available during the initial configuration of the Pickup Group. Figure 11-11 shows the Call Pickup Group Configuration page.

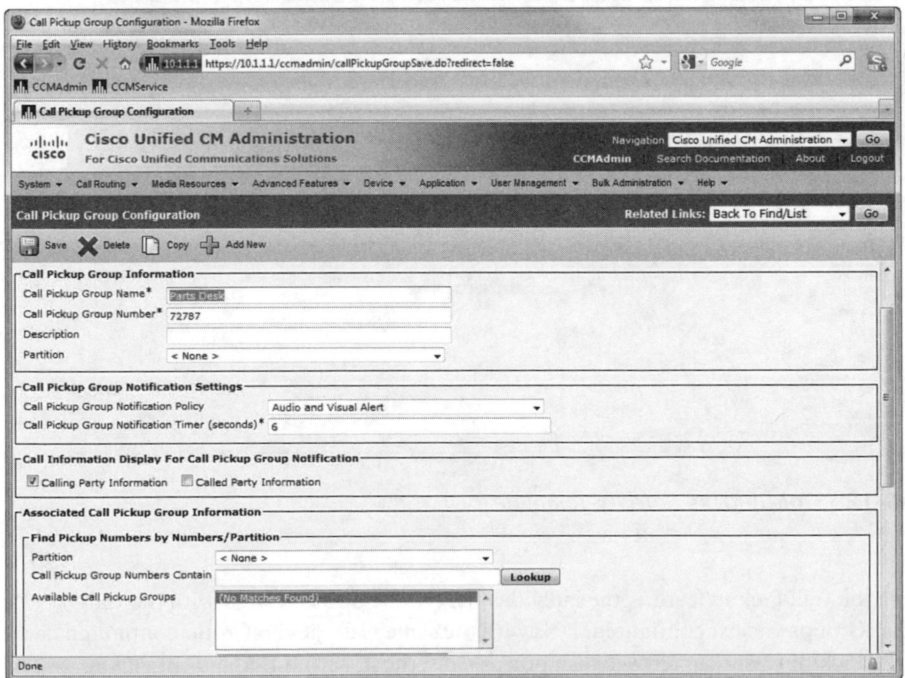

Figure 11-11 *Call Pickup Group Configuration*

To use the Pickup feature, a softkey must be added to the phone(s). The following steps outline the process of modifying and adding a softkey template to enable Call Pickup:

1. To configure the softkey template, navigate to **Device > Device Settings > Softkey Template.**
2. Select, add, or copy a softkey template.
3. From the **Related Tasks** pull-down, select **Configure Softkey Layout.**
4. Add the **Pickup, Group Pickup**, or **Other Group Pickup** softkeys as desired. (These keys can be selected in the **Off Hook** or **On Hook** call states.)
5. Click **Save.**
6. Apply the modified softkey template to phones or Device Profiles as desired.

Figure 11-12 Shows the Softkey Template Configuration Page with the On Hook call state template options being configured.

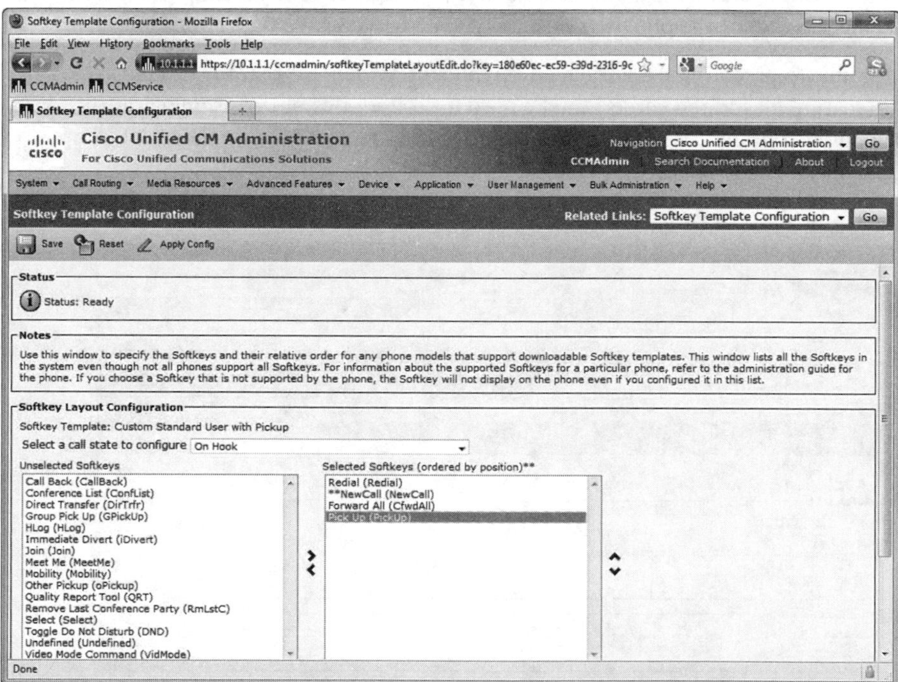

Figure 11-12 *Configuring Softkey Template for Barge*

To use the Call Pickup feature, the individual DNs must be associated with the Call Pickup Groups we just configured. DNs with the same Call Pickup Group configured can use the Pickup feature to answer each other's calls; those with different Call Pickup Groups can use Group Pickup or Other Group Pickup if their Call Pickup Groups are pre-associated. Figure 11-13 illustrates the association of a DN with a Call Pickup Group.

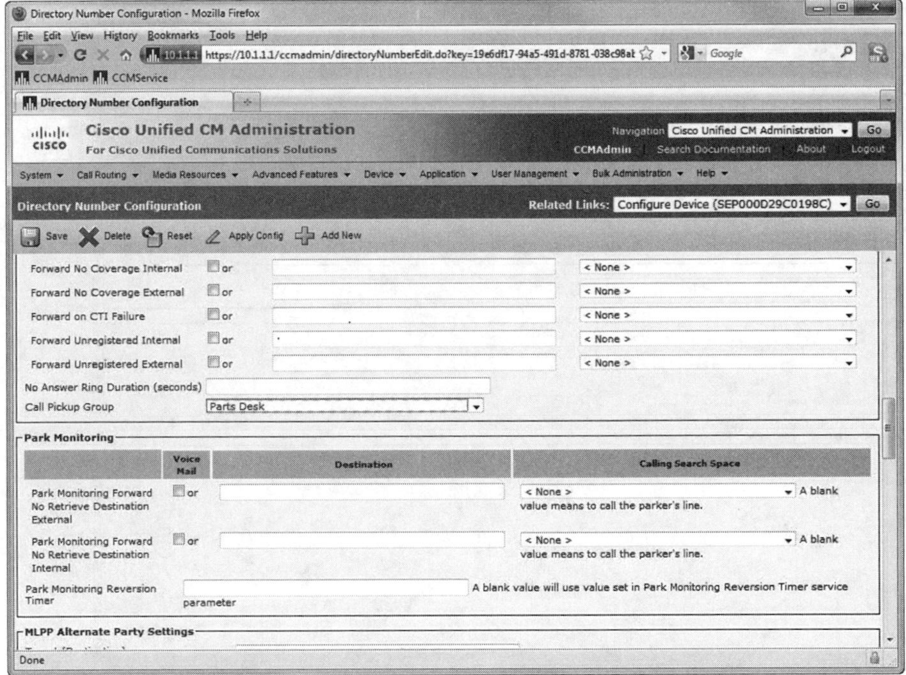

Figure 11-13 *Call Pickup Group Association to DN*

Configuring Call Park and Directed Call Park

The Call Park feature allows a user to park a call at a reserved DN, and retrieve it from any IP Phone. Directed Call Park is similar, with the added features of requiring a prefix code to retrieve the call and the ability to specify a different reversion number. The following steps configure the Call Park and Directed Call Park features:

1. In CM Administration, navigate to **Call Routing** > **Call** Park.
2. Click **Add New**.
3. Specify either an individual DN or a range of DNs to be used for Call Park. The number can be partitioned if desired by selecting a custom Partition. A number or range is associated with the CUCM server you select from the pull-down; if you are associating Call Park slots to multiple servers, ensure that the number ranges do not overlap between servers.
4. Click **Save**.

Note: A Call Park number range is defined with the same wildcard used in Route Patterns; for example, the range of 880X defines ten Call Park slots numbered 8800 through 8809.

Figure 11-14 shows Call Park configuration.

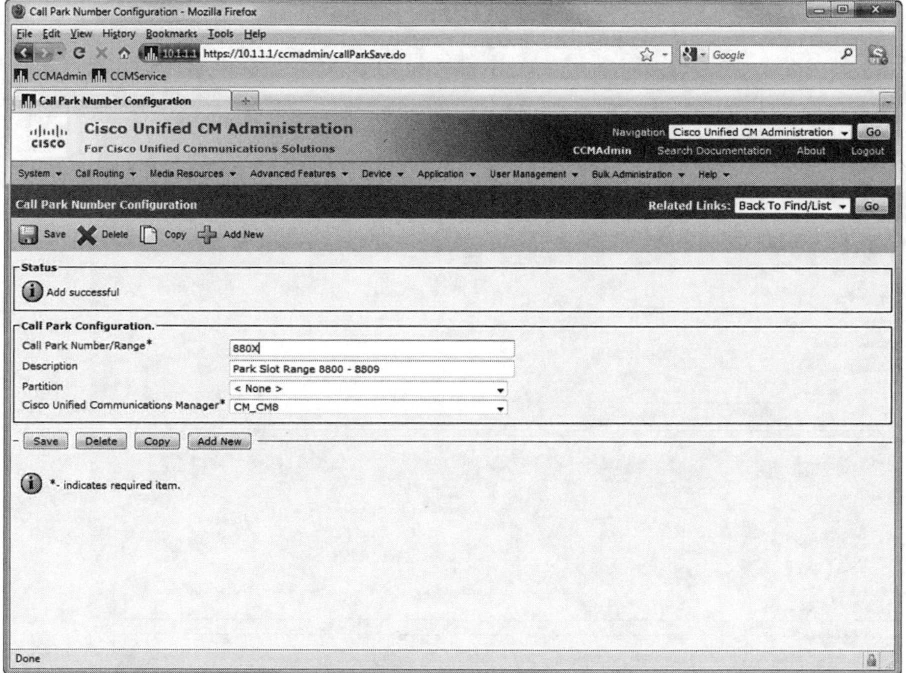

Figure 11-14 *Call Park Configuration*

Configuring Directed Call Park is similar. Follow these steps:

1. In CM Administration, navigate to **Call Routing > Directed Call Park.**
2. Click **Add New.**
3. Enter a unique number or range and specify a Partition if desired.
4. Specify the **Reversion Number** (the DN to which a call will be forwarded if not picked up from the park slot before the Call Park Reversion timer [60 seconds by default] expires).
5. Specify the **Reversion Calling Search Space** to allow the phone to find the reversion number specified above, if it is not in the normal CSS of the phone or line.
6. Specify the **Retrieval Prefix**, which is the code the person picking up the parked call must dial in order to retrieve it.
7. Click **Save.**

Figure 11-15 shows Directed Call Park configuration.

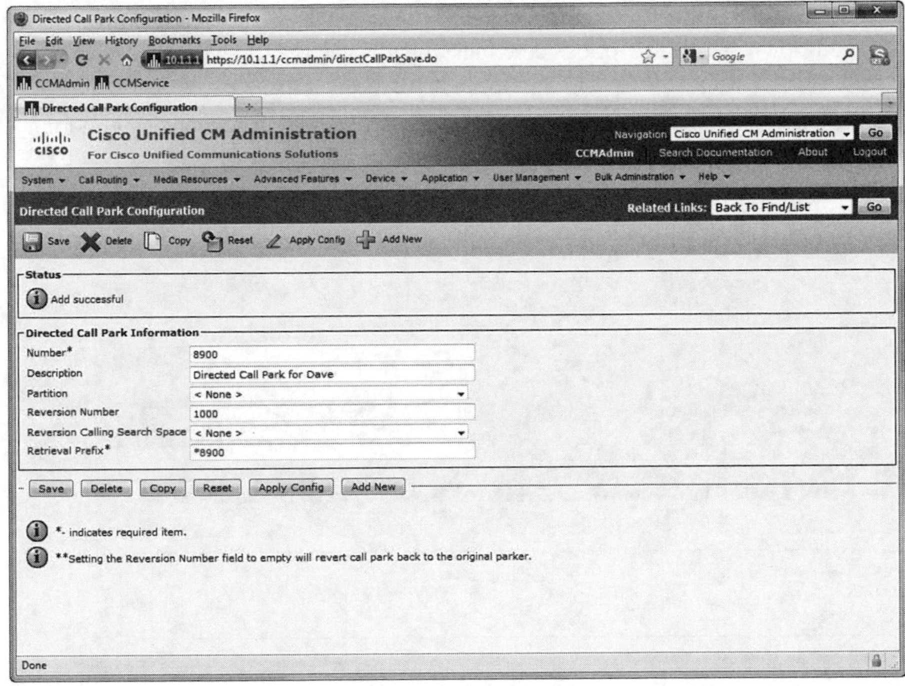

Figure 11-15 *Directed Call Park Configuration*

Configuring Call Hunting

To configure Call Hunting, groups of DNs are associated with Line Groups that specify the hunting behavior. Line Groups are added to Hunt Lists, which select the order of hunting through the Line Groups. A Hunt Pilot number is associated with a Hunt List and serves as the dialed number trigger of the hunting system. To configure Call Hunting, follow these steps:

Create Line Groups

1. Create DNs and associate them with phones.
2. In CM Administration, navigate to **Call Routing > Route/Hunt > Line Group.**
3. Click **Add New.**
4. Enter a Line Group Name.
5. Specify the **RNA Reversion Timeout** (the number of seconds each DN in the Line Group will ring before the No Answer trigger is reached).
6. Select the **Distribution Algorithm: Top Down** (each new call starts with the DN at the top of the list), **Circular** (each new call begins starts at the next DN in the last after the one used by the previous call), **Broadcast** (all DNS in the Line Group ring simultaneously), or **Longest Idle Time** (the DN that has been in the On Hook state for the longest rings).

7. Select the **Hunt** option for each call state (No Answer, Busy, and Not Available) from the pull-down. The Hunt options determine if and when the call will move from the current Line Group to the next Line Group in the Hunt List.
8. Add DNs to the Line Group. The order in which the DNs are listed may be important depending on the earlier choice of distribution algorithm.
9. Click Save.

Figure 11-16 shows the Line Group Configuration page.

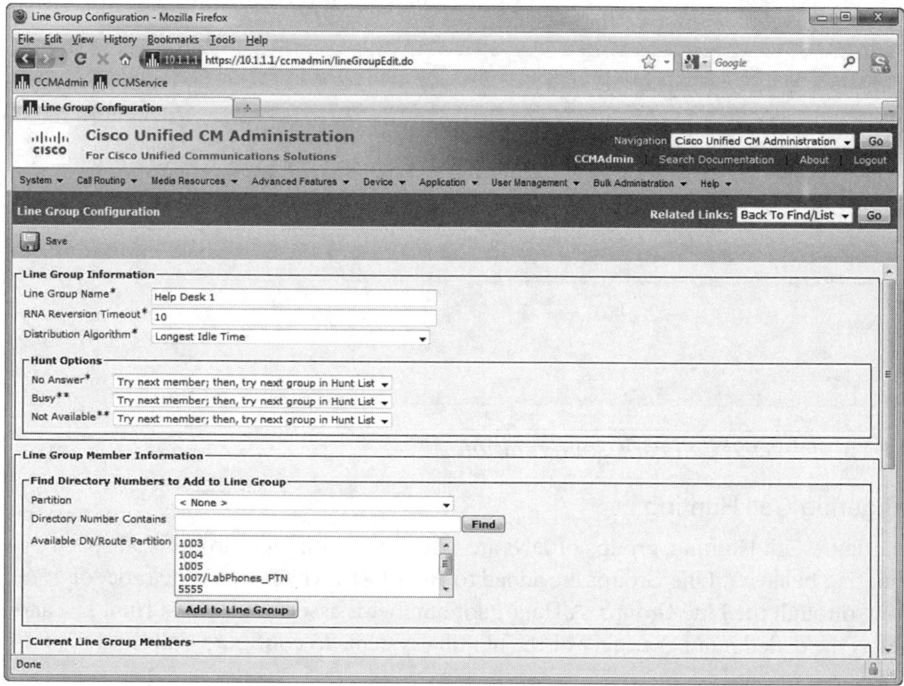

Figure 11-16 *Line Group Configuration*

Create Hunt Lists

1. Navigate to **Call Routing > Route/Hunt > Hunt List.**
2. Click **Add New.**
3. Specify a **Name.**
4. Set the **CUCM Group** in order to provide CUCM redundancy for the Hunt List.
5. Click **Save.** The Hunt List configurations now appear on the page.
6. **Add line groups** to the Hunt List. The Hunt List is always top-down processed, so the order of the Line Groups is important; use the arrows to adjust the order if needed.
7. Click **Save** when the desired Line Groups appear in the correct order.

Create a Hunt Pilot

1. Navigate to **Call Routing > Route/Hunt > Hunt Pilot.**
2. Click **Add New.**

3. Enter a Hunt Pilot number. This behaves the same way as a Route Pattern; you can specify any string you like, including a valid PSTN or DID number.
4. Set a **Partition**, if desired, to control access to the Hunt Pilot.
5. In the **Hunt List** field, select the Hunt List that should be accessed by dialing this Hunt Pilot number.
6. Specify an **Alerting Name**, which displays on phones receiving calls as part of the hunting system.
7. Set **Hunt Forwarding** options to control where calls that cannot be handled by the hunting system are sent. The **Use Personal Preferences** checkbox ignores the configured settings, instead using the CFNC setting of the station that forwarded the call to the Hunt Pilot.
8. Click **Save**.

Figure 11-17 shows the Hunt Pilot Configuration page.

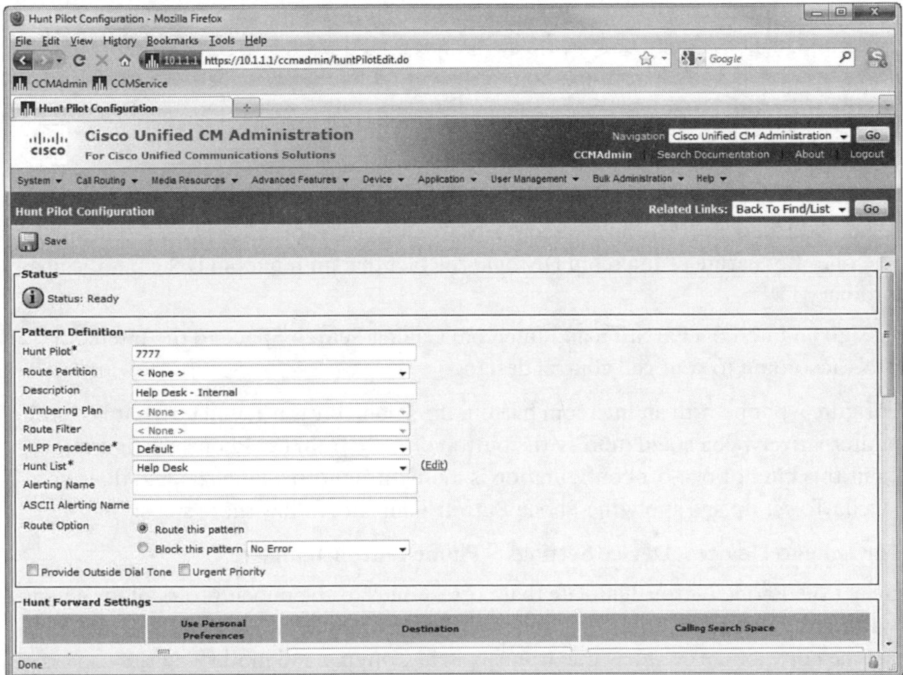

Figure 11-17 *Hunt Pilot Configuration*

> **Note:** The Phone Button Template and/or Softkey Template can be configured to include the HLog key, which allows a user to log into or out of a Hunt Group. The CallManager Service parameter Hunt Group Logoff Notification specifies a ringtone that plays to alert a user that there is an incoming call that would ring their phone if it were not logged out of the hunting system. This is useful as a reminder to users that they are logged out if they return to their desk after an absence.

Configuring Intercom Features

As previously discussed, the Intercom feature uses special Intercom DNs, Partitions, and Calling Search Spaces. The following steps walk through setting up the Intercom feature:

1. Navigate to **Call Routing > Intercom > Intercom Route Partition.**
2. Click **Add New**. In the Configuration Page, enter the name and comma-separated description of the new Intercom Partition. You can create up to 75 Partitions at once on this page.
3. Click **Save.**
4. Navigate to **Call Routing > Intercom > Intercom Calling Search Space**. Click **Find**, and note that an Intercom CSS has been automatically created as a result of creating the Intercom Partition (the auto-named entry will be *<partition_name>*_GEN). The auto-generated CSS automatically includes the Intercom Partition just created. You may use the auto-generated CSS, modify it, or create custom Intercom CSSs as desired.

Note: A custom Intercom CSS is really only necessary if an Intercom button needs to support multiple (dialed instead of speed dialed) targets, and access to some of those targets must be limited. If you are creating point-to-point Intercom lines, there is no need to customize the Intercom CSS.

5. Navigate to **Call Routing > Intercom > Intercom Directory Number.**
6. Click **Add New** to create an Intercom DN. You must create at least two, because of the one-way nature of Intercom DNs and the fact that an Intercom DN cannot call an ordinary DN.
7. Assign an Intercom **Partition** and Intercom **Calling Search Space** to the Intercom DNs according to your call control design.

To configure a phone with an Intercom button, the Phone Button Template must be modified. (Alternatively, you could modify the button directly from the Phone configuration page, but this kind of one-off configuration is more difficult to administer and does not scale well.) To set up and apply the Phone Button Template, follow these steps:

1. Navigate to **Device > Device Settings > Phone Button Template.**
2. Select the Phone Button Template that corresponds to the phone/protocol for which you want to configure Intercom. You may modify the stock profile, copy it and modify the copy, or create a new one from scratch. Copying and modifying the copy is the recommended action.
3. In the Phone Button Template configuration window, add the Intercom feature to the desired button appearance.
4. Click **Save.**

Figure 11-18 shows the Phone Button Template configuration.

Now that we have the DNs and template set up, we can assign the template to the phones:

1. Apply the template to a phone by selecting the modified template from the pull-down in the phone's configuration page.

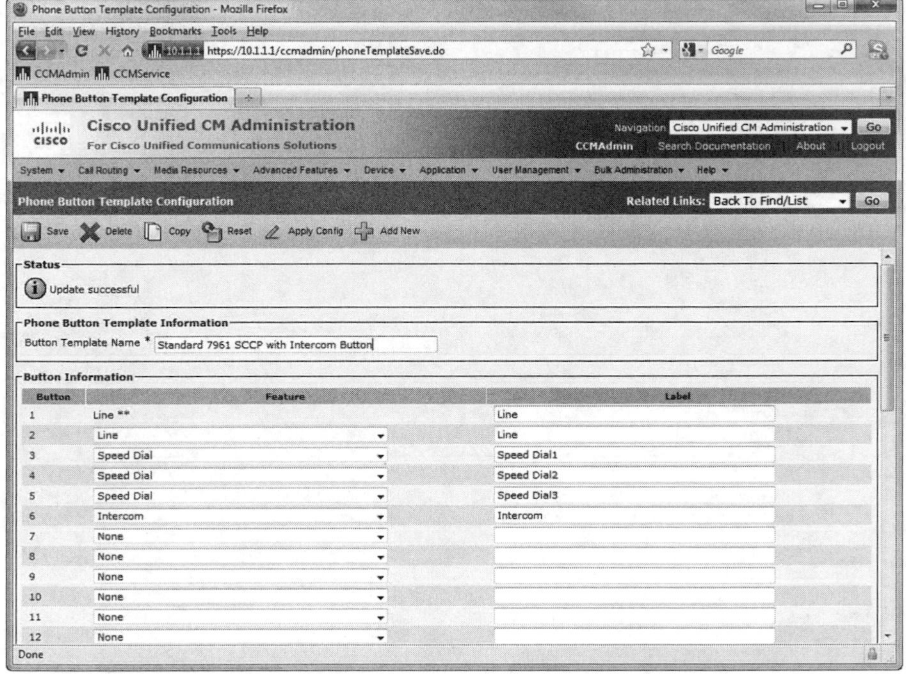

Figure 11-18 *Adding Intercom DN to a Phone Button Template*

2. Configure a button on the phone with an Intercom DN, Intercom Partition, and Intercom CSS.

Figure 11-19 illustrates the Intercom button configuration.

Configure CUCM Native Presence

As previously described, there are two implementations of CUCM native Presence. The first, simplest implementation is BLF Speed Dial; the second, more involved implementation is Presence-Enabled Call Lists, which uses Presence Groups and a special Subscribe CSS.

Configuring BLF Speed Dials

To add a BLF Speed Dial to a phone, it is recommended that you modify the Phone Button Template; it is possible to modify the individual phone's buttons, but doing so creates administrative burden and does not scale well. To set up BLF Speed Dial, follow these steps:

1. Navigate to **Device > Device Settings > Phone Button Template.**
2. Select, copy, or create the appropriate template for the phone model/protocol.
3. Add the **Speed Dial BLF** feature to one or more of the available buttons.
4. Apply the template to the phones.
5. In the phone configuration page, select an available Add a new BLF SD button and configure the **Destination** DN and display **Label.**

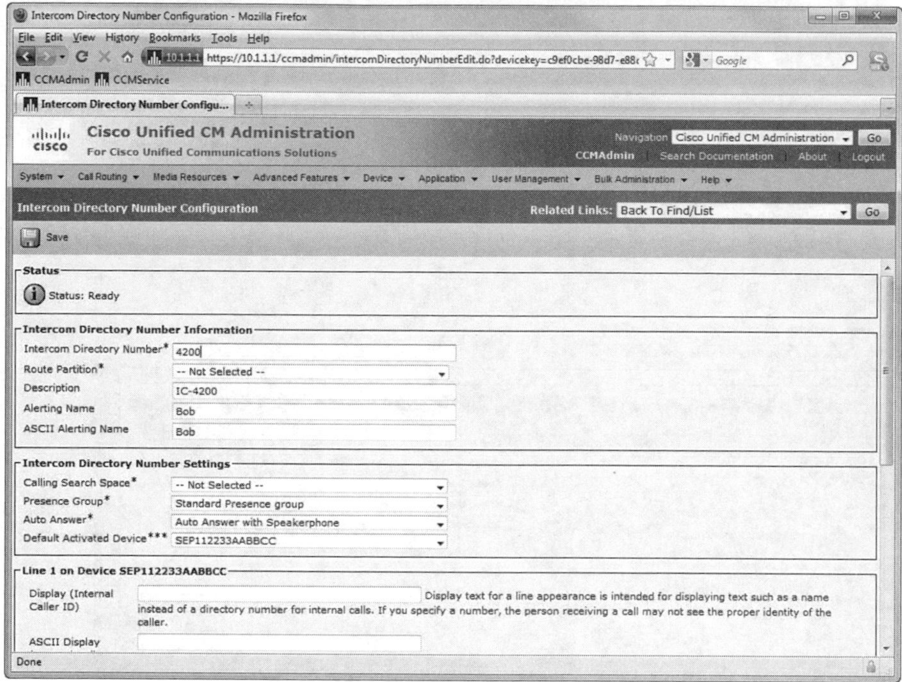

Figure 11-19 *Intercom Button Configuration*

Figure 11-20 illustrates the configuration of a BLF Speed Dial.

Configuring Presence-Enabled Call Lists

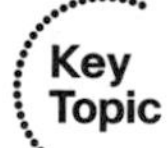

As described earlier, the CUCM system can track the on hook status of phones both through BLF Speed Dials and Presence-Enabled Call Lists. Presence-Enabled Call Lists use the device's configured Subscribe CSS and/or Presence Group subscription policies to determine whether a device can watch a DN's Presence status:

- If the Subscribe CSS applied does not include the Partition of the DN being watched, Presence status is unavailable.
- If the Inter-Presence Group Subscription setting is denied between the two groups, Presence status information is unavailable.
- If both the Subscribe CSS and Presence Groups are used together, both must allow the subscription in order for Presence status to be watched.

Configuring Presence-Enabled Call Lists is somewhat more complex than BLF Speed Dials, but it allows a greater flexibility, precision, and scalability when large numbers of devices need to watch large numbers of DNs. To configure Presence-Enabled Call Lists, follow these steps:

1. Navigate to **System > Enterprise Parameters.** Scroll down to **Enterprise Parameters Configuration.**
2. Set **Enable BLF for Call Lists** to **Enabled.**

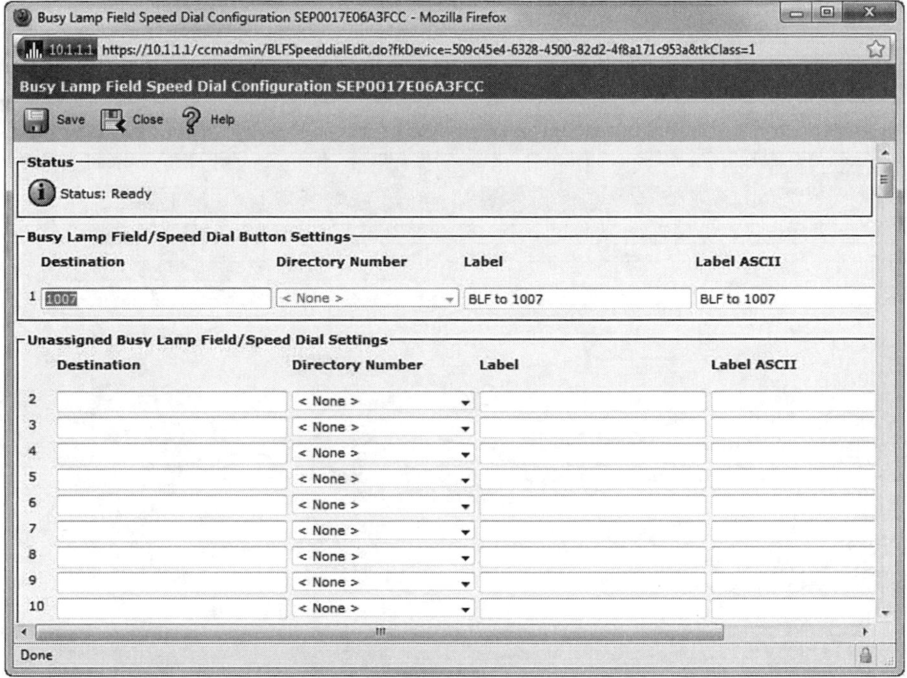

Figure 11-20 *Configuring a BLF Speed Dial*

Figure 11-21 illustrates the Enterprise BLF for Call Lists configuration.

3. If using a custom Subscribe CSS is desired, create new Subscribe CSSs. It is generally recommended to create custom Subscribe CSSs and use existing Partitions; the overall Class of Control plan should be well thought out prior to making these changes.
4. Apply the appropriate CSS to phones and SIP trunks as required.

Configuring Custom Presence Groups

Because all devices and DNs are part of the Standard Presence Group by default and all devices can watch all DNs within the same Presence Group, the configured Subscribe CSS and existing Partitions may provide adequate control over Presence subscriptions. If a more complex design is required, it may be necessary to set up custom Presence Groups and define custom inter-Presence Group subscriptions, as follows:

1. Navigate to **System > Presence Group.**
2. Click **Add New** and configure a **name.**
3. Set the **Presence Group Relationship** to each other Presence Group (**Allow Subscription** or **Disallow Subscription**) to control whether or not this group can watch the Presence status of other groups. Setting it to **System Default** references the **Default Inter-Presence Group Subscription** Enterprise Parameter discussed next. Each subscription is one-way: For example, Executives may be allowed to watch Employee Presence status, but Employees may not be allowed to watch Executive status. Figure 11-22 shows a Presence Group Configuration page.

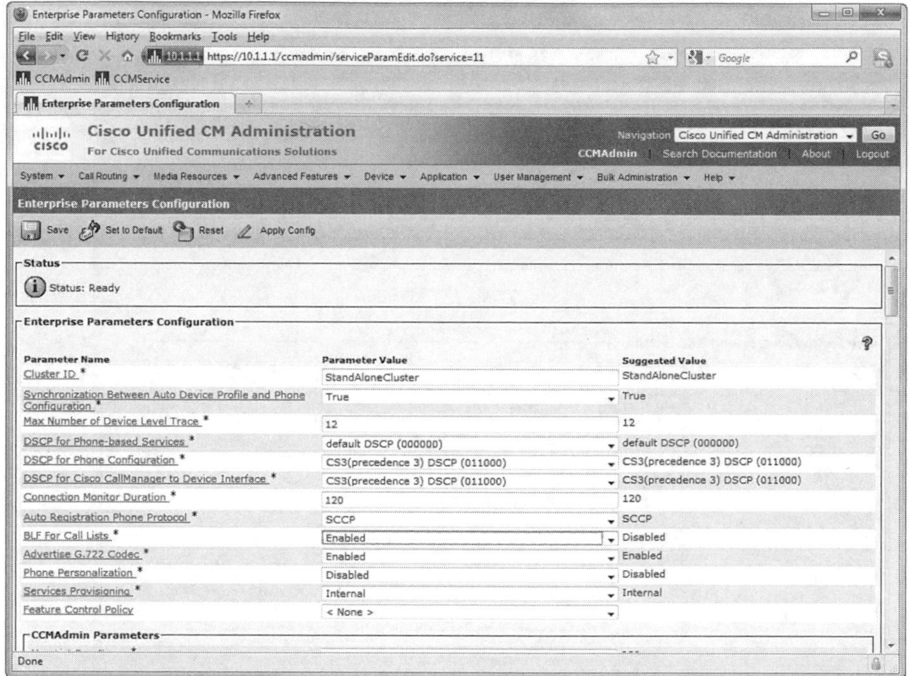

Figure 11-21 *BLF for Call Lists Enterprise Parameter*

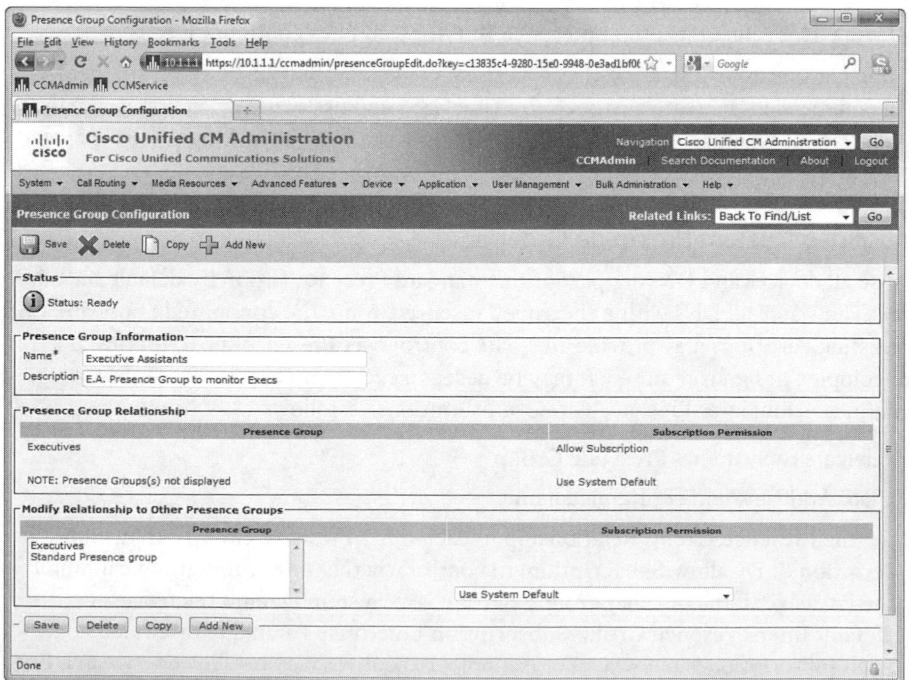

Figure 11-22 *Presence Group Configuration*

4. Navigate to **System > Service Parameters.** Select the server you want to configure, and then the **Cisco CallManager Service.** Scroll down to **Clusterwide Parameters (System-Presence).**

5. Set the default **Inter-Presence Group Subscription** policy to either **Allow** or **Disallow,** as desired. (The **Use System Default** setting mentioned references this parameter setting.)

6. Assign Presence Groups to DNs and Phones. Remember that Phones watch DNs. The Presence Groups assigned to the phone and the DN, the Inter-Presence Group Subscription Setting, the Subscribe CSS of the watcher, and the Partition of the DN all interact to determine whether Presence information is available to the watcher.

Figures 11-23 through 11-25 show the Presence Group configuration for a phone, a DN, and a SIP trunk.

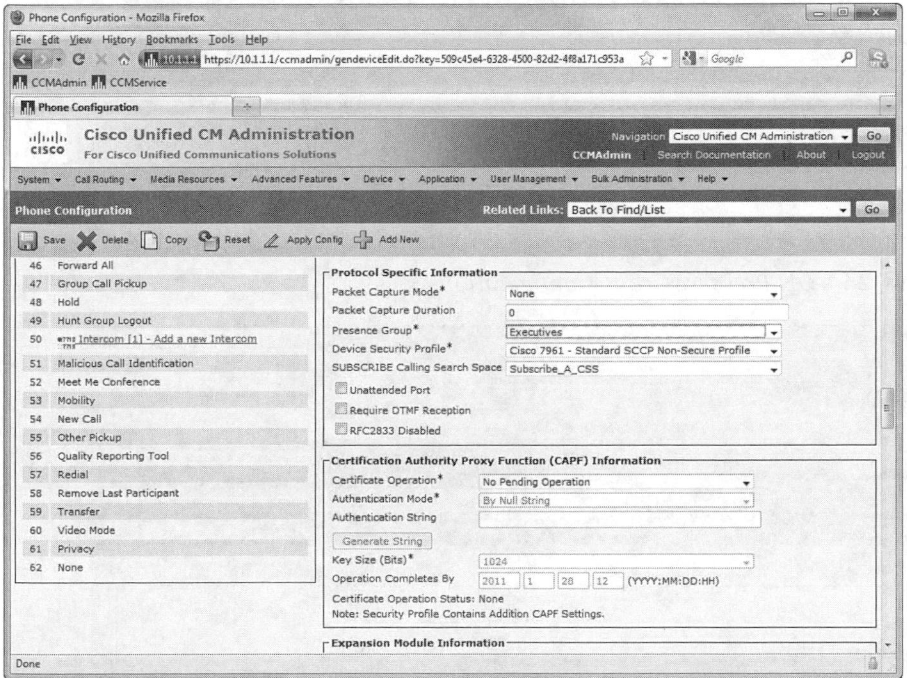

Figure 11-23 *IP Phone Presence Group Configuration*

Note: Phones are watchers that monitor the Presence status of Presence entities (such as DNs and SIP trunks). The Presence Group assignment and Inter-Presence Group Subscription Setting control whether or not the watcher can see the Presence status of the Presence entity. A SIP trunk, however, is both a watcher and a Presence entity, but only one Presence Group can be assigned to a SIP trunk. This single Presence Group is applied to both sending and receiving Presence subscriptions. Keeping that in mind, make sure that the Presence Group assigned to a SIP trunk has the correct permissions to watch and be watched by (or not) the other Presence Groups in the system.

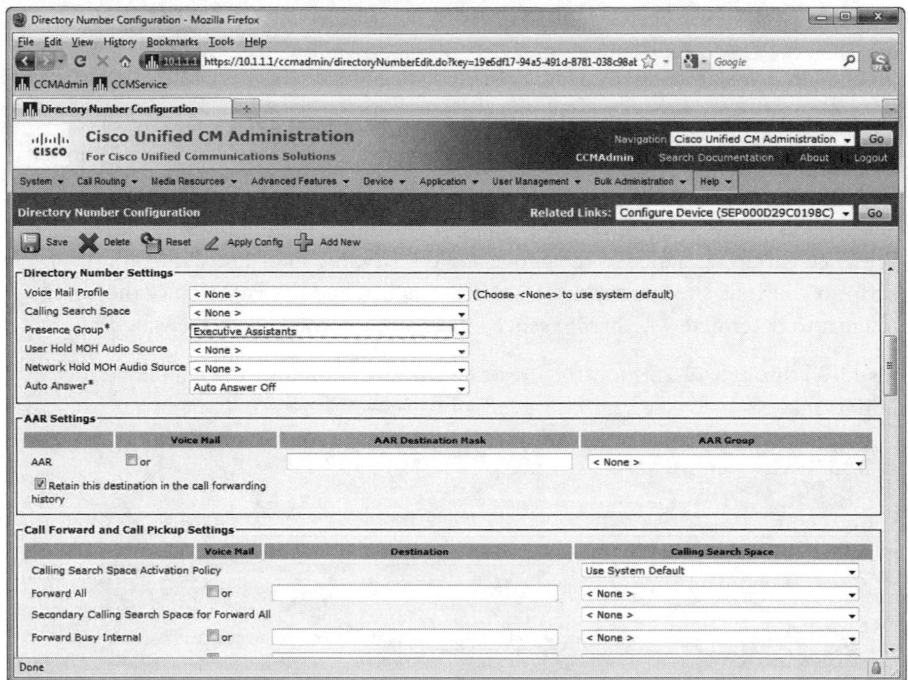

Figure 11-24 *DN Presence Group Configuration*

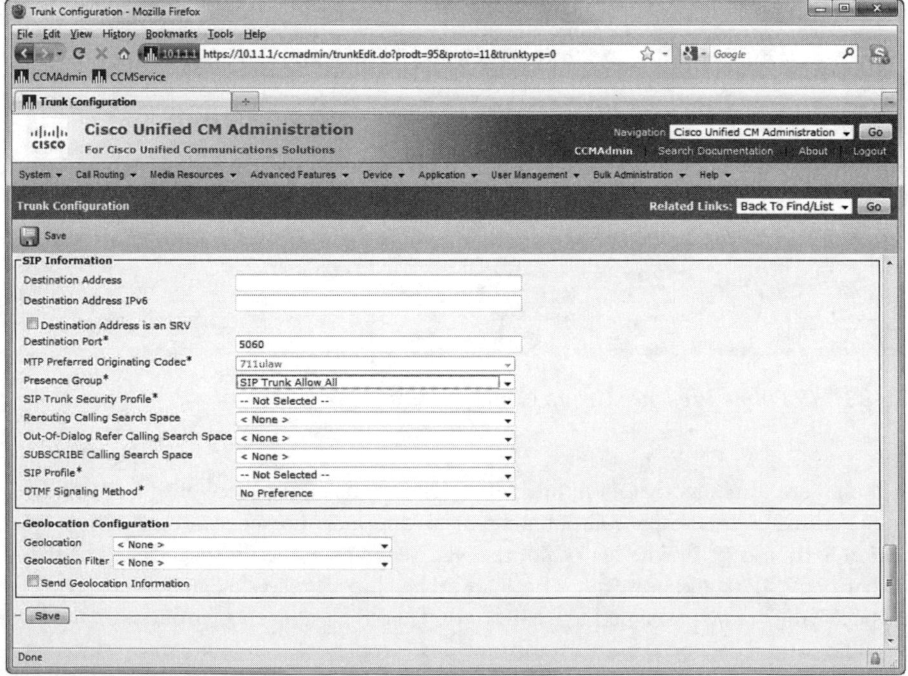

Figure 11-25 *SIP Trunk Presence Group Configuration*

Exam Preparation Tasks

Review All the Key Topics

Review the most important topics in the chapter, noted with the key topics icon in the outer margin of the page. Table 11-2 lists and describes these key topics and identifies the page number on which each is found.

Table 11-2 *Key Topics for Chapter 11*

Key Topic Element	Description	Page Number
Paragraph	Extension Mobility in CUCM	292
Paragraph	Enable Extension Mobility in CUCM	293
Paragraph	Intercom	303
Paragraph	Native Presence	303
Paragraph	Presence architecture	303
Paragraph	Configure Call Hunting	311
Paragraph	Configure Intercom	314
Paragraph	Configure Presence-Enabled Call Lists	316
Note	Presence configuration for SIP trunks	319

This chapter includes the following topics:

- **Understanding CUCM Mobility Features:** This section describes the Mobile Connect and Mobile Voice Access (MVA) features in CUCM.
- **Implementing Mobility Features in CUCM:** This section outlines the procedures for implementing Mobile Connect and MVA.

CHAPTER 12

Enabling Mobility Features in CUCM

The explosion in mobile technology has made communication possible from a huge array of devices, including home phones, cellular phones, WiFi-enabled smartphones, tablets, laptops, desktops, and specialized wireless IP phones. It has never been easier to stay in touch. But all this has created its own set of problems: Managing all the different methods of communication, all the phone numbers and voicemail boxes takes increasingly more time and creates more confusion.

In a business environment (and increasingly in a personal context), this confusion creates inefficiencies that actually impair communications instead of facilitating them. That might mean a loss of valuable business or simply missing an important call. The Unified Mobility feature set allows a person to be reached at a single number, and to place calls from any device and have all calls appear to come from that same number, creating a consistent point of voice contact and greatly simplifying the management of voice communications with an individual.

This chapter defines, describes, and reviews the implementation steps for both of these mobility features, explaining the advantages, drawbacks, and integration of each with the Cisco Unified Communications Manager (CUCM) architecture.

"Do I Know This Already?" Quiz

The "Do I Know This Already?" quiz allows you to assess whether you should read this entire chapter or simply jump to the "Exam Preparation Tasks" section for review. If you are in doubt, read the entire chapter. Table 12-1 outlines the major headings in this chapter and the corresponding "Do I Know This Already?" quiz questions. You can find the answers in Appendix A, "Answers Appendix."

Table 12-1 *"Do I Know This Already?" Foundation Topics Section-to-Question Mapping*

Foundation Topics Section	Questions Covered in This Section
Mobile Connect	1–3, 6
Mobility (General)	4–5
Mobile Voice Access	7–9

1. Which of the following best describes Mobile Connect?
   a. It has the capability to have multiple IP phones ring when a single DN is called.
   b. It has the capability to enable users to dial in from the PSTN and be greeted by an auto-attendant that allows them to enter the extension of the person they are trying to reach.
   c. It has the capability for IP phone users to forward calls to their mobile phones.
   d. It has the capability to have multiple remote PSTN devices ring simultaneously with their enterprise IP phone.
2. Which of the following are valid steps in the configuration of Mobile Connect? (Choose all that apply.)
   a. Add the Mobility softkey to softkey templates.
   b. Activate Mobility for end users.
   c. Associate users to their IP phones.
   d. Configure Remote Destination Profiles.
   e. Associate Remote Destinations with Remote Destination Profiles.
   f. Configure and apply Access Lists.
   g. Obtain additional licenses for Mobile Connect.
3. Which of the following is true of Access Lists?
   a. Access Lists are required for Mobility features to function.
   b. Access Lists are configured on the IOS gateways to filter calls at the interface.
   c. Access Lists cannot be empty.
   d. An empty Access List applied to an Allowed Calls filter allows no calls.
   e. An empty Access List applied to a Disallowed Calls filter allows no calls.
4. What is the maximum number of Remote Destination Profiles that can be configured for a user?
   a. 1
   b. 5
   c. 10
   d. Unlimited
5. What is the maximum number of Remote Destinations that can be configured for a user?
   a. 1
   b. 5
   c. 10
   d. Unlimited

6. What is the correct order of processing during a Mobile Connect call, if the caller has dialed the user's IP phone number?

   a. Remote Destination, Remote Destination Profile, Ring Schedule, Access List
   b. Remote Destination Profile, Remote Destination, Ring Schedule, Access List
   c. Access List, Ring Schedule, Remote Destination Profile, Remote Destination
   d. Ring Schedule, Access List, Remote Destination, Remote Destination Profile

7. Which component is not part of a Mobile Voice Access configuration?

   a. Cisco Unity Connection Auto-Attendant
   b. Cisco IOS H.323 VXML gateway
   c. Mobile Voice Access Media Resource
   d. End user configured for Mobile Voice Access

8. How does an H.323 gateway route calls inbound for the MVA service to the CUCM server hosting the service?

   a. One dial peer matching the MVA access PSTN number pointing to the CUCM server.
   b. Two dial peers: One matching the MVA access PSTN number with **incoming called-number** configured, and one matching the MVA access PSTN number pointing to the CUCM server.
   c. A static route redirecting all HTTP calls to the CUCM server.
   d. All H.323 configuration is dynamically created by the CUCM server via TFTP download.

9. Which IOS dial peer configuration command associates the dial peer matching the MVA access PSTN number with the VXML application hosted by the CUCM server running the MVA service?

   a. incoming called-number 4085555000
   b. session target ipv4:10.1.1.1
   c. service mva
   d. service mva http://10.1.1.1:8080/ccmivr/pages/IVRMainpage.vxml

Foundation Topics

Understanding CUCM Mobility Features

CUCM incorporates a range of Mobility features that allow a user to interact with their Unified Communications devices and applications regardless of where he happens to be. The goal is to extend the ability to communicate with customers or colleagues using their enterprise IP phone number, both for inbound and outbound calls, in a seamless and flexible way. The following sections describe the features and configuration of some of the Mobility capabilities of CUCM.

Describe Mobile Connect

Mobile Connect is often called Single Number Reach: A user's IP phone number becomes the single number by which all the various other devices that person uses can be reached, including home phones, mobile phones, Internet-based VoIP numbers, and so on. The benefit is that a single point of voice contact is published for simplicity and consistency, whereas a range of devices can actually take calls, which provides maximum flexibility and reachability for the person almost regardless of where they may be or which communication method they may have available to them.

The user experience is simple, but powerful; if he receives a call at his business number, his IP phone rings. In addition, all the other devices configured for Mobile Connect ring at the same time. Whichever device is answered receives the extended call, and all other devices stop ringing.

Suppose the user answers the call on his mobile phone while on the way to his office. When he gets to his desk, he has the option of picking up the call at his IP phone by pressing a softkey. The call is seamlessly transferred to the desk phone, and the caller may not even realize it has happened.

Likewise, the user can be in a call on his IP phone and redirects it to his mobile device as he leaves the office, again without the other party knowing that it has happened (except, of course, the possibility of a change in background noise or voice quality on a cell phone).

If the user calls a colleague's IP phone by dialing his Direct Inward Dial (DID) from his mobile phone, CUCM recognizes the Automatic Number Identification (ANI) (the Caller ID) of the user as matching a Remote Destination Profile, which has a shared line with the user's directory number (DN). The call to the colleague's IP phone is presented as being from the DN of the user's IP phone. This functionality also allows the user to call in to the Cisco Unity Connection voice-messaging system and have Easy Message Access to his personal mailbox (which is associated with the DN on his IP phone).

Note: Most implementations of Mobile Connect use access codes for remote destinations. If this is the case, some digit manipulation of the incoming ANI may be necessary to match the entire Remote Destination Profile number pattern. Alternatively, the CallManager Service Parameter **Matching Call ID with Remote Destination** can be set to

Partial Match, which causes CUCM to make the closest match with a Remote Destination Profile, starting with the least-significant digit of the ANI.

Unified Mobility Architecture

Mobile Connect uses Remote Destination Profiles to configure virtual phones that share several configuration settings with the user's primary IP phone. In effect, the Remote Destination Profiles act as phones with shared lines; when the primary number rings, the shared lines also ring, but the system configuration redirects the call out to the PSTN to ring the other devices.

Remote Destination Profiles are configured with many of the same settings as the physical IP phone, including a Partition, Device Pool, Calling Search Space (CSS), User and Network Music on Hold (MoH), and, of course, the same DN. The profile also includes a Rerouting Calling Search Space to allow the system to route calls to the device, even if that call would normally be restricted by the CSS of the IP phone.

Up to ten Remote Destinations can be defined per user. Immediate complexities emerge with respect to the behavior of the Mobile Connect feature: How long should the system wait before ringing the Remote Destination Profile phones? How long should the remote devices be made to ring? How long must the remote device ring before the call can be picked up on it? The adjustment of these timers can greatly improve the utility of the Mobile Connect feature, as well as avoid unwanted behavior, such as calls going to voicemail too soon or possibly going to some other voicemail (such as the personal voicemail at the user's home phone).

Access Lists

Not to be confused with IOS router access control lists, these Access Lists give both administrators and users control over which calls will ring which Remote Destination Profile devices, and at what time of day.

Access lists filter calls based on the Caller ID; they can be configured to allow certain number patterns (called a *white list*) or to block certain number patterns (called a *black list*). The callers are identified using three types of match:

- **Not available:** The Caller ID is not provided.
- **Private:** The Caller ID is not displayed.
- **Directory number:** The Caller ID matches a specific number, or a wildcard-defined range of numbers (using the digits 0 through 9, *, #, and the wildcards of X and !).

Time-of-Day Access

Each Remote Destination Profile can be configured with a schedule that controls when it should be included in the set of remote device that will ring. By default, all remote devices ring. The time zone of the actual remote device should be specified so that the time-of-day rules apply correctly.

For incoming calls, the Time-of-Day rules are processed first. If the time-of-day rule allows the call to be extended to the Remote Destination Profile, any configured Access Lists are processed next. If the Access List allows the call, it is extended to the Remote Destination device.

Key Point: Avoid using empty Access Lists. An empty Access List selected in a white list causes no calls to be routed; an empty Access List in a black list causes all calls to be routed to the remote destination device.

Mobile Voice Access

Mobile Voice Access (MVA) provides the same single-number consistency and flexibility for outbound calls from users: By accessing the CUCM system from their mobile device, they can instruct the system to place calls and have the call appear to be from their IP phone using their primary IP phone number (assumed to be a DID, but could be the main business number). As a result, their physical location has no impact on their voice communication consistency. Whether they are sitting in front of their IP phone or using their mobile phone from the golf course, the calls they place all appear to come from their primary business number.

To use the feature, the user dials in to a specific PSTN DID to access the MVA service. A specially configured VoiceXML gateway routes calls to an Interactive Voice Response (IVR) application that guides the user through his MVA session. The IVR app provides security by prompting the user to authenticate with their User ID (optional) and PIN. Once successfully authenticated, the IVR prompts the user for the number they want to dial. The user enters the PSTN number and the system places the call, using the ANI of the user's IP phone. The user can switch between his mobile device and his IP phone during the call. The called party sees the Caller ID as that of the user's IP phone, providing the single-number consistency and recognition, as well as ease of callback from a call list.

Figure 12-1 shows the basic call flow in MVA.

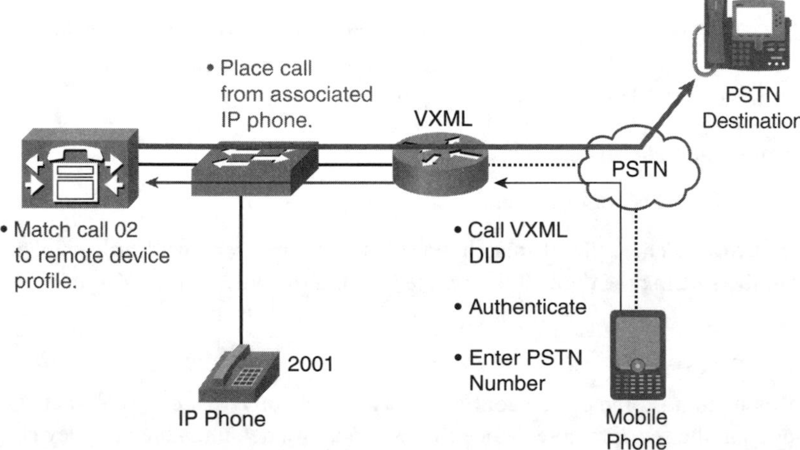

Figure 12-1 *Mobile Voice Access Basic Call Flow*

Implementing Mobility Features in CUCM

The configuration of the various Mobility features in CUCM is not difficult, but it is repetitive and potentially time-consuming, especially when many users need to be configured. This section outlines the configuration tasks for the Mobility features described earlier.

Configuring Mobile Connect

Many different components interact to allow Mobile Connect to function. The basic steps to configure Mobile Connect are as follows:

1. Configure Softkey Templates to include the Mobility key.
2. Configure user accounts for Mobility.
3. Configure IP phones to support Mobility features.
4. Create Remote Destination Profiles and assign them to each user.
5. Add Remote Destinations to Remote Destination Profiles.
6. Configure ring schedules for each Remote Destination.
7. Configure Access Lists.
8. Apply Access Lists to Remote Destinations.
9. Configure Service Parameters.

Each of the following sections describes these configuration steps in detail.

Step 1: Configure Softkey Templates

To use Mobile Connect, the user activates a softkey on his IP phone. Complete the following tasks to add the softkey to the phone(s) that will use this feature:

1. Navigate to **Device** > **Device Settings** > **Softkey Template**.
2. Select, copy, and modify or ad a new template.
3. In the **Related Tasks** pull-down, select **Configure Softkey Layout** and click **Go**.
4. Add the **Mobility** softkey to the **OnHook** and **Connected** call states lists. Click **Save** after each move.

Figure 12-2 shows the Softkey Template with the Mobility key added.

Step 2: Configure User Accounts for Mobility

Individual user accounts must be enabled for Mobility, and some settings tuned for optimal functionality. Complete the following steps to set up the user accounts for Mobility:

1. Navigate to **User Management** > **End User** and select a user.
2. Check the **Enable Mobility** checkbox.
3. Set the **Remote Destination Limit** (maximum 10).
4. Set the **Maximum Wait Time for Desk Pickup** timer. This is how much time (in milliseconds) is allowed for the user to pick up a call that was redirected from the remote device to the IP phone. The default is 10,000 ms (10 sec) with a maximum of 30,000 ms (30 sec).

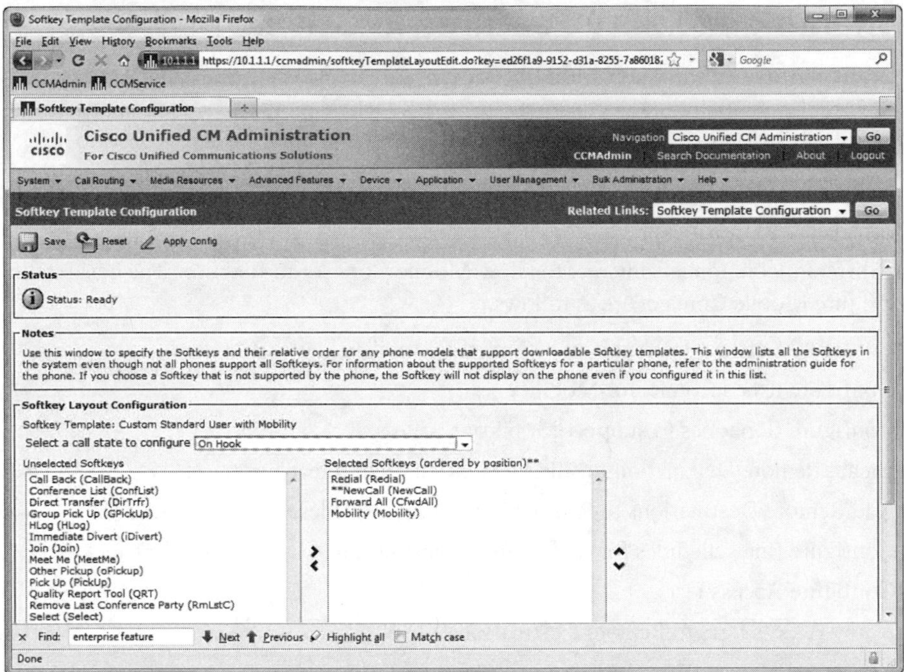

Figure 12-2 *Softkey Template with Mobility Key Added*

Figure 12-3 shows the End User Configuration page for Mobile Connect.

Step 3: Configure the IP Phone to Support Mobility Features

The users' IP phones must be configured to link the user configuration and softkey template. To do this, complete the following steps:

1. Assign the Softkey Template to which you previously added the Mobility key.
2. Set the **Owner User ID** to the appropriate Mobility-configured user.

Step 4: Create Remote Destination Profiles

The following steps create the Remote Destination Profiles, link them to the user accounts, and ensure that calls can reach the remote numbers:

1. Navigate to **Device > Device Settings > Remote Destination Profile.**
2. Configure a name.
3. Select the **User ID** to be associated with this profile.
4. Select the **Rerouting Calling Search Space.** This CSS will redirect calls to remote devices and, therefore, must provide access to the remote devices' phone numbers.

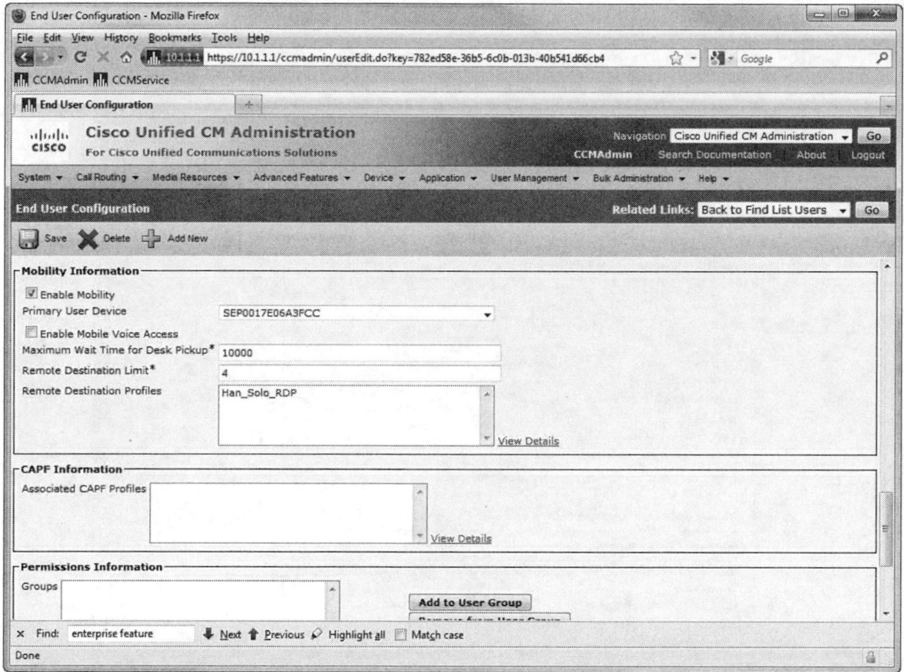

Figure 12-3 *End User Configurations for Mobile Connect*

Figure 12-4 illustrates a Remote Destination Profile configuration.

Step 5: Add Remote Destinations to Remote Destination Profiles

These steps link the Remote Destinations to the Remote Destination Profiles:

1. Navigate to **Device > Remote Destination.**
2. Click **Add New.**
3. Enter a **Name.**
4. Set the **Destination Number**, just as it would be dialed from an IP phone, including any access codes. This entry must be a PSTN number.
5. Associate the appropriate **Remote Destination Profile** for the user. Once configured, this cannot be changed; to change it, delete the Remote Destination and re-create it with the desired setting.
6. Check the **Mobile Phone** checkbox to allow manual handoff of calls from the IP phone using the Mobility softkey.
7. Check the **Enable Mobile Connect** checkbox to include this Remote Destination in the set of those that will ring when the IP phone shared line rings.
8. In the **Association Information** area, select one or more of the shared lines on the Remote Destination Profile.

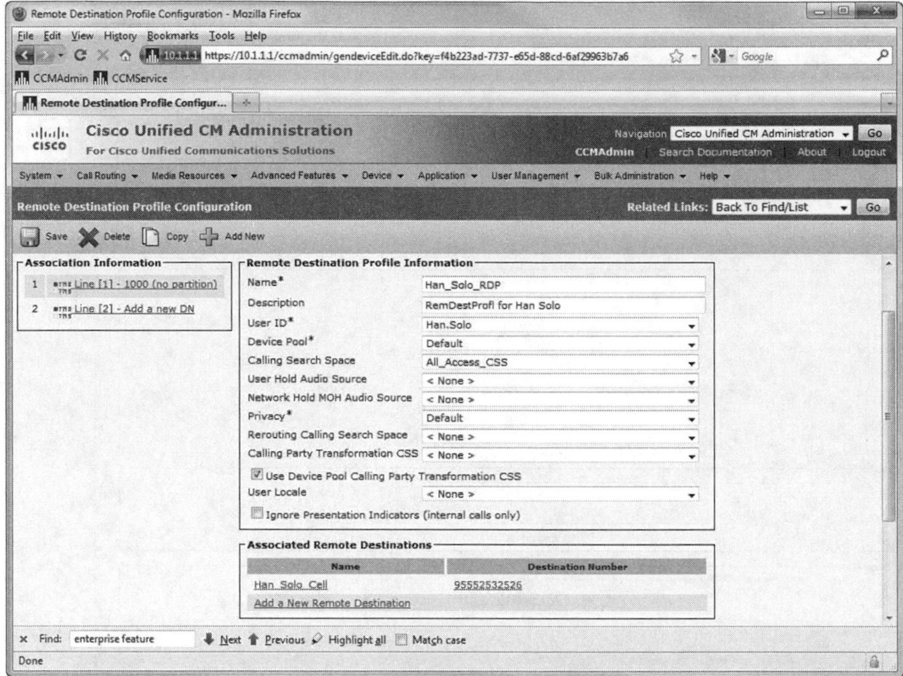

Figure 12-4 *Remote Destination Profile Configuration*

Figure 12-5 illustrates adding a Remote Destination to a Remote Destination Profile.

Step 6: Configure Ring Schedules for Each Remote Destination

The next steps tune the functionality of the Remote Destinations by limiting the times of day when they will ring:

1. Set the days and times when this remote device should ring.
2. Select the correct time zone of the remote device.

Step 7: Configure Access Lists

These steps configure Access Lists to limit which numbers can or cannot ring Remote Destinations:

1. Navigate to **Call Routing** > **Class of Control** > **Access List.**
2. Click **Add New.**
3. Configure a **Name** for the list.
4. Set the **Owner** User ID from the pull-down. This entry should be the Mobile Connect user to whom the Access List applies.
5. Choose either **Allowed** or **Blocked** to set the function of the list.
6. Click **Save.**

When the screen refreshes, in the **Access List Member** area, click **Add Member.**

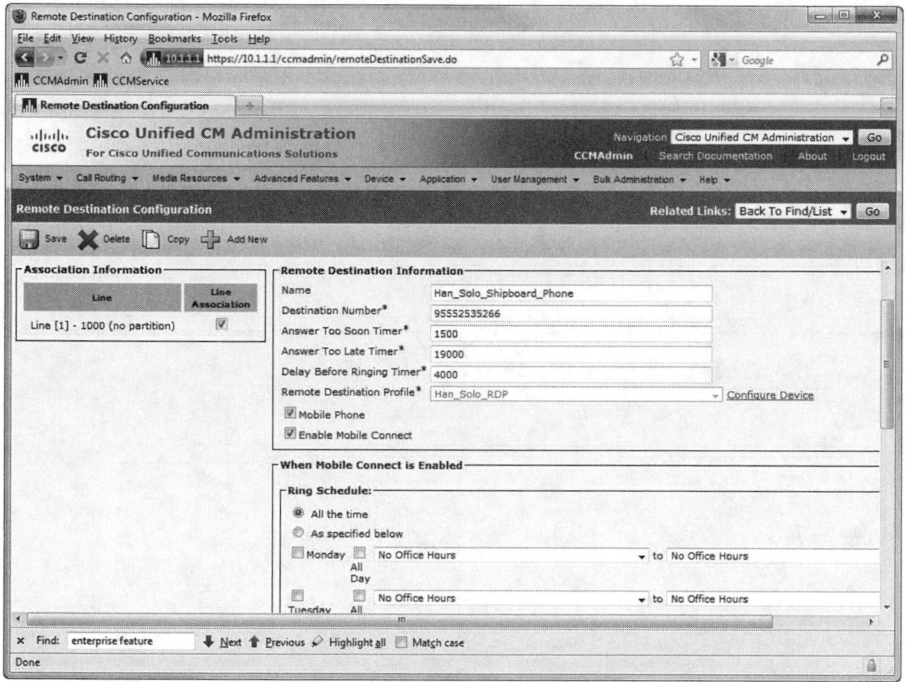

Figure 12-5 *Adding Remote Destinations to a Remote Destination Profile*

Figure 12-6 shows an Access List configuration to block certain calls (a Disallowed list or blacklist).

- In the **Access List Member Detail** page, select **Filter Mask:**
  - **Directory Number:** To enter a specific ANI number or wildcard pattern.
  - **Private:** To filter based on calls with Caller ID not displayed.
  - **Not Available:** To filter based on calls without Caller ID.

- In the **DN Mask** field, you can enter a specific digit string (for example, 5558675309) or use the wildcards X (matching a single dialed digit) and ! (matching any number of digits) to represent multiple strings in one entry (much like Route Patterns).

Figure 12-7 illustrates a Filter Mask configuration to block a specific Caller ID number.

Step 8: Apply Access Lists

These Access Lists configured previously must be applied to the Remote Destinations, as described in the following steps:

1. Navigate to **Device > Remote Destination** and select a Remote Destination.
2. Select either the **Always ring this destination** or one of **Ring this destination if the caller is in** (Allowed) or **Do not ring this destination if the caller is in** (Blocked) radio button. You must select one or the other; selecting both is not an option.

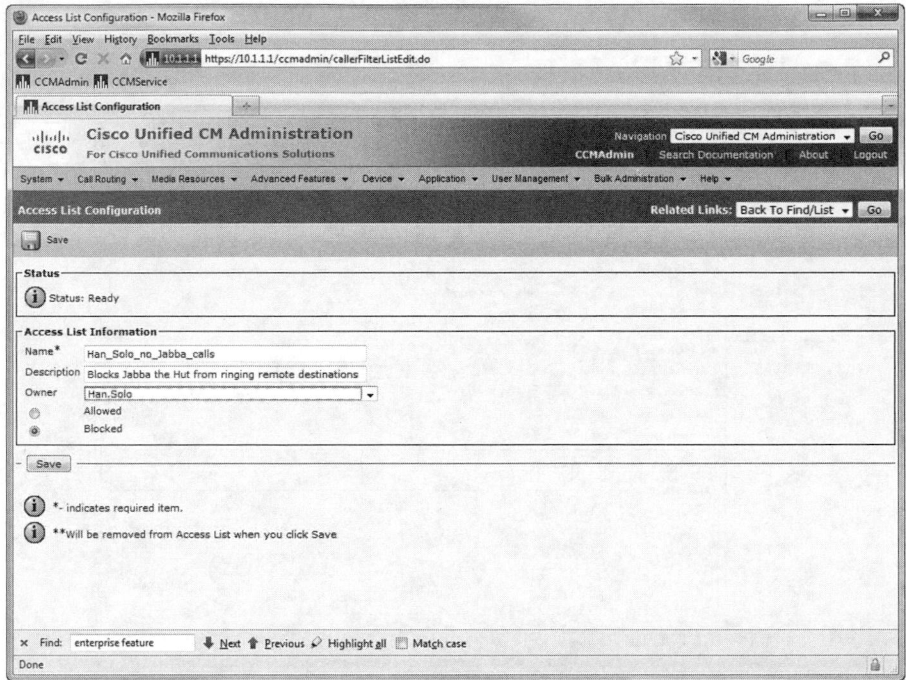

Figure 12-6 *Access List Configuration*

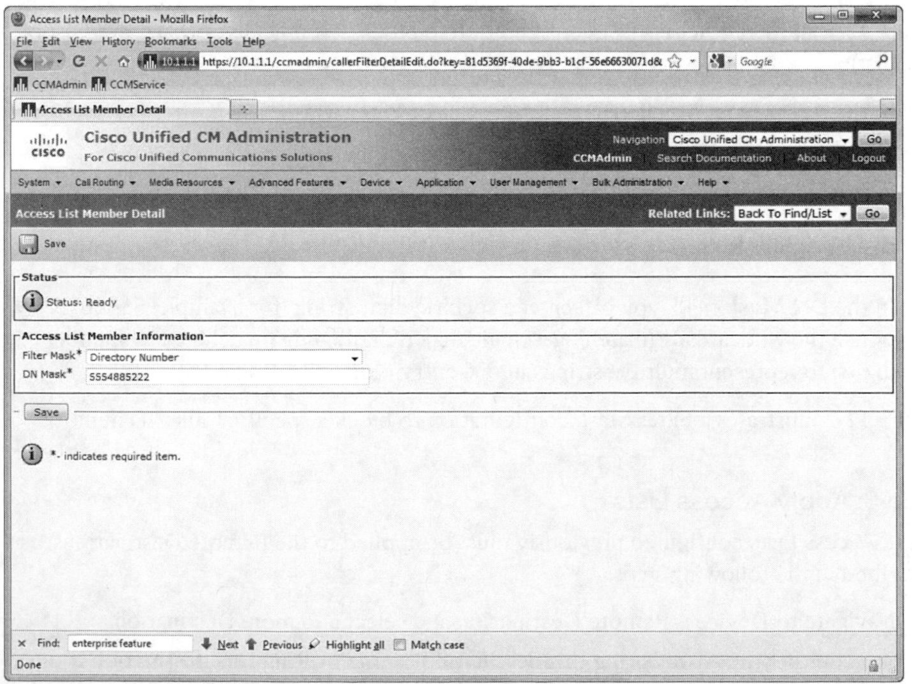

Figure 12-7 *Configuring a Filter Mask for an Access List*

3. In the pull-down next to the ring selection, select the Access List that provides the desired filter.
4. Click **Save.**

Figure 12-8 shows an Access List applied to a Remote Destination.

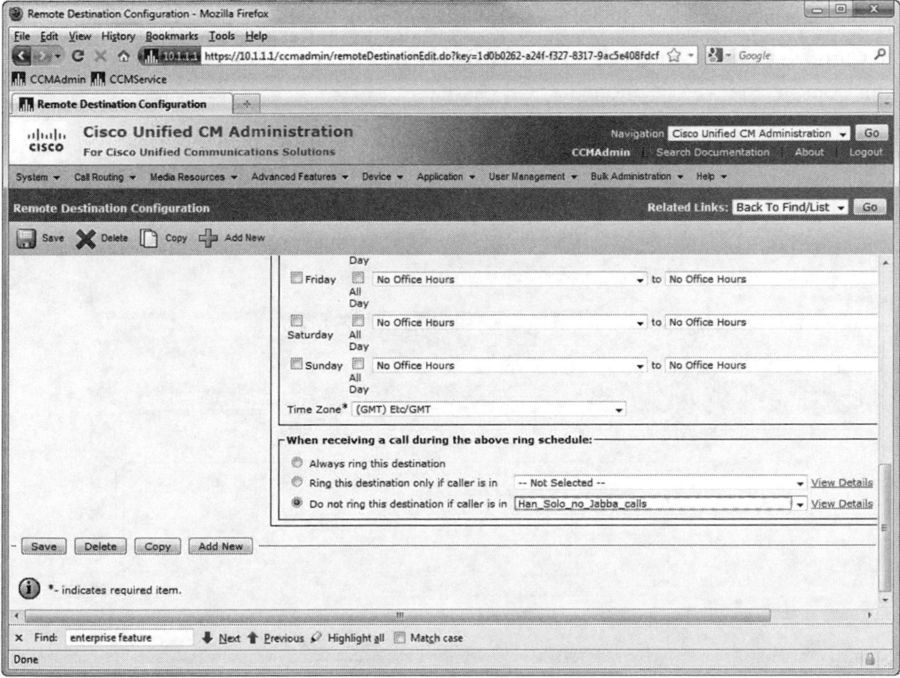

Figure 12-8 *Applying an Access List to a Remote Destination*

Step 9: Configure Service Parameters

Certain service parameters can be tuned to customize the behavior of the Mobility features, as discussed in these steps:

1. Navigate to **System > Service Parameters.** Select the **Server** you want to configure from the pull-down.
2. Select the **CallManager** service from the **Service** pull-down.
3. Scroll down to the **Clusterwide Parameters (System - Mobility)** section.
4. In the **Inbound Calling Search Space for Remote Destination** field, choose either **Trunk or Gateway Inbound Calling Search Space** (the default, which uses the CSS of the trunk or gateway that is routing the inbound call from the Remote Destination) or **Remote Destination Profile + Line Calling Search Space** (which uses the combined line and Remote Destination Profile CSS).
5. In the **Matching Caller ID with Remote Destination** field, select either **Complete Match** (the default, which requires the incoming Caller ID to exactly match the

Remote Destination number) or **Partial Match**, which allows you to specify how many digits of the Caller ID to match, starting with the least significant digit.

6. Scroll down to the **Clusterwide Parameters (Feature - Reroute Remote Destination Calls to Enterprise Number)** section.
7. Set **Reroute Remote Destination Calls to Enterprise Number** to **True** (the default is **False**) to cause direct calls to a Remote Destination number to be extended to the IP phone number, allowing the user to take advantage of Mobility features.
8. Set **Ignore Call Forward All on Enterprise DN** to **True** to route calls to Remote Destinations, even if the IP phone has CFA active.

Figure 12-9 shows some of the Service Parameters configuration for Mobility.

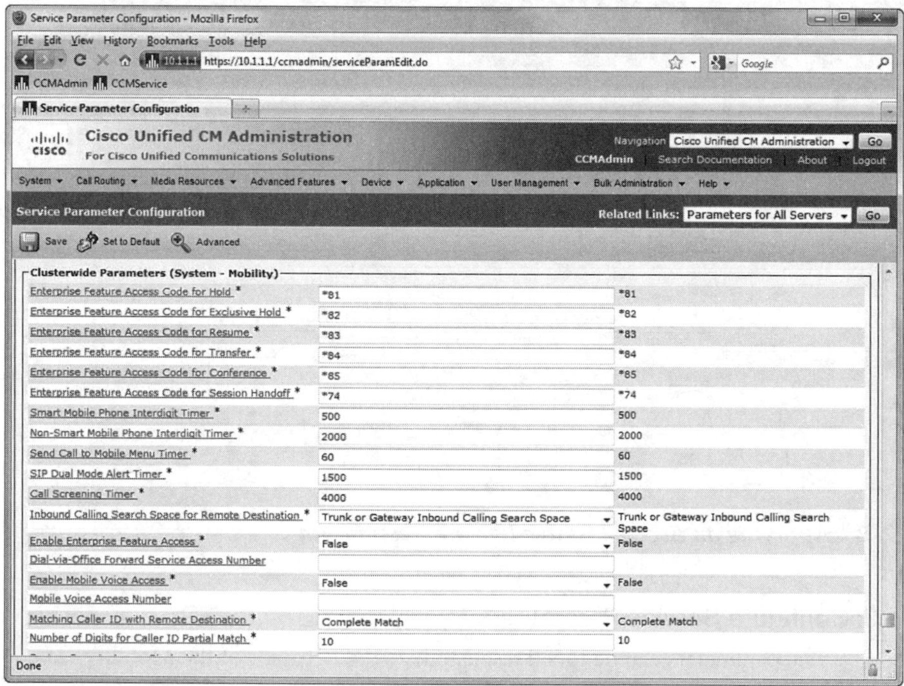

Figure 12-9 *Service Parameters Configuration for Mobility*

Configuring MVA

The basic steps to configure MVA are as follows:

1. Activate the MVA service.
2. Configure service parameters.
3. Enable MVA for each user.
4. Configure the MVA media resource.
5. Configure the MVA VXML application at the IOS gateway.

These steps are described in more detail in the following sections.

Step 1: Activate the MVA Service

Before the MVA feature can function, the service must be activated, as described in the following:

1. Navigate to **Unified Serviceability > Tools > Service Activation.**
2. Select the Cisco Unified Mobile Voice Access Service.
3. Click **Save.**

Step 2: Configure Service Parameters

With the MVA service active, you can now enable it for the cluster, as shown in the following steps.

1. Navigate to **Unified CM Administration > System > Service Parameters.**
2. Select the server you want to configure, and select the **Cisco CallManager Service.**
3. Scroll down to **Clusterwide Parameters (System - Mobility).**
4. Set the **Enable Mobile Voice Access** value to **True.**
5. Modify other system parameters, such as access codes, if desired.

Figure 12-10 shows the Cisco Unified Mobile Voice Access Service activated.

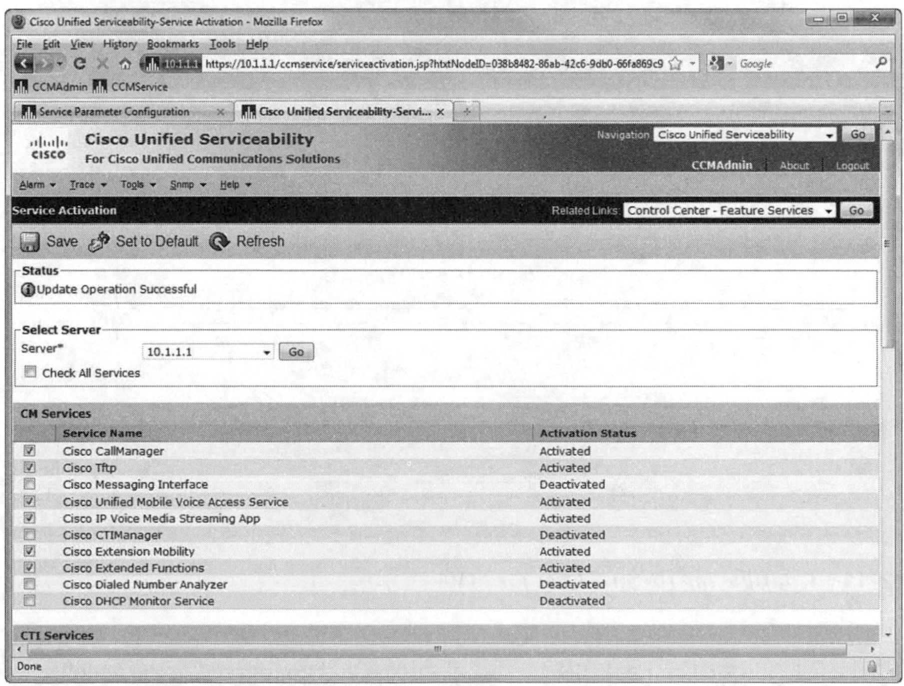

Figure 12-10 *Activating the MVA Service*

Note: Enabling the MVA service is required to enable MVA globally, but it is also necessary to activate MVA for each user for it to actually function.

Step 3: Enable MVA for Each User

As noted previously, it is not sufficient to activate the MVA service and enable MVA for the cluster; you must now enable MVA for each user, as described in the following steps:

1. Navigate to the user configuration page for the user(s) for whom you want to enable MVA.
2. Scroll down to the **Mobility Information** section.
3. Check the **Enable Mobile Voice Access** box.
4. Verify that the **Remote Destination Profile** listed is configured correctly to provide authentication.

Figure 12-11 shows the End User Configuration page for MVA.

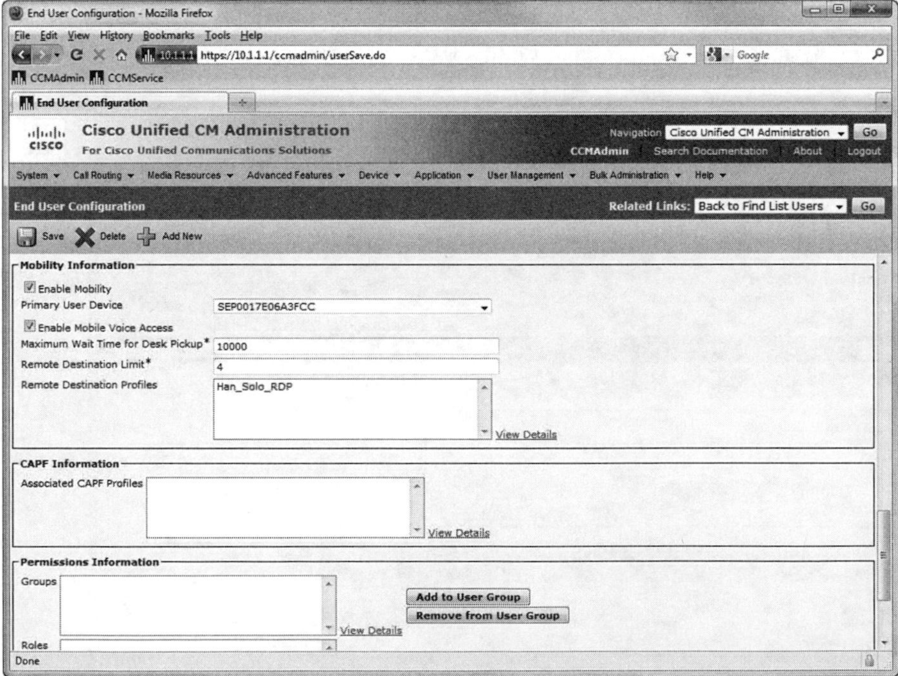

Figure 12-11 *Configuring the End User for MVA*

Note: Other Mobility-related parameters were discussed and configured in the Mobile Connect sections of this chapter.

Step 4: Configure the MVA Media Resource

The MVA media resource is automatically created when the MVA service is activated. To configure options, follow these steps:

1. Navigate to **Media Resources > Mobile Voice Access.**
2. Enter a number for the **Mobile Voice Access Directory Number**. This number is the internal number to which the H.323 MVA gateway will forward calls it receives on the PSTN number for MVA access. On the gateway, a dial peer must be configured that matches the PSTN MVA number to the MVA CUCM server.
3. Assign a **Partition** if desired. This partition must be in the CSS of the MVA gateway.
4. Set additional **Locales**, if desired, to provide users with MVA IVR service in multiple languages.

Figure 12-12 shows the MVA media resource configuration page.

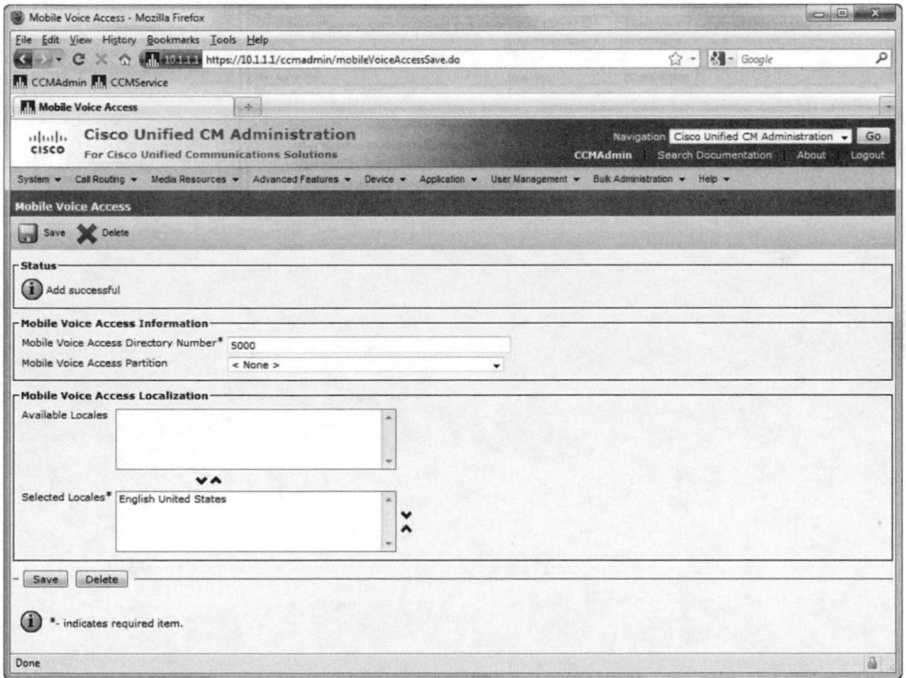

Figure 12-12 *MVA Media Resource Configuration*

Step 5: Configure the MVA VXML Application at the IOS Gateway

The H.323 gateway configuration must include the following statements. The output shown here is annotated to explain what the commands do:

```
! Define the MVA Application and URL
application
 service mva http://10.1.1.1:8080/ccmivr/pages/IVRMainpage.vxml
```

```
dial-peer voice 50001 pots
! Associate the MVA application to this dial peer
 service mva
! Match the PSTN MVA access number to this inbound dial peer
 incoming called-number 4085555000
 direct-inward-dial
dial-peer voice 50002 voip
! Match the PSTN MVA access number to this outbound dial peer
 destination-pattern 4085555000
! Identify the CUCM server running the MVA service VXML app referenced above
 session target ipv4:10.1.1.1
 dtmf-relay h245-alphanumeric
 codec g711ulaw
 no vad
```

Exam Preparation Tasks

Review All the Key Topics

Review the most important topics in the chapter, noted with the key topics icon in the outer margin of the page. Table 12-2 describes these key topics and identifies the page number on which each is found.

Table 12-2 *Key Topics for Chapter 12*

Key Topic Element	Description	Page Number
"Unified Mobility Architecture" section	Understanding the components of Mobile Connect and MVA	327
"Mobile Voice Access" section	Understanding MVA	328
List	Steps required to configure Mobile Connect	329
List	Steps required to configure MVA	336
CLI output	IOS configuration to support MVA	339

Definition of Key Terms

Define the following key terms from this chapter, and check your answers in the Glossary:

Mobile Connect, Mobile Voice Access (MVA)

This chapter includes the following topics:

- **Describe Cisco Unity Connection:** This section describes the capabilities and features of the CUC application with CUCM integration.
- **Describe Cisco Unity Connection Users and Mailboxes:** This section describes the core components and related configuration requirements of the CUC system.
- **Implement Cisco Unity Connection Users and Mailboxes:** This section reviews the implementation of users and mailboxes in CUC.

CHAPTER 13

Voicemail Integration with Cisco Unity Connection

Voicemail is a ubiquitous feature of a modern business phone system. The Cisco Unity Connection product certainly qualifies as voicemail, but has so many related and advanced features that calling it simply voicemail is at least inaccurate and probably an injustice. This chapter reviews the features and capability of the CUC application, the core components and systems, and the configuration and implementation of basic services for voice messaging.

"Do I Know This Already?" Quiz

The "Do I Know This Already?" quiz allows you to assess whether you should read this entire chapter or simply jump to the "Exam Preparation Tasks" section for review. If you are in doubt, read the entire chapter. Table 13-1 outlines the major headings in this chapter and the corresponding "Do I Know This Already?" quiz questions. You can find the answers in Appendix A, "Answers Appendix."

Table 13-1 *"Do I Know This Already?" Foundation Topics Section-to-Question Mapping*

Foundation Topics Section	Questions
Describe Cisco Unity Connection	1–6
Describe Cisco Unity Users and Mailboxes	7–8
Implement Cisco Unity Connection Users and Mailboxes	9–10

1. What is the maximum number of mailboxes supported on a single CUC server?
   a. 3,000
   b. 7,500
   c. 10,000
   d. 20,000

2. Which of the following is *not* deployed as a Linux appliance?
   a. Cisco Unity Connection 8.x
   b. Cisco Unified Contact Center 8.x
   c. Cisco Unified Communications Manager 8.x
   d. Cisco Emergency Responder
   e. Cisco Unified Presence

3. Which of the following are supported phone-system integrations in CUC? (Choose all that apply.)
   a. CUCM using SCCP
   b. IP PBX using SIP
   c. Digital PBX using PIMG
   d. PBX using AMIS
   e. PBX using analog DTMF

4. Which of the following represents the correct order of processing of components when a user presses his Messages button?
   a. Voicemail Pilot, Voicemail Profile, Hunt Group, Hunt List, Line Group, Voicemail Port
   b. Voicemail Profile, Voicemail Pilot, Hunt Pilot, Hunt List, Line Group, Voice Mail Port
   c. Voicemail Profile, Voicemail Port, Voicemail Pilot, Hunt Pilot, Hunt List, Line Group
   d. Hunt Pilot, Hunt List, Line Group, Voicemail Profile, Voicemail Pilot, Voicemail Port

5. Which of the following is *not* a call handler type in CUC?
   a. System Call Handler
   b. Directory Call Handler
   c. Interview Call Handler
   d. Holiday Call Handler

6. Which of the following would be processed by Direct Routing Rules in CUC? (Choose two.)
   a. Bill dials Max's phone, but Max is on the phone, so it goes to voicemail.
   b. Max dial's Keisa's phone, but Keisa is away from her desk, so it goes to voicemail.
   c. Keisa dials Angie's phone, which is forwarded to Angie's cell phone, and Keisa gets Angie's cellphone voicemail.
   d. Max sees his message lamp and presses the Messages button to check his voicemail.
   e. Bill calls the company number from home and hears the auto-attendant opening greeting.

7. Where can the Maximum Message Length be set? (Choose three.)
   a. Class of Service
   b. User Template
   c. User Account Settings
   d. Mailbox Quotas
   e. Message Store Settings
8. Guy is the CUC administrator. He uses AXL to import user accounts from the CUCM database into CUC. What is the result?
   a. Usernames and aliases of users imported via AXL cannot be changed in CUC.
   b. User passwords cannot be changed in CUC.
   c. Active Directory regularly syncs users to CUC.
   d. User accounts can no longer be created in the local CUC database.
9. Guy notices that some of the user accounts he wants to import from CUCM into CUC using AXL do not appear in the list of users found via AXL. What might be the problem?
   a. The LDAP Manager account password is incorrect.
   b. The DirSync Service is not activated.
   c. The missing user accounts are not configured with a Primary Extension in the CUCM database.
   d. The missing user accounts had the Do Not List In Directory setting checked.
10. Pete wants to be able to call into the CUC Auto-Attendant number from his mobile phone and be prompted to enter his PIN for quick access to his mailbox. What should he ask the CUC administrator to do for him?
    a. Configure a special dial peer on the gateway router to send calls from Pete's number to an Attempt Sign-In conversation on CUC.
    b. Tell Pete about the hidden key press during the Opening Greeting that will send him to the Attempt Sign-In conversation.
    c. Buy a different voicemail system, because CUC cannot do this.
    d. Add an alternate extension with the ten-digit ANI of Pete's mobile phone.

Foundation Topics

Describe Cisco Unity Connection

Cisco Unity Connection (CUC) is a full-featured voice-messaging and voice-recognition system, providing universal access to calls and messages as part of a Unified Communications solution. Up to 20,000 users can be hosted by a single CUCv8.x server (assuming the most powerful hardware platform). Voice recognition capabilities allow speech-activated commands to be used by both internal and external callers. A built-in IMAP server allows e-mail access to voice messages, and a clientless, web-based interface provides the same capability from any compatible browser with web access to the CUC server. The following sections discuss CUC.

Overview of Cisco Unity Connection

CUC is one of the five Unified Communication products currently offered as a Linux appliance. (The others are Cisco Unified Communications Manager [CUCM], Cisco Unified Presence [CUP], Cisco Emergency Responder, and Unified Contact Center Express.) The data and message store databases are held locally on the server, both using an instance of the Informix Database Service application, just as the other Unified Communications applications do.

CUC supports integration with a variety of traditional PBX systems that support either native IP functionality or a digital TDM circuit that can be connected via PBX or T1 IP Media Gateway (PBX IP Media Gateway [PIMG] or T1 IP Media Gateway [TIMG]). CUC users can be manually configured, bulk imported from a Comma-Separated Values (CSV) file, imported from a CUCM server database, or synchronized directly from a Lightweight Directory Access Protocol (LDAP) system. Password authentication can be redirected to the LDAP system for single sign-on functionality.

CUC can integrate with a Microsoft Exchange server, using Web-Based Distributed Authoring and Versioning (WebDAV) to provide calendar and journal information for integration with Cisco Unified MeetingPlace and Cisco Unified MeetingPlace Express, Microsoft Exchange 2003, and personal call routing rules within CUC itself. Calendar integration services for Microsoft Exchange 2007 are handled using the web service's application programming interface (API).

CUC provides a traditional Telephone User Interface (TUI) for interaction over Dual-Tone Multi-Frequency (DTMF) phones, a Voice User Interface (VUI) for hands-free interaction, and the IP phone service-based Voice View Express to see voice-message headers on the IP phone screen or in the Cisco Unified Personal Communicator.

Single-Site and Multisite Deployment Considerations

The simplest deployment of the CUC application is as a single-site model, with one building or campus accessing a single CUC server (or active-active redundant pair). The advantages of design simplicity, a single codec for all calls, and a greatly reduced implementation task list make this an attractive option.

If there are multiple locations (or will be in the future), a multi-site deployment may be a better choice; although users can call across the IP WAN to check or leave voice messages (or use other features) in a single-site deployment, doing so can put a significant extra load on WAN bandwidth and transcoder resources. This is especially true as the number of users in the system increases. Locating additional servers in branch locations can greatly reduce the impact of these problems while providing the same seamless functionality as in a single-site model.

CUC Integration Overview

Integration in this context refers to interoperation with a PBX- or IP-based telephone system. CUC supports a variety of integrations using SCCP, SIP, or PIMG/TIMG. Multiple phone systems are supported concurrently; CUCM and CME can be supported using SCCP or SIP, a SIP-capable PBX will integrate using SIP, and a variety of digital PBX products can be supported using a PIMG or TIMG device that converts a digital TDM circuit to a SIP trunk.

CUC Integration with CUCM Using SCCP

Key Topic

A Voicemail Port Wizard is available in CUCM 8.x that simplifies the integration of CUC with CUCM. The wizard requests user input to correctly set up the system, and then generates voicemail ports in CUCM and adds them to a Line Group. The administrator must manually configure the Hunt List and Hunt Pilot to support the Line Group.

The Hunt Pilot is referenced by a Voicemail Pilot, which is itself referenced by a Voicemail Profile. Figure 13-1 illustrates the architecture of the voicemail integration on the CUCM side.

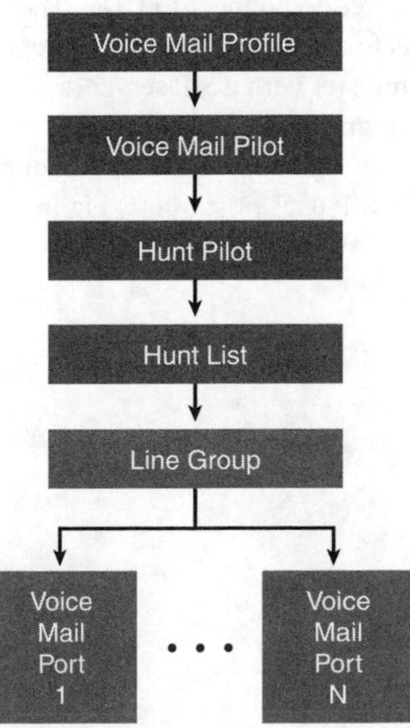

Figure 13-1 *SCCP Voicemail Integration Components in CUCM*

Default entries for the voicemail profile and voicemail pilot exist, which are used by all users of the CUCM system; these may be used and customized, or others may be added for other integrations and used by different subsets of users. The Voicemail Port Wizard has all but eliminated a common problem in CUCM-to-CUC integrations with SCCP: Often, administrators would forget one or more of the critical steps of creating the Voicemail Pilot and linking it to the Hunt Pilot, and creating the Voicemail Profile and linking it to the Voicemail Pilot. Miss one of these, and the Messages button on the phones won't work, even though you can dial the CUC Hunt Pilot number directly and reach CUC.

On the CUC server, a set of ports (the number of available ports being limited by the server hardware capacity, and then by licensing) is defined, and each port is configured for various call-behavior options, including whether the port should answer calls, perform Message Waiting Indicator (MWI) or Message Notification, and other settings. Call routing within CUC can be controlled by (among other things) the phone system or the port group.

MWI uses a separate and unique Directory Number (DN) for MWI On and MWI Off. The DNs must be configured (and match) in both CUCM and CUC. One of the tricks you can play on your co-workers is to dial the MWI On DN from their phone; their MWI light will come on (dial MWI off and it turns off, too). This practical joke is also an effective way to test MWI functionality.

An integration using Skinny Client Control Protocol (SCCP) may be secured using digital certificates and SCCP over port 2448. (Nonsecure SCCP uses port 2000.)

CUC Integration Using SIP

The Session Initiation Protocol (SIP) integration components are slightly different from SCCP: Instead of the voicemail pilot pointing to a Hunt Pilot, it points to a Route Pattern, which in turn points to a SIP trunk. The SIP trunk is configured to connect to CUC. The number of ports is not defined on the CUCM server as it is for SCCP integration; rather, they are only defined in CUC. Each port is configured to register with a SIP server (which is the CUCM server). A significant difference with a SIP integration is that there are no separate DNs for MWI On/Off; instead, SIP itself handles the signaling of the MWI lamp state. SIP can also be secured using port 5061. (Nonsecure SIP uses port 5060.) Figure 13-2 illustrates the SIP integration components on CUCM.

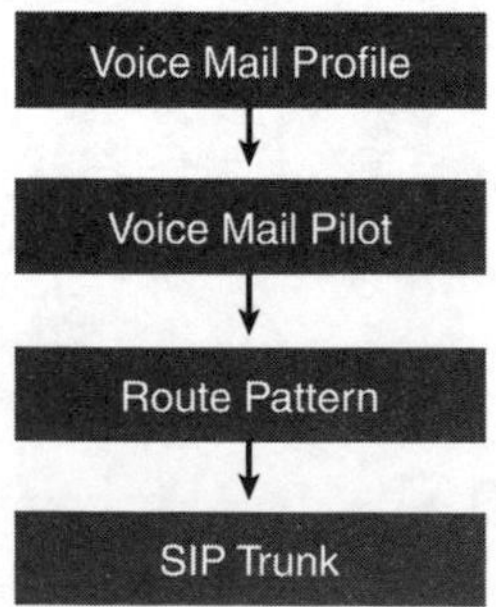

Figure 13-2 *SIP Voicemail Integration Components in CUCM*

CUC Features

This section describes many of the system-level features and settings of CUC.

System Settings

The installation and configuration of CUC includes many system settings. Because the ICOMM exam scope is relatively limited, we describe only a few here.

General Configuration

The General Configuration page includes the defaults for the system **Time Zone**, **Language**, and **Maximum Greeting Length**.

Roles

The Roles page lists the eight administrative roles defined in the application. An administrative role gives (and limits) administrative capability to users. Table 13-2 lists these roles and brief descriptions of their functions.

Table 13-2 *CUC Roles and Descriptions*

Role	Description
Audio text administrator	Administers call handlers, directory handlers, and interview handlers.
Greeting administrator	Manages call handler recorded greetings via TUI.
Help desk administrator	Resets user passwords, unlocks user accounts, and views user settings.
Mailbox access delegate account	Accesses all messages via messaging APIs.
Remote administrator	Administers the database using remote management tools.
System administrator	Top-level connection administrator; accesses all connection administrative functions, reports, and tools for server and users.
Technician	Accesses functions that enable management of system and phone system integration settings, views all system and user settings, runs all reports and diagnostic tools.
User administrator	Administers users; accesses all user administration functions, user reports, and user administration tools.

Enterprise Parameters and Service Parameters

Equivalent to the pages of the same name in CUCM, these pages define and tune system and service parameters, such as what users can see and configure on the User Web Pages, Quality of Service (QoS) settings for CUC-generated traffic, and so on.

LDAP

These pages define the integration with an LDAP system to provide user synchronization and optional authentication.

Call Handlers

All inbound calls to CUC are handled by a series of call handlers. The three basic types of call handlers are as follows:

- **System Call Handlers** are used for greetings and can be customized to offer user input options ("For Sales, Press 2...") and automation, such as playing a different greeting when the business is closed.
- **Directory Handlers** allow a caller to search the CUC directory for the user they want to contact. Different directories can be defined based on location, distribution list membership, and so on.
- **Interview Handlers** provide the caller with recorded information, and then ask questions and record the caller's answers in a single message. Interview Handlers can be used for telephone-based reporting for almost any purpose, such as automating job applications.

Three System Call Handlers are defined by default: Goodbye Call Handler, Opening Greeting, and Operator Call Handler. Opening Greeting is what outside callers (those without a voicemail box on the CUC server) hear; it is expected, of course, that the greeting will be customized by the business.

Call Routing

Two primary call routing criteria are built in to CUC: The application identifies Direct calls and Forwarded calls. A Direct call is when the caller dials the CUC system directly, either by pressing the Messages button on his IP phone (or dialing the voicemail pilot manually) or by dialing the public switched telephone network (PSTN) Direct Inward Dial (DID) of the CUC Auto-Attendant.

The system examines the information presented in the call as it is routed to the CUC port that answers it. The information available to CUC for its decision making includes the following:

- Calling number
- Called number
- Forwarding station
- Phone system/port
- Schedule

Two rules are defined under each category of call routing rule. The following sections describe these rules.

Figure 13-3 illustrates the call routing actions and rule criteria within CUC.

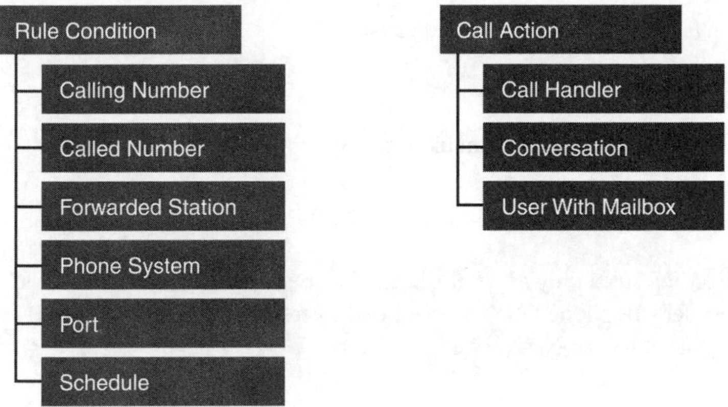

Figure 13-3 *Call Routing Actions and Rule Criteria*

Direct Routing Rules

For calls placed directly to CUC, the following two default rules apply:

- **Attempt Sign-In:** If the calling number is recognized as the primary DN associated with a voicemail box, the call is sent to the Attempt Sign-In Conversation, and the caller is prompted to enter his PIN to log in to his voicemail box.
- **Opening Greeting:** If the calling number is not associated with a voicemail box on the CUC server, the call is sent to the Opening Greeting.

Additional rules can be defined administratively (for example, routing calls to the business' customer help number to a specific call handler); the rules are processed top-down for each call, so the order of the rules is critical to their behavior.

Forwarded Routing Rules

For calls that are forwarded to CUC (typically because the user was on the phone or did not answer their phone), the following two default rules apply:

- **Attempt Forward:** If the forwarding station is associated with a voicemail box on the CUC system, the forwarding phone user's personal greeting is played.
- **Opening Greeting:** If the forwarding station is not associated with a voicemail box on the CUC system, the Opening Greeting is played.

Call Routing Rule Filters

When defining custom call routing rules (whether for Direct or Forwarded calls), the following filters can be applied (singly or combined within a single rule):

- Calling number
- Called number

- Voicemail port
- Phone system
- Forwarding station (applies only to forwarded calls)
- Schedule

The use of these rule capabilities provides administrators with a powerful customized call-routing capability.

Key Point: CUC Call Routing rules only apply if the call has been answered by CUC; for example, a PSTN call to a user's IP phone DID will not be answered by CUC unless the IP phone forwards it because of a Busy or No Answer condition.

Distribution Lists

Distribution Lists (DL) provide a simple way to send a voice message to a group of users. Two types of DL can be configured: System DLs are managed by the administrator and can be made available to all users or a subset of users as required. Private DLs are managed and maintained by an individual user and are usable only by the user who made them. The administrator can limit how many private lists a user can create and how many members can be in each.

Authentication Rules

To set the security level for access to the CUC system, authentication rules for **Voicemail** (for TUI access via PIN) and **Web Application** (for access to the User Web Pages) can be customized. Authentication rules specify how many failed login attempts can occur before the account is locked out, how long the account is locked out, the minimum number of characters in a password, how often the password must be changed, and so on. The default Authentication Rules apply to all users. You can create new Authentication Rules with customized settings, and apply them to User Templates or even individual users. In this way, you could allow a four-digit password for a new User Template while keeping a five-digit password for another.

Dial-Plan

CUC incorporates the concepts of Partitions and Search Spaces in a similar way to CUCM. Objects that can receive calls, such as a user mailbox or call handler, are assigned a Partition; objects that can place or transfer calls are assigned a Search Space. The object being called must be in one of the Partitions listed in the Search Space of the object making the call. A default Search Space, containing the default Partition, allows all objects to reach all other objects until the administrator customizes the system as needed.

Using this mechanism, it is possible to create a directory handler for the Vancouver office, and limit the Search Space of the directory handler to include only the Vancouver

Partition. In this way, searching the Vancouver directory would list only users assigned to the Vancouver Partition.

Describe Cisco Unity Connection Users and Mailboxes

The system components identified previously are important, but without users and mailboxes, system functionality is limited. The following sections outline the basics of adding users and mailboxes to the system.

User Templates

User Templates, as the name implies, provide a pattern used in the creation of new user accounts. Most of the required configuration information common to all the new users can be entered in the template, and then the individual user-specific information is combined with the template to create (potentially many) new user accounts with speed and accuracy. The template settings are applied as the user is created; changing the template does not retroactively change the user accounts that were created using that template.

Two default User Templates are created at install: one for administrators and one for users. These can be modified if desired, and as many custom templates as needed can be created.

The following sections highlight some of the settings on the User Template Pages (of which there are many, so not all settings are discussed).

User Template Basics

The basic elements of User Template configuration are summarized in the following sections:

- **Name:** The name of the template, such as Managers. The Alias and Display Name generation for user accounts can also be specified; the default is first name followed by last name.
- **Phone:** In this section, the Dial-Plan (Partition/Search Space), Class of Service (CoS), and Schedule are defined.
- **Location:** Geographic location information, language localization, and time zone are set here.

Key Topic

Key Point: CoS in CUC (not to be confused with Layer 2 Quality of Service [QoS] marking or CUCM Class of Control) is a simple and powerful method of assigning and restricting user privilege. The CoS defines greeting and message length timers, licensed feature access, advanced feature access, alternate extension definition, Private DL number, and membership limits and call transfer abilities.

An unlimited number of CoS can be defined, providing exactly the combination of abilities and features for as many sets of users as needed.

Figure 13-4 illustrates the basic components of a CoS.

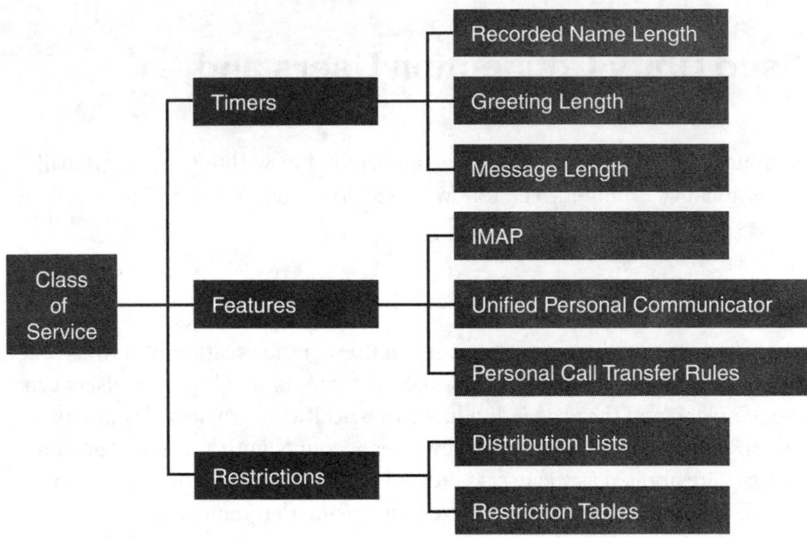

Figure 13-4 *CoS Components*

Password Settings

On this page, the administrator can lock and unlock the account, control when and if the password must be changed, and set the **Authentication Rule.** All these can be set for both the **Voicemail** password and the **Web Application** password.

Roles

Roles define one of the default administrative capability assignments, as described in Table 13-2.

Transfer Rules and Greetings

Three transfer rules are defined by default. The **Standard** rule cannot be modified and is active by default. The **Alternate** rule can be modified to be active according to a different schedule or a specific end date. The **Closed** rule takes effect during defined closed hours.

These rules are applied to determine the behavior of the user's mailbox, including which greeting is played to callers. The available greetings are the following:

- **Alternate:** Used for personalization of the voicemail box with a custom greeting.
- **Busy:** Plays when the user's extension is busy.
- **Internal:** Plays only to internal (on-net) callers.
- **Holiday:** Plays when the Holiday schedule is in effect.
- **Closed:** Plays when the Closed schedule is in effect.
- **Standard:** Plays the default "*<user name>* is not available..." greeting.

Call Actions

The administrator can select what action the system takes after the greeting has played. Typically, **Take Message** would be selected, but **Hang Up** and **Restart Greeting** are available options.

Other Call Actions include sending the call to any configured call handler or to a user mailbox.

Message Settings, Message Actions, and Caller Input

When the caller leaves a message, he can be allowed to edit it or not, allowed to mark it as Urgent, or have the system set all messages as Urgent or Normal. Messages can also be marked as secure, which can be used to limit where the message can be delivered. (For example, a secure message can be restricted from delivery to an IMAP client.)

After the message has been left, the administrator can set what action the system takes, with the default being say goodbye and hang up. Other options include sending the caller to any configured call handler or to a user mailbox.

During the conversation with CUC, the caller may be allowed to press a key to perform a configured action (such as logging in to the Greetings Administrator TUI). The default key presses are zero (0) for the operator and asterisk (*) to log in to the mailbox.

TUI Settings

The TUI user experience can be customized to speed up or slow down the conversation with CUC, make it louder or quieter, change how long the system will wait for key presses, and customize the order of playback of different message types, among others.

CUC End Users

Creating a new end user requires many configuration parameters; if using User Templates to create new users, most of these settings are pulled from the template. The unique individual settings that must be configured include the **Alias** (the unique user ID), **Name**, **Mailbox Store**, **Extension**, and **Alternate Extensions**.

Extension and Call Forward Options

Key Topic

The **Extension** number is a required entry. This number should be the primary DN on the user's IP phone; when she presses the Messages button, it is the caller ID that CUC uses to determine if she is a mailbox owner and if so, prompts her to log in. Likewise, if a call to a user's phone is forwarded to CUC because the user was on the phone or did not answer, CUC uses the caller ID to determine which mailbox greeting to play to the caller.

The call forward options on the IP phone (discussed earlier in this book) can change the behavior of CUC; for example, a different greeting can be played for internal versus external callers.

Voice Messaging with SRST and AAR

In the event the IP WAN fails, calls can be rerouted over the PSTN using Automated Alternate Routing (AAR) or Survivable Remote Site Telephony (SRST). If the CUC server is normally reached over the WAN from by branch user, in the event of an outage, these systems reroute the call to CUC over the PSTN, too. When the call arrives at CUC, the ten-digit PSTN caller ID will not be recognized as a mailbox owner unless it is added as an alternate extension for the user (the same goes for each other user on the system as well).

Voicemail Box

When creating the user mailboxes, the administrator can choose whether to list the user in the directory or not, record the voice name (the spoken version of the username; CUC speaks the name in the configuration page if the name is not recorded), and record a greeting. The administrator can also require the user to go through these steps at his next login.

Private Distribution Lists

Each user is permitted to create up to 99 private DLs, each with a maximum of 999 members. Lower limits can be set in the CoS or individually per user. Private DLs are visible only to the user who created them (and to administrators).

Notification Devices

In addition to the MWI lamp on the IP phone, users can be notified of new messages by way of up to three PSTN numbers (mobile phone, home phone, and pager) and e-mail. Toll call control is handled by restriction tables that define what numbers CUC can call for message notification. When CUC calls the configured number, it informs the user that there is a new voice message and prompts the user to authenticate before playing the message.

User Creation Options

There are several ways to create or import user accounts into CUC:

- **Manual creation:** Creates users one at a time. All user data is maintained locally in the CUC database.
- **Bulk administration:** Creates many users at once from data in a .csv file. All user data is maintained locally in the CUC database.
- **Migration from Cisco Unity:** Users can be migrated and imported using the Consolidated Object Backup and Restore Application Suite (COBRAS) tool. COBRAS helps administrators migrate users from a Cisco Unity system to a Cisco Unity Connection system. The users can be imported with or without their mailboxes.
- **Import from CUCM:** Synchronizes the CUC user database with an existing CUCM database. Some user data is maintained in CUCM and copied to CUC, but CUC-specific data is locally maintained in the CUC database.

- **Import from LDAP:** Synchronizes the CUC user database with an existing LDAP user database. Some user data is maintained in LDAP and copied to CUC, but CUC-specific data is locally maintained in the CUC database. Optionally, web-password authentication can be redirected to LDAP to provide a single point of administration and single sign-on for user passwords.

CUC Voicemail Boxes

A voicemail box is typically associated with each user (one per user in most scenarios). The mailbox is held in a database store that may be synchronized between two CUC servers in an Active-Active redundant pair. The user mailbox may be moved to another store if necessary.

Message Aging Policy and Mailbox Quotas

To control disk space utilization by voicemail box storage, administrators can set message aging policies that move read messages to the Deleted Items folder after a specified number of days (disabled by default). Messages in the Deleted Items folder are automatically permanently deleted after 15 days by default (configurable).

User storage quotas can be configured to warn users when their mailbox nears the maximum allowed size (warning at 12 MB by default). Users are prevented from sending new messages when their mailbox reaches 13 MB (configurable to any appropriate value), and they cannot send or receive messages if their mailbox reaches 14 MB by default (also configurable).

12 MB of disk space is approximately 200 minutes of recorded messages using the G.729 codec, and about 25 minutes using G.711.

Implement Cisco Unity Connection Users and Mailboxes

In the following sections, the concepts introduced previously are put into action. What follows are the basics of implementation of users and mailboxes in CUC. There are, of course, many other possible steps that are not included here, both for clarity and to stay within the scope of the CCNA Voice exam.

Configure End User Templates

Templates are a powerful and useful way to speed up and simplify the creation of users. You may modify an existing template or create a new one to meet requirements. Although many configurations are available, the ICOMM course is limited in scope and examines only the following Edit menu entries:

- User Template Basics
- Password Settings
- Roles
- Message Settings and Actions

- Phone Menu
- Playback Settings
- Message Notification

Remember that all the following sections reference editing a User Template, and consequently will set configurations for all user accounts created with the template. Some settings obviously need to be unique per user and can either be included as part of the user creation steps or configured after the accounts are created. You can create as many User Templates as you need. To begin, navigate to **Templates > User Templates** and select an existing template (or create a new one).

User Template Basics

The following points review most of the elements found on the User Template Basics screen. You may or may not need to change them all; each template will be at least slightly different in your environment:

- **Alias:** This is a required field; in this case, it is the name of the template itself.
- **Display Name: This** is a required field; this is the template name as it appears in the Find/List Templates page.
- **Display Name Generation:** Choose one of **First Name, then Last Name** or **Last Name, then First Name.**
- **Outgoing Fax Server:** Select the correct fax server for the users, if any.
- **Partition:** Select the appropriate Partition for the users.
- **Search Scope:** Select the appropriate Search Scope (Search Space) for the users.
- **Phone System:** Select the phone system for the users. Most deployments will have only one available, but CUC supports multiple integrations so this setting can be important.
- **Class of Service:** The CoS controls what features and capabilities the users can access.
- **Active Schedule:** Select the schedule that will be applied to the users. Schedules can be used to affect when the Standard or Closed greetings play, as well as what after-greeting actions CUC takes.
- **Set for Self-enrollment at Next Login:** Check this box to force the user to go through the tutorial and record his voice name and greetings the next time he logs in to CUC.
- **List in Directory:** Select this checkbox to list the users in the CUC directory; doing so allows outside callers to search for, find, and then call users.
- **Time Zone:** Select Use System Default Time Zone or choose the appropriate time zone for the users. Select the time zone with care, because it will modify CUC behavior for displaying the time a message was received and for message notification.
- **Language:** Set the language that CUC uses to play instructions to users and for text-to-speech. This setting does not apply to the voice recognition conversation.

Figure 13-5 shows part of the User Template Basics page.

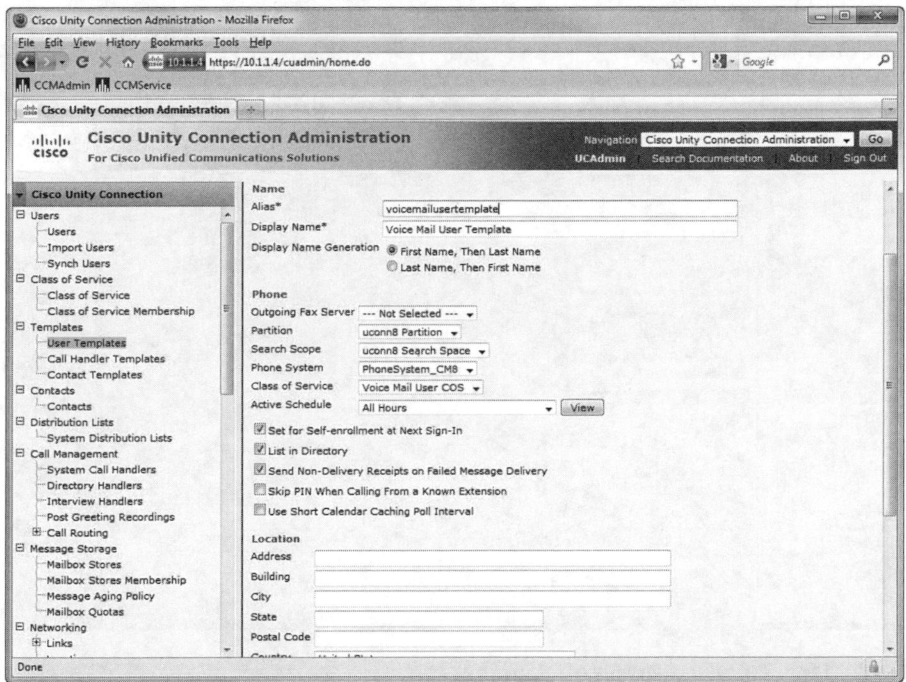

Figure 13-5 *User Templates Basics Page*

Password Settings

CUC uses two separate passwords for each user: The Web Application password is used for logins to CUC web pages (including the User Web Pages, and Administration pages if the user has the privilege to access them); the voicemail password is actually the PIN, which is used for TUI logins. To begin, select which password you want to configure. On the page that opens, select or modify the following options:

- **Locked by Administrator:** Select this if you want to prevent the user from logging in to CUC.
- **User Cannot Change:** Select this to prevent the user from changing his password. This is recommended if more than one user will access the voicemail box; it is also recommended in this case to select the Does Not Expire checkbox.
- **User Must Change at Next Login:** Select this to force the user to change his password the next time he logs in to CUC.
- **Authentication Rule:** Set the Authentication Rule that will be applied to the user account's password.

Figure 13-6 shows the Password Settings page.

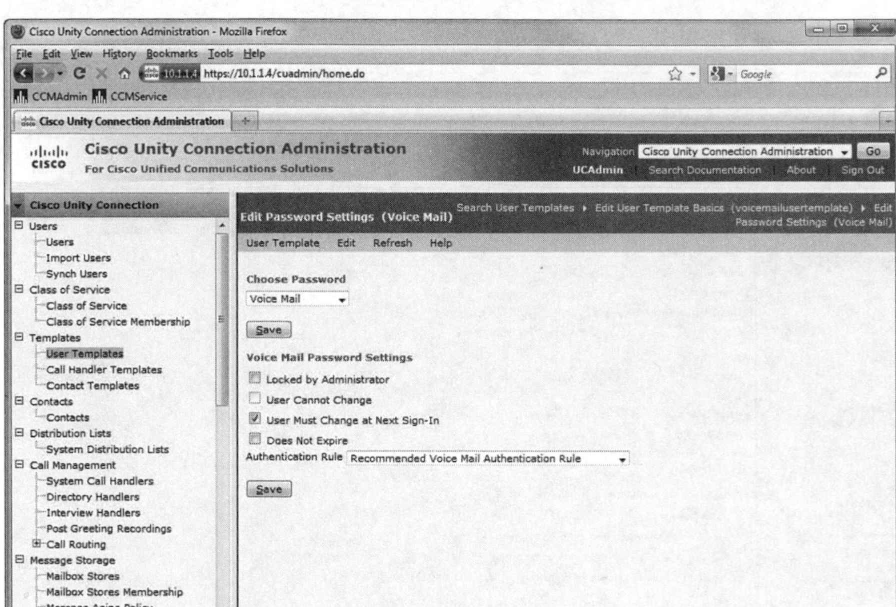

Figure 13-6 *User Template: Password Settings*

Roles

By default, no roles are selected for user accounts. Assign the appropriate role(s), if these users need administrative privilege. Table 13-2 described the role functions.

Message Settings

On the Message Settings page, you can select or modify the following as appropriate for your environment:

- **Maximum Message Length:** Limits how long a single message can be (the default is 300 sec).
- **Callers Can Edit Messages:** Select to prompt users to listen to, add to, re-record, or delete a message they have just left.
- **Language That Callers Hear:** Choose one of the installed languages. This setting affects the language of system recordings, such as "Record your message at the tone." If Inherit Language from Caller is selected, when an IP phone calls, the user locale setting (in CUCM) of that phone determines what language will be used (as long as that language is installed on CUC).
- **Unidentified Callers Message Urgency:** If the caller leaving the message is not a mailbox owner on the system, he is classified as an Unidentified Caller. Messages left by these callers can all be set to Normal or Urgent, or the system can be set to Ask

Callers, giving them the choice. By default, when a user logs into their mailbox, CUC plays messages marked Urgent first. Some organizations use this behavior to improve customer service, by forcing all outside callers' messages to be marked Urgent. Setting the Unidentified Callers Message Urgency here (in the User Template) will apply the chosen setting to all users created with this template; you can override this setting at the individual user's account under the Phone menu.

Figure 13-7 shows part of the Message Settings page.

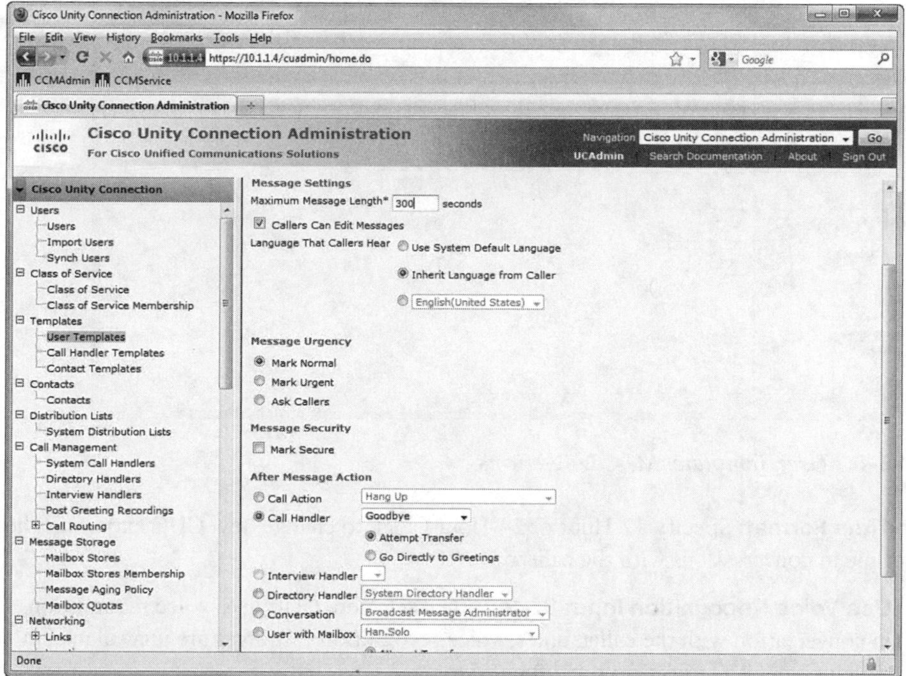

Figure 13-7 *User Template: Message Settings*

Message Actions

For each type of message (Voicemail, E-mail, Fax, and Delivery Receipt), set the action CUC will take: Accept, Reject, Relay, or Accept and Relay.

Figure 13-8 shows the Message Actions page.

Phone Menu

Under the Phone menu, you can select or modify the following settings:

- **Touchtone Conversation Menu Style:** Select **Full** to have CUC play detailed instructions for callers or **Brief** for shorter, less-detailed instructions.
- **Conversation Volume:** Adjusts the loudness of CUC conversations.
- **Conversation Speed:** Adjusts the playback speed of CUC conversations.

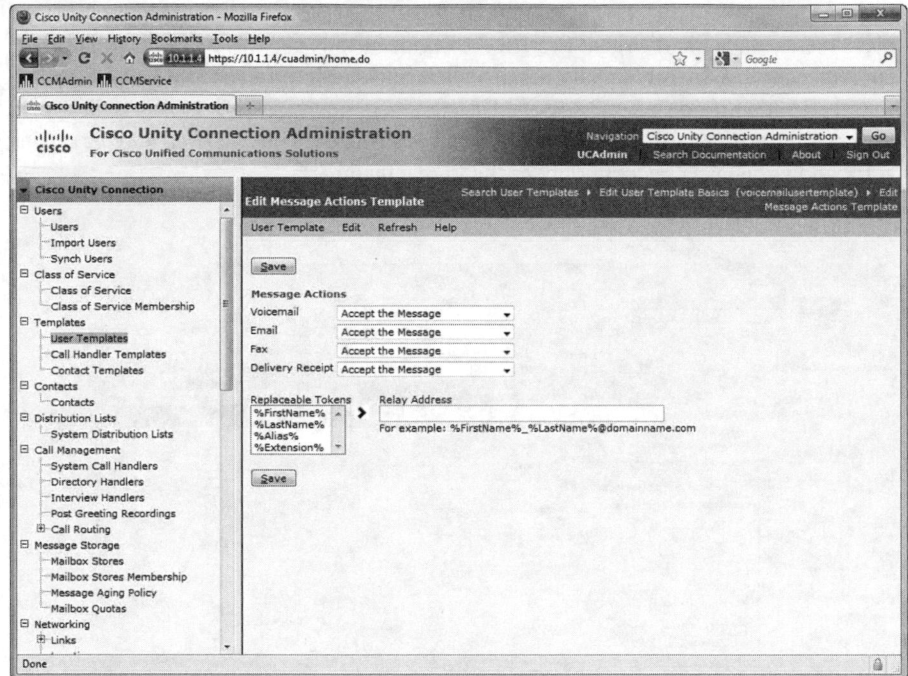

Figure 13-8 *User Template: Message Actions*

- **Time Format:** Specify 12-Hour or 24-Hour Clock to change how CUC expresses the time in conversations with the caller.
- **Use Voice Recognition Input Style:** When selected, CUC uses voice recognition in conversation with the caller, unless voice-recognition resources are unavailable; in which case, CUC reverts to touchtone style.
- **Touchtone Conversation Style:** Select the keypad emulation (to make the key press functions the same as other voice-messaging systems; this is normally done if you want to provide a smoother transition for users from a legacy voice-messaging system to CUC).
- **Enable Message Locator:** Enables users to search messages by user name, extension, or caller ID.

Figure 13-9 shows part of the Phone Menus page.

Playback Message Settings

Under the Playback Message Settings menu, the following settings can be selected or modified:

- **Message Volume:** Sets the volume for message playback.
- **Message Speed:** Sets the speed at which messages are played back.

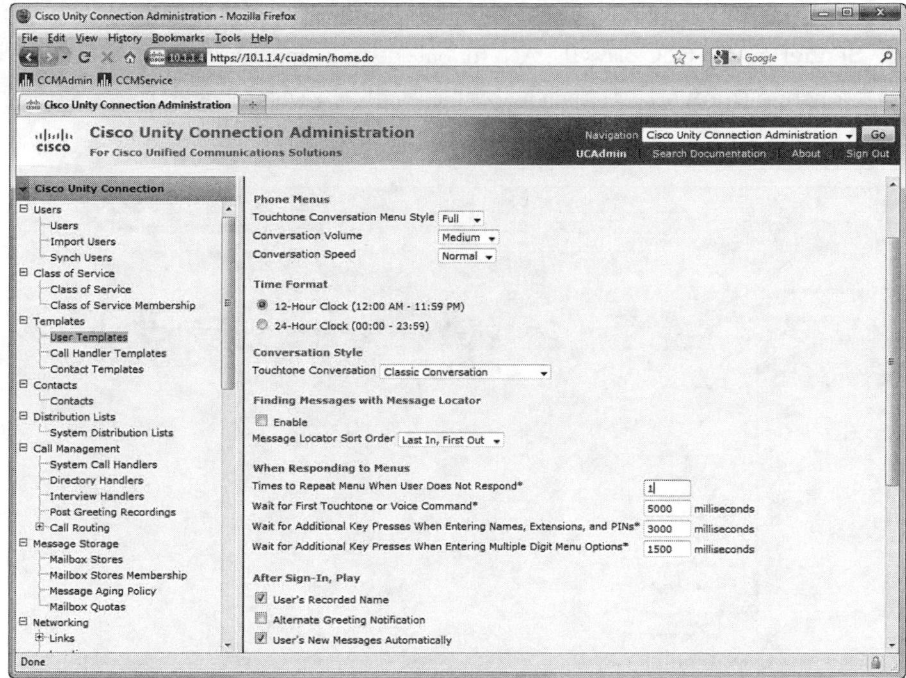

Figure 13-9 *User Template: Phone Menus*

- **For New Messages, Play:** Check each box as desired to hear message counts for Totals (all new messages), E-mails, Faxes, and Receipts.
- **For Saved Messages, Play:** Check this box to have CUC announce the number of saved messages.
- **Before Playing Messages, Play:** Check the **Message Type Menu** box to hear a menu of key press options to hear messages of each type.
- **New Message Play Order and Saved Message Play Order:** Use this set of preference lists to set the order in which CUC plays messages to the users. To hear e-mails and faxes, the user must be assigned a CoS that has the **Access to E-mail in Third-Party Message Stores** and **Fax** features enabled. If there are fax messages, CUC announces just the sender, date, and time (it does not read the fax body).
- **Before Playing Each Message, Play:** Check each box as desired to hear any (or all) of the following:
  - **Sender's Information:** Recorded name or ANI for internal callers; ANI for outside callers is not played.
  - **Include Extension:** In conjunction with the **Sender's Information** checkbox, selecting this checkbox causes CUC to play the extension of an internal caller and the recorded name (if available).
  - **Message Number:** CUC announces the sequential number of messages in the mailbox as it plays them.

- **Time the Message Was Sent:** CUC plays the timestamp of each message.
- **Sender's ANI:** CUC plays the ANI for outside caller messages.
- **Message Duration:** CUC announces the length of each message.

Figure 13-10 shows part of the Playback Settings page.

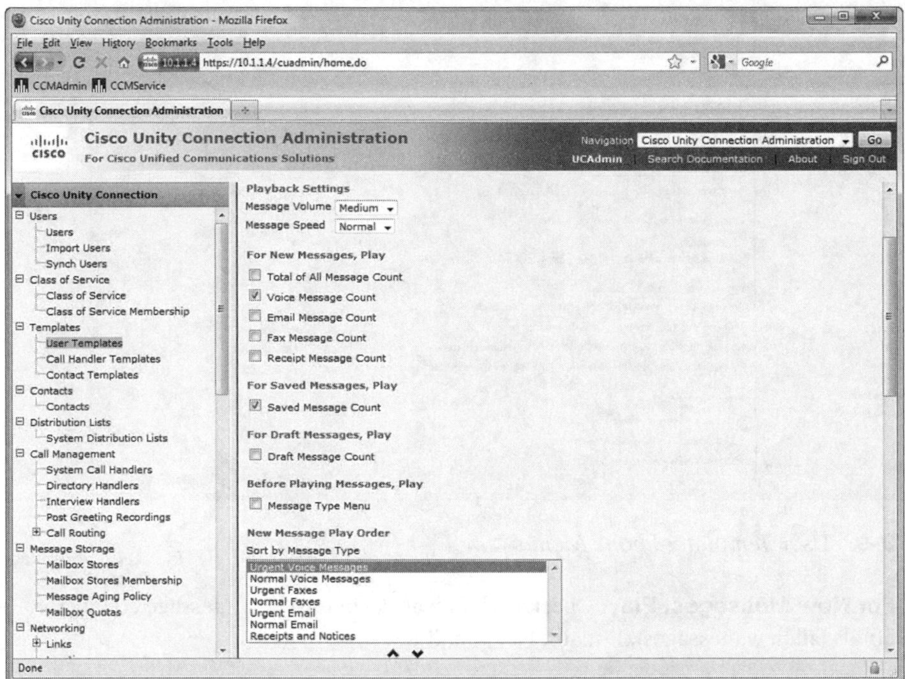

Figure 13-10 *User Template: Playback Settings*

Notification Devices

Message notification should not be confused with MWI; notification refers to CUC making a phone call or sending an e-mail to advise a user that he has a new message. CUC supports notification on a **Pager**, **Work Phone**, **Home Phone**, and **Mobile Phone** and via **SMTP** by default. Additional notification devices may be added.

For each device, set the following as needed:

- **Enable:** Check this box to allow CUC to send notifications to this device.
- **Display Name:** Modifies the name as needed.
- **Delay Before First Notification Event:** Sets the number of minutes CUC waits after a new message is left before it sends a notification using this device.
- **Notification Repeat Interval:** Sets how frequently CUC resends the notification.
- **Notify Me Of:** Select the checkboxes to have CUC send notifications for **All Messages**, **All Voice Messages**, **Fax Messages**, **Calendar Appointments**, **Calendar Meetings**, and for each type, whether to send for **Urgent Only**.

- **Pager/Phone Settings:** Sets the **Phone Number** CUC should call. Modify the extra settings to add extra digits, wait times, and so on to ensure the call completes correctly. All these settings should be applied at the individual user account rather than at the template.

Configure CUC End Users

Adding users to CUC can be done manually (one at a time), via import from CUCM or LDAP, or by using the Bulk Administration Tool (BAT). In each case, the goal is to provide individually specific information per user to complete the common information provided by the different User Templates previously configured.

In the following sections, some possible configuration steps have been omitted in the interest of relevancy to ICOMM.

Manual Process

To manually create a new user, navigate to **Users > Users** and create a new user. Select the appropriate **User Type** (typically **User with Mailbox**) and select or enter the following:

1. From the **Based on Template drop-down,** select the appropriate User Template for this user to supply the common configurations.
2. Add the **Alias** (which must be unique), **First Name**, **Last Name**, and **Display Name.**
3. Choose the appropriate **Mailbox Store**, if there is more than one in use.
4. Add the **Extension** of the user, which is typically the primary DN associated with his IP phone.
5. Click **Save**.

Figure 13-11 shows part of the User Configuration Basics page.

Alternate Extensions and Names

An Alternate Extension can be provided to users to allow them to call in to CUC from a number other than their primary Extension and gain access to their mailbox without having to go through the Opening Greeting. The following steps outline the configuration of Alternate Extensions:

1. In the User Configuration page, from the **Edit** menu, select **Alternate Extensions** and click **Add New.**
2. Set the **Type of Phone** from the drop-down (**Work**, **Home**, **Mobile**).
3. Provide a **Display Name.**
4. Set the correct **Phone Number**. This is the Caller ID (ANI) number of the phone as it will appear to CUC; typically, this is the full PSTN number, but be aware of the possibility that CUCM may be stripping or modifying digits before they are sent to CUC.

   Figure 13-12 shows the Alternate Extension page.
5. From the **Edit** menu, select **Alternate Names.**

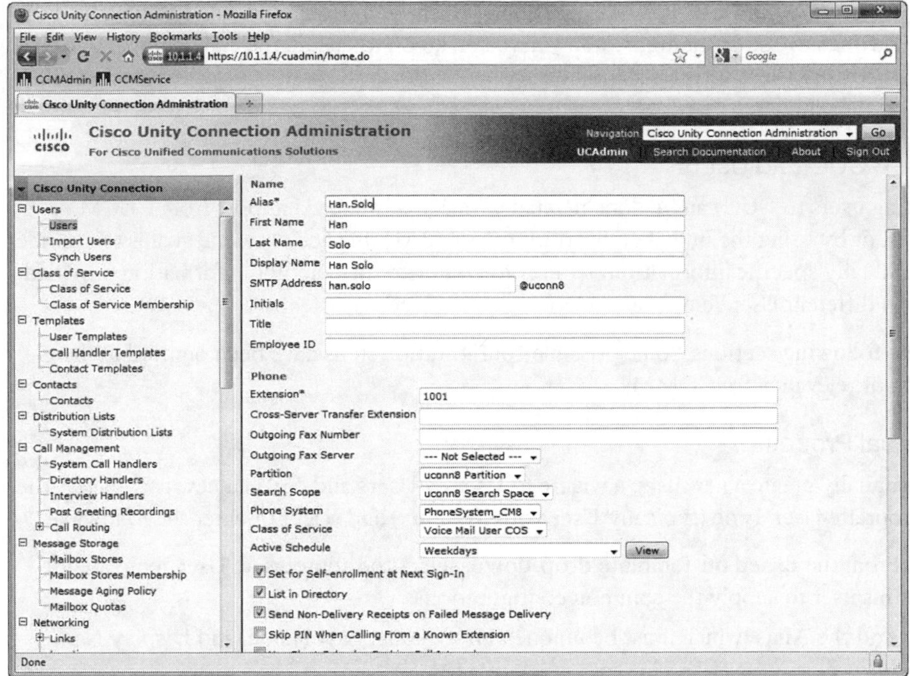

Figure 13-11 *User Configuration Basics*

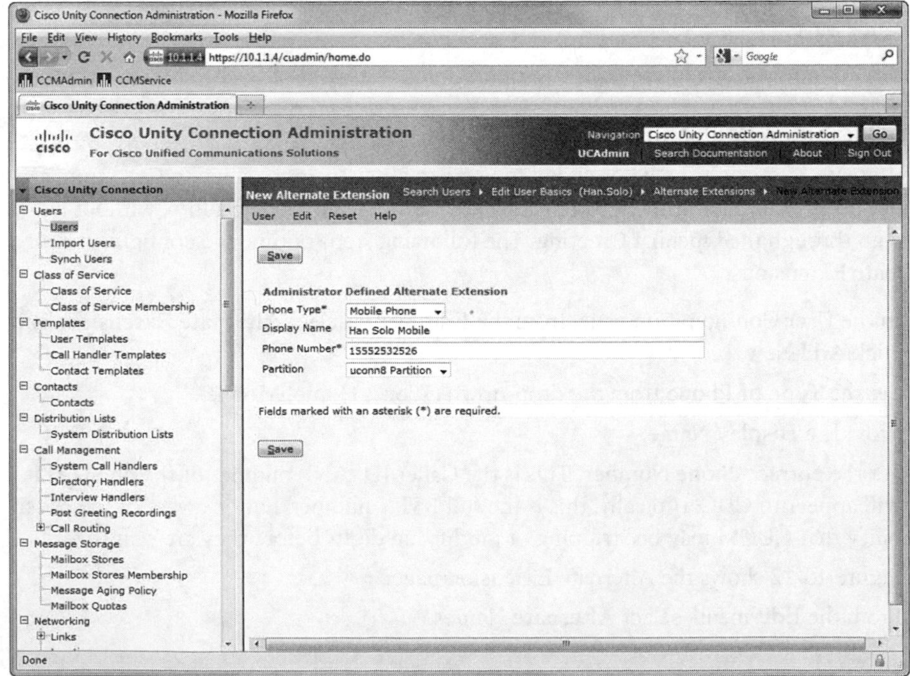

Figure 13-12 *User Configuration: Alternate Extensions*

An Alternate Name allows the administrator to define nicknames, familiar names, or phonetic name spellings to allow callers to search for users by a name other than the entry in their user page main configuration. For example, if Jedediah is the first name entry, but callers search for J.D. because that is his nickname, the administrator can add the alternate name of J.D., or even Jaydee if voice recognition is in use.

Private DLs

The administrator can add Private DLs on behalf of the user, or the user can add them from his Personal Communications Assistant (PCA) web pages. (Note that the administrator interface also opens the PCA, logged in as the user.) Each list (up to 99 can be created) needs a unique name and can contain up to 999 members. These maximums can be limited in the CoS.

Figure 13-13 shows the PCA view of Private Lists.

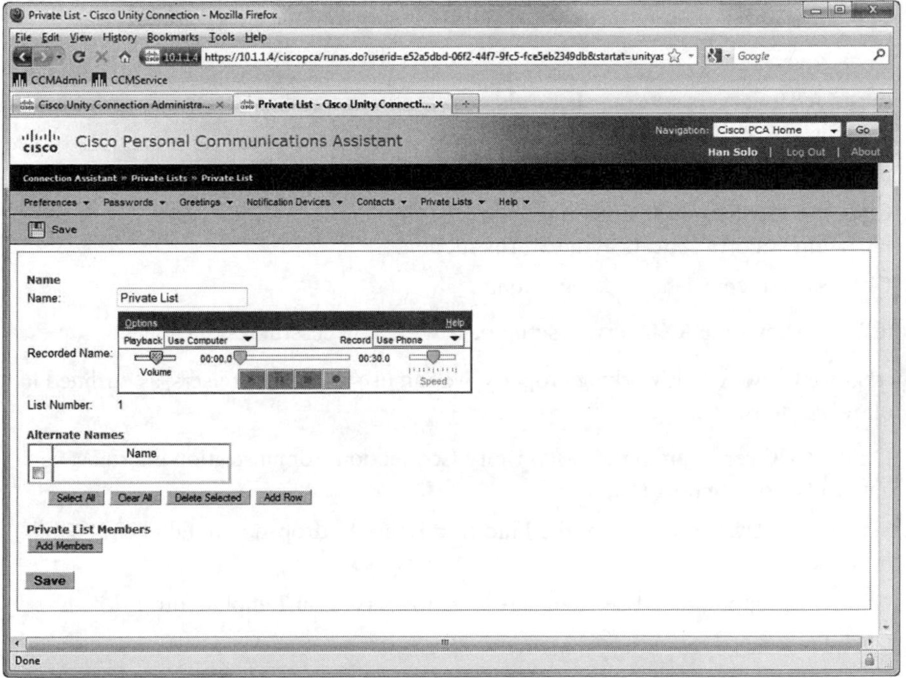

Figure 13-13 *PCA: Private Lists*

Importing End Users in to CUC

Performing a user import is one of the quickest and simplest ways to create user accounts in CUC. By using fully configured User Templates, the work of creating users is much faster and more accurate. Depending on the implementation, there may be an added benefit if the user accounts are synchronized with another system, which reduces the replication of administrative tasks associated with user maintenance.

Importing Users from CUCM

If the CUCM server already has a fully populated user database, those entries may be easily imported and synchronized in the CUC database. To enable this capability, perform the following actions on both the CUC and CUCM server:

1. From the **Cisco Unified Serviceability** interface, under the **Tools** menu, select **Service Activation.**
2. Select the **Cisco AXL Web Service** and click **Save.**
3. In **Cisco Unity Connection Administration**, navigate to **Telephony Integrations > Phone System.**
4. Click the name of the CUCM server from which you want to import users.
5. Under the **Edit** menu, click **Cisco Unified Communications Manager AXL Servers.**
6. In the **Edit AXL Servers** page, under **AXL Server Settings** enter the **User Name** and **Password** of the account that CUC will use to login to the CUCM AXL server.
7. Click **Save.**
8. In the **AXL Servers** section, click **Add New.**
9. Enter the **IP Address** and **Port** of the CUCM server. (CUCM 8.x supports SSL, so use port 8443 or 443.)
10. Click the **Test** button. **Test message successfully sent to AXL server** <ip_address:Port> appears in the **Status** section.
11. Click **Save** to complete the integration.

Figure 13-14 shows the AXL Servers setup page with a successful test message.

Now that we know AXL is working properly, we can use it to import users, as outlined in the following steps:

1. On the CUC server, from the Cisco Unity Connection Administration interface, navigate to **Users > Import Users.**
2. Select the CUCM server from the **Find End Users In** drop-down. Filter the search if needed.
3. Choose the appropriate User Template from the **Based on Template** drop-down.
4. Click **Find.**
5. Select the user(s) to import from the list, and then click **Import Selected.**

Note: CUCM users must have a primary extension defined or the users will not appear on the Import Users page in CUC.

Figure 13-15 shows an import of users from the CUCM AXL server.

Importing Users from LDAP

LDAP is a standards-based user, password, and privilege database system. Multiple other applications can access a LDAP directory to determine if a particular username and password are valid and have access privileges to a particular resource.

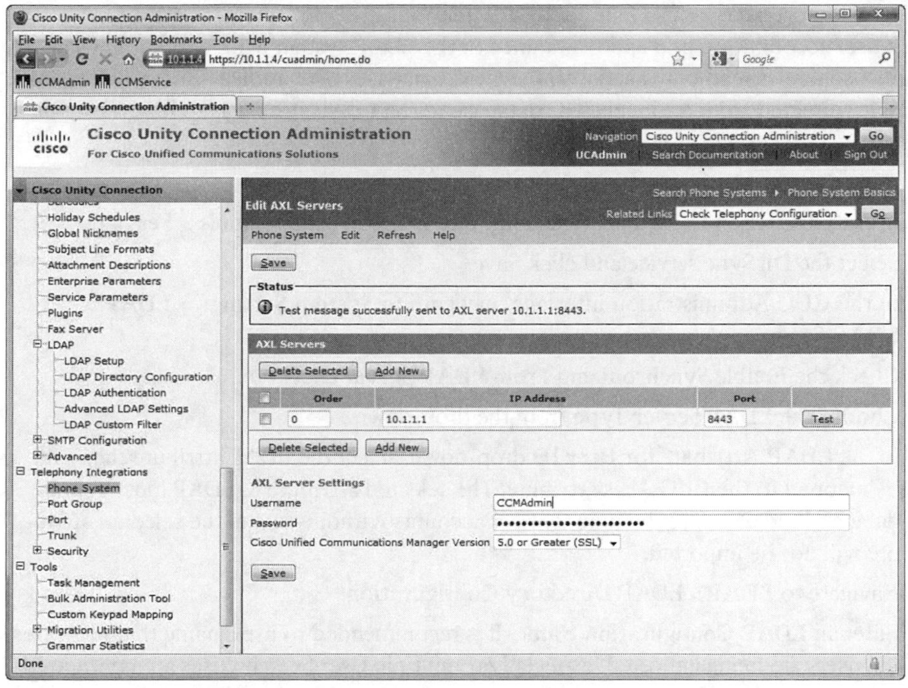

Figure 13-14 *AXL Server Edit Page: Successful Test*

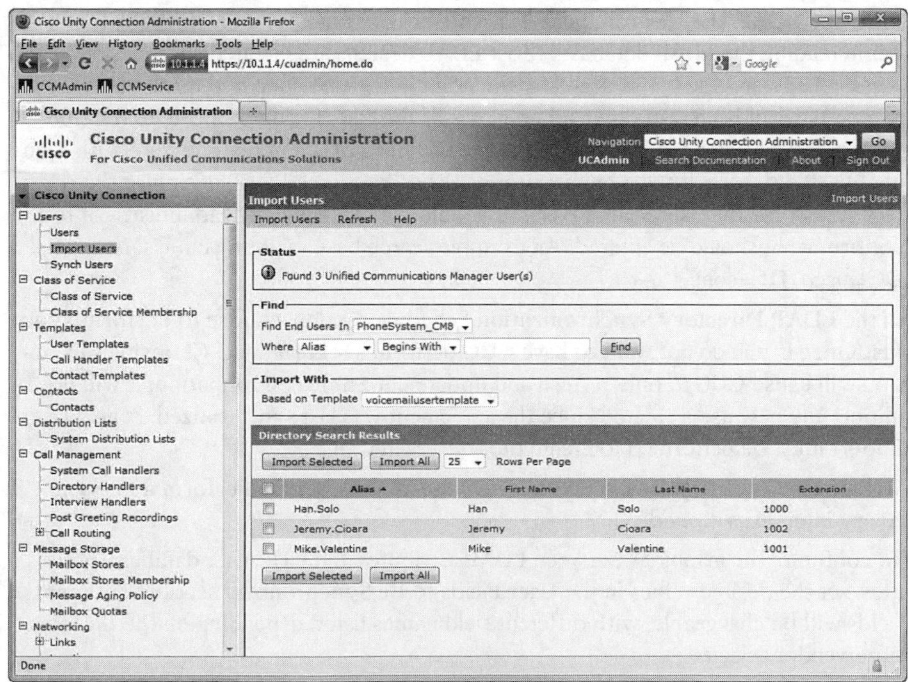

Figure 13-15 *User Import from CUCM Using AXL*

CUC can import users from a number of different vendors' implementations of LDAP. These user accounts are held and maintained in the LDAP system, with certain fields (attributes) copied as read-only entries to the CUC database. User authentication can be optionally redirected from the local CUC database to the LDAP system as well.

To enable LDAP synchronization, follow these steps:

1. In the CUC Cisco Unified Serviceability interface, navigate to **Tools > Services**.
2. Select the **DirSync** service and click **Save**.
3. In the CUC Administration interface, navigate to **System Settings > LDAP > LDAP Setup**.
4. Check the **Enable Synchronizing From LDAP Server** checkbox.
5. Choose the **LDAP Server Type** from the drop-down.
6. In the **LDAP Attribute for User ID** drop-down, select the LDAP attribute that will be mapped to the CUC Alias attribute. The selected attribute in LDAP must contain data, and the data must be unique. User accounts without data in the selected attribute will not be imported.
7. Navigate to **LDAP > LDAP Directory Configuration**.
8. Enter an **LDAP Configuration Name**. It is recommended to use a name that identifies the users are being imported, especially if multiple User Search Bases are configured.
9. Enter the **LDAP Manager Distinguished Name** and **LDAP Password**: This is the LDAP account and password that CUC uses to read and import the LDAP database.
10. Enter the **LDAP User Search Base**: This entry defines the point at which CUC will begin reading the LDAP database. Most LDAP designs are hierarchical tree structures; CUC starts the LDAP search at the point in the tree specified by the User Search Base and can read down all branches of the tree. It cannot move up the tree from that point, nor can it cross to other branches. CUC can only integrate with a single LDAP database. If the administrator does not know the LDAP design or the correct syntax for the User Search Base, he should contact the LDAP administrator to confirm what should be entered. An example search base might be cn=Users, DC=cisco, DC=com.
11. In the **LDAP Directory Synchronization Schedule** section, choose to **Perform Sync Just Once** if you do not want to have CUC perform a regular sync. Choosing this option will cause CUC to only refresh and update current user information; it will not import any new users created since the agreement was last synchronized. A new User Import must be performed to create those users in CUC.
12. To have CUC synchronize on a regular scheduled basis, set the **Perform a Re-sync Every** interval as desired.
13. To configure the mappings between LDAP attributes and CUC user database attributes, set the desired values in the **User Fields to Be Synchronized** section. Different fields will be changeable, with different field names listed depending on the LDAP type/vendor selected.

Figure 13-16 shows part of the LDAP Directory Configuration page.

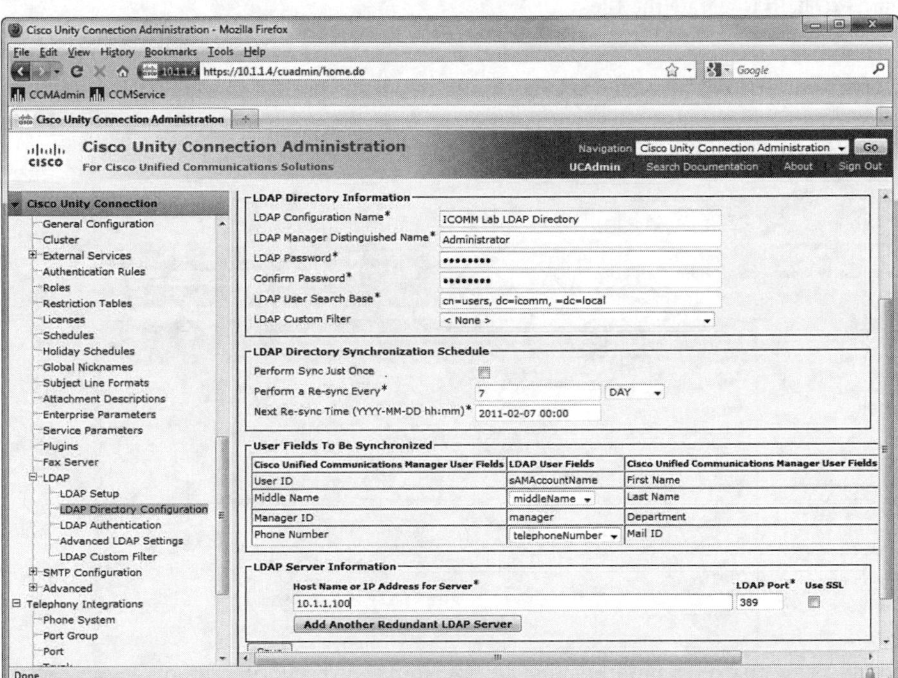

Figure 13-16 *LDAP Directory Configuration Page*

Bulk Administration Import of CUC Users

Importing users using the Bulk Administration Tool is a fast and easy way to create multiple accounts if the required user information can be formatted as a CSV file. To bulk import users, follow these steps:

1. Navigate to **Tools > Bulk Administration Tool.**
2. Under **Select Operation**, select **Create.** (Note that you can also **Update, Delete,** or **Export.**)
3. Under **Select Object Type**, choose **Users With Mailbox.**
4. Under **Override CSV Fields When Creating User Accounts**, select **User Template Yes** and select the desired **User Template.**
5. Under **Select File**, browse to the .csv file that contains the user import information.
6. Specify a **Failed Objects Filename.**
7. Click **Submit.**

Note: The formatting of the .csv is critical to the success of the import. The easiest way to get a correctly formatted file is to perform a BAT Export of a single user. The .csv file then has all the necessary information you need to create the import file correctly. This is different from the CUCM BAT tool, which includes an Excel template file for download with macros to help to create the file.

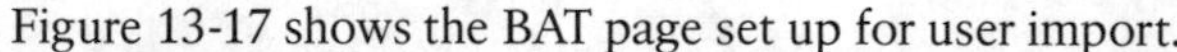

Figure 13-17 shows the BAT page set up for user import.

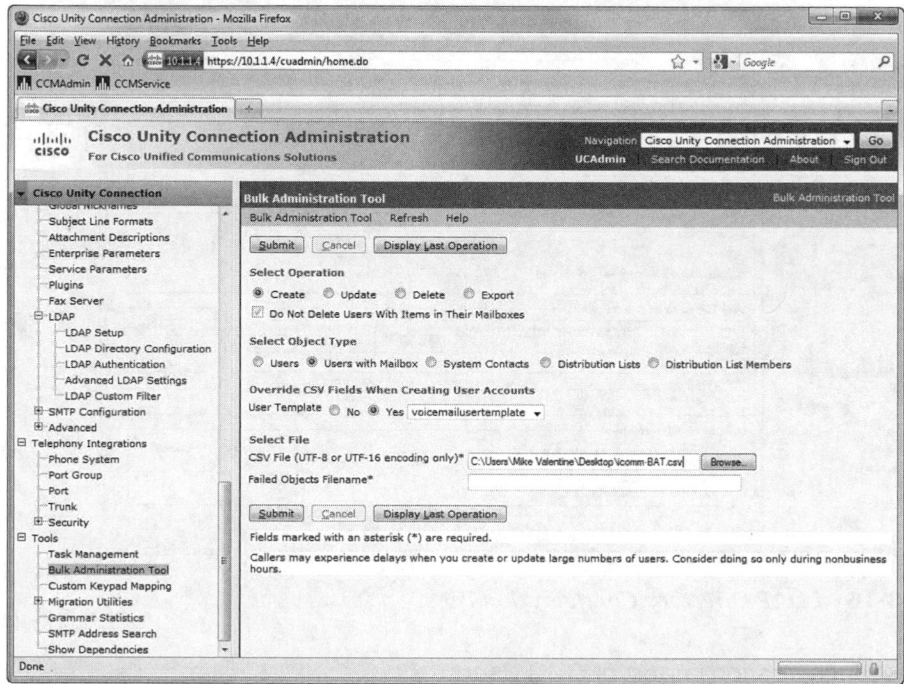

Figure 13-17 *BAT: User Import*

Managing the CUC Message Store

The size and some details of the mailbox store can be checked by navigating to **Message Storage > Mailbox Stores** and selecting the store you want to check. The display page provides information on the size of the store, the number of mailboxes, when it was created, and allows you to set the **Maximum Size Before Warning** value to determine when CUC begins sending warnings about the store size. (When 90% of the configured value is reached, warnings are logged; at 100%, errors are logged.)

Figure 13-18 shows the Edit Mailbox Store page.

Mailbox Stores Membership

Additional Message Stores can be created if additional space is required. Users can be easily moved to the new store by navigating to **Message Storage > Mailbox Stores Membership**.

Select the user(s) you want to move, select the database to move them to, and click **Move Selected Mailboxes**.

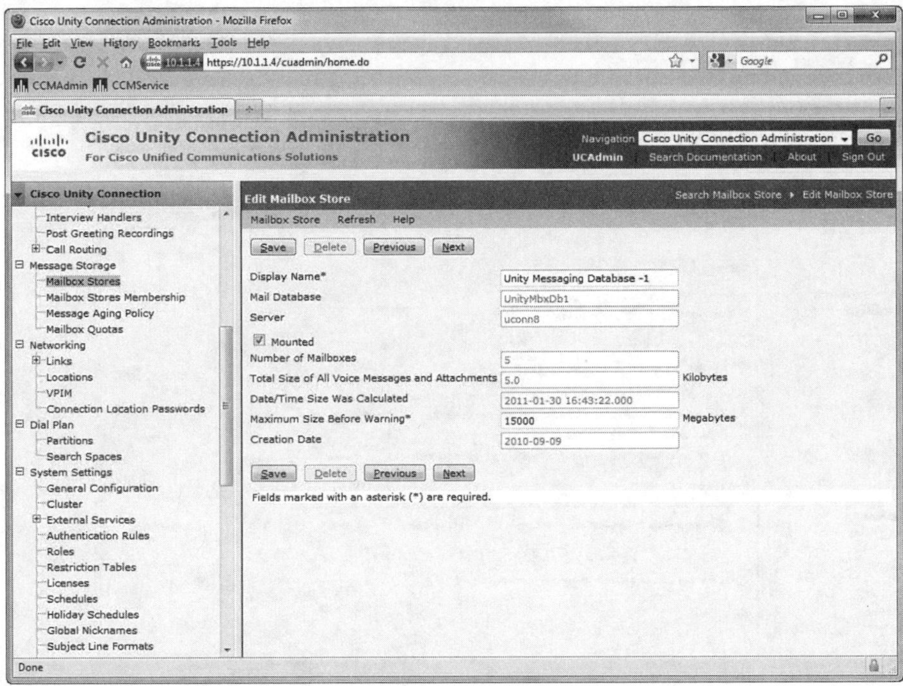

Figure 13-18 *Edit Mailbox Store Page*

Message Aging Policy

The default Message Aging Policy deletes items in the Deleted Items folder after 15 days. To modify the policy, navigate to **Message Storage > Message Aging Policy**, and then select **Default System Policy.** (Note that you may create custom aging policies if you want.) You may disable the policy by deselecting the **Enabled** checkbox; the default is enabled. You may also choose to modify the following:

- Under the **Message Aging Rules Based on When the Message Was Last Modified** heading, you may select whether or not to move New Messages to the Saved folder, and Saved Messages to the Deleted Items folder, and after how many days for each.
- Under the **Secure Message Aging Rules Based on When the Message Was Created** heading, you may choose to permanently delete secure touched messages or all secure messages that reach a specified age.

Figure 13-19 shows the Message Aging Policy page for the Default System Policy.

Mailbox Quotas

Setting strict mailbox size limits early in the deployment is a good idea. To begin, navigate to **Message Storage > Mailbox Quotas.** Here, you can change the **System-Wide Mailbox Quotas,** including the **Warning, Send,** and **Send/Receive** thresholds. These quotas may be overridden in the User Template or individually per user.

Figure 13-20 shows the System-Wide Mailbox Quotas page.

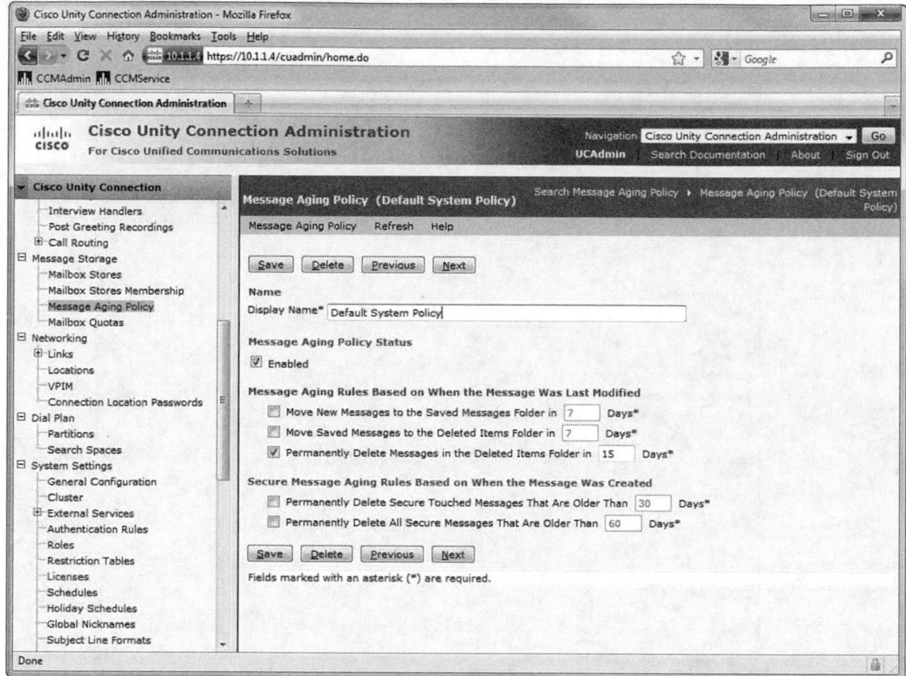

Figure 13-19 *Message Aging Policy Page*

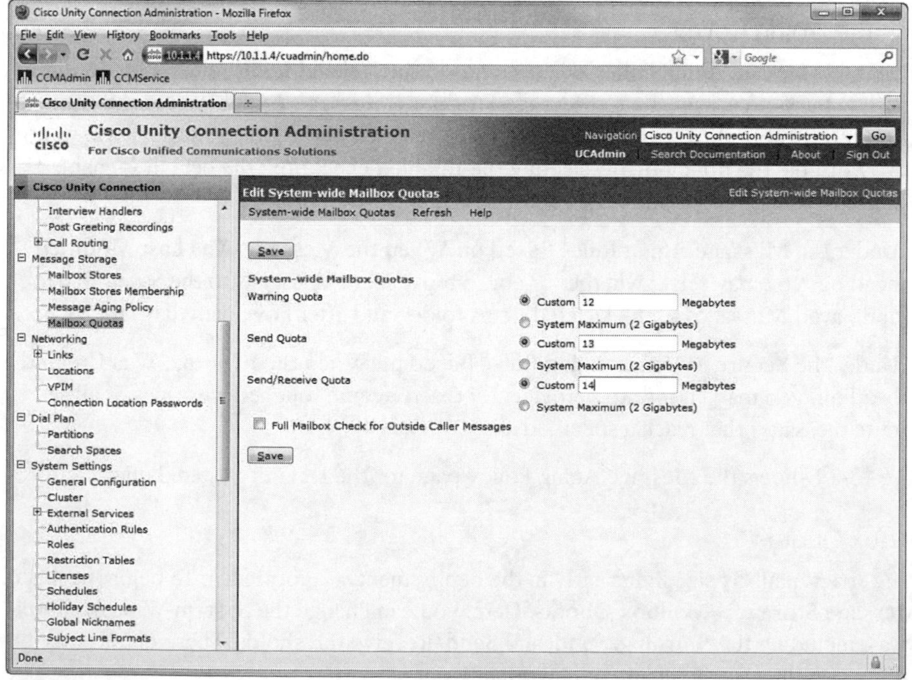

Figure 13-20 *System-Wide Mailbox Quotas Page*

Exam Preparation Tasks

Review All the Key Topics

Review the most important topics in the chapter, noted with the key topics icon in the outer margin of the page. Table 13-3 describes these key topics and identifies the page number on which each is found.

Table 13-3 *Voicemail Integration with Cisco Unity Connection Key Topics*

Key Topic Element	Description	Page Number
Paragraph	CUC integration with CUCM using SCCP	347
Paragraph	Call routing	350
Key Concept Box	CoS in CUC	353
Paragraph	Extension and call forward options	355
Paragraph	User creation options	356
Paragraph	Configure CUC end users	365
Paragraph	Importing end users in to CUC	367
Paragraph	Importing users from LDAP	368
Paragraph	Bulk administration import of CUC users	371

Definitions of Key Terms

Define the following key terms from this chapter, and check your answers in the Glossary:

integration, call handler, class of service (CoS), Administrative Extensions for XML (AXL), call routing rules

This chapter includes the following topics:

- **Describe Cisco Unified Presence Features:** This section discusses Cisco Unified Personal Communicator, its features, operating modes, and protocol utilization. The basic capabilities of the Cisco Unified Communications Manager IP Phone Service and IP Phone Messenger are also reviewed.
- **Describe Cisco Unified Presence Architecture:** This section identifies the protocols and interactions associated with CUPS integrations to other Unified Communications applications and features. Cisco Unified Personal Communicator call flows in both deskphone and softphone mode are described, along with regulatory compliance and QoS considerations.
- **Enabling Cisco Unified Presence:** This section reviews the steps required to configure end users so that they can use Cisco Unified Personal Communicator in either desk phone control mode or softphone mode.

CHAPTER 14

Enabling Cisco Unified Presence Support

Cisco Unified Presence Server (CUPS) extends the native Presence capabilities of Cisco Unified Communications Manager to include multiple communications methods and status settings, as well as important enterprise features relating to compliance with legislation regarding retention of communications. This chapter introduces CUPS and briefly describes some of the implementation considerations in its deployment.

"Do I Know This Already?" Quiz

The "Do I Know This Already?" quiz allows you to assess whether you should read this entire chapter or simply jump to the "Exam Preparation Tasks" section for review. If you are in doubt, read the entire chapter. Table 14-1 outlines the major headings in this chapter and the corresponding "Do I Know This Already?" quiz questions. You can find the answers in Appendix A, "Answers Appendix."

Table 14-1 *"Do I Know This Already?" Foundation Topics Section-to-Question Mapping*

Foundation Topics Section	Questions
Describe Cisco Unified Presence Features	1–6
Describe Cisco Unified Presence Architecture and Call Flows	7–8

1. CUPS uses standards-based protocols for call control signaling and Instant Messaging chat. Which of the following are protocols used by CUPS for these purposes? (Choose two.)

   a. SCCP

   b. SSH

   c. SIP/SIMPLE

   d. XMPP

2. Which external communications and presence applications can CUPS interact with? (Choose all that apply.)
   a. Cisco Unity Connection 8.x
   b. Microsoft Exchange Server 2007
   c. Google Talk
   d. Cisco Unified MeetingPlace
   e. Cisco WebEx
3. Which of the following is *not* a capability of CUPC?
   a. Voice Calling
   b. Video Calling
   c. IM Chat
   d. Visual Voicemail
   e. Persistent Chat
4. CUPC can operate in two modes. What are they?
   a. SIMPLE Telephony User Node (STUN) mode
   b. Deskphone mode
   c. Autonomous mode
   d. Softphone mode
5. Which of the following is true of the Client Services Framework?
   a. It is a standards-based communication platform that enables cross-platform Presence signaling.
   b. It is a Windows Server software that enables features and functionality for Cisco Unified Communications.
   c. It is a core component of CUPC and enables integration with Microsoft Office Communicator.
   d. Multiple CSF clients can be installed on a single workstation, enabling simultaneous cross-platform communications.

6. Group Chat is a feature of CUPS. If group chat rooms and discussions need to be available for future reference after all participants log off, which two steps are required?

   a. Obtain additional licenses for Persistent Chat.
   b. Install a WebEx server to record group chats.
   c. Install a third-party IM-retention compliance application on the CUPS server.
   d. Enable Persistent Chat on the CUPS server.
   e. Provision a PostgreSQL-compliant external database.

7. What components are required for CUPC to operate if the user is traveling with his laptop? (Choose three.)

   a. CSF registers as a SIP device with CUCM using SIP.
   b. User account associated with CSF device in CUCM.
   c. CUPC downloads config file from TFTP.
   d. Cisco IP Communicator must be installed on the client workstation.

8. How can calendar information in Microsoft Exchange be integrated into CUPS Presence signaling so that Free/Busy status in an Outlook calendar can be mapped to the Available/Away Presence status?

   a. Using WebDAV
   b. Using SIP/SIMPLE
   c. Using CTIQBE
   d. Using XMPP

Foundation Topics

Describe Cisco Unified Presence Features

Cisco Unified Presence Server (CUPS) extends the basic Presence capability of Cisco Unified Communications Manager (CUCM) (on-hook, off-hook, or unknown) to include advanced capability and availability signaling. The availability and capability of colleagues and business partners to communicate (Available, Busy, and Away status for various devices) is immediately visible and can be additionally enhanced using their calendar status. Enterprise Instant Message (IM) capability is available both within and extended beyond the enterprise to external contacts, including Group Chat, Persistent Chat, and compliance features, such as IM logging and IM history.

The primary goal of CUPS is to reduce communication delays by providing instant and useful information about the capability and availability of the person you are trying to contact. Users can indicate whether they are available, and by which methods, including desk phone, mobile phone, IM, or conferencing application. This signaling capability can be extended to existing applications to allow contact with the right individual to be quickly established. This helps resolve support or customer issues more quickly, eliminating phone tag and the delays associated with e-mail exchanges.

CUPS is tightly integrated with CUCM, which provides call control and native Presence signaling (on-/off-hook status). CUPS itself provides a central collection point for user capabilities and status by way of standards-based signaling using Session Initiation Protocol (SIP), SIP for Instant Messaging and Presence Leveraging Extensions (SIMPLE), and Extensible Messaging and Presence Protocol (XMPP). A variety of client interfaces are available, including Cisco Unified Personal Communicator (CUPC), which provides a rich and tightly integrated user experience.

Cisco Unified Personal Communicator

CUPC provides a single interface for the most commonly used Cisco Unified Communications tools. From the CUPC application, users can initiate voice calls (either through deskphone control or in softphone mode). Contacts are listed with their enhanced Presence status, providing click-to-communicate functionality by phone call, video call, chat, group, chat, and collaboration interaction with a simple interface to transition from one medium to another. Chat is enabled using the Jabber Extensible Communications Platform (XCP) using XMPP as the chat protocol.

CUPC Operating Modes

CUPC operates in two modes:

- **Deskphone mode:** CUPC can control the user's desk phone to place calls. The IP phone must be registered with CUCM and associated with the user. CUPC uses Computer Telephone Integration Quick Buffer Encoding (CTIQBE) for IP phone control. Figure 14-1 illustrates the protocol interaction when CUPC is in deskphone mode.

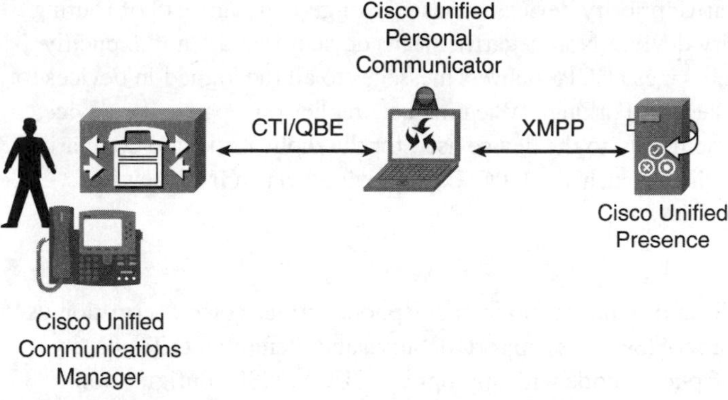

Figure 14-1 *CUPC in Deskphone Mode*

- **Softphone mode:** When the IP phone is not available or the user is away from his desk, CUPC activates the associated softphone based on the Cisco Unified Client Services Framework (CSF), which registers the softphone with CUCM as a SIP device (see Figure 14-2). The CUCM administrator must create the CSF device in order to enable this functionality.

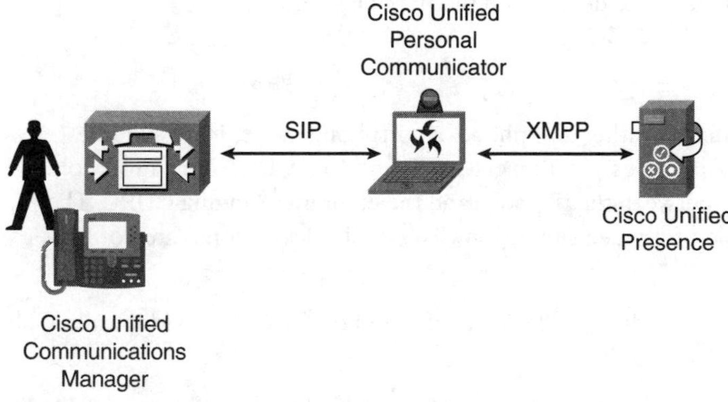

Figure 14-2 *CUPC in Softphone Mode*

Enterprise Instant Messaging

CUPC provides TLS-secured Chat and Group Chat capability. Ad-Hoc Group Chat sessions are stored in memory on the CUPC server. The Persistent Chat feature enables group chat rooms where the conversation persists even when all participants have left the chat session. If Persistent Chat is enabled, it requires an external database to store the chat rooms and conversations. Offline IM allows chat messages to be sent to users who are currently offline; these messages are stored in the local IDS database on the CUPS server.

The Jabber XCP message routing system has been modified in the CUPS implementation to improve functionality when users are available on multiple devices. Messages are routed to all the non-negative priority devices the user is logged into, instead of routing only to the highest-priority device. (Non-negative means devices that are not explicitly blocked from receiving messages.) CUPS delivers messages to all the logged-in devices for a user (this behavior is called IM Forking). When the user replies on a particular device, subsequent messages are sent only to the device used for the reply. Backward compatibility for SIP-only Presence clients (such as CUPC 7.x) is provided via an IM gateway.

Voice Calls

CUPC supports voice calls in deskphone mode or softphone mode. Voice encryption using Secure Real-Time Protocol (SRTP) is supported. Survivable Remote Site Telephony (SRST) is supported in softphone mode with appropriate CUCM/CSF configuration. CUPC supports the following codecs in softphone mode:

- G.711 a-law, mu-law
- G.722 Wideband
- G.729A, G.729B
- iLBC (Internet Low Bandwidth Codec)
- iSAC (Internet Speech Audio Codec)

Codec support in deskphone mode depends on the IP phone model.

Video Calls

CUPC supports video calling in either deskphone or softphone mode. In deskphone mode, CUPC uses Cisco Audio Session Tunnel (CAST) and Cisco Discovery Protocol (CDP) for communication between the IP phone and the computer running CUPC. The desk phone must be enabled for video support in CUCM. SIP desk phones are not supported by CAST.

In softphone mode, CUPC uses the CSF device configured on CUCM for video calls. CSF devices are video-enabled by default.

Video conferencing is supported, using a video-conference bridge accessible by the Media Resource Group List (MRGL) assigned to the IP phone or CSF device.

Integration Support

CUPC can integrate with many Cisco Unified Communications applications, providing a wide range of rich features. CUPC can provide visual voicemail when integrated with Cisco Unity or Unity Connection. Users can control their mailboxes and listen to, send, and delete messages using CUPC.

Click-to-call functionality is supported from CUPC and other applications, such as Microsoft Outlook, Word, and Excel. Presence indicators for contacts can also be viewed in Outlook. If Cisco Unified MeetingPlace is integrated in the system, a call in CUPC can be easily escalated to a conference.

CUPC System Requirements

Refer to the Cisco documentation at www.cisco.com/en/US/products/ps6844/index.html for current system requirements for Windows and Macintosh operating systems (OS).

Cisco Unified Client Services Framework

The Cisco Unified Client Services Framework (CSF) provides the foundation of all Unified Communications client software. It is part of CUPC, but it also extends the functionality of Microsoft Outlook and WebEx Connect. The core functionality of the CSF is voice and video, secure communication with CUCM, and communication with text (IM) servers such as CUPS. CSF provides audio and video call control and advanced features, such as visual voicemail support.

Only one CSF client can be installed at one time on a client PC; for example, CUPC and Cisco Unified Communications Integration for Microsoft Office Communicator cannot co-reside on the same client computer.

Cisco Unified Communications Manager IP Phone Service

The Cisco Unified Communication Manager IP Phone (CCMCIP) Service was originally used to provide authentication, directory services, and help for end users. It has been adapted for use by CUPC and CSF clients to retrieve a list of devices on which the user can be reached when they log in. Figure 14-3 illustrates CUPC's use of the CCMCIP service.

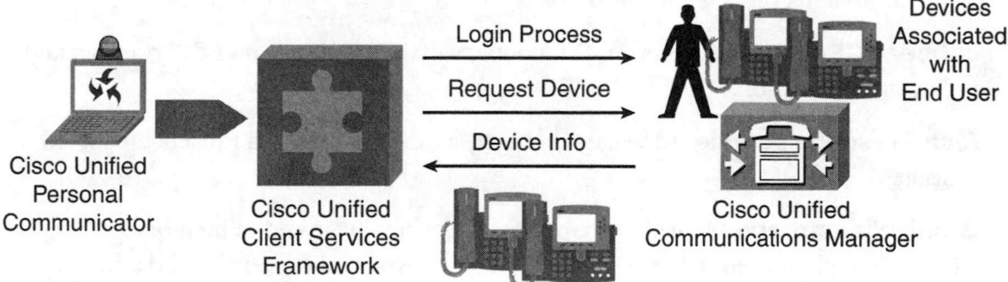

Figure 14-3 *CCMCIP Interaction*

Cisco IP Phone Messenger

Cisco IP Phone Messenger (IPPM) is an IP phone service that allows a user to build a contact list and see the Presence status of those contacts, all from their IP desk phone. The service also allows users to read and clear IM messages from others, respond to the sender by calling them, or send preconfigured text messages directly from the phone. (The text of these messages can be customized by the CUCM administrator.) Users can also change their Presence status using the IP phone.

The background components of IPPM include XML-over-HTTP signaling between the IP phone running the service and CUPS, and XMPP between CUPS and CUPC on the user's workstation. SIP is used for communication with SIP proxy or registrar servers. Figure 14-4 illustrates protocol interactions when using IPPM.

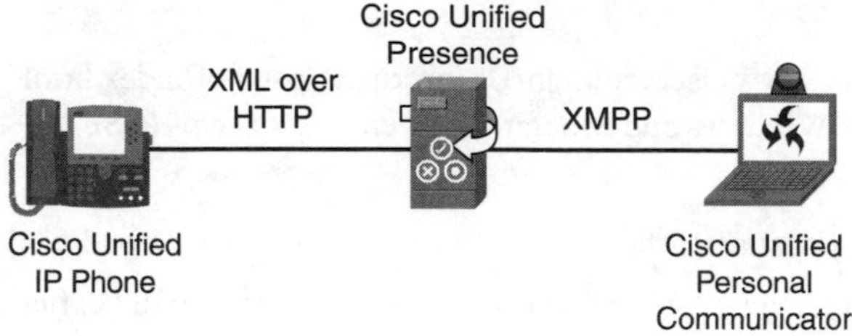

Figure 14-4 *Cisco IPPM Protocol Interaction*

Describe Cisco Unified Presence Architecture

CUPS is a standards-based application that uses several protocols to provide functionality and feature richness:

- SIP, SIMPLE, and XMPP to provide generic Presence and federation functionality
- Simple Object Access Protocol (SOAP) to access the CUCM database via Cisco Unified Communications System XML and CUPC configuration profiles
- Computer Telephone Interface Quick Buffer Encoding (CTIQBE) for CTI integration for remote call control with Microsoft Office Communicator
- HTTP to provide IP Phone Messenger (IPPM) on Cisco IP phones

The core components of CUPS are as follows:

- **Jabber XCP:** Provides Presence, IM, contacts listing, message and call routing, and policy and federation.
- **Rich Presence service:** Manages Presence state gathering and Presence-enabled routing.
- **Group chat storage:** Ad-Hoc group chats are normally stored in memory on the CUPS server; Persistent Chat and message archiving require an external database.

Integration with Microsoft Office Communications Server

CUPS interoperates with Microsoft Live Communications Server 2005 or Microsoft Office Communications Server 2007 using SIP. The Microsoft Office Communicator client and associated IP phone interoperate to provide click-to-dial, phone control, and Presence capability.

Integration with LDAP

CUPS can be integrated with Lightweight Directory Access Protocol (LDAP) (including Microsoft Active Directory), allowing users to log in with their LDAP credentials and synchronize their Presence status with their Outlook/Exchange calendar. The LDAP directory can be searched from the CUPC interface. CUPS can communicate with Exchange using Outlook Web Access (a Web Distributed Authoring and Versioning interface provided by Exchange). The CUPS user database is synchronized with CUCM, which in turn may synchronize its user database with LDAP. CUPS can redirect user authentication to LDAP, providing a single sign-on experience for users. Figure 14-5 shows CUPS integration with LDAP.

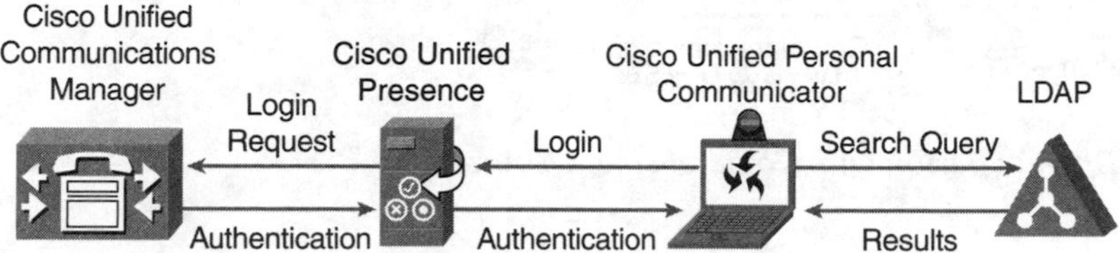

Figure 14-5 *CUPS Integration with LDAP*

Integration with Cisco Unity Connection

CUPS integration with CUC provides the user with the ability to sort, view, listen to, and delete voicemail messages, and call the sender of voicemail messages, all from the CUPC interface. Presence information for the sender is displayed, allowing the user to select how to communicate with the sender or escalate to a call, message, or conference. Interaction with the CUC mailbox is via Internet Message Access Protocol (IMAP) and requires a voicemail profile to be configured on the CUPS server. Figure 14-6 illustrates CUPS interaction with CUC.

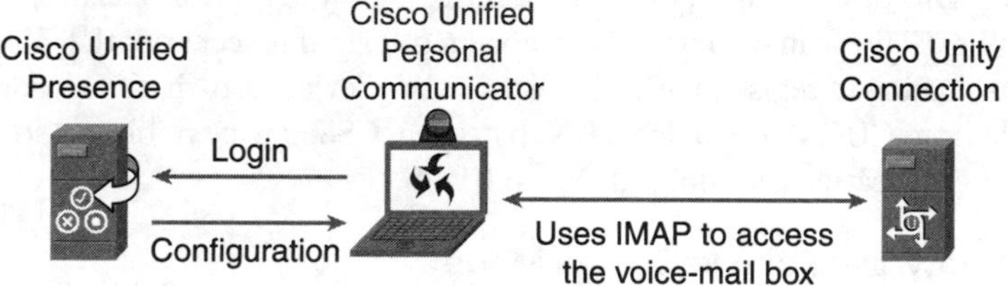

Figure 14-6 *CUPS Integration with CUC*

Integration with Conferencing Resources

CUPS uses WebEx or MeetingPlace for its conferencing capability. The use of WebEx requires a local Meeting Center server. Communication with conferencing server is via Macromedia Flash using HTTP or HTTPS for transport. Figure 14-7 shows CUPS integration with conferencing resources.

Integration with Calendar Resources

CUPS can integrate with Microsoft Exchange 2003 or 2007 (Active Directory 2003 or 2008 is required) to provide access to calendar status (Free/Busy/Out of Office) and map that status to a Presence status (Available, Busy, Away). A special Exchange account (with membership in the View-Only Administrators Groups in Exchange and Receive-As permissions on all user mailboxes) must be configured for CUPS to inspect user calendars. Figure 14-8 shows CUPS interaction with calendar resources.

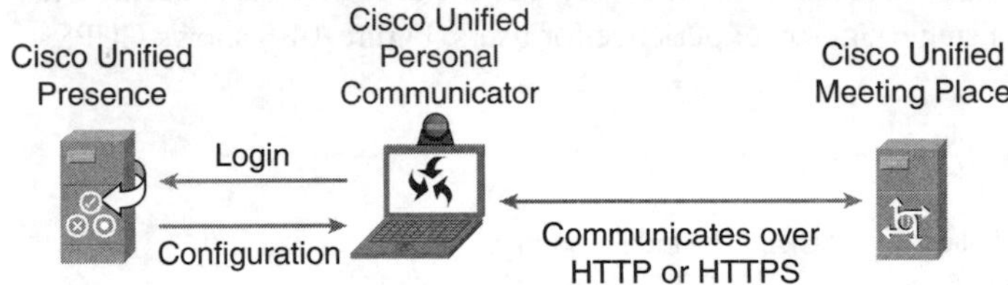

Figure 14-7 *CUPS Integration with Conferencing Resources*

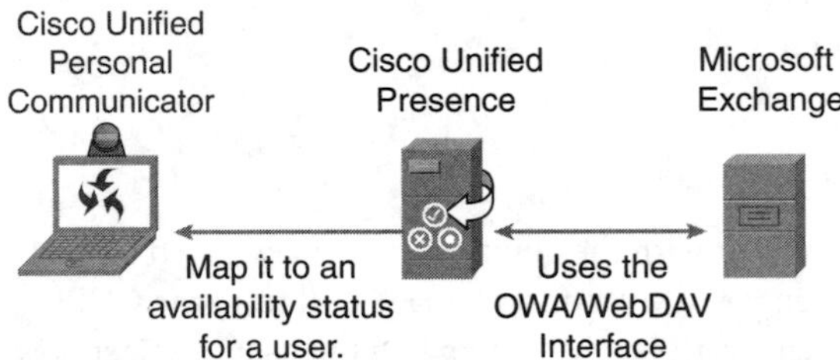

Figure 14-8 *CUPS Interaction with Calendar Resources*

Architecture and Call Flow: Softphone Mode

If the user is away from his desk (using a laptop while travelling is the typical scenario) or no IP phone is available, CUPC can operate in softphone mode. In this scenario, the CSF framework uses SIP signaling to register with CUCM as a CSF device. It then downloads the configuration file from CUCM, obtaining a DN, partition, CSS, device pool, and so on. XMPP is used for chat features, sending all IMs to CUPS.

Architecture and Call Flow: Deskphone Control Mode

When using CUPC in deskphone control mode, CUPC registers with CUCM and downloads its configuration file. It then logs in with the user-provided credentials. The CCMCIP service provides CUPC with the list of user-associated controlled devices. CUPC uses CTIQBE to control the IP phone. (If the user has multiple phones associated with his account, he can select the phone CUPC should control.) CUPC uses XMPP for chat features, sending all IMs to CUPS.

Compliance and Persistent Chat

Persistent Chat refers to group chat messages that are preserved when all group chat participants have left the chat session, so that the information can be referenced later. For Ad-Hoc chats, these messages are preserved in server memory. Persistent Chat requires an external database store; a separate database store is required for each CUPS server configured for Persistent Chat. The database instances can run on the same server but do not have to. Interaction with the PostgreSQL database is via Open Database Connectivity (ODBC).

Regulatory compliance may require that IM chats be preserved. CUPS can provide this capability by using a PostgreSQL external database. Third-party compliance applications may provide more features than a simple PostgreSQL database store; capabilities such as inline virus scanning of IMs and antispam measures for IM are among the more common features. If a third-party compliance application is in use, all chats are sent through the compliance server; this means that if the server is not available to CUPS, no IMs can be sent. Figure 14-9 illustrates CUPS integration with an external PostgreSQL server.

Figure 14-9 *CUPS Integration with External Compliance Server*

CUPS and QoS Considerations

Because Quality of Service (QoS) is so vital to successful Unified Communications deployments, some measures must be taken to ensure that CUPC traffic is appropriately processed by QoS mechanisms. CUPC marks traffic outbound from the user workstation with values appropriate for voice, video, and signaling traffic. Normally, all traffic coming from the user workstation is untrusted and marked down to a low QoS value by the first QoS-enabled device that handles it; this behavior must be modified by specifying the port ranges that CUPC uses and applying the appropriate QoS markings to that traffic. This requires that the network device (switch, router, firewall, and so on) have the capability to classify traffic based on port number, mark the traffic as appropriate for the QoS environment, and apply a QoS policy to forward the traffic to specific destination addresses with appropriate bandwidth and delay guarantees. It is possible that the OS on the workstation running CUPC may not support the QoS marking outbound, but this is increasingly unlikely in modern LAN environments.

Table 14-2 lists the protocols transmitted by CUPC, along with their port numbers and a brief description.

Table 14-2 *Protocols, Port Numbers, and Descriptions of CUPC Outbound Traffic*

Port Number	Protocol	Description
69	UDP	Connects to the TFTP server to download the TFTP file.
80	TCP HTTP	Connects to services such as Cisco Unified MeetingPlace for meetings, or Cisco Unity or Cisco Unity Connection for voicemail features.
143	IMAP (TCP/TLS)	Connects to Cisco Unity or Cisco Unity Connection to retrieve and manage the list of voice messages for the user, and the voice messages themselves.
389	TCP	Connects to the LDAP server for contact searches.

Table 14-2 *Protocols, Port Numbers, and Descriptions of CUPC Outbound Traffic*

Port Number	Protocol	Description
443	TCP HTTPS	Connects to services such as Cisco Unified MeetingPlace for meetings, or Cisco Unity or Cisco Unity Connection for voicemail features.
636	LDAPS	Connects to the secure LDAP server for contact searches.
993	IMAP (SSL)	Connects to Cisco Unity or Cisco Unity Connection to retrieve and manage the list of voice messages for the user, and the voice messages themselves.
2748	TCP	Connects to the CTI gateway, which is the CTIManager component of Cisco Unified Communications Manager.
5060	UDP/TCP	Provides Session Initiation Protocol (SIP) call signaling.
5061	TCP	Provides secure SIP call signaling.
5222	TCP (XMPP)	Connects to the Cisco Unified Presence server for availability status and IM features.
7993	IMAP (TLS)	Connects to Cisco Unity Connection to retrieve and manage the list of secure voice messages for the user, and the secure voice messages themselves.
8191	TCP	Connects to the local port to provide SOAP web services.
8443	TCP	Connects to the Cisco Unified Communications Manager IP Phone (CCMCIP) server to get a list of currently assigned devices.
16384-32766	UDP	Sends RTP media streams for audio and video.

Enabling Cisco Unified Presence

This section reviews the steps required to configure end users so that they can use Cisco Unified Personal Communicator (CUPC) in either desk phone control mode or softphone mode.

Enabling End Users for Cisco Unified Personal Communicator in CUCM

The steps required to enable end users for CUPC must be carried out on CUCM, CUPS, and the CUPC client. The steps required for CUCM are summarized in the following list:

1. Assign license capabilities in CUCM.
2. Configure end users in CUCM.
3. Associate the directory numbers with the end users in CUCM.
4. Create a Cisco Unified Client Services Framework (CSF) device.
5. Associate the CSF device with the end user in CUCM.

These summary steps are detailed in the following sections.

Step 1: Assign License Capabilities in CUCM

CUCM controls which end users can use CUPS and CUPC. In CUCM Administration, navigate to **System > Licensing > Capabilities Assignment.** Find the user for whom you want to enable Presence and CUPC and click on the username. On the page that opens, select **Enable CUP (Cisco Unified Presence)** to enable CUP for that user. Select **Enable CUPC (Cisco Unified Personal Communicator)** to enable CUPC for that user. You can also use the **Bulk Assignment** button, which allows you to select multiple users from the list and assign these two capabilities to all the selected users at once.

Step 2: Configure End Users in CUCM

Key Topic

To configure desk phone control for a user, in CUCM Administration, navigate to **User Management > End User** and select the user you want to configure. In the **Device Information** section on the End User configuration page, click the **Device Association** button and select the user's IP Phone so that it appears in the **Controlled Devices** pane. (You should also verify that the **Allow Control of Device from CTI** checkbox is selected on that device's configuration page; it is selected by default.) If the user uses Extension Mobility, you must move the appropriate Device Profile to the **CTI Controlled Device Profiles** pane.

Key Point: If you do not check the **Allow Control of Device from CTI** checkbox, the user cannot control his desk phone while using CUPC, and he cannot make any calls!

Because different IP Phone models support CTI in different ways, the end user must be added to one or more groups for the CTI functions to work correctly on the various phone models the user might have access to. On the End User configuration page, scroll down to the **Permissions Information** section. Use the following guidelines to add the user to the appropriate group or groups:

- **For all IP Phone models:** Add the user to the **Standard CTI Enabled** group.

- **For IP Phone 69XX series models:** Add the user to the **Standard CTI Allow Control of Phones supporting Rollover Mode** group.
- **For IP Phone 89XX and 99XX series models:** Add the user to the **Standard CTI Allow Control of Phones supporting Connected Xfer and conf** group.

Step 3: Associate the Directory Numbers with the End Users in CUCM

Navigate to **Device > Phone**, select the user's IP Phone, and go to the Directory Number Configuration. Select the **Allow Control of Device from CTI** checkbox. If using Extension Mobility, click to enable the same box on the appropriate Device Profile configuration page for the user.

Next, scroll down to the **Users Associated with Line** section and click the **Associate End Users** button. Select the user associated with this DN and click **Add Selected.** Click **Apply Config.**

Step 4: Create a Cisco Unified CSF Device

At this point, you need to set up CUPC in softphone mode. Navigate to **Device > Phone** and click **Add New**. Select the appropriate **Phone Type** for the version of CUCP in use. The Key Concept following Step 4 provides the detail you need to know to select the correct Phone Type.

Next, select the **Allow Control of Device from CTI** checkbox. Then, select a **Device Security Profile** followed by a **SIP Profile.** On the user's Directory Number configuration page, associate this new CSF device with the same DN as on the user's primary IP Phone. Verify that the user is associated with the DN.

Key Concept: Keep this in mind with respect to the version of CUPC in use:

1. If using CUPC version 8.0+, select **Cisco Unified Client Services Framework** as the Phone Type:
   - CSF devices can be named anything you like; there isn't a naming convention, but there is a limit of 15 characters (letters and numbers only).
2. If using CUPC version 7.0, select **Cisco Unified Personal Communicator** as the Phone Type. CUPC version 7.0 devices must follow a specific naming convention:
   - The CUPC v7.0 device name must start with the letters "UPC," followed by a derivation of the username.
   - The CUPC v7.0 device name can contain only numbers and uppercase letters; it can be a maximum of 12 characters after the "UPC" prefix.
   - For example: The CUPC 7.0 device name for the user ID "Mike.Valentine" would be "UPCMIKEVALENTIN."

This might seem counterintuitive at first. Just remember that the CSF is the newer way of doing things, so the newer version of CUPC uses CSF. The newer version has no naming restrictions; again, newer is easier and fancier.

Step 5: Associate the CSF Device with the End User in CUCM

On the End User configuration page, under the **Device Information** section, add the newly created CSF device to the **Controlled Devices** list by clicking the **Device Association** button. Click **Save.**

Enabling End Users for CUPC in Cisco Unified Presence

CUPC has a tight integration with CUPS. On the CUPS server, there are several configurations that must be built to support CUPC features, such as accessing voice mail messages from the CUPC client, desk phone control mode, LDAP directory lookups, and CCMCIP-based user device information. Some features might not be necessary in some deployments. The following list summarizes these features and the CUPS configurations required to support them:

- **Access personal voice mail using CUPC:** CUPC can retrieve and process voice messages from a voice messaging server. There are several interconnected steps that must be configured to support the feature in CUPC, as follows:
  1. **Define the Mailstore:** In CUPS, navigate to **Application > Cisco Unified Personal Communicator > Mailstore** and select **Add New.** Configure a **Name** and **IP address** for the voice messaging server, optionally modifying the port and protocol as appropriate for the messaging server deployment. This particular configuration allows the use of IMAP to retrieve voice messages.
  2. **Define the Voicemail Server:** Navigate to **Application > Cisco Unified Personal Communicator > Voicemail Server.** Click **Add New.** The **Server Type** can be **Unity** or **Unity Connection.** Set the appropriate **IP Address,** and then click **Save.**
  3. **Define the Voicemail Profile:** Navigate to **Application > Cisco Unified Personal Communicator > Voicemail Profile.** Select the previously configured **Mailstore** and **Voicemail Server.**
- **Allow Desk Phone Control:** CUPC can control the user's desk phone using CTI. To enable this function, perform the following steps:
  1. Navigate to **Application > Cisco Unified Personal Communicator > CTI Gateway Server.** Click **Add New.** Enter a **Name** and **IP Address.**
  2. Navigate to **Application > Cisco Unified Personal Communicator > CTI Gateway Profile.** Verify that the auto-created entry exists.
- **Allow LDAP Directory Lookups:** CUPC can access the local LDAP directory to provide a list of users that can be clicked-to-dial:
  1. Navigate to **Application > Cisco Unified Personal Communicator > LDAP Server.** Click **Add New.** Enter a **Name** and **IP Address,** and then click **Save.**
  2. Navigate to **Application > Cisco Unified Personal Communicator > LDAP Profile.** Configure a **Name** and the LDAP account and password used to access the LDAP system. In the **Search Context** field, enter the LDAP search string (in the same syntax as used for LDAP Sync on CUCM) that contains the CUPC users. From the **Primary LDAP Server** drop-down, select the LDAP server configured in Step 1.

- **Define the CCMCIP Profile:** The CCMCIP service has been adapted to allow CUPC to discover all the devices on which a particular user is logged in and available, presenting the user with this information in the CUPC interface:
  1. Navigate to **Application > Cisco Unified Personal Communicator > CCMCIP Profile**, and then click **Add New**.
  2. Enter the IP Address of the **Primary CCMCIP Host**, and optionally add a **Backup CCMCIP Host** as well.

What you have just done is create a set of application profiles that allows CUPC to connect and communicate with the various resources and applications on the Unified Communications network. Other application profiles can be created for things like conferencing applications.

Each application profile must be assigned to users in order for them to be able to use it. This can be done in two ways: The simplest way is to enable the **Make this the default <*profile_type*> Profile for the system** box on each profile configuration page. If this is done before any users are synchronized from CUCM, then all new user accounts will use these default profiles without requiring any further configuration. The other way is to associate users individually to the profiles as (or after) the profiles are created. This is done on each profile page or from the End User configuration page in CUPS. This manual method can be used at any time to modify an individual user's application profile set.

The last configurations needed are on the CUCP application itself. Each user should customize his CUPC client with the correct credentials for each application in his profile set. Applications for which no profile has been defined for this user is not listed in his CUPC client and is not accessible,

Troubleshooting CUPC

Table 14-3 provides a few examples of issues that can arise with CUPC, along with things to check to fix the problems.

Table 14-3 *CUPC Troubleshooting Quick Reference*

Symptom	Things to Check
Error on starting CUPC: "The selected device is not available."	■ Verify that devices are registered in CUCM. ■ Verify that the end user is associated with IP Phone in CUCM. ■ Verify that the CCMCIP Profile is associated with the user in CUPS. ■ Verify that the device and DN can be controlled by CTI in CUCM.
User can't make calls using CUPC in softphone mode.	■ Verify that the user is associated with the CSF device in CUCM. ■ Verify that the CSF device is registered in CUCM. ■ Check for correct DN, partition, and CSS.

Table 14-3 *CUPC Troubleshooting Quick Reference*

Symptom	Things to Check
Users are not shown as on the phone during an active call.	■ Verify that the SIP trunk between CUCM and CUPS exists and is correctly configured. ■ Verify that the user is associated with the line (check the configuration of IP Phone, Device Profile, or CSF as appropriate) in CUCM.
User cannot log on to CUPC.	■ Verify that the user account is not locked. ■ Verify the correct server IP address in CUPC (the user may have changed it). ■ If using the host name instead of the IP address, verify that DNS is available and correctly configured. ■ Verify the license capabilities assignment in CUCM.
User cannot add contacts; search returns no results.	■ Verify that the user is associated with the correct LDAP profile in CUPS. ■ Verify LDAP Search Context in CUPS.
User cannot control the IP Phone 9971.	■ Verify that the IP Phone is associated with the user in CUCM. ■ Verify that the Allow Control of Device from CTI box is checked on the device configuration page in CUCM. ■ Verify that the user is a member of both the **Standard CTI Enabled** and **Standard CTI Allow Control of Phones supporting Connected Xfer and Conf** groups.

Exam Preparation Tasks

Review All the Key Topics

Review the most important topics in the chapter, noted with the key topics icon in the outer margin of the page. Table 14-4 describes these key topics and identifies the page number on which each is found.

Table 14-4 *Key Topics for Chapter 14*

Key Topic Element	Description	Page Number
Paragraph	CUPS fundamentals	380
Paragraph	CUPC operating modes	380
Paragraph	Enterprise Instant Messaging	381
Paragraph	Client services framework	383
Paragraph	Presence architecture	384
Table 14-2	CUPC port numbers and descriptions	387
Section	Configure End Users in CUCM	389
Key Concept Box	CUPC Device Naming Conventions	390
Paragraph	CUPS Application Profiles	392

Definitions of Key Terms

Define the following key terms from this chapter, and check your answers in the Glossary:

CTIQBE, XMPP, SIMPLE, CSF, Persistent Chat, compliance, IPPM, CCMCIP

- **Troubleshooting:** This section walks you through a general troubleshooting process you can use to approach almost any network-related issue.
- **Troubleshooting Common CME Registration Issues:** One of the most common issues you'll encounter in Cisco VoIP is an IP phone that continually boot cycles. This section discusses these issues and provides an approach to solving them.
- **Troubleshooting Dial-Plan and QoS Issues:** When a phone call fails or starts crackling during a call, people on staff have no problem letting you know that they want you to do something about it.

CHAPTER 15

Common CME Management and Troubleshooting Issues

If it worked right the first time every time, none of us would have jobs! This chapter discusses how to handle questions and troubleshoot Cisco Unified Communication Manager Express (CME). The chapter is divided into the three major areas of troubleshooting typically encountered on production networks: IP phone registration issues, dial-plan issues, and Quality of Service (QoS) issues.

"Do I Know This Already?" Quiz

The "Do I Know This Already?" quiz allows you to assess whether you should read this entire chapter or simply jump to the "Exam Preparation Tasks" section for review. If you are in doubt, read the entire chapter. Table 15-1 outlines the major headings in this chapter and the corresponding "Do I Know This Already?" quiz questions. You can find the answers in Appendix A, "Answers Appendix."

Table 15-1 *"Do I Know This Already?" Foundation Topics Section-to-Question Mapping*

Foundation Topics Section	Questions Covered in This Section
Troubleshooting	1–2
Troubleshooting Common CME Registration Issues	3–7
Troubleshooting Common Dial-Plan and QoS Issues	8–10

1. You are planning a structured troubleshooting approach for an IP phone registration issue. You just defined the problem; what is your next step?

   a. Consider the possibilities.

   b. Gather the facts.

   c. Create an action plan.

   d. Implement an action plan.

2. You just finished resolving an outage issue in the voice network. Which of the following should you do as a follow-up measure? (Choose three.)
   a. Reboot the devices to ensure the issue does not reappear.
   b. Document the solution.
   c. Document the root cause of the issue.
   d. Document the changes made to the system.
   e. Document the next change window for follow-up.
3. You are troubleshooting an IP phone registration issue. You verify that the IP phone is receiving an IP address with Option 150 from the DHCP server. What should the phone do next after this point?
   a. Reboot with a new configuration.
   b. Contact the TFTP server.
   c. Register with the CME router.
   d. Update its firmware.
4. A Cisco IP Phone is plugged into an Ethernet wall jack. The phone does not respond to the connection. What should your first area of troubleshooting be for this situation?
   a. Verify that the voice VLAN is assigned to the port.
   b. Enable CDP on the interface.
   c. Verify PoE configuration.
   d. Ensure that you are using Category 6 Ethernet cable.
5. You believe one of your Cisco IP Phones has not been assigned to the correct voice VLAN. What symptom is typically seen when this occurs?
   a. The IP phone has unforeseen call restrictions or permissions.
   b. The IP phone displays "invalid VLAN" on the screen.
   c. The IP phone continually reboots.
   d. The PC attached to the IP phone loses network connectivity.
6. After a Cisco IP Phone determines its voice VLAN configuration via CDP, what does it do?
   a. The phone reboots in the new VLAN.
   b. The phone sends out a DHCP request tagged with the voice VLAN number.
   c. The phone sends out an untagged DHCP request.
   d. The phone queries for a TFTP server in the new VLAN.

7. An IP phone boots and displays "Registration Rejected" on the screen. What is the most likely cause of the issue?

   a. The CME router has no appropriate ephone configuration.

   b. An ACL is blocking access to the CME router.

   c. The TFTP server is not serving the correct files.

   d. The MAC address of the phone is in the disallow list.

8. What command can you enter to watch CME process calls as digits are dialed?

   a. **show dialpeer voice**

   b. **debug dialpeer voice**

   c. **debug voip dialpeer**

   d. **debug voice dialed**

9. What is the one-way delay requirement Cisco recommends to achieve high-quality voice calls?

   a. 100 ms

   b. 150 ms

   c. 200 ms

   d. 250 ms

10. What command can you use to verify QoS operations and packet statistics on a specific interface of your CME router?

   a. **show run**

   b. **show qos interface**

   c. **show service-policy interface**

   d. **show policy-map interface**

Foundation Topics

Troubleshooting

When troubleshooting, employing a consistent and systematic methodology saves time and helps prevent errors that might make the situation worse. The sequence of steps described in this section is supported by Cisco best practices as one model for effective troubleshooting. Figure 15-1 illustrates the troubleshooting sequence. This same model and process is also seen in Chapter 16, "Management and Troubleshooting of Cisco Unified Communications Manager," because it also applies to Cisco Unified Communications Manager (CUCM) troubleshooting.

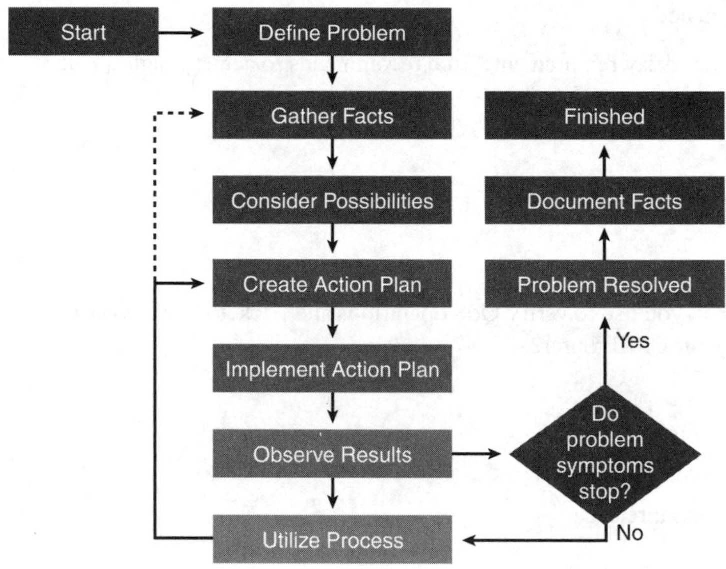

Figure 15-1 *Troubleshooting Methodology*

The steps illustrated in Figure 15-1 are described here:

1. **Define the problem:** Analyze the problem and create a clear problem statement. Define the symptoms and probable causes. Compare current conditions to a baseline "normal" condition.
2. **Gather facts:** Collect and consider command outputs and user statements. Eliminate possible causes to reduce the number of potential issues. Ask: When did this problem occur? What changed before the problem started? It is intermittent or constant? Is there a pattern to the intermittence? Are there any error messages? Does anyone else have the problem?
3. **Consider possibilities:** Based on the facts gathered in Step 2, identify a short list of likely causes. This may be a quick, almost intuitive process, or it might require significant research and discussion.

4. **Create action plan:** Beginning with the most likely cause on the list from Step 3, define a plan of action to correct the problem. The plan should modify only one variable at a time, so that the effect of that one change can be easily evaluated.
5. **Implement an action plan:** Execute the commands or make the changes decided in Step 4. Do not improvise; follow the plan. If new information or ideas come up, you may want to start at Step 3 again. At each step, make sure that the action taken does not make the problem worse—if it does, undo the change. Minimize the impact of changes on production systems, especially limiting the duration of security vulnerabilities, such as temporarily removing access lists to see if they are the problem.
6. **Observe results:** Has the problem been solved?

6a. **Unsolved:** If the problem is not solved, undo the change implemented in Step 5 and return to Step 4.

6b. **Solved:** If the problem is solved, document the solution, root cause, and the changes made to the system.

Troubleshooting Common CME Registration Issues

I'm sure many of you have had the experience where a friend or family member calls and says, "My computer is not working." Instinctively, your mind begins to try to narrow down this broad statement by asking probing questions, such as, "Are there lights on the PC? Does the monitor display an image? Do applications open? Do you smell smoke?" Based on the answers to these questions, you can begin deducing a plan of action.

In the same sense, troubleshooting IP telephony issues requires you to ask probing questions to help narrow down the many possible issues to a focused troubleshooting process. For example, imagine you receive a call from a user who states, "My IP phone is not working." You might ask the probing questions, "How are you talking to me now? Does the IP phone show your extension number on the screen? Is there any dial tone when you lift the handset? What happens when you dial? What messages are on the screen right now? Does it seem like the phone is rebooting? When did you notice the phone stop working?"

All these questions are geared to focus your troubleshooting into one of the following areas:

- Issues relating to the network
- Issues relating to the IP phone configuration
- Issues relating to CME configuration

Just like any data device, an IP phone relies on the network and supporting servers to operate correctly; without them, the IP phone will not function. The absolute biggest thing for you to know when troubleshooting network-related issues is the boot process of the Cisco IP Phone (see Figure 15-2). This was covered in earlier chapters, but we add it here for discussion.

1. The 802.3af Power over Ethernet (PoE) switch sends a small DC voltage on the Ethernet cable, detects an unpowered 802.3af device, and supplies power to the line.
2. The switch delivers voice VLAN information to the Cisco IP Phone using Cisco Discovery Protocol (CDP).

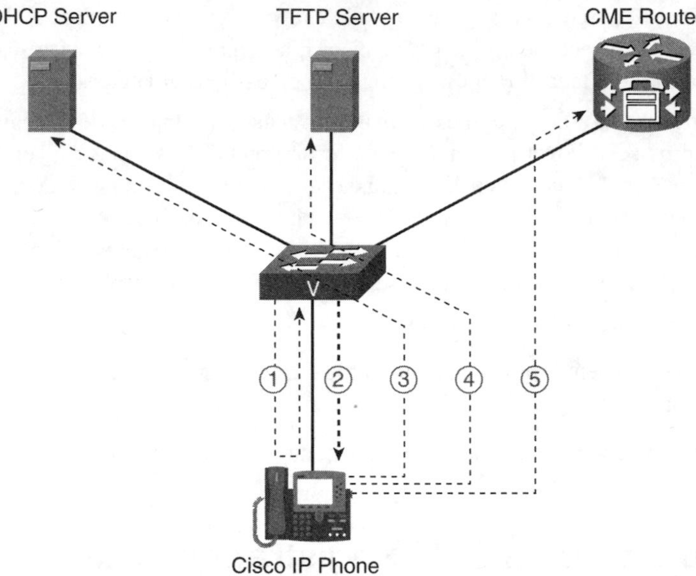

Figure 15-2 *Cisco IP Phone Boot Process*

3. The IP phone sends a DHCP request on its voice VLAN. The DHCP server replies with IP addressing information, including DHCP Option 150, which directs the IP phone to the TFTP server.
4. The IP phone contacts the TFTP server and downloads its configuration file and firmware.
5. Based on the IP address listed in the configuration file, the IP phone contacts the call processing server (the CME router, in this case), which supports VoIP functions.

Knowing this boot process virtually plans your troubleshooting process for you! Anytime a Cisco IP Phone gets stuck in the boot process (for example, it is unable to get a DHCP-assigned address, reach a TFTP server, or register with the CME router), it reboots itself and tries again. You can get many clues to how far the phone is getting in the boot process by carefully watching the messages on the screen. The IP phone tells you when it attempts to configure a voice VLAN, obtain an IP address, or contact a TFTP server. By watching the screen, you can identify the step where the phone stops in the boot process and begin your troubleshooting from that point.

Tip: Keep in mind that it is normal for a newly installed IP phone to reboot several times because of firmware and configuration updates.

Here's the process you should use for troubleshooting network-related issues:

Issue 1: Verifying PoE

Obviously, if the IP phone does not receive power, nothing is going to work correctly. You can quickly diagnose this issue by asking the user if her phone is displaying anything on the screen. If not, here are your areas of focus:

- **Check the physical connections:** Verify that the IP phone is securely plugged in, verify all patch panel connections, and verify the Ethernet cable used has all eight pins functioning properly (because PoE uses the four pins not used by data).
- **Check the PoE switch:** Ensure that the switch is online and operational, verify that the PoE clients have not exhausted the switch power supply by using the **show power inline** command, and use a **show run** command to verify that the port does not have the command **power inline never** (which disables PoE).
- **Check the IP phone:** Verify that the IP phone does not have expansion modules causing it to exceed PoE capabilities, move the IP phone to a different port to see if it receives power, try a different IP phone on the same port to see if it receives power, and be sure that the IP phone and PoE switch support a compatible PoE standard (Cisco inline power, 802.3af, and 802.3at).

Tip: Before you move too far into the troubleshooting process, it's always best to reset the phone's configuration settings to factory default. Although the various IP Phone models have different locations to reset the configuration, all of them will have the option. Often, a bogus configuration item will end up stuck in one of the key fields that you may not initially see. Resetting the phone configuration gives you a clean starting ground to work from.

Issue 2: Voice VLAN Assignment

If an IP phone is assigned to the wrong VLAN, it may not be able to contact the necessary supporting servers (DHCP, TFTP, CME, and so on). You can diagnose this and the following issues when you receive a report that the IP phone is continually rebooting:

- **Check the switch configuration:** Ensure that you correctly configured the voice VLAN by viewing the running-configuration (**show run**), verifying the interface operating mode by entering a **show interface** <*interface_name*> **switchport** command, verifying you created the voice VLAN on the switch (**show vlan**), verifying that the voice VLAN is added to all applicable trunk connections (**show interface trunk**), and verifying CDP is enabled on the port connected to the IP phone (because the voice VLAN is sent to the IP phone using CDP).
- **Check the IP phone:** If you have physical access to the IP phone, navigate to the network settings page (using the Settings button on the phone). Verify that the IP phone has received the voice VLAN configuration, verify the IP phone has an IP address, and access the phone IP address in a web browser and view the log files for more clues.

Issue 3: DHCP Server

Receiving an IP address through DHCP goes hand-in-hand with the voice VLAN assignment (Issue 2). If a Cisco IP Phone is not assigned to the correct VLAN, it may not receive an IP address from the DHCP server, or if it does, the DHCP options for the pool may not be correct. After you verify the voice VLAN configuration, you can use this process to troubleshoot the DHCP process:

- **Check the DHCP helper-address:** If you are using a centralized DHCP server, ensure a router (or L3 switch) supporting the voice VLAN is forwarding DHCP requests to a proper server. (You can find the **helper-address** command under the router interface connected to the VLAN.)
- **Check the DHCP server:** Verify that the DHCP server has an IP address pool created for the voice VLAN devices, ensure that the pool has not run out of IP addresses, verify that DHCP option 150 (TFTP Server) is properly configured and assigned to the pool, connect other test devices (laptop or PC) in the voice VLAN, and ensure these devices receive IP addresses.
- **Check the IP phone:** If you have physical access to the IP phone, navigate to the network settings page (using the Settings button on the phone). Verify that the IP phone received an IP address from the appropriate subnet, verify all applicable DHCP options (subnet mask, default gateway, TFTP server) are filled in, and attempt to ping the phone from another subnet to ensure routing works (assuming no access control lists [ACL] block this communication).

Keep in mind that it's easy to mix up DHCP-related troubleshooting with other phone system issues. Because a phone experiencing communication issues constantly reboots, there are times when the phone does not have an IP address (which can send you down a wrong track of troubleshooting). Before you pull your hair out focusing on DHCP issues, try statically assigning an IP configuration to the phone and see if the phone registers successfully with CME. If it does register successfully, the problem is most likely related to DHCP or VLAN issues. If not, the problem is more likely related to routing, TFTP, or CME issues.

Issue 4: TFTP Server

The TFTP server is a critical part of the IP phone boot process because it supplies the phone firmware and configuration file with the base settings the phone should use for operation (and the IP address of the CME server for registration). Although CME supports using an external TFTP server to store all this data, most CME deployments simply use the flash memory and dynamic RAM of the router to store these files. Take the following steps to troubleshoot TFTP communication:

- **Check routing configuration:** If the TFTP server is on a different subnet than the IP phone, validate that data is able to route between the two subnets by placing a test devoice (such as a laptop or PC) in the voice VLAN and testing connectivity to the TFTP server (by transferring files via TFTP).
- **Check the TFTP server:** Verify that the TFTP server is operational and serving files, validate the firmware for the IP phone model in question exists on the TFTP

server as well as a specific configuration file for the phone (the configuration file should have the MAC address of the IP phone in the filename), and verify that you entered the **create cnf-files** command from telephony-service configuration mode to generate the necessary configuration files on the TFTP server.

- **Check the IP phone:** If you have physical access to the IP phone, navigate to the network settings page (using the Settings button on the phone). Verify that the IP phone is configured (either statically or via DHCP) for the appropriate TFTP server IP address.

Issue 5: CME Server

The final troubleshooting step is to investigate the CME server itself. The most common CME issue encountered is a "Registration Rejected" error message on the IP phone. Seeing this error is actually fantastic news: It means that the IP phone is moving through the entire boot process but fails when trying to register with the CME router. Almost 100% of your focus should be on the ephone settings. A registration rejected message almost always appears because the IP phone has not been properly configured in CME. First, validate that an ephone entry exists in your CME configuration for the IP phone in question. If so, verify that the MAC address entered for the ephone matches the MAC address of the IP phone. Don't trust the sticker on the outside of the IP phone! Verify the MAC address directly from the phone settings or the Cisco switch (viewing the dynamic MAC addresses learned on the port connecting to the IP phone).

If the MAC address appears correct in the CME configuration, try enabling auto-registration in CME (type **auto-reg-ephone** under the telephony-service configuration). This should allow the phone to register without any extension assignment. You can then validate the MAC address of the IP phone by entering the **show ephone** command. If you want to get into the nitty-gritty, issue a **debug ephone detail** command, and you'll be able to watch the IP phone registration process line by line. Be careful with this command, because it might become overwhelming (to both you and your CME router) if you have many phones registering at the same time.

Troubleshooting Dial-Plan and QoS Issues

Although this topic is a bit outside the current scope of the CCNA Voice exam, it is extremely advantageous to you if you are able to troubleshoot basic dial-plan and QoS issues. These issues occur after the IP phone successfully registers with the CME router and attempts to place calls. Symptoms that arise range from call failure when dialing (fast busy/reorder tone) to static, distortion, or dropped calls after the call is connected. The former issue is typically related to a misconfigured dial-plan, while the latter issue is typically related to QoS.

Dial-Plan Issues

To troubleshoot issues related to the dial-plan on the CME router, you must first focus on the dial peers. Remember, the dial-peer configuration builds the routing table for your voice calls. If you configured it inaccurately or incompletely, calls will not complete. Although you can use many commands to troubleshoot calls, two key commands rise to the surface: **show dial-peer voice summary** and **debug voip dialpeer**.

Similar to the **show ip interface brief** command, using **show dial-peer voice summary** enables you to see a table view of all the dial peers that exist on your voice gateway. If calls fail as they are dialed, this is usually the best place to start. Use this command to verify that the expected dial peers exist, have the correct destination pattern configured, and point to a port or IP address that is reachable from the CME router. Example 15-1 shows the output of this command.

Example 15-1 *show dial-peer voice summary Command Output*

```
CME_A# show dial-peer voice summary
dial-peer hunt 0
             AD                                    PRE PASS                 OUT
TAG    TYPE  MIN  OPER PREFIX    DEST-PATTERN      FER THRU SESS-TARGET     STAT PORT
20005  pots  up   up             1500$             0                         50/0/20
20006  pots  up   up             1501$             0                         50/0/21
20007  pots  up   up             1502$             0                         50/0/22
20008  pots  up   up             1503$             0                         50/0/23
20009  pots  up   up             1504$             0                         50/0/24
20010  pots  up   up             1505$             0                         50/0/25
20011  pots  up   up             1506$             0                         50/0/26
20012  pots  up   up             1507$             0                         50/0/27
20013  pots  up   up             1508$             0                         50/0/28
20014  pots  up   up             1509$             0                         50/0/29
1101   pots  up   up             1101              0                      up   0/0/0
1102   pots  up   up             1102              0                      up   0/0/1
1200   voip  up   up             91..........      0   syst ipv4:67.215.241.250
1201   voip  up   up             9[^1]..[2-9].... 0   syst ipv4:67.215.241.250
```

In this example, if you tried to call the destination 916025551212, you could identify a match on dial peer 1200.

Now, verifying that a dial peer matches a dialed string and actually completing a call can be two different things. If you verify the dial peer and still receive reorder tones, you can use the **debug voip dialpeer** command. This command shows digits as they are dialed, as shown in Example 15-2.

Example 15-2 *debug voip dialpeer Command Output*

```
CME_A# debug voip dialpeer
Mar 31 16:07:13.195: //-1/xxxxxxxxxxxx/DPM/dpAssociateIncomingPeerCore:
   Calling Number=1500, Called Number=, Voice-Interface=0x8905DD70,
   Timeout=TRUE, Peer Encap Type=ENCAP_VOICE, Peer Search Type=PEER_TYPE_VOICE,
   Peer Info Type=DIALPEER_INFO_SPEECH
Mar 31 16:07:13.195: //-1/xxxxxxxxxxxx/DPM/dpAssociateIncomingPeerCore:
   Result=Success(0) after DP_MATCH_ORIGINATE; Incoming Dial-peer=20005
Mar 31 16:07:13.195: //-1/xxxxxxxxxxxx/DPM/dpMatchSafModulePlugin:
   dialstring=NULL, saf_enabled=0, saf_dndb_lookup=0, dp_result=0
Mar 31 16:07:16.047: //-1/C64C50C58929/DPM/dpMatchPeersCore:
   Calling Number=, Called Number=9, Peer Info Type=DIALPEER_INFO_SPEECH
Mar 31 16:07:16.047: //-1/C64C50C58929/DPM/dpMatchPeersCore:
   Match Rule=DP_MATCH_DEST; Called Number=9
Mar 31 16:07:16.047: //-1/C64C50C58929/DPM/dpMatchPeersCore:
   Result=Partial Matches(1) after DP_MATCH_DEST
Mar 31 16:07:16.051: //-1/C64C50C58929/DPM/dpMatchSafModulePlugin:
   dialstring=9, saf_enabled=1, saf_dndb_lookup=0, dp_result=1
Mar 31 16:07:16.051: //-1/C64C50C58929/DPM/dpMatchPeersMoreArg:
   Result=MORE_DIGITS_NEEDED(1)
Mar 31 16:07:19.551: //-1/C64C50C58929/DPM/dpMatchPeersCore:
   Calling Number=, Called Number=91, Peer Info Type=DIALPEER_INFO_SPEECH
Mar 31 16:07:19.551: //-1/C64C50C58929/DPM/dpMatchPeersCore:
   Match Rule=DP_MATCH_DEST; Called Number=91
Mar 31 16:07:19.551: //-1/C64C50C58929/DPM/dpMatchPeersCore:
   Result=Partial Matches(1) after DP_MATCH_DEST
Mar 31 16:07:19.551: //-1/C64C50C58929/DPM/dpMatchSafModulePlugin:
   dialstring=91, saf_enabled=1, saf_dndb_lookup=0, dp_result=1
Mar 31 16:07:19.551: //-1/C64C50C58929/DPM/dpMatchPeersMoreArg:
   Result=MORE_DIGITS_NEEDED(1)
Mar 31 16:07:21.159: //-1/C64C50C58929/DPM/dpMatchPeersCore:
   Calling Number=, Called Number=916, Peer Info Type=DIALPEER_INFO_SPEECH
Mar 31 16:07:21.159: //-1/C64C50C58929/DPM/dpMatchPeersCore:
   Match Rule=DP_MATCH_DEST; Called Number=916
Mar 31 16:07:21.159: //-1/C64C50C58929/DPM/dpMatchPeersCore:
   Result=Partial Matches(1) after DP_MATCH_DEST
Mar 31 16:07:21.159: //-1/C64C50C58929/DPM/dpMatchSafModulePlugin:
   dialstring=916, saf_enabled=1, saf_dndb_lookup=0, dp_result=1
Mar 31 16:07:21.159: //-1/C64C50C58929/DPM/dpMatchPeersMoreArg:
   Result=MORE_DIGITS_NEEDED(1) DPM/dpMatchPeersCore:
Mar 31 16:07:21.359: //-1/C64C50C58929/DPM/dpMatchPeersCore:
   Calling Number=, Called Number=9160, Peer Info Type=DIALPEER_INFO_SPEECH
Mar 31 16:07:21.359: //-1/C64C50C58929/DPM/dpMatchPeersCore:
   Match Rule=DP_MATCH_DEST; Called Number=9160
```

```
Mar 31 16:07:21.359: //-1/C64C50C58929/DPM/dpMatchPeersCore:
   Result=Partial Matches(1) after DP_MATCH_DEST
Mar 31 16:07:21.359: //-1/C64C50C58929/DPM/dpMatchSafModulePlugin:
   dialstring=9160, saf_enabled=1, saf_dndb_lookup=0, dp_result=1
Mar 31 16:07:21.359: //-1/C64C50C58929/DPM/dpMatchPeersMoreArg:
   Result=MORE_DIGITS_NEEDED(1)
<...output omitted...>
Mar 31 16:07:23.843: //-1/C64C50C58929/DPM/dpMatchPeersCore:
   Calling Number=, Called Number=916025551212, Peer Info
Type=DIALPEER_INFO_SPEECH
Mar 31 16:07:23.843: //-1/C64C50C58929/DPM/dpMatchPeersCore:
   Match Rule=DP_MATCH_DEST; Called Number=916025551212
Mar 31 16:07:23.843: //-1/C64C50C58929/DPM/dpMatchPeersCore:
   Result=Success(0) after DP_MATCH_DEST
Mar 31 16:07:23.843: //-1/C64C50C58929/DPM/dpMatchSafModulePlugin:
   dialstring=916025551212, saf_enabled=1, saf_dndb_lookup=0, dp_result=0
Mar 31 16:07:23.843: //-1/C64C50C58929/DPM/dpMatchPeersMoreArg:
   Result=SUCCESS(0)
   List of Matched Outgoing Dial-peer(s):
     1: Dial-peer Tag=1200
```

The key areas in Example 15-2 are highlighted. Notice at the beginning of the debug output that the CME router matches an incoming dial-peer when the IP phone with extension 1500 goes off-hook (incoming dial peer 20005 matched). Then, CME analyzes each digit as it is dialed from the IP phone. The first digit dialed is a 9. The debug output shows Partial Matches(1), indicating this dialed string partially matches one or more dial peers. The output then continues along this path until the IP phone has dialed a complete string (916025551212). At this point, CME realizes it has a full match on outgoing dial peer 1200.

This **debug** command can be useful to watch the CME router go through the dial peer matching process in real time. This is where you might talk a user through dialing a number that is failing and watch how CME handles the digits as the user dials them.

QoS Issues

Troubleshooting QoS is a different skill set and strategy than troubleshooting dial-plan issues. Instead of working with the routing table for voice, you're working with the voice traffic itself.

You might have heard the saying before, "If you have no goal, you'll hit it every time." Before we can troubleshoot voice quality issues, we need to have a goal we're shooting for. Table 15-2 shows the Cisco recommended parameters for high-quality voice calls.

Table 15-2 *Requirements for High-Quality VoIP Calls*

Parameter	Requirement
End-to-end (one-way) delay	150 ms or less
Jitter	30 ms or less
Packet loss	1% or less

The full definition of each of these parameters was discussed in Chapter 6, "Understanding the CME Dial-Plan." This is half of the puzzle: We know how fast and consistent our voice traffic must travel across the network. The other half of the puzzle is how *much* voice traffic must travel across the network. You can find this out based on two factors: the voice CODEC you are using for the calls and how many concurrent calls you plan to support. Table 15-3 shows the average bandwidth usage for the two most popular codecs used in a Cisco VoIP network.

Table 15-3 *Average Bandwidth Utilization for G.711 and G.729A*

Codec	Packetization Interval	Bandwidth Per Call
G.711	20 ms	80 kbps
G.711	30 ms	74 kbps
G.729A	20 ms	24 kbps
G.729A	30 ms	19 kbps

Keep in mind that these are simply average values. After you develop the ninja skills needed for CCNP Voice, you can calculate these values down to the bit-level for your specific environment. The methods you need to do this are covered in the *CVOICE Authorized Self-Study Guide*, which you can find at www.ciscopress.com/bookstore/product.asp?isbn=1587055546.

Tip: The packetization interval represents how much audio is included per packet. The larger your packetization interval, the more audio data you put into each packet. The more audio data you put in each packet, the less packets you send (thus, the slight bandwidth savings by choosing 30 ms packetization intervals). The default packetization interval on Cisco routers is 20 ms.

Now, we can put the two puzzle pieces together. We know the requirements for high-quality audio and we know how much bandwidth each call consumes. Now, we need to provision QoS to ensure our switches and routers are able to guarantee priority queuing for our voice call bandwidth.

After you configure QoS, keep a proactive eye on the network to ensure that it meets quality standards. Although there are many sophisticated (and expensive!) tools available to help measure and monitor the voice traffic crossing the network, you actually get a basic monitoring utility each time you purchase a Cisco IP Phone. Whenever a phone is on an active call, you can press the question-mark button (help menu) twice to retrieve call statistics, as shown in Figure 15-3.

The call statistics include:

- Codec
- Packet size
- Received and transmitted packets
- Average and maximum jitter
- Lost packets

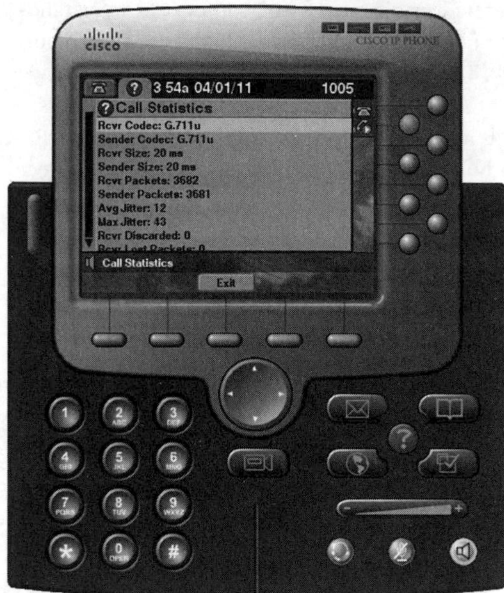

Figure 15-3 *Gathering Call Statistics*

You can also retrieve these statistics by accessing the built-in web server of the IP phone (by entering the phone's IP address into a web browser while it is on an active call).

Although a plethora of QoS troubleshooting techniques are available (and discussed in the CVOICE book previously mentioned), we highlight only one of them here. When you deploy QoS on a router, there are a variety of methods and configurations you can use. However, when it comes time to apply these methods to your router, you use an interface command known as **service-policy**. This command applies a policy-map to your interface and engages QoS similar to the **ip access-group** command applying an access-list to an interface and engaging security. You can verify your QoS statistics by using the **show policy-map interface** command. Example 15-3 shows the output of this command.

Example 15-3 *show policy-map interface Command Output*

```
Router# show policy-map interface serial3/1 output
 Serial3/1
  Service-policy output: VoicePriority
    Class-map: voice (match-all)
      0 packets, 0 bytes
      5 minute offered rate 0 bps, drop rate 0 bps
      Match: ip precedence 5
      Weighted Fair Queueing
        Strict Priority
        Output Queue: Conversation 264
```

```
      Bandwidth 128 (kbps) Burst 3200 (Bytes)
      (pkts matched/bytes matched) 14231/3903582
      (total drops/bytes drops) 1253/391292
```

This output allows you to see the success or failure of the QoS policy applied to the outbound direction of Serial 3/1. Notice the QoS policy is named VoicePriority. It grants strict priority to the first 128 kbps of voice traffic. Notice the number of packets dropped (1253). This indicates a problem; this is roughly 8 percent of your voice packets. On an actual network, this could cause degradation issues of active calls.

The drops are most likely caused by one of two things: a congested WAN link or voice traffic exceeding the provisioned amount. A congested WAN link simply means that there is not enough bandwidth to send the calls in the time required, so the packets are dropped. The latter issue seems almost identical to the first. That is, if you have too many voice calls, won't you congest the WAN and not be able to get all the packets through? Although this may be true, the more likely cause is the Strict Priority queuing configuration. Strict Priority ensures that the VoIP traffic gets the first and fastest 128 kbps available. However, once the voice traffic tries to exceed this amount, the router begins dropping the traffic (even if there is bandwidth available on the WAN link). This ensures that you know exactly how much voice traffic is going over your WAN connections. The data traffic won't have to fight for the scraps of bandwidth left over.

Note: All these troubleshooting guidelines also apply to phones registering with CUCM.

Exam Preparation Tasks

Review All the Key Topics

Review the most important topics in the chapter, noted with the key topics icon in the outer margin of the page. Table 15-4 describes the key topics and the page numbers on which each is found.

Table 15-4 *Key Topics for Chapter 15*

Key Topic Element	Description	Page Number
List	Cisco best practice troubleshooting guidelines	400
List	Boot process for Cisco IP Phones registering with CME	401
Table 15-2	QoS requirements for high-quality voice calls	408
Table 15-3	Average bandwidth utilization for popular Cisco codecs	409

Definitions of Key Terms

Define the following key term from this chapter, and check your answer in the Glossary:

packetization interval

This chapter includes the following topics:

- **Describe How to Provide End-User Support for Connectivity and Voice Quality Issues:** This section reviews the basic troubleshooting method as it applies to a Unified Communications environment.
- **Describe CUCM Reports and How They Are Generated:** This section reviews the content of the built-in reports for CUCM and how to create them.
- **Describe CUCM CAR Reports and How They Are Generated:** This section reviews CUCM Call Detail Record reports and how they are created.
- **Describe Cisco Unified RTMT:** This section reviews the RTMT and how to use it for system monitoring.
- **Describe the Disaster Recovery System:** This section summarizes the features and uses of the native backup and restore service in CUCM.

CHAPTER 16

Management and Troubleshooting of Cisco Unified Communications Manager

Cisco Unified Communications Manager (CUCM) is a large and complex application and, in most deployments, it is considered "mission critical," which means that when trouble happens, it is important to be able to find out what is wrong and fix it quickly. A solid troubleshooting methodology will help you keep your head when everyone around you is losing theirs.

"Do I Know This Already?" Quiz

The "Do I Know This Already?" quiz allows you to assess whether you should read this entire chapter or simply jump to the "Exam Preparation Tasks" section for review. If you are in doubt, read the entire chapter. Table 16-1 outlines the major headings in this chapter and the corresponding "Do I Know This Already?" quiz questions. You can find the answers in Appendix A, "Answers Appendix."

Table 16-1 *"Do I Know This Already?" Foundation Topics Section-to-Question Mapping*

Foundation Topics Section	Questions
Describe How to Provide End-User Support for Connectivity and Voice Quality Issues	1–3
Describe CUCM Reports and How They Are Generated	4
Describe CUCM CAR Reports and How They Are Generated	5–8
Describe Cisco Unified RTMT	9
Describe the Disaster Recovery System	10

1. Cisco defines a series of steps in the process of troubleshooting. Which of the following lists those steps in the correct order?

   a. Gather Facts.
      Define the Problem.
      Consider Possibilities.
      Create Action Plan.
      Implement Action Plan.
      Observe Results.

b. Create Action Plan.
   Gather Facts.
   Define the Problem.
   Observe Results.
   Consider Possibilities.
   Implement Action Plan.

c. Consider Possibilities.
   Gather Facts.
   Define the Problem.
   Create Action Plan.
   Implement Action Plan.
   Observe Results.

d. Define the Problem.
   Gather Facts.
   Consider Possibilities.
   Create Action Plan.
   Implement Action Plan.
   Observe Results.

2. Rob is having problems with his IP Phone. It isn't working. Rob says he has tried "fiddling with a few things on the phone." Where should you begin troubleshooting?
   a. Examine CUCM to see if the phone is registered.
   b. Verify that the DHCP server is active, reachable from Rob's phone, and has addresses available.
   c. Verify that the TFTP service is running on the CUCM.
   d. Verify that the local settings on Rob's phone are correct.
   e. Verify that the switch configuration is correct.

3. Greg is an end user in the Engineering department. He has done some reading on the Internet, and has learned that there is a Unified Reporting tool he can use to run reports. However, he phones you to tell you that he can't run any reports because he is denied access to the web page. What action will allow Greg to run reports?
   a. Modify the permissions on the Unified Reporting web pages to allow Greg access.
   b. Give Greg a copy of Crystal Reports and the Platform Administration account.
   c. Install the RTMT on Greg's PC.
   d. Make Greg a member of the Standard CCM Super Users Group.
   e. Tell Greg he is not allowed to run reports.

4. The CAR Reporting tool allows three types of users to access reports: Administrators, Managers, and Users. What defines a report user as a Manager as opposed to just a User?
   a. Their account is referenced in the Manager User ID field of another User's account.
   b. Their account is a member of the Standard CAR Manager Users Group.
   c. The **Manager** checkbox is selected in the CAR User Configuration page.
   d. Managers log in to a different CAR Reports tool than users do.

5. Aunt Beru wants to find out who on her team makes the longest phone calls. Which CAR report should she run?

   a. Bills > Department
   b. Bills > Individual
   c. Top N > By Duration
   d. Top N > By Charge

6. Greg is a CAR Reports User. Can he run the Top N > By Charge report?

   a. No
   b. Yes

7. Janice is upset because the custom IP Phone service she commissioned is not being adopted by many users. Which CAR report will determine how many phones are subscribed to her custom service?

   a. Cisco IP Phone Services (with the custom service selected)
   b. Top N > Service Subscriptions
   c. Cisco Unified Communications Manager Assistant > Manager Call Usage
   d. Top N > By Number of Calls

8. You are the CUCM Administrator. A couple months ago, management asked you to implement Client Matter Codes to track employees' personal calls. What report can you run to provide a list of calls made with the CMC assigned to personal calls?

   a. Traffic > Summary
   b. Traffic > Summary By Extension
   c. Forced Authorization Code / Client Matter Code > Client Matter Code
   d. Forced Authorization Code / Client Matter Code > Authorization Level

9. Luke has been a CUCM administrator for two years. He is trying to use his RTMT to look at system summary information for the new CUC server. He complains that the menu is not visible. What should Luke do to make the CUC menu visible?

   a. Under Edit > Preferences, check the CUC box under System Menu.
   b. Install the Linux version of RTMT on his PC.
   c. Download and install the RTMT from the CUC server.
   d. Luke must be made a member of the Standard RTMT Administrators Group.

10. What two storage options are available to the Disaster Recovery System?

   a. Local disk file and TFTP
   b. Local disk file and FTP
   c. Local tape drive and FTP
   d. Local tape drive and local disk file
   e. Local tape drive and SFTP

Foundation Topics

Describe How to Provide End-User Support for Connectivity and Voice Quality Issues

Troubleshooting a complex application like CUCM can be challenging. The scope of CCNA Voice (ICOMM) requires us to keep it relatively simple, so the following sections should by no means be considered a comprehensive CUCM troubleshooting manual. In this chapter, you learn how to troubleshoot basic IP Phone registration issues and look at the extensive reporting capabilities of the CUCM system, which helps you monitor the health and well-being of the servers in your deployment.

Note that the following section on troubleshooting methodology is duplicated in Chapter 15, "Common CME Management and Troubleshooting Issues." It appears again here because it is important, and it makes it easier to refer to as you read this chapter.

Troubleshooting

When troubleshooting, employing a consistent and systematic methodology saves time and helps prevent errors from making the situation worse. The sequence of steps described here is supported by Cisco best practices as one model for effective troubleshooting. Figure 16-1 illustrates the troubleshooting sequence.

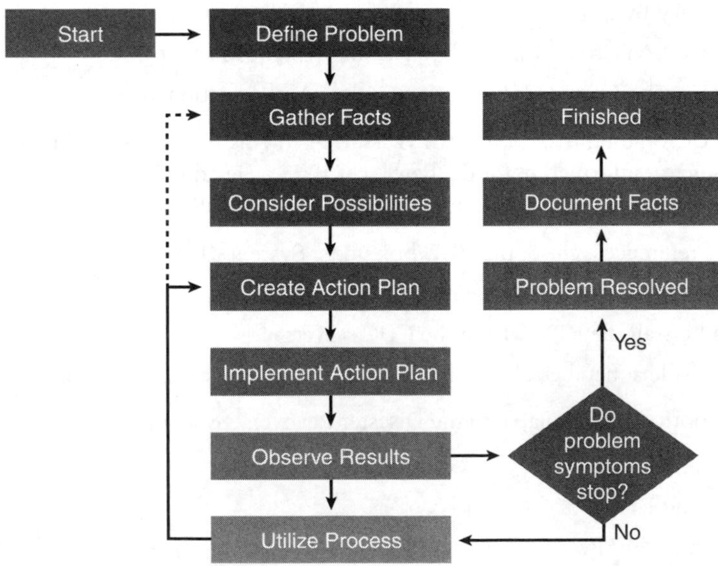

Figure 16-1 *Troubleshooting Methodology*

Here are the steps illustrated in Figure 16-1:

1. **Define the Problem:** Analyze the problem and create a clear problem statement. Define the symptoms and probable causes. Compare current conditions to a baseline "normal" condition.
2. **Gather Facts:** Collect and consider command outputs and user statements. Eliminate possible causes to reduce the number of potential issues. Ask: When did this problem occur? What changed before the problem started? It is intermittent or constant? Is there a pattern to the intermittence? Are there any error messages? Is anyone else having the problem?
3. **Consider Possibilities:** Based on the facts gathered in Step 2, identify a short list of likely causes. This may be a quick, almost intuitive process, or it may require significant research and discussion.
4. **Create Action Plan:** Beginning with the most likely cause on the list from Step 3, define a plan of action to correct the problem. The plan should modify only one variable at a time, in order that the effect of that one change can be easily evaluated.
5. **Implement Action Plan:** Execute the commands or make the changes decided in Step 4. Do not improvise: Follow the plan. If new information or ideas come up, you may want to start at Step 3 again. Make sure at each step that the action taken does not make the problem worse—if it does, undo the change. Minimize the impact of changes on production systems, especially limiting the duration of security vulnerabilities, such as temporarily removing access lists to see if they are the problem.
6. **Observe Results:** Has the problem been solved?

7a. **Unsolved:** If the problem is not solved, undo the change implemented in Step 5 and return to Step 4.

7b. **Solved:** If the problem is solved, document the solution, the root cause, and the changes made to the system.

8. **You are done.**

A list like that (which you need to memorize) needs a good mnemonic. Here's one for you:

Define, Gather, Consider, Create, Implement, Observe: *"Dogs Gobble Cookies, Cats Irritate Owners."*

Troubleshooting IP Phone Registration Problems

The IP Phone registration process is deceptively complex, and several discrete areas may cause problems—and, of course, if one step fails, subsequent steps fail, too. That being the case, you can use a Divide and Conquer methodology: Quickly check which steps have succeeded, and then start troubleshooting from the first point of failure. IP phone registration problems can be categorized according to the following points of failure:

- Local to the IP Phone
- VLAN or switch mismatches
- DHCP problems

- TFTP problems
- CUCM registration problems

Let's examine a scenario in which the phone is not registering. In this scenario, the phones are supposed to use DHCP. Each point that follows describes a possible place where problems may be happening. Best practices dictate that we follow these troubleshooting procedures in the order that follows; there will be times when experience or specific knowledge allows us to skip to a later procedure:

- **Local to the IP Phone:** The IP Phone itself can display its current configuration and settings, which can quickly indicate which part of the sequence has failed. Press the **Settings** button, and then select Network Configuration from the displayed list. Scroll down to **IP Configuration** and verify that the phone has received an IP address (in the correct subnet), subnet mask, default gateway, and the correct TFTP server address. If the entries are absent or incorrect, verify that the phone is configured to use DHCP by pressing **Settings > Network Configuration**, and then scrolling down to **DHCP Enabled** and verifying that it is set to **Yes.** If all that is correct, but the phone is still not receiving an address from DHCP, move on to the next step.
- **VLAN or switch mismatches:** Verify that the switch is correctly configured to support IP Phones. The switch should have a voice VLAN defined, and if there is a PC connected to the phone, a separate access VLAN. (See Chapter 3, "Understanding the Cisco IP Phone Concepts and Registration," for the configurations.) Verify that the VLAN numbers are correct. If the switch configuration is correct, move on.
- **DHCP problems:** Verify that the DHCP server is running and that it has not run out of IP addresses. Make sure that the DHCP scopes (subnets or pools) are correct with respect to the IP address range being assigned, the subnet mask, default gateway, and Option 150 (TFTP server IP address). On the IP Phone, navigate to **Settings > Network Configuration.** Verify that the **DHCP Server** entry lists the IP address of the correct DHCP server. Check that an IP address has been assigned, and if so, that it is in the correct range. Verify that the **TFTP Server 1** address entry is correct.

Note: If the DHCP server is on a remote subnet from the IP phones, the local router blocks the DHCP broadcasts by default. Use the **ip helper-address <ip_address>** command on the local router to allow it to forward DHCP requests to the IP of the DHCP server.

- **TFTP problems:** As it boots up, the phone queries the TFTP server (at the address learned via DHCP) for its configuration file. The filename it asks for is called SEP<mac>.cnf.xml. If the phone has been successfully added to the CUCM or Cisco Unified Communication Manager Express (CME) application, the file will exist and will be downloaded to the phone. If the phone has never been added to the application before, the file will not be there. The phone will then ask for the file called XMLDefault.cnf.xml. This default file is always available. If the phone is not getting its config file, verify that the phone has the correct TFTP address in the **Network Configuration** list. Verify that the TFTP service is running on the server at that IP. You can check on the status of the TFTP process on the phone by pressing Settings > Status Messages;

example messages include File Not Found:SEP<mac address>.cnf.xml, TFTP Timeout: SEP<mac address>.cnf.xml, and SEP<mac address>.cnf.xml.

- **CUCM registration problems:** The TFTP download file contains the IP address of the CUCM server with which it is supposed to register. Check **Settings > Device Configuration > Unified CM Configuration** to verify that the Unified CM IP address listed is correct. There may be a backup and tertiary server IP listed, depending on how the cluster is configured. Verify that the Cisco CallManager Service is running on the server(s) listed; you may also need to verify that autoregistration is correctly set up (only if you are actually using Autoregistration, of course).

If all the previous are verified as correct, there may be a problem with the phone itself; that type of troubleshooting is beyond the scope of CCNA Voice.

Deleting Unassigned Directory Numbers Using the Route Plan Report

When using Autoregistration, one of the more common issues encountered is that the phones fail to register, displaying "Error DB Config" or a similar message on the IP Phone screen. The source of the problem is that the range of Directory Numbers (DNs) allocated for Autoregistration has been used up; Autoregistration is working, but there are no more DNs available to assign to the phones. This situation arises because it is normal practice to change the DN of an autoregistered phone (which is assigned sequentially from the range defined for Autoregistration) to its "real" production DN. An odd thing happens to the Autoregistration-assigned DN: It is not released back to the available range, but instead is marked as "Unassigned" and held in "database limbo." Unassigned Autoregistration DNs are not visible unless you go looking for them, so it is not obvious that they are the source of your problem.

The fix is simple: Either add to the range of Autoregistration DNs, or use the following steps to "reclaim" them so they can be re-used on newly-registering phones:

1. Navigate to **System > Route Plan Report.**
2. Set the first filter (the left-most field) to **"Unassigned DNs."**
3. Click **Find.**
4. Delete all of the listed unassigned DNs. This can be done easily by selecting all the listed DNs using the checkboxes to their left and then clicking **Delete Selected.**

The DNS are now released back to the Autoregistration range as available for assignment.

These steps are also used to "clean up" the database after modifications to your Partitions design. Another interesting behavior of the CUCM database is that when a DN is assigned to a different Partition, it still exists in the previous one but is flagged as unassigned. These unassigned DNs can create confusion, because they *do* appear in lists as selectable (for example, when building a Line Group)—but they do not function because they are not assigned to any device.

You should be familiar with using the Route Plan Report to delete unassigned DNs as a routine maintenance task.

Describe CUCM Reports and How They Are Generated

CUCM includes numerous reporting tools. This section reviews how to generate and access those reports, and how to use them for troubleshooting, maintenance, and system analysis.

The Cisco Unified reporting tool pulls information from a range of sources, and formats the data into a single simplified output. The report tool alerts the user if the report job will cause performance issues for the server or take an excessive amount of time.

The reports pull data from the Publisher and the subscribers, including the following sources:

- Real-Time Monitoring Tool (RTMT) counters
- Call Detail Records (CDR) and the CDR Administration and Reporting database
- CUCM database
- Disk files (traces and logs)
- Prefs settings
- CLI
- Real-Time Information Server (RIS)

The reporting system uses the Cisco Tomcat service and Remote Procedure Calls (RPC) to the other servers via HTTPS. Make sure the Tomcat service is running and that HTTPS traffic can reach and return from the servers.

Generating Reports

Access the Cisco Unified Reporting tool from the navigation drop-down at the top right of the CUCM Administration interface or directly via the URL https://<ip_address>/cucreports. By default, the only users with the necessary privilege to view the reports are member of the CCM Super Users Group, the only member of which by default is the CUCM application administration account defined at install.

The reports are accessed under the System Reports menu, as shown in Figure 16-2.

Select the desired report from the menu list. The Reporting tool stores one of each previously created report for later access; if the report selected has not been run before, a message is displayed, indicating that the report does not exist and the user should generate a new report. The Generate a New Report link is directly below this message.

If an old copy of the report exists, a message is displayed to that effect, and a link to the report is listed. Bear in mind that the old report contains old information; check the time/date stamp on the report to be sure that the report is recent enough to be valid. To recreate the report, click the icon at the top right to run it again.

Figure 16-3 shows a sample report (in this case, a Device Counts Summary). The green checkmark at the top indicates that the report ran successfully. The other buttons to the right side of the page allow you to upload an XML report that is stored on your local workstation to keep it on the CUCM server, download the report to your local workstation, or run it again.

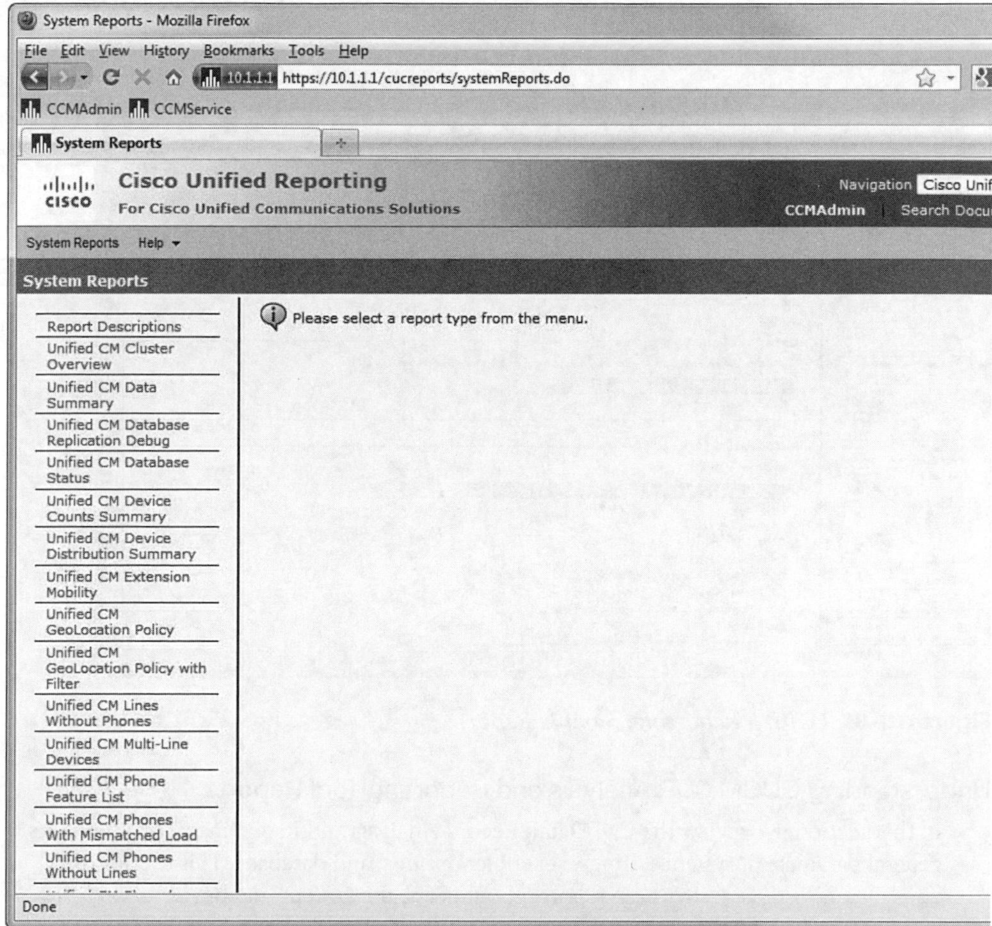

Figure 16-2 *Cisco Unified Reporting: System Reports Menu*

Analyzing Reports

The reports you generate can be used as part of the following tasks:

- **Troubleshooting:** For gathering facts, considering possibilities, and observing results.
- **Maintenance:** Find configuration or load mismatches, or summarize system information.
- **System Analysis:** List phones by type, feature, or several other filters.

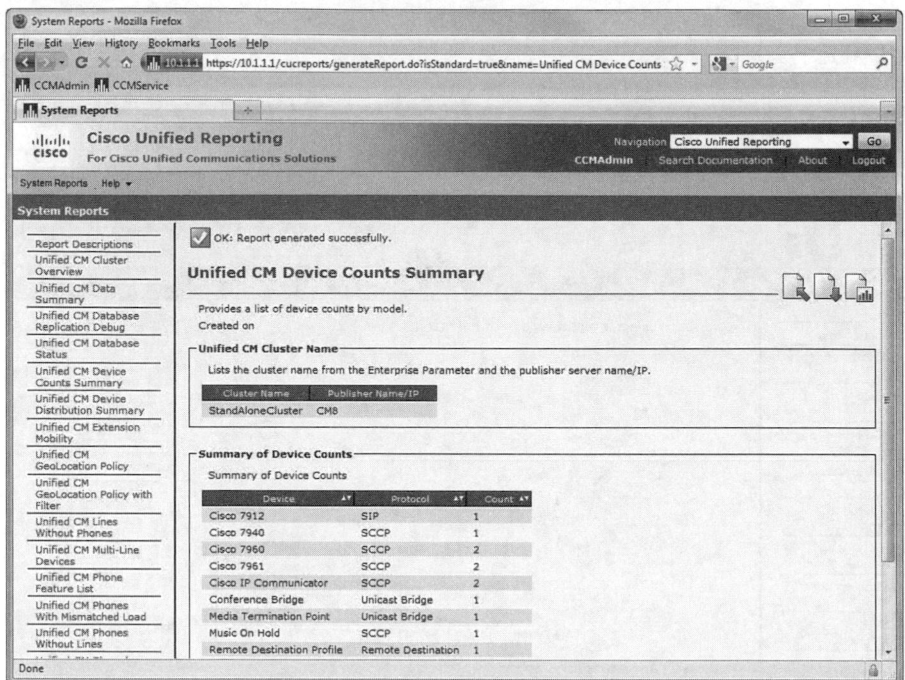

Figure 16-3 *Unified Reporting: Sample Report*

Understanding CUCM CDR Analysis and Reporting Tool Reports

In this section, we review the Call Detail Record Analysis and Reporting (CAR) tool—its general parameters, system settings, scheduler options, and database. This section also goes over how to generate these reports about users, the system, and devices.

Every call that CUCM processes can be logged. These logs, called Call Detail Records (CDR) and Call Management Records (CMR), contain information about the call and the voice quality metrics for the call. These CDRs are stored as flat files on the subscriber servers and uploaded to the CDR/CAR database on the Publisher at regular intervals (the interval can be administratively set). In addition to providing useful information for internal administrative purposes, the CDR database can be used by third-party billing applications to prepare internal or external billing reports.

Administrators can pull reports manually using the web interface at https://<ip_address>/car or set up reporting jobs to occur automatically. The option to load CMRs in addition to CDRs is determined administratively; CMRs are not loaded by default. Figure 16-4 shows the CAR Reports tool.

Activate CAR-Related Services

To use CAR, the **Cisco CAR Web Service** must be activated:

1. From the Unified Serviceability page, navigate to **Tools > Service Activation.** Select the **Cisco CAR Web Service** and click **Save.**

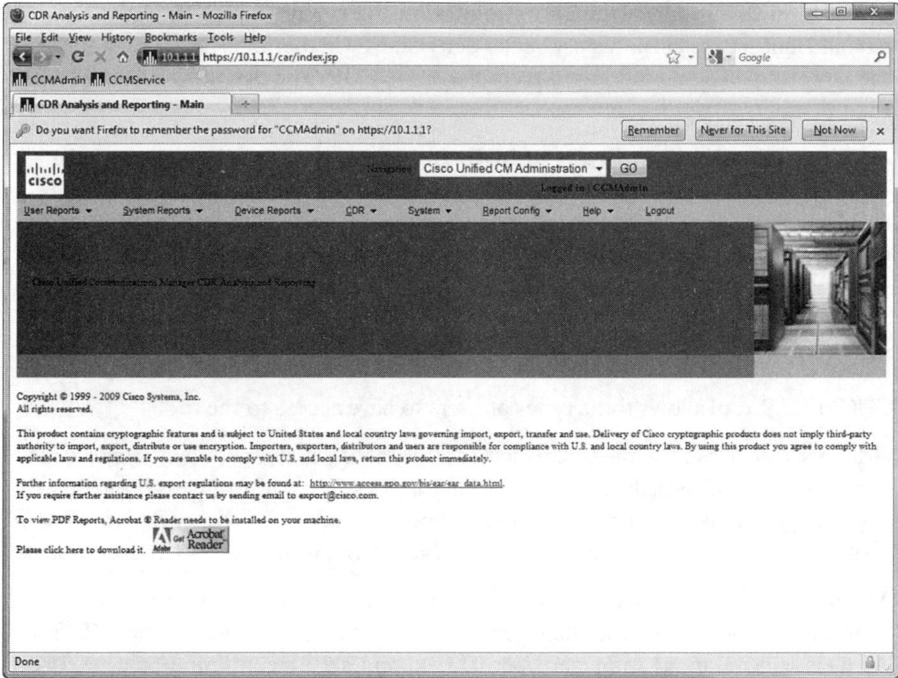

Figure 16-4 *CAR Report Tool*

2. If an external billing server is to be used, you must also activate the **Cisco SOAP-CDROnDemand Service.** (SOAP stands for Simple Object Access Protocol.)

Configure CDR Service Parameters

In the CM Administration interface, navigate to **System > Service Parameters.** Select a server from the first drop-down, and then select the **Cisco CallManager** service. Click the **Advanced** button at the top of the page to display all parameters. Adjust the following as required:

- **CDR Enabled Flag:** This setting determines whether CDRs will be generated. This must be set on all servers. The default is **False** (CDRs not collected).
- **CDR Log Calls with Zero Duration Flag:** Setting this parameter to **True** causes CUCM to generate CDRs for calls that never connect or that last less than 1 second. Calls that are unsuccessful (including calls that result in reorder tone or that fail because of a busy trunk) are always logged regardless of this setting. The default setting is **False.**
- **Call Diagnostics Enabled:** This parameter enables the logging of CMRs. You have the choice to never generate CMRs, to generate them only if CDRs are also being generated, or to generate CMRs whether CDRs are being collected or not. The default value is **Disabled.**
- **Display FAC in CDR:** This parameter controls whether the Forced Authorization Code used to make a call will be included in the CDR. The default value is **False.**

- **Show Line Group Member DN in finalCalledPartyNumber CDR Field:** For calls to Call Hunting systems, this parameter determines whether the Hunt Pilot number or the DN that picked up the calls is recorded in the CDR. The default is False (the Hunt Pilot number is recorded, not the DN).
- **Add Incoming Number Prefix to CDR:** This parameter controls whether the incoming prefix (several are defined in the service parameters) is added to the calling party number in CDRs. Prefixes added to an inbound call are always recorded in CDRs; this setting controls whether prefixes are added to CDRs if they are added to outbound calls. The default value is **False.**

CAR Tool Users

The CUCM CAR tool allows three types of users to have access to the tool:

- **Administrators** can use the tool to access all reporting features to build reports for system performance analysis, load balancing verification, and troubleshooting purposes. Any end user or application user can be given administrator access to the CAR tool by making them a member of the **Standard CAR Admin Users** Group.
- **Managers** can generate reports for users, departments, and Quality of Service (QoS). Managers are defined by the **Manager User ID** field in CM Administration: If User A's ID is selected in the **Manager User ID** field on User B's configuration page, User A can run Manager reports on User B. Managers can set up automatic reports to be delivered to their configured mail ID.
- **Users** can generate a billing report for their calls. Users can generate reports for a specific date range and can have the system e-mail the report. Users must be members of the **standard CCM End Users Group** to access these reports, and the devices they want to generate reports for must be associated with their account.

CDR and CMR Architecture

CDRs are generated by CUCM servers that are processing calls (those that are running the CallManager Service). The CDRs contain information about the called and calling numbers, the time/date stamp for connect and disconnect, and why the call was disconnected. CMRs contain information about latency, jitter, packet loss, and the amount of data sent during the call. Each call may generate several CDRs and CMRs. The call processing nodes collect the CAR data (the collective term for CDRs and CMRs) in a local log and periodically upload them to the CDR repository node using SFTP. The repository node runs the CDR Repository Manager Service, which is responsible for maintaining the CDR and CMR files, managing the disk space used by the CDR data, and sending the files to (up to) three configured locations (typically, these are third-party billing servers).

CAR System Parameters

The CAR system requires some setup to function most effectively. The following tasks need to be completed:

- Configure **Mail Parameters** so the CAR system can send reports and alerts by e-mail.

- Configure the dial-plan to match local calling pattern so that CDRs are interpreted correctly (for example, 4-digit calls are classified as On Net and 10-digit calls are local). The default settings are based on the NANP.
- Configure the **gateways** in the CAR tool. Gateways configured in CUCM are automatically added and deleted, but the area codes local to the gateway must be added so that the reports can determine which calls are long distance.
- Set the **COMPANY_NAME** value (maximum of 64 characters) as desired; this name appears in the header of CAR reports.
- Set up the **CAR Scheduler** to control which types of records are loaded, at what intervals, and for how long. The aim is to allow sufficient loading time while not impacting the server performance. Loading the CDR and CMR file should not be confused with the actual generation of the CDRs and CMRs; loading refers to the Repository node pulling the raw CDR/CMR data from the call processing nodes into the CDR/CAR database.
- Configure CDR/CAR database purging. **Automatic Purge** is on by default; you can adjust the **High** and **Low Water Mark** values to control when the purging begins and stops, based on the percentage utilization of the available database disk space. Which records to purge is based on the age of the records. **Manual Purging** is also available.
- Configure Automatic Report Generation. Choose which reports should be automatically generated and optionally select whether you would like the reports emailed. The interval for report generation is fixed (daily, weekly, or monthly), but the start time can be customized using the Scheduler interface.

Exporting CDR and CMR Records

CDR and CMR files can be downloaded from the server as a CSV file, typically to be imported into a billing application. The steps are as follows:

1. From the CAR tool, navigate to **CDR > Export CDR/CMR.**
2. Set the **From Date** and **To Date** values.
3. In **Select Records**, check **CDR Records**, **CMR Records**, or both.
4. Click **Export File.**
5. In the new window, choose either **CDR Dump** or **CMR Dump**, and then click **Save As** to select a file location on the workstation.
6. Selecting the **Delete File** checkbox causes the CAR tool to delete the downloaded records from the CAR database. This is recommended to prevent the database size from ballooning unnecessarily.

Generating CDR Reports

As mentioned previously, users, Managers, and Administrators can use the CAR report tool to generate different reports. This section discusses the reports in more detail.

The following reports are available to users, Managers, and Administrators, as indicated in each description:

- **Bills**
  - **Individual:** Available to users, Managers, and Administrators. Provides call information for a specified date range in summary or detail format. Report can be viewed or e-mailed.
  - **Department:** Available to Managers and Administrators. Provides call and QoS information in summary or detail format, for all users who report to the Manager or to selected users.
- **Top N**
  - **By Charge:** Available to Managers and Administrators. The report lists the top number of users in order of call charges for the specified time period (N is the number of users in the list). Reports can also be configured to report by charges to the destination of the calls, or by all calls, which lists the top N calls that incurred the most charges.
  - **By Duration:** Available to Managers and Administrators. Reports list users sorted by duration of call during the specified time period. The report can also list top calls in order of duration by destination or all calls by duration.
  - **By Number of Calls:** Available to Managers and Administrators. Reports list users by number of calls or extensions by number of calls.
- **Cisco Unified Communications Manager Assistant**
  - **Manager Call Usage:** Available to Administrators. Reports list summary or detail information for call completion by Managers using Cisco Unified Communications Manager Assistant. Reports can list the calls Managers made for themselves, calls that assistants handled for Managers, or calls handled by both for Managers.
  - **Assistant Call Usage:** Available to Administrators. Reports can list the calls assistants made for themselves, calls that assistants handled for Managers, or calls handled by assistants for both assistants and Managers.
- **Cisco IP Phone Services:** Available to Administrators, this report shows selected IP Phone services, the number of users subscribed to the services, and the utilization percentage for each.

Example Report Generation

The following steps illustrate how to generate a report using the CAR tool. As an example, we run a report that tells us what users have made the most calls:

1. Navigate to the CAR Reports tool at https://<ip_address>/car, as shown in Figure 16-4.
2. Select **User Reports > Top N > By Number of Calls**, as shown in Figure 16-5.
3. The next screen allows you to define the parameters for the report, including call types, user types, and date range, as shown in Figure 16-6.

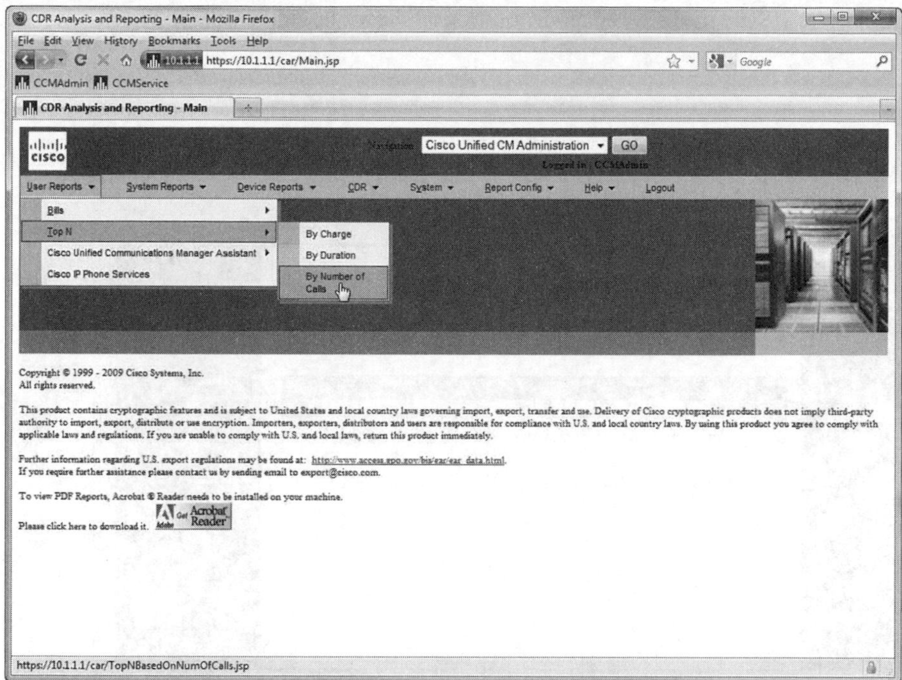

Figure 16-5 *Selecting a User Report: Top N by Number of Calls*

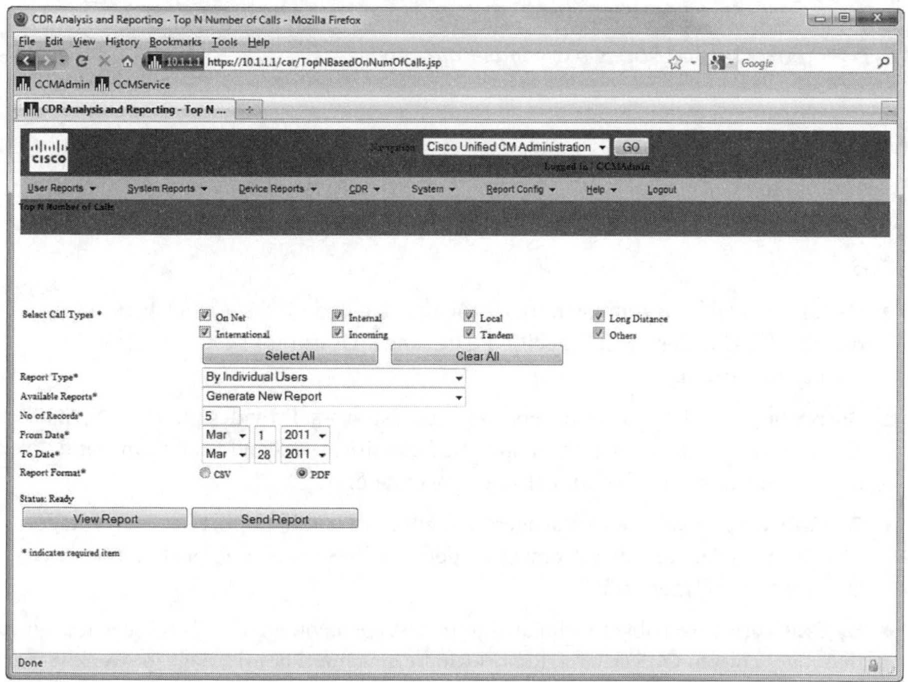

Figure 16-6 *Top N By Number of Calls Parameters Selection*

4. Click View Report to see the report output, as shown in Figure 16-7.

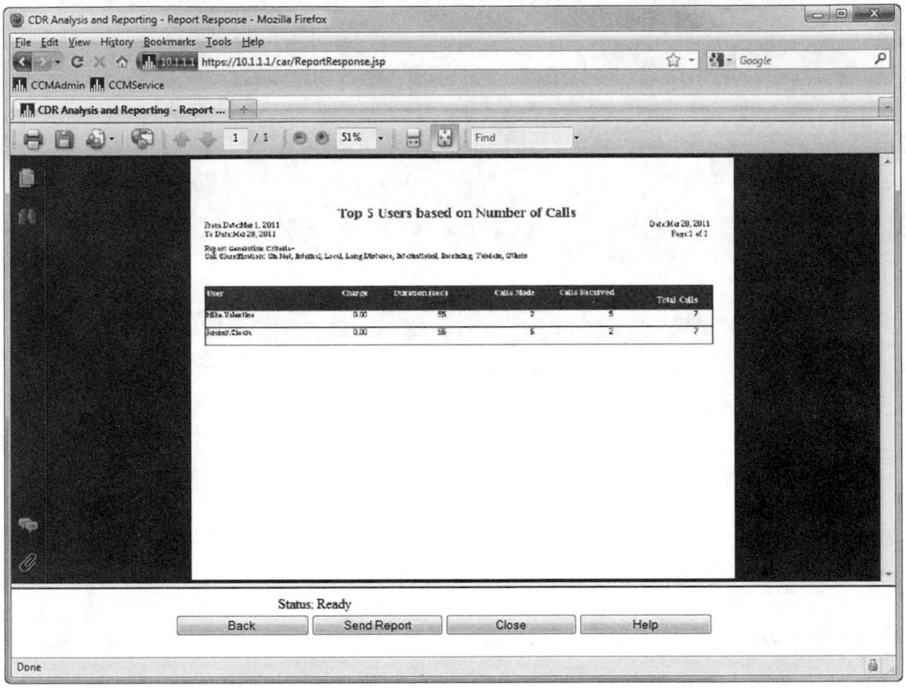

Figure 16-7 *Report Output: Top N by Number of Calls*

Generating System Reports

The CAR tool provides several system reports in addition to the user reports just listed. The following sections summarize the available system reports:

- **QoS**
    - **Detail:** Available to administrators. Provides detailed QoS statistics for calls handled by CUCM during the specified date range. Useful for system-wide voice-quality monitoring.
    - **Summary:** Available to managers and administrators. Provides pie-chart format showing QoS ratings for calls of specified classifications and time frame, and includes a summary table for calls per QoS grade.
    - **By Gateway:** Available to managers and administrators. Report lists percentage of calls per selected gateways meeting defined QoS criteria. Report can be generated hourly, daily, or weekly.
    - **By Call Type:** Available to administrators. Lists percentage of calls by selected type that meet chosen QoS criteria. Report can be generated hourly, daily, or weekly.

- **Traffic**
  - **Summary:** Available to administrators. Displays call volume for selected call types and QoS categories for a specified time frame. Useful for displaying the number of calls made hourly/daily/weekly.
  - **Summary by Extension:** Available to administrators. Displays call volume per specified extension(s) and call types during the selected time frame.
- **Forced Authorization Code/Client Matter Code (FAC/CMC)**
  - **Client Matter Code:** Available to administrators. Lists called and calling numbers, call duration, and call classification for specified time period.
  - **Authorization Code Name:** Available to administrators. Lists called and calling numbers, call time stamps, duration, and call classification for specified time period by FAC name (includes authorization level).
  - **Authorization Level:** Available to administrators. Lists called and calling numbers, call time stamps, duration, and call classification for a specified time period by FAC authorization level (includes FAC name).
- **Malicious Call Details:** Available to administrators. Displays details for calls tracked by the Malicious Caller Identification (MCID) service for the specified time period.
- **Precedence Call Summary:** Available to administrators. The report lists (in bar graph format) summary information for calls that were preempted by the selected MLPP levels for the specified time period.
- **System Overview:** Available to administrators. Provides high-level information about the CUCM network.
- **CDR Error:** Available to administrators. Lists statistics for errors encountered during CDR data transfer.

Generating Device Reports

The CAR tool provides several reports to monitor loading and performance of CUCM-related devices, such as gateways and conference bridges. These device reports include the following:

- **Gateway:** **Detail**, **Summary**, and **Utilization** reports display gateway utilization according to various call and gateway criteria.
- **Route Patterns and Hunt Groups:** Includes the following reports:
  - **Route/Line Group Utilization**
  - **Route Pattern/Hunt Pilot Utilization**
  - **Hunt Pilot Summary**
  - **Hunt Pilot Detail**

- **Conference Bridge:** **Conference Call Detail** and **Conference Bridge Utilization** reports monitor conference resources.
- **Voice Messaging:** The **Voice Messaging Utilization** report estimates the percent utilization of voice-messaging devices.

Describe Cisco Unified RTMT

The Cisco Unified Real-Time Monitoring Tool (RTMT) allows administrators to collect, view, interpret, and monitor the various counters, trace files, and logs generated by CUCM, Cisco Unity Connection (CUC), and Cisco Unified Presence (CUP).

The RTMT is a client application installed on an administrative workstation. The software can be downloaded from the CUCM, CUC, CUP, and Cisco Unified Contact Center Express (CUCCX) servers. The RTMT for each server product is specific to that server product, with the exception that the RTMT version for CUCM and for CUC are interchangeable. Only one instance of RTMT can be installed on one workstation. RTMT uses HTTPS to connect to Unified Communications servers and monitor system performance, device status, device discovery, CTI applications, and voice-messaging ports.

End users (or application users) must be added to the standard CCM Admin Users and Standard RealtimeAndTraceCollection groups to use RTMT. They can log into RTMT using their User ID and password.

The administrative capabilities of RTMT include the following:

- Monitor predefined system health objects
- Generate e-mail alerts for objects that fall below or exceed defined threshold values
- Collect and view trace files from different services
- View syslog messages
- Configure and monitor performance counters

RTMT Interface

The RTMT GUI includes the following menus and options (plus several others not listed):

- **File:** Save, restore, and delete RTMT profiles, monitor Java Virtual Machine (JVM) information, access the report archive, access the Unified reporting tool, log off, or exit.
- **Edit:** Set up categories for table format views, set polling rates for performance counters and devices, show/hide Quick Launch Channel, and edit trace settings for RTMT.
- **Window:** Close current (or all) RTMT windows.
- **Application:** Provides links to administration, serviceability, and application-specific interfaces, depending on which RTMT is in use.

When RTMT is in use, the RTMT menu is divided into three submenus:

- **System:** Allows monitoring of platform health, including CPU and memory and disk utilization. Administrators can set up and monitor various performance counters, alerts, and traces, and access the Trace & Log Central tool and syslog viewer.

  - **CallManager:** If RTMT is connected to a CUCM server, administrators can view summary information about the server, search for devices, and monitor services.

  or

  - **Unity Connection:** If RTMT is connected to a CUC server, administrators can use the Port Monitor Tool and view statistics and summaries applicable to CUC.

  or

  - **CUP:** If RTMT is connected to a CUP server, administrators can view summary information applicable to the CUP application.

- **Analysis Manager:** If the RTMT is connected to a CUCM server, the administrator can display configuration and licensing summaries, and use the Call Path Analysis tool.

Monitoring CUCM with RTMT

The sections include examples that show some of the ways in which RTMT can monitor a CUCM server.

CallManager Summary

The CallManager Summary view shows graphs for registered phones, calls in progress, and active MGCP gateway ports and channels.

Gateway Activity

The Gateway Activity view displays a summary of calls in progress for a specific type of gateway (MGCP FXS/FXO/T1/PRI or H.323). Information on the number of completed calls per gateway type (per server or per cluster) can also be displayed.

Note: This information is listed per gateway *type*, not per gateway.

Device Search

Administrators can search for phones, gateway devices, H.323 devices, CTI devices, voice-messaging devices, media resources, hunt lists, and Session Initiation Protocol (SIP) trunks. For each type of device, the administrator can search by status (registered, unregistered, rejected, any status and devices that are only configured in the database. Additionally, the search can be limited to a specific model of device or (for phones) specific protocol. The search results are presented in table format, one row per device and one column for each criterion selected for display.

Database Summary

The Database Summary view shows the replication status, number of replicates created, the number of change notification requests queued in the database and in memory, and the total number of connection clients. For each server in the cluster, it can also display the current replication status.

Describe the Disaster Recovery System

The Disaster Recovery System (DRS) allows administrators to perform scheduled backups or manual backups of the CUCM and CDR/CAR databases. It also backs up and restores its own configuration settings, so in the event of a restore, the DRS does not have to be totally reconfigured.

The DRS includes GUI and CLI user interfaces, a backup scheduler, tape or SFTP backup storage, and a distributed system architecture for backup and restore functions. Backups must be restored to the same version of the application. (The DRS cannot be used as an upgrade/downgrade mechanism, although it can be used to migrate from a physical server to a VMware server.)

The DRS architecture features a Local Agent on each server in the cluster (which performs the backup and restore operations), and a Master Agent on the Publisher. The Master Agent does the following:

- Stores system-wide component registration information.
- Maintains the schedule of backup tasks and sends the tasks to the Local Agents as scheduled.
- Stores backups on a local tape drive or a remote SFTP server.
- Interfaces with the administrator via the DRS web page.

The DRS web interface is accessed at https://<ip_address>/drf or by using the drop-down navigation selection at the top-right of the CUCM administration page. By default, only the Platform Administration account has access to the DRS, but other accounts can be given the necessary privilege.

The DRS is a common feature of all Linux-based Unified Communication applications, but different components are backed up based on the application. Table 16-2 lists the components that can be backed up for CUCM, CUC, and CUP.

Table 16-2 *Components Backed Up by DRS*

CUCM	CUP	CUC
Platform	Platform	Platform
Cisco License Manager	Cisco License Manager	Cisco License Manager
Trace Collection Tool	Trace Collection Tool	Trace Collection Tool
Syslog	Syslog	Syslog
CUCM DB	CUP DB	CUC DB
TFTP/MoH Files	XCP Data	Mailbox Store
CDR/CAR Data	CUP Data	Greetings

Using the DRS

The DRS is a simple interface. The following sections outline its use.

Set Up a Backup Device

Before any backups can happen, you must create a backup device by following these steps:

1. In the DRS, navigate to **Device > Backup Device.**
2. Provide a name for the backup device being created.
3. Specify whether this device is a local tape drive or a network directory (meaning SFTP server).
4. Select the tape drive device or provide the IP address, SFTP root path, and SFTP account the DRS should use.
5. Click **Save** (up to ten backup devices can be created).

Create a Scheduled Backup

Now that we have a backup device, we can schedule a backup to use it by following these steps:

6. Navigate to **Backup > Scheduler.**
7. Click **Add New.**
8. Provide a name for the schedule.
9. Select the previously defined backup device this job should use.
10. Select the features to back up. Depending on the application, these may be:
    - CUCM: CCM, CDR_CAR
    - CUC: CONNECTION_DATABASE, CONNECTION_GREETINGS_VOICE-NAMES, CONNECTION_MESSAGES_UNITYMBXDB1, CUC
    - CUP: CUPS, CUP
11. Define the schedule for the backup.
12. Enable the scheduled job.

If desired, a manual backup can be started by navigating to **Backup > Manual Backup** and performing the same steps, except that instead of defining a schedule, simply start the backup job.

Whether the backup is scheduled or manual, understand that the process is resource-intensive, and it is recommended that they be run during times of low demand on the server if possible.

The status of the backup job(s) can be observed by navigating to **Backup > Current Status.** A list of the components of the backup job and the completion percentage for each component is presented. Components that are complete show a link to the log file.

Perform a Restore

The purpose of having backups is to be able to restore our data when necessary. The following steps describe the basic restore process:

1. Navigate to **Restore > Restore Wizard.**
2. Select the device that holds the backup file from which you wish to restore.
3. Select the correct backup file from the list available on that device.
4. Select the feature(s) you want to restore. It should be self-evident that if a feature was not backed up, it cannot be restored!
5. If the restore is coming from an SFTP server, you may select the optional **File Integrity Check**, which ensures that the restored data is not corrupted. Doing so takes significant server and network resources and slows down the restore process.
6. Select the server(s) that should be restored. If the Publisher (first node) is selected for restore, the DRS automatically restores the database on the subscribers (subsequent nodes). However, in either case, all existing data is overwritten by the restore.
7. Monitor the restore progress by navigating to **Restore > Status.**

Administrators should be familiar with the content of the *Disaster Recovery System Administration Guide* for their version(s) of software and should practice restore scenarios in a lab environment.

Exam Preparation Tasks

Review All the Key Topics

Review the most important topics in the chapter, noted with the key topics icon in the outer margin of the page. Table 16-3 describes these key topics and identifies the page number on which each is found.

Table 16-3 *Key Topics for Chapter 16*

Key Topic Element	Description	Page Number
List	Troubleshooting steps	419
Topic	IP Phone registration troubleshooting	419
Topic	Generating CDR reports	427
Topic	Generating system reports	430
Topic	Describe Cisco Unified RTMT	432
Topic	Describe the Disaster Recovery System	434

Definitions of Key Terms

Define the following key terms from this chapter, and check your answers in the Glossary:

troubleshooting, Call Detail Record (CDR), Call Management Record (CMR), Call Detail Record Analysis and Reporting (CAR), Disaster Recovery System (DRS)

This chapter includes the following topics:

- **Generating and Accessing Cisco Unity Connection Reports:** This section reviews how to create and locate CUC Reports.
- **Analyzing Cisco Unity Connection Reports:** This section examines the content of CUC Reports and how to interpret them.
- **Troubleshooting and Maintenance Operations Using Cisco Unity Connection Reports:** This section provides guidance and examples for using CUC Reports for the troubleshooting and maintenance of CUC.

CHAPTER 17

Monitoring Cisco Unity Connection

Cisco Unity Connection (CUC) includes a variety of built-in reports to help administrators track the server's health and performance, as well as the activities of the users. This chapter introduces the use and content of many of these reports.

"Do I Know This Already?" Quiz

The "Do I Know This Already?" quiz allows you to assess whether you should read this entire chapter or simply jump to the "Exam Preparation Tasks" section for review. If you are in doubt, read the entire chapter. Table 17-1 outlines the major headings in this chapter and the corresponding "Do I Know This Already?" quiz questions. You can find the answers in Appendix A, "Answers Appendix."

Table 17-1 *"Do I Know This Already?" Foundation Topics Section-to-Question Mapping*

Foundation Topics Section	Questions
Generating and Accessing Cisco Unity Connection Reports	1–6
Analyzing Cisco Unity Connection Reports	7–8
Troubleshooting and Maintenance Operations Using Cisco Unity Connection Reports	9–10

1. Cisco Unity Connection provides two built-in reporting interfaces. Name them. (Choose two.)

   a. Cisco Unified Reporting
   b. Cisco Unity Connection Reporting
   c. Cisco Unity Connection Serviceability Reports Tool
   d. Cisco Unified Serviceability Reports Archive

2. Which of the following is *not* a Cisco Unity Connection Serviceability Reports Tool report?

   a. Server Report
   b. Unused Voice Mail Accounts Report
   c. User Lockout Report
   d. Port Activity Report
   e. Users Report

3. Which of the following are Cisco Unified Serviceability Reports Archive reports? (Choose two.)

   a. Mailbox Store Report
   b. Alert Report
   c. System Configuration Report
   d. Server Report
   e. Message Traffic Report

4. What service must be activated in order to begin the collection of the Cisco Unified Serviceability Reports Archive report data?

   a. Cisco Serviceability Reporter
   b. Cisco UXL Web Service
   c. Cisco Reports Harvester
   d. Cisco Unity Connection Serviceability Harvester

5. How many reports are available in the Cisco Unified Serviceability Reports Archive?

   a. 19
   b. 2
   c. 19 per day
   d. 2 per day of data collection

6. You decide to increase the number of entries that can be held in the Audit Log. Where can this be done?

   a. It cannot be done.
   b. It is done by saving old log files to a syslog server and removing them from the original log location.
   c. It is done in the Service Parameters configuration page.
   d. In CUC Administration, navigate to System Settings > Advanced > Reports.

7. Ermeniglio decides to start using the Serviceability Reports archive. He starts the correct service, waits until the next day, and opens the Alerts Report. He is confused to discover that the report shows no alerts were generated the previous day. What could be the issue?

   a. The service sometimes "sticks" and should be restarted.

   b. No alerts were generated.

   c. The Display Alert Threshold setting must be lowered to Critical or below.

   d. Ermeniglio must use the RTMT to see alerts.

8. Bob says he wants to see the details of the alerts shown in the Alerts Report in the Serviceability Archives. What should you do?

   a. Open the RTMT for CUC, go to Alert Central, and right-click any alert to see the details.

   b. Open RTMT, go to the Server tab, and look for the alert list.

   c. Select the **Include Detail** checkbox when generating the Alerts Report.

   d. Wait 30 seconds to make sure Bob is real, and then download the Event Log from the CUCM server.

9. The previous CUC administrator recently quit. You have been hired to take her place. You want to get a sense of how well-maintained the CUC server is. How can you find out if any accounts are still active but not in use?

   a. Run the Weekly Diagnostics Report.

   b. Run the Security Report.

   c. Run the Unused Voice Mail Accounts Report.

   d. Run the User Lockout Report.

10. Several days a week, usually between 8:00 a.m. and 9:00 a.m., users complain of MWI problems. Some have new messages, but their lamp is not lit; others complain that their lamp is on, but there are no new messages. You suspect that the server is not making MWI calls. How can you quickly check whether or not this is the case?

   a. Run the Port Simulator utility and send a test message.

   b. Run the Port Status Monitor in the afternoon.

   c. Come in early on Thursday, and try leaving a message for a user to see if the lamp comes on.

   d. Run the Port Activity Report.

Foundation Topics

Generating and Accessing Cisco Unity Connection Reports

There are two main interfaces for generating and viewing reports in CUC: in Cisco Unity Connection Serviceability under **Tools > Reports**, and in Cisco Unified Serviceability under **Tools > Serviceability Reports Archive.**

Cisco Unity Connection Serviceability Reports

The CUC Serviceability application provides 19 different reports that assist the administrator in monitoring and understanding the status and behavior of the CUC application. Figure 17-1 shows the CUC Serviceability Reports page with all 19 reports listed.

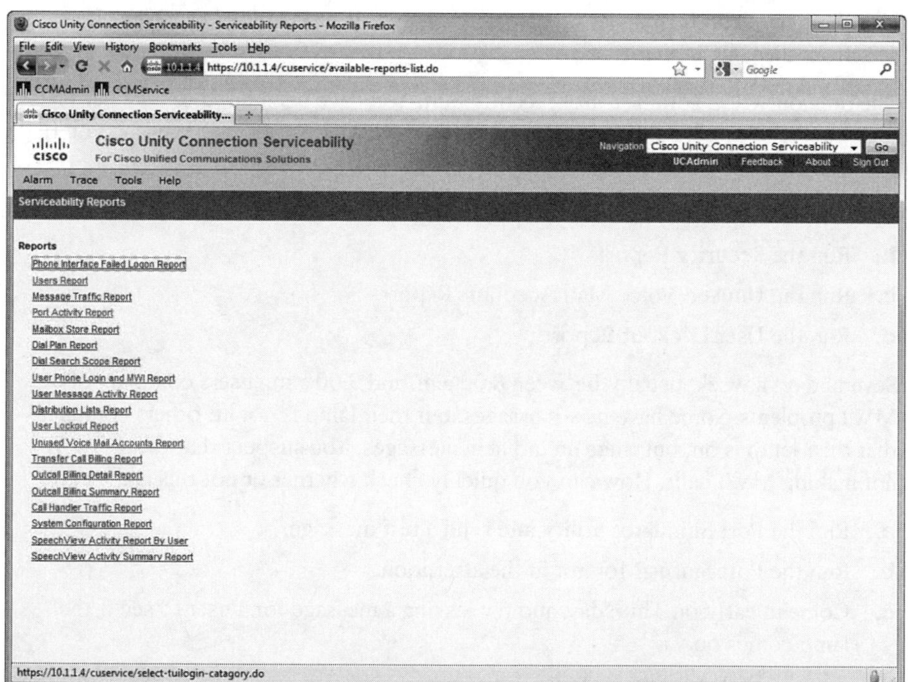

Figure 17-1 *CUC Serviceability Reports Page*

To access these reports, open the CUC Serviceability application and navigate to Tools > Reports. The list of reports includes the following:

- Phone Interface Failed Logon Report
- Users Report

- Message Traffic Report
- Port Activity Report
- Mailbox Store Report
- Dial Plan Report
- Dial Search Scope Report
- User Phone Login and MWI Report
- User Message Activity Report
- Distribution Lists Report
- User Lockout Report
- Unused Voice Mail Accounts Report
- Transfer Call Billing Report
- Outcall Billing Report
- Outcall Billing Summary Report
- Call Handler Traffic Report
- System Configuration Report
- SpeechView Activity Report by User
- SpeechView Acitivity Summary Report

Let's take the example of running the Users Report. From the CUC Serviceability Reports page, click **Users Report.** The screen shown in Figure 17-2 appears.

The following selections can be made to customize the report (shown in Figure 17-2):

Run This Report For:

- Select Class:
  - User
  - Distribution List
  - COS
- User:
  - All Users
  - Selected User

File Format:

- Web Page
- Comma-Delimited File
- PDF File

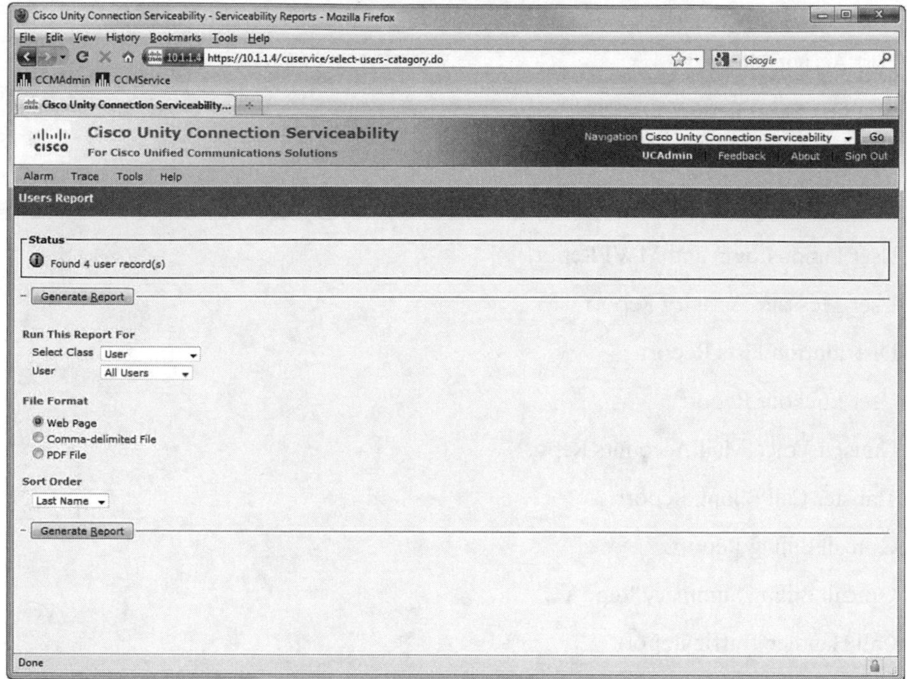

Figure 17-2 *Users Report Configuration Page*

Sort Order:

- Last Name
- First Name
- Extension
- COS

After you customize your selections, click **Generate Report.**

Figure 17-3 shows a sample output of the Users Report.

The Users Report includes the following information, as shown in Figure 17-3:

- Last Name, First Name, and Alias
- Location
- Home Mail Server
- Billing ID, CoS, and Extension
- Account Lockout Status
- Personal Call Transfer Rules Enabled/Disabled Status

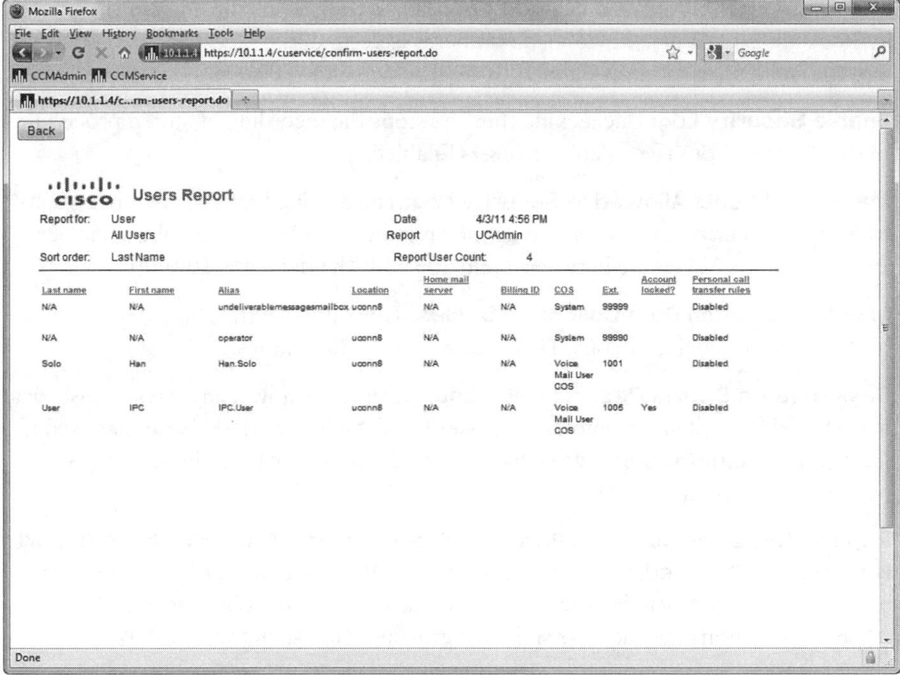

Users Report

Report for: User, All Users — Date: 4/3/11 4:56 PM — Report: UCAdmin

Sort order: Last Name — Report User Count: 4

Last name	First name	Alias	Location	Home mail server	Billing ID	COS	Ext.	Account locked?	Personal call transfer rules
N/A	N/A	undeliverablemessagesmailbox	uconn8	N/A	N/A	System	99999		Disabled
N/A	N/A	operator	uconn8	N/A	N/A	System	99990		Disabled
Solo	Han	Han.Solo	uconn8	N/A	N/A	Voice Mail User COS	1001		Disabled
User	IPC	IPC.User	uconn8	N/A	N/A	Voice Mail User COS	1005	Yes	Disabled

Figure 17-3 *Sample Output: Users Report*

The other reports in the CUC Serviceability Reports list are used in a similar way. Each report provides insight into the current status, configuration, and utilization of the CUC server.

Cisco Unified Serviceability: Serviceability Reports Archive

CUC includes a built-in reporting system to monitor server health and performance. These reports are accessed through the Unified Serviceability web application (not to be confused with the CUC Serviceability application). To access these reports, you must first activate the **Cisco Serviceability Reporter** service. In Unified Serviceability, navigate to **Tools > Service Activation.** Select the **Cisco Serviceability Reporter** service and click **Save.**

The Serviceability Reporter collects data from log files and populates the Serviceability Reports Archive, which stores report information and makes it available on a daily basis. It is a CPU-intensive service, so consider whether activating it will negatively impact your server health or performance. The type and amount of data collected can be tuned in the CUC Administration interface. Navigate to **System Settings > Advanced > Reports** to modify the following:

- **Enable Audit Log:** Unchecking this box stops the logging of stored procedures. The default setting is Enabled.

- **Maximum Events Allowed in Audit Log:** This setting limits the number of entries in the Audit Log. When the defined number is exceeded, the oldest entries are overwritten. Values are between 1 and 100,000. The default is 100,000.
- **Enable Security Log:** Unchecking this box stops the recording of stored procedures to the Security Log. The default setting is Enabled.
- **Maximum Events Allowed in Security Log:** This setting limits the number of entries in the Security Log. When the defined number is exceeded, the oldest entries are overwritten. Values are between 1 and 100,000. The default is 100,000.
- **Minutes Between Data Collection Cycles:** This value controls how frequently report data is gathered from logs. The default is every 30 minutes.
- **Days to Keep Data in Reports Database:** Determines how many days of historical data should be kept in the reports database. The default is 90 days. Note that even if the report specifies a date range of more than 90 days in the past, the report will still be limited to this setting's value.
- **Reports Database Size (as a Percentage of Capacity) After Which the Reports Harvester Is Disabled:** Sets the maximum percentage of disk space the reports database may take up. When the value is reached, the CUC Report Harvester service is stopped, preventing the database size from growing. The default value is 80%.
- **Maximum Records in Report Output:** Limits the number of records presented in the report output. The allowed range is from 5,000 to 30,000; the default is 25,000. Some reports impose their own restrictions due to the size of the report (for example, User Message Activity is limited to 25,000 records).
- **Minimum Records Needed to Display Progress Indicator:** This value determines whether running a report will cause a message to pop up before the report is generated, and a progress bar to be shown as it runs. The idea is to provide a warning that the selected report is large and may impact server performance. The allowed range is from 1 to 10,000. The default is 2,500.

Figure 17-4 shows the Report Configuration page.

Analyzing Cisco Unity Connection Reports

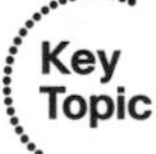

In the Cisco Unified Serviceability web application, navigate to **Tools > Serviceability Reports Archive**. Click the month for which you want to view reports, and then click the specific date. Figure 17-5 shows the Serviceability Reports Archive list page.

There are two Serviceability Archive Reports available to view (assuming enough time has elapsed to provide data to collect it).

The **Alerts Report** displays the following:

- Number of alerts per severity in the cluster
- Number of alerts per server
- Top ten alerts in the cluster

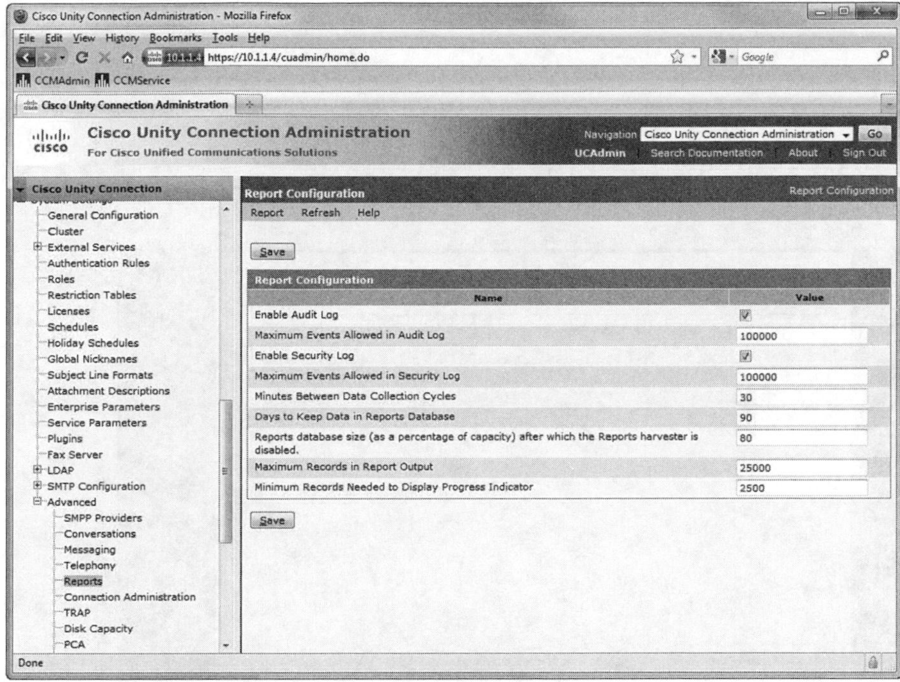

Figure 17-4 *CUC Administration Report Configuration Page*

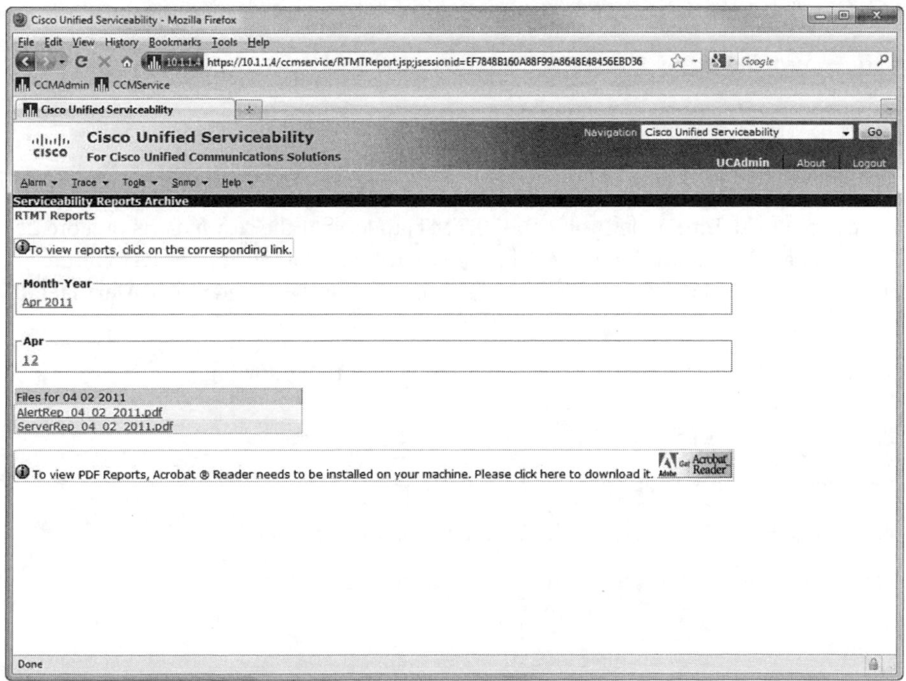

Figure 17-5 *Serviceability Reports Archive List Page*

Figure 17-6 shows one of the pages in the Alerts Report.

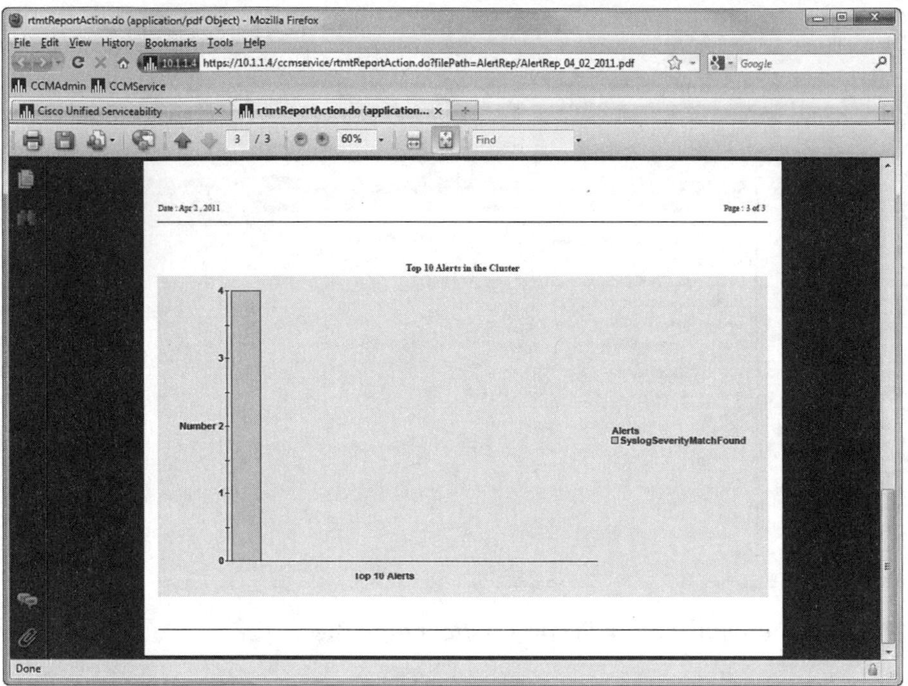

Figure 17-6 *Alerts Report*

Because these reports are just summaries, showing no details of the alerts (only what severity level and to which server they are attributed), it is likely that administrators will want to use the Real-Time Monitoring Tool (RTMT) to look at the server alerts in more detail. By opening **Alert Central** in RTMT (be sure to use the CUC version of RTMT), administrators can view the list of alerts, right-click any one of them, and select **Alert Detail.** A window pops up to show the details of the logged alert.

The Server Report provides statistics (in graph format) for the following:

- Percentage CPU per Server
- Percentage Memory Usage per Server
- Percentage Hard Disk Usage of the Common Partition per Server
- Percentage Hard Disk Usage of the Spare Partition per Server

Figure 17-7 shows the Server Report.

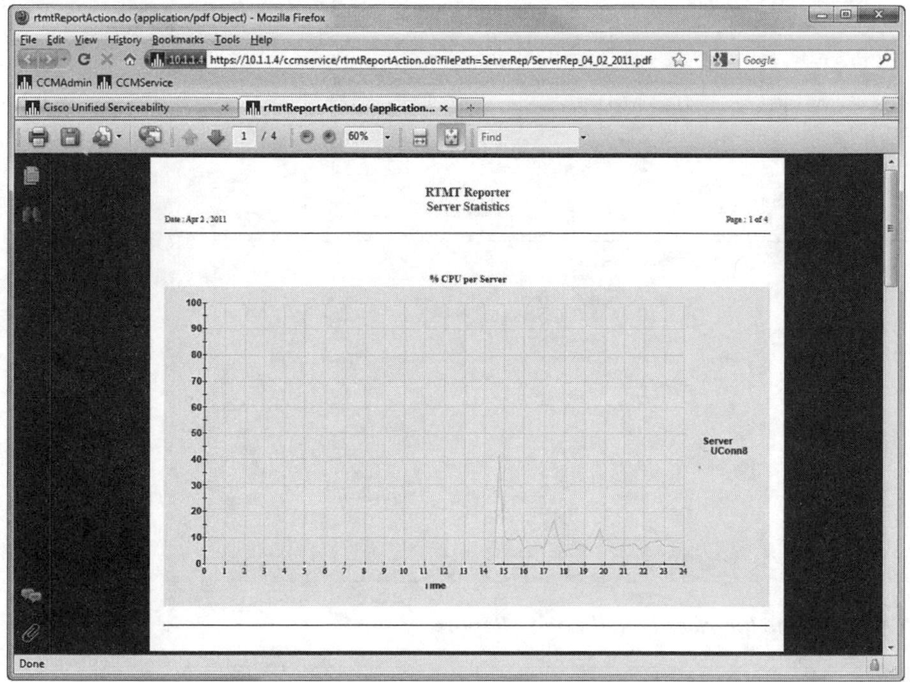

Figure 17-7 *Server Report*

Troubleshooting and Maintenance Operations Using Cisco Unity Connection Reports

The CUC Serviceability Reports provide insight into what is happening on the server. For example:

- An administrator might run the **Phone Interface Failed Logon** Report to see if there are a significant number of failed logins for a given time period. If there are, the next question is whether these failed attempts to log in are a user issue (which could be resolved by talking with the user) or evidence of an attempt to hack into the user's mailbox. Figure 17-8 shows the Phone Interface Failed Logon Report.
- The **User Lockout Report** (often run in conjunction with the **Failed Login** Report) provides a quick list of which accounts are locked out, why, and when they were locked. The administrator can then contact the locked-out users and take any corrective action required; then, in the CUC Administration web application, the administrator can navigate to **Users > Users**, select the affected user, navigate to **Edit > Password Settings**, and click **Unlock Password** to unlock the user's account. Figure 17-9 shows the User Lockout Report.

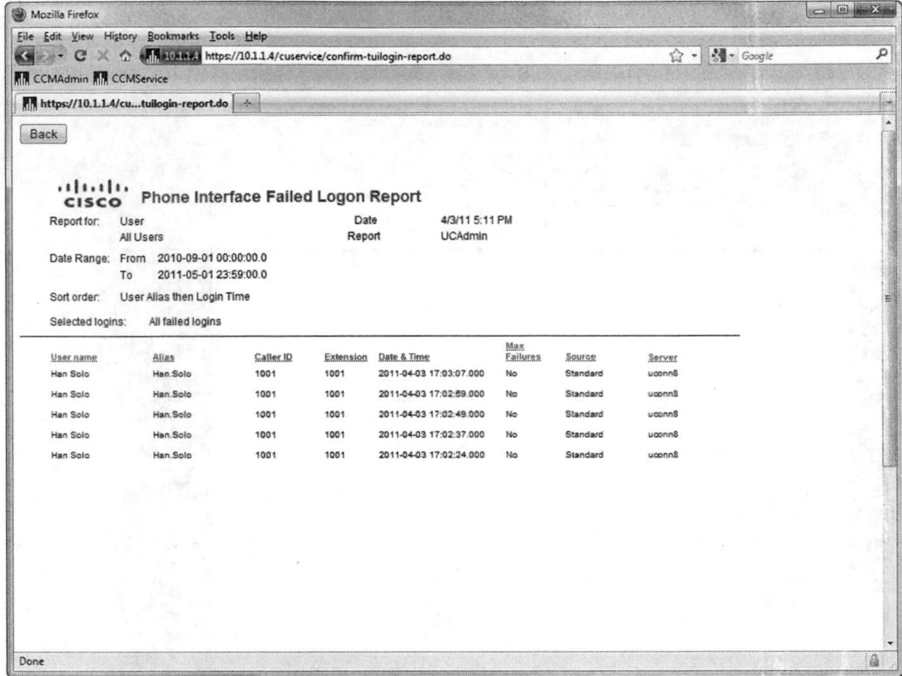

Figure 17-8 *Phone Interface Failed Logon Report*

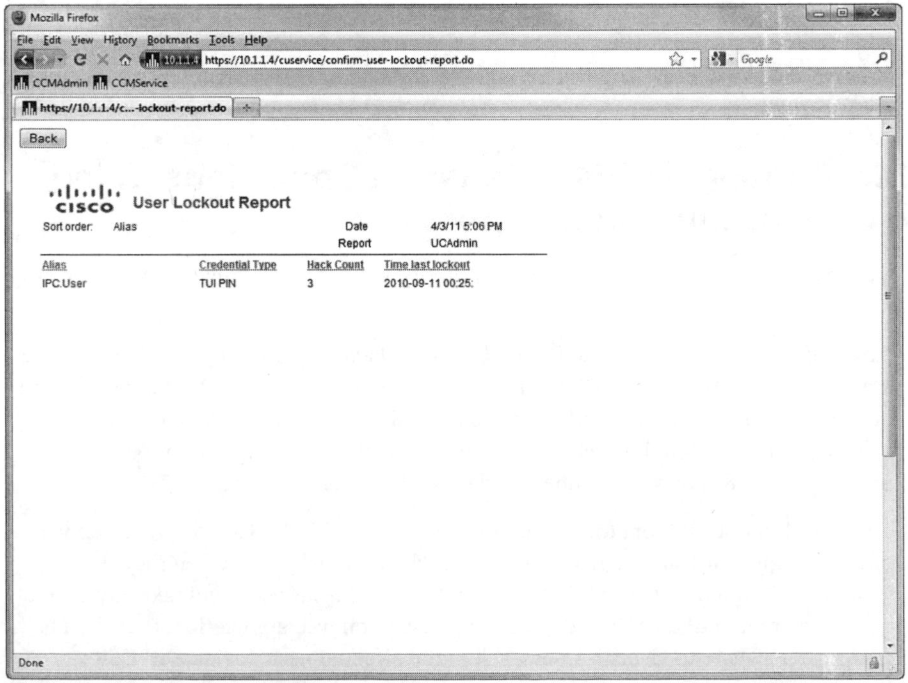

Figure 17-9 *User Lockout Report*

- The **Port Activity Report** shows the following statistics for each voicemail port on the server:
  - Port Name
  - Inbound Calls
  - Outbound MWI
  - Outbound AMIS
  - Outbound Notification
  - Outbound TRAP
  - Port Total

The administrator can determine whether all ports are active and usable and may be able to determine the cause of an MWI problem (perhaps because no ports have been assigned to perform MWI only). Figure 17-10 shows the Port Activity Report.

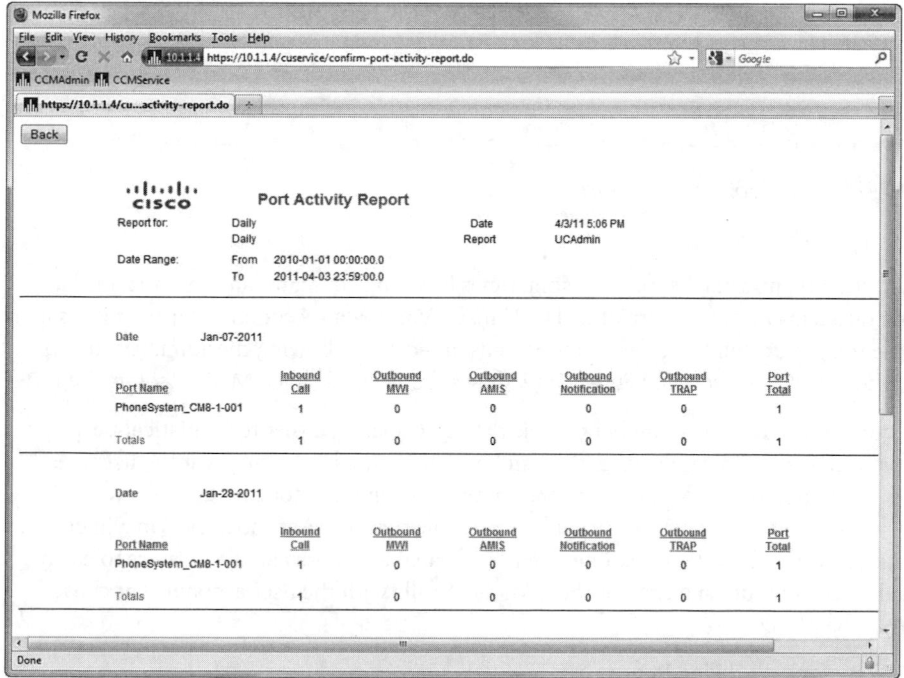

Figure 17-10 *Port Activity Report*

Reports to Support Routine Maintenance

Close monitoring of the mailbox stores help prevent running out of disk space and spot any other issues before they cause a service interruption. The **Mailbox Store Report** provides a summary view of the current size, last error condition, and status of the mailbox store. Figure 17-11 shows the Mailbox Store Report.

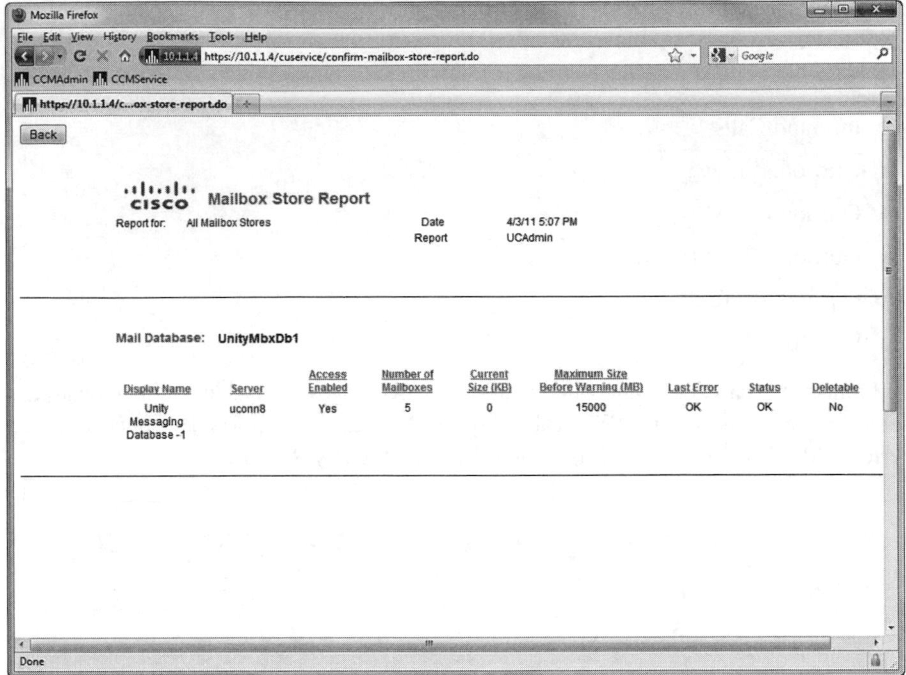

Figure 17-11 *Mailbox Store Report*

In a larger organization, employees sometimes leave the company, and the voicemail administrators may not be aware of it. The **Unused Voice Mail Accounts Report** makes it simple to spot accounts that have not been used recently, allowing the administrator to take the appropriate action. Figure 17-12 shows the Unused Voice Mail Accounts Report.

It is common to track the number of calls that CUC makes, either for statistical purposes or for actual cost billing purposes. Remember that CUC can place calls at the user's request (assuming their CoS allows it), and these calls can incur toll charges. Likewise, if CUC performs Message Notification, it is possible that some of those calls may incur toll charges as well. There are three Billing reports available to allow administrators to easily correlate the number and time of these kinds of calls with the user account that caused them to be placed:

- **Transfer Call Billing Report** lists:
  - Name, extension, and billing ID of the user
  - Date/time stamp for the call
  - Called number
  - Transfer result (Connected, Ring No Answer, Busy, or Unknown)

- **Outcall Billing Detail Report** is sorted by day, user extension, and lists:
  - Name, extension, and billing ID of the user
  - Date/time stamp for the call
  - Called number
  - Transfer result (Connected, Ring No Answer, Busy, or Unknown)
  - Duration of the call (in seconds)
- **Outcall Billing Summary Report** is sorted by date, name, extension, and billing ID, and shows the dial-out time in seconds for each hour of the day.

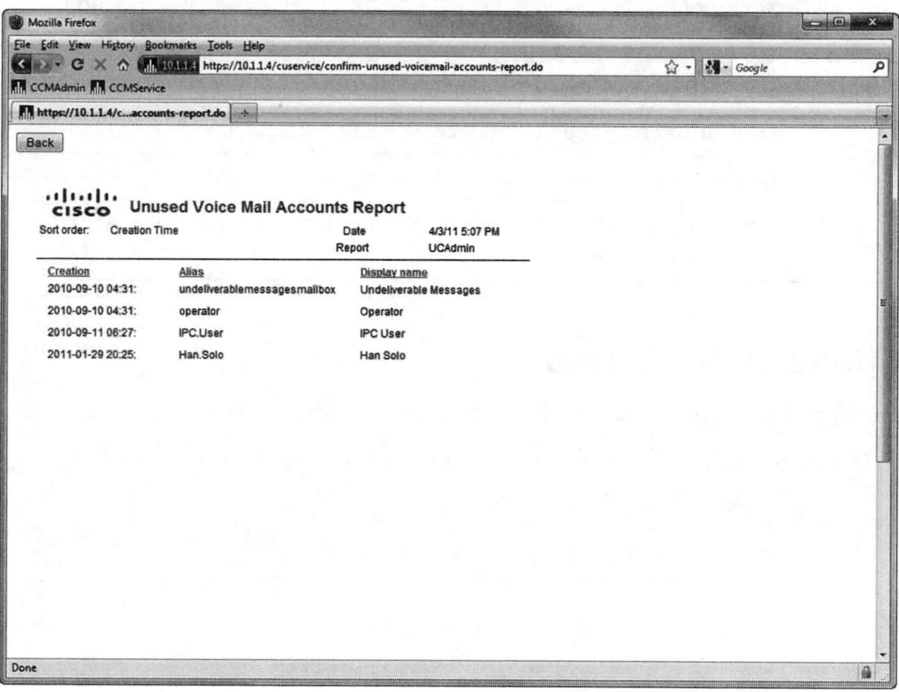

Figure 17-12 *Unused Voice Mail Accounts Report*

Exam Preparation Tasks

Review All the Key Topics

Review the most important topics in the chapter, noted with the key topics icon in the outer margin of the page. Table 17-2 describes these key topics and identifies the page number on which each is found.

Table 17-2 *Key Topics for Chapter 17*

Key Topic Element	Description	Page Number
Sub-topic	Cisco Unity Connection Serviceability Reports	442
Sub-topic	Cisco Unified Serviceability: Serviceability Reports Archive	445
Topic	Analyzing Cisco Unity Connection Reports	446
Topic	Troubleshooting and maintenance operations using Cisco Unity Connection Reports	449

Definitions of Key Terms

Define the following key terms from this chapter, and check your answers in the Glossary:

Cisco Unified Serviceability Reports, Cisco Unified Serviceability Archives Reports

The first 17 chapters of this book cover the technologies, protocols, design concepts, and considerations required to be prepared to pass the 640-461 ICOMM Exam. Although these chapters supply the detailed information, most people need more preparation than simply reading the first 17 chapters of this book. This chapter details a set of tools and a study plan to help you complete your preparation for the exams.

This short chapter has two main sections. The first section lists the exam-preparation tools useful at this point in the study process. The second section lists a suggested study plan now that you completed all the earlier chapters of this book.

CHAPTER 18

Final Preparation

Tools for Final Preparation

This section lists some information about the available tools and how to access the tools.

Pearson Cert Practice Test Engine and Questions on the CD

The CD in the back of this book includes the Pearson Cert Practice Test engine—software that displays and grades a set of exam-realistic multiple-choice questions. Using the Pearson Cert Practice Test engine, you can either study by going through the questions in Study mode or take a simulated (timed) CCNA Voice exam.

The installation process requires two major steps. The CD has a recent copy of the Pearson Cert Practice Test engine. The practice exam—the database of CCNA Voice exam questions—is not on the CD.

Install the Software from the CD

The software-installation process is routine compared with other software-installation processes. To be complete, the following steps outline the installation process:

Step 1. Insert the CD into your PC.

Step 2. The software that automatically runs is the Cisco Press software to access and use all CD-based features, including the exam engine and the CD-only appendices. From the main menu, click the option to **Install the Exam Engine.**

Step 3. Respond to window prompts as with any typical software-installation process.

The installation process gives you the option to activate your exam with the activation code supplied on the paper in the CD sleeve. This process requires that you establish a Pearson website login. You need this login to activate the exam, so please register when prompted. If you already have a Pearson website login, there is no need to re-register; just use your existing login.

Activate and Download the Practice Exam

After the exam engine is installed, you should activate the exam associated with this book (if you did not do so during the installation process), as follows:

Step 1. Start the Pearson Cert Practice Test (PCPT) software from the Windows **Start** menu or from your desktop shortcut icon.

Step 2. To activate and download the exam associated with this book, from the **My Products** or **Tools** tab, select the **Activate** button.

Step 3. At the next screen, enter the Activation Key from paper inside the cardboard CD holder in the back of the book. Once entered, click the **Activate** button.

Step 4. The activation process downloads the practice exam. Click **Next**, and then click **Finish.**

After the activation process is completed, the **My Products** tab should list your new exam. If you do not see the exam, make sure you selected the **My Products** tab on the menu. At this point, the software and practice exam are ready to use. Simply select the exam and click the **Use** button.

To update a particular exam you have already activated and downloaded, simply select the **Tools** tab and select the **Update Products** button. Updating your exams ensures that you have the latest changes and updates to the exam data.

If you want to check for updates to the PCPT exam engine software, simply select the **Tools** tab and select the **Update Application** button. This ensures that you are running the latest version of the software engine.

Activating Other Exams

The exam software-installation process, and the registration process, has to happen only once. Then, for each new exam, only a few steps are required. For example, if you buy another new Cisco Press Official Cert Guide or Pearson IT Certification Cert Guide, extract the activation code from the CD sleeve in the back of that book—you don't even need the CD at this point. All you have to do is start the exam engine and perform Steps 2 through 4 from the previous list.

Premium Edition

In addition to the free practice exam provided on the CD-ROM, you can purchase additional exams with expanded functionality directly from Pearson IT Certification. The Premium Edition of this title contains an additional two full practice exams and an eBook (in both PDF and ePub format). In addition, the Premium Edition title also has remediation for each question to the specific part of the eBook that relates to that question.

Because you purchased the print version of this title, you can purchase the Premium Edition at a deep discount. There is a coupon code in the CD sleeve that contains a one-time use code and instructions for where you can purchase the Premium Edition.

To view the premium edition product page, go to http://www.pearsonitcertification.com/title/9780132748322.

Cisco Learning Network

Cisco provides a wide variety of CCNA Voice preparation tools at a Cisco Systems website called the Cisco Learning Network. This site includes a variety of exam-preparation tools, including sample questions, forums on each Cisco exam, learning video games, and information about each exam.

To reach the Cisco Learning Network, go to www.cisco.com/go/learnnetspace, or just search for "Cisco Learning Network." You need to use the login you created at www.cisco.com. If you don't have such a login, you can register for free. To register, simply go to www.cisco.com, click **Register** at the top of the page, and supply the required information.

Chapter-Ending Review Tools

Chapters 1–17 each have several features in the Exam Preparation Tasks at the end of the chapter. You might have already worked through these in each chapter. It can also be useful to use these tools again as you make your final preparations for the exam.

Suggested Plan for Final Review/Study

This section lists a suggested study plan from the point at which you finish reading through Chapter 17 until you take the 640-461 ICOMM Exam. Certainly, you can ignore this plan, use it as is, or just take suggestions from it.

The plan uses six steps:

Step 1. Review key topics and DIKTA questions:

You can use the table that lists the key topics in each chapter or just flip the pages looking for key topics. Also, reviewing the Do I Know This Already? questions from the beginning of the chapter can be helpful for review.

Step 2. Use the PCPT engine to practice:

The PCPT engine on the CD can be used to study using a bank of unique exam-realistic questions available only with this book.

Step 3. Review weak topic areas:

Working through the practice exam will reveal topics you may want to review and study. Take the time to fill in the gaps in your knowledge and then attempt the practice exam again. However, be careful about taking the practice exam too many times because you may gain a false sense of confidence simply because you begin to memorize the answers to the practice test. Be sure you truly understand the content before attempting the real exam.

Step 4. Schedule and take the 640-461 ICOMM Exam:

Put your knowledge to the test in the real certification exam environment! While we wish you the best of luck, please keep in mind that Cisco certification exams are some of the most difficult in the industry, and there are many factors beyond just technical knowledge that weigh in on the final exam results.

Step 5. Study based on your results:

When you leave the certification testing center, you will know the exam results. If you passed, congratulations! If not, write down areas or topics you remember struggling with on the exam. In your studies, focus 90% of your time on these content areas and 10% of your time brushing up on the other areas.

Step 6. Repeat Step 4 and 5 until fully certified!

Using the Exam Engine

The PCPT engine on the CD includes a database of questions created specifically for this book. The PCPT engine can be used either in Study mode or Practice Exam mode, as follows:

- **Study mode:** Most useful when you want to use the questions for learning and practicing. In Study mode, you can select options like randomizing the order of the questions and answers, automatically viewing answers to the questions as you go, testing on specific topics, and many other options.
- **Practice Exam mode:** Presents questions in a timed environment, providing you with a more exam-realistic experience. It also restricts your ability to see your score as you progress through the exam and view answers to questions as you are taking the exam. These timed exams not only allow you to study for the actual 640-461 ICOMM Exam, they help you simulate the time pressure that can occur on the actual exam.

When doing your final preparation, you can use Study mode, Practice Exam mode, or both. However, after you see each question a couple of times, you will likely start to remember the questions, and the usefulness of the exam database may go down. So, consider the following options when using the exam engine:

- Use this question database for review. Use Study mode to study the questions by chapter, just as with the other final review steps listed in this chapter. Plan on getting another exam (possibly from the Premium Edition) if you want to take additional simulated exams.
- Save the question database, not using it for review during your review of each book part. Save it until the end, so you will not have seen the questions before. Then, use Practice Exam mode to simulate the exam.

Picking the correct mode from the exam engine's user interface is obvious. The following steps show how to move to the screen from which to select Study or Practice Exam mode:

Step 1. Click the **My Products** tab if you are not already in that screen.

Step 2. Select the exam you want to use from the list of available exams.

Step 3. Click the **Use** button.

By taking these actions, the engine should display a window from which you can choose **Study Mode** or **Practice Exam Mode.** When in Study mode, you can further choose the book chapters, limiting the questions to those explained in the specified chapters of this book.

Summary

The tools and suggestions listed in this chapter are designed with one goal in mind: to help you develop the skills required to pass the 640-461 ICOMM Exam. This book has been developed from the beginning to not just tell you the facts, but help you learn how to apply the facts. No matter what your experience level leading up to when you take the exams, it is our hope that the broad range of preparation tools, and even the structure of the book, will help you pass the exam with ease. I hope you do well on the exam.

APPENDIX A

Answers Appendix

Chapter 1

1. A
2. B
3. A
4. B and C
5. C
6. D
7. D
8. C and E
9. B
10. C
11. C
12. B and D

Chapter 2

1. A
2. A and B
3. A
4. C
5. B
6. A
7. D
8. C
9. B
10. D
11. A
12. B
13. B and C

Chapter 3

1. C
2. A and C
3. A and C
4. C
5. A
6. B
7. C
8. D
9. A
10. C
11. A
12. A and B

Chapter 4

1. A, D, and E
2. B
3. B
4. A
5. D
6. A
7. C

Chapter 5

1. A
2. A, E, and G
3. A
4. C
5. D
6. B
7. C
8. B
9. A
10. B

Chapter 6

1. A
2. D
3. C
4. E
5. D
6. A and C
7. B
8. C
9. B
10. B
11. C

Chapter 7

1. B
2. C
3. A
4. D
5. D
6. C
7. D
8. B
9. A
10. A
11. B and C
12. A and C
13. A

Chapter 8

1. C
2. C and D
3. B
4. B
5. F
6. A, C, and E
7. C
8. A, B, C, and D
9. C
10. B

Chapter 9

1. D
2. A
3. B
4. B
5. D
6. C
7. B
8. D
9. C
10. A

Chapter 10

1. B and C
2. B and D
3. A
4. E
5. E
6. A
7. F
8. E and F
9. B, D, E, and F
10. D
11. C
12. C

Chapter 11

1. A
2. B
3. A, C, and D
4. B
5. D
6. A, D, E, and F
7. B
8. D
9. A, C, and D
10. D

Chapter 12

1. D
2. A, B, C, D, E, and F
3. D
4. A
5. C
6. B
7. A
8. B
9. C

Chapter 13

1. D
2. B
3. A, B, and C
4. B
5. D
6. D and E
7. A, B, and C
8. A
9. C
10. D

Chapter 14

1. C and D
2. A, B, C, D, and E
3. E
4. B and D
5. C
6. D and E
7. A, B, and C
8. A

Chapter 15

1. B
2. B, C, and D
3. B
4. C
5. C
6. B
7. A
8. C
9. B
10. D

Chapter 16

1. D
2. D
3. D
4. A
5. C
6. A
7. A
8. C
9. C
10. E

Chapter 17

1. C and D
2. A
3. B and D
4. A
5. D
6. D
7. B
8. A
9. C
10. D

Over time, reader feedback allows Cisco Press to gauge which topics give readers the most problems when taking the exams. To assist readers, authors may create new materials clarifying and expanding upon those troublesome exam topics. This additional content about the exam will be posted as a PDF document on this book's companion website, www.ciscopress.com/title/9781587204173.

This appendix provides you with updated information if Cisco makes minor modifications to the exam upon which this book is based. When Cisco releases an entirely new exam, the changes are usually too extensive to provide in a simple updated appendix. In those cases, you need to consult the book's new edition for updated content.

This appendix attempts to fill the void that occurs in any print book. In particular, this appendix does the following:

- Mentions technical items that might not have been mentioned elsewhere
- Covers new topics if Cisco adds new content to the exam over time
- Provides a way to get up-to-the-minute current information about content for the exam

APPENDIX B

640-461 CCNA Voice Exam Updates, Version 1.0

Always Get the Latest at the Companion Website

You are reading the version of this appendix that was available when your book was printed. However, given that the main purpose of this appendix is to be a living, changing document, it is important that you look for the latest version online at the book's companion website. To do so:

Step 1. Browse to www.ciscopress.com/title/9781587204173.

Step 2. Select the Updates option under the More Information box.

Step 3. Download the latest Appendix B document.

Note: The downloaded document has a version number. Comparing the version of this print Appendix B (version 1.0) with the latest online version of this appendix, you should do the following:

- **Same version:** Ignore the PDF that you downloaded from the companion website.
- **Website has a later version:** Ignore the Appendix B in your book and read only the latest version that you downloaded from the companion website.

If no appendix is posted on the book's website, that means that there are no updates to post, and version 1.0 is still the latest version.

Technical Content

The current version of this appendix does not contain any additional technical coverage.

Glossary

802.1Q An industry-standard trunking protocol that allows traffic for multiple VLANs to be sent between switches.

802.3af Power over Ethernet (PoE) Industry-standard method of supplying power over an Ethernet cable to attached devices.

administration via telephone (AVT) system Gives an administrator an easy way to record custom prompts and to quickly record and enable the AA alternate greeting via a telephone connection.

Administrative Extensions for XML (AXL) Method by which user accounts in a CUCM database may be replicated and synchronized with the user account database in CUC.

analog signal A method of signaling that uses properties of the transmission medium to convey sound characteristics, such as using electrical properties to convey the characteristics of voice.

application (CUCM) Includes CM Administration, Unified Serviceability, Cisco Extension Mobility, and so on. Each application has resources that roles are permitted or restricted from viewing and/or editing.

auto-assignment A feature that allows the Cisco Unified CME router to distribute ephone-dns to auto-registered IP phones.

auto-attendant A common application of Interactive Voice Response (IVR) that allows callers to use automated menus to navigate to specific areas of your company.

auto-registration A Cisco Unified CME feature that allows Cisco IP phones to register with the CME router without an existing ephone configuration; auto-registration is turned on by default.

automated attendant (AA) Provides a business with the ability to answer and direct incoming phone calls to the appropriate person within the business without requiring human intervention.

Automatic Number Identification (ANI) Describes caller ID information delivered to a voice-processing device. Closely related to Dialed Number Identification Service (DNIS), which identifies dialed number information.

Call Detail Record (CDR) A log of specific information about a phone call that can be used for billing, analysis, reporting, and troubleshooting.

Call Detail Record Analysis and Reporting (CAR) A built-in reporting system that allows administrators to easily generate reports from the CDR/CMR database.

call handler A software element that answers a call, typically plays an informational greeting recording, may offer user input options for navigation, and may allow transfer of the call to other call handlers or user extensions.

Call Management Record (CMR) A log of voice quality statistics about IP phone calls.

call park A Cisco Unified CME feature that allows you to park a call on hold in a virtual "parking spot" until it can be retrieved.

call pickup A Cisco Unified CME feature that allows you to answer another ringing phone in the network.

call routing rules One of the primary mechanisms that CUC uses to analyze and direct calls to the appropriate call handler, conversation, or extension. Two basic types are available: Direct, for calls dialed directly to CUC, and Forwarded, for calls forwarded because of Busy or RNA events.

channel associated signaling (CAS) A method of digital signaling in which signaling information is transmitted using the same bandwidth as the voice.

Cisco Configuration Assistant (CCA) The GUI tool created by Cisco to provision, manage, troubleshoot, and maintain the Smart Business Communications System.

Cisco Discovery Protocol (CDP) Protocol that allows Cisco devices to discover other, directly attached Cisco devices. Switches also use CDP to send voice VLAN information to attached IP phones.

Cisco Emergency Responder Dynamically updates location information for a user based on the current position in the network and feeds that information to the emergency service provider if an emergency call is placed.

Cisco Inline Power Cisco-proprietary, prestandard method of supplying power over an Ethernet cable to attached devices.

Cisco IOS license A license from Cisco that allows a router to run the IOS software; most newly purchased routers come with an IOS license.

Cisco Smart Assist The name commonly associated with the group of wizard-like features integrated into the Cisco Configuration Assistant that simplifies the provisioning and maintenance of the SBCS suite.

Cisco Unified Communications An architecture that seeks to minimize the differences between the way we would like to communicate and the way we have to communicate given time, device, and location constraints.

Cisco Unified Communications 500 (UC500) The small business call processing platform that is able to support up to 48 users.

Cisco Unified Communications Manager The multiserver call processing platform that is able to support up to 30,000 users per cluster.

Cisco Unified Communications Manager Business Edition The single-server call processing platform that is able to support up to 500 users and includes integrated voicemail.

Cisco Unified Communications Manager Express (CME) The call processing platform that is able to support up to 250 users (depending on router hardware).

Cisco Unified Communications Manager IP Phone Service (CCMCIP) Used to list user-associated devices that can be used for communication.

Cisco Unified Contact Center Express A call center application that is able to support up to 300 agents.

Cisco Unified MeetingPlace Provides a multimedia conference solution that gives you the capability to conference voice, video, and data into a single conference call.

Cisco Unified Mobility Allows users to have a single contact phone number that they can link to multiple devices.

Cisco Unified Presence Provides status and reachability information for the users of the voice network.

Cisco Unified Serviceability Archives Reports Two reports per day of collected data: one for Alerts and one for Server statistics (CPU, hard-disk utilization).

Cisco Unified Serviceability Reports A set of 19 built-in reports to provide administrators with information about the configuration, utilization, and status of the CUC application.

Cisco Unity The unified messaging platform that is capable of supporting up to 7,500 users per server and redundant server configurations.

Cisco Unity Connection The single-server unified messaging platform that is capable of supporting up to 7,500 users.

Cisco Unity Express The unified messaging platform that is integrated into a Cisco Unified CME router; capable of supporting up to 250 users.

Cisco Unity Express administrator A subscriber that is a member of the administrators group.

Cisco Unity Express Advanced Integration Module (AIM-CUE) An entry-level hardware platform for Cisco Unity Express, providing up to 50 mailboxes, 14 hours of storage, and either four or six ports that can be used for simultaneous voice sessions, depending upon the model of Cisco ISR router it is installed in.

Cisco Unity Express automated-attendant (AA) script A collection of software steps that defines each action to be performed on a received call.

Cisco Unity Express CLI The command-line interface used to configure and administer Cisco Unity Express.

Cisco Unity Express custom scripting The act of modifying the default Cisco Unity Express AA script to match a business need.

Cisco Unity Express Editor A PC software application that is used to create Cisco Unity Express custom scripts.

Cisco Unity Express Enhanced Network Module (NME-CUE) The high-end hardware platform for Cisco Unity Express, providing up to 250 mailboxes, 300 hours of storage, and 24 ports that can be used for simultaneous voice sessions.

Cisco Unity Express GUI Provides subscribers and administrators with a web interface to use and manage Cisco Unity Express features and functions.

Cisco Unity Express greeting/prompt A recorded message played to a caller.

Cisco Unity Express Network Module (NM-CUE) The lower-midlevel hardware platform for Cisco Unity Express, providing up to 100 mailboxes, 100 hours of storage, and eight ports that can be used for simultaneous voice sessions.

Cisco Unity Express Network Module with Enhanced Capability (NM-CUE-EC) The upper-midlevel hardware platform for Cisco Unity Express, providing up to 250 mailboxes, 300 hours of storage, and 16 ports that can be used for simultaneous voice sessions.

Cisco Unity Express password Used to authenticate subscribers via the GUI.

Cisco Unity Express PIN Used to authenticate subscribers via the TUI.

Cisco Unity Express subscriber A user account configured in Cisco Unity Express.

Cisco Unity Express telephony user interface (TUI) Provides subscribers and administrators with a telephone interface to use and manage Cisco Unity Express features and functions.

Class of Restriction (COR) The method used to implement calling restrictions in the CME environment. An inbound COR list assigns privileges, whereas an outgoing COR list restricts calling.

class of service (CoS) Group of settings that provides or restricts access to licensed features, advanced features, or user capabilities.

Client Services Framework (CSF) The client-side foundations for CUPC and other CUC-integrated applications, such as Microsoft Outlook.

common channel signaling (CCS) A method of signaling in which information is transmitted using a separate, dedicated signaling channel.

community A group of devices managed by the Cisco Configuration Assistant via its IP address.

compliance As it relates to CUPS, compliance refers to the preservation of IMs on an external PostgreSQL database or compliance server.

Computer Telephone Interface Quick Buffer Encoding (CTIQBE) The protocol used for CUPC desk phone control.

dial-peer Logical configuration used to define dial plan information on a Cisco router.

Dialed Number Identification Service (DNIS) Describes dialed number information delivered to a voice processing device. Closely related to Automatic Number Identification (ANI), which identifies caller ID information.

Direct Inward Dial (DID) A voice configuration that allows users from the PSTN to dial directly into an individual phone at an organization without passing through a receptionist or automated attendant application.

directed pickup A method used with call pickup to answer a phone directly by dialing the extension number of the ringing phone.

Disaster Recovery System (DRS) The built-in Unified Communications platform backup and restore utility.

dual-tone multifrequency (DTMF) A type of address signaling in which the buttons on a telephone keypad use a pair of high and low electrical frequencies to generate a signal each time a caller presses a digit.

Dynamic Trunking Protocol (DTP) Allows switches to dynamically negotiate trunk links.

E.164 An international numbering plan created by the ITU and adopted for use on the PSTN.

Ear and Mouth (E&M) Analog interface type that acts as a trunk to a PBX system.

ephone A configuration in the CME router that represents a single IP phone (or IP telephony device).

ephone-dn A configuration in the CME router that represents a single directory number (DN).

extended super frame (ESF) A modern T1 signaling method that sends 24 T1 frames at a time.

Extensible Messaging and Presence Protocol (XMPP) Used for instant messaging with CUPS.

feature license A license dictating the number of IP phones a Cisco Unified CME router is able to support.

feature ring Causes an IP phone to ring with three consecutive pulses; configured by using the f button separator.

Foreign Exchange Office (FXO) Analog interface type that connects to a telephone carrier central office (CO) or PBX system; FXO ports receive dial tone from the attached device.

Foreign Exchange Office (FXO) ports Analog interfaces that allow you to connect a VoIP network to legacy telephony networks such as the PSTN or a PBX system.

Foreign Exchange Station (FXS) Analog interface type that connects to a legacy analog device (station); FXS ports provide dial tone to the attached device.

Foreign Exchange Station (FXS) ports Analog interfaces that allow you to connect a legacy analog telephony device to a VoIP network.

G.711 Uncompressed audio codec consuming 64 kbps of bandwidth.

G.726 Compressed audio codec consuming 32 kbps of bandwidth.

G.728 Compressed audio codec consuming 16 kbps of bandwidth.

G.729 Compressed audio codec consuming 8 kbps of bandwidth.

general delivery mailbox (GDM) A mailbox that is shared by a group of subscribers.

glare An instance in which a user picks up a phone and connects unexpectedly to an incoming call.

ground start signaling A method of signaling that relies on grounding wires connecting to a device to signal a new call; typically used in PBX systems to avoid glare.

H.323 Protocol suite created by the ITU-T to allow multimedia communication over network-based environments.

H.450.2 Industry-standard method of transferring calls without hairpinning.

H.450.3 Industry-standard method of forwarding calls without hairpinning.

hairpinning A problem that occurs when a call is transferred or forwarded from one IP phone to another that keeps the audio path established (or hairpinned) through the original IP phone; this tends to cause QoS issues with the call.

integrated messaging Provides access to voicemail message via an e-mail client and allows a subscriber to treat voicemail messages similarly to e-mail messages.

integration Configuration of a voicemail system (CUC) to interact with a call agent (CUCM). An integration provides MWI, call forward to personal greeting, and easy message access.

Interactive Voice Response (IVR) An automated system that provides a recorded, automated process to callers accessing your voice network.

Internet Low Bitrate Codec (iLBC) Compressed audio codec consuming 15.2 kbps of bandwidth.

Internet telephony service provider (ITSP) Provides VoIP trunk connectivity to the PSTN to provide a cost savings over traditional telephony service providers (TSP).

Inter-Switch Link (ISL) A Cisco-proprietary trunking protocol, which has been replaced by the industry-standard 802.1Q.

IP Phone Messenger (IPPM) An interface to handle IMs on the desk IP Phone.

key system A system that allows a company to run a private, internal voice network; key systems are usually used in smaller companies and provide shared-line extensions to all devices, although many key systems now provide unique extensions to all devices.

LAN expansion port The 10/100BASE-TX port in the UC520 that is automatically configured in the security configuration as the external system interface. This port is used to connect to a DSL or cable router for Internet access.

live record Enables a subscriber to record a live call and have that call delivered into the subscriber's mailbox.

live reply Enables a subscriber to use the received caller identification number (ANI) and place a phone call to that caller during voicemail message playback.

local directory The directory that is built automatically by the CME router as you define caller ID information for the ephone-dns.

local group pickup A method used with call pickup to answer a ringing phone from within the local group of an IP phone.

local loop The PSTN link between the customer premises (such as a home or business) and the telecommunications service provider.

loop start signaling A method of signaling that relies on connecting the tip and ring wires connecting to a device to complete an electrical circuit; typically used in devices connecting to the PSTN.

mailbox caller features Mailbox features that Cisco Unity Express offers to a caller, where the caller may or may not be a subscriber configured on Cisco Unity Express.

mailbox subscriber features Mailbox features that Cisco Unity Express offers to a configured subscriber.

mean opinion score (MOS) A subjective method of determining voice quality; listeners hear a phrase read over a voice network and rate the quality of the audio on a scale of 1 to 5.

Media Convergence Server (MCS) The server hardware platforms that support Cisco Unified Communications Manager software.

Media Gateway Control Protocol (MGCP) Voice signaling protocol created by the IETF; allows voice gateways to be controlled by a centralized call agent in client/server fashion.

message notification Feature used to generate a call, send an e-mail, or send a page to the subscriber when a new message has arrived in their mailbox.

Message Waiting Indicator (MWI) Provides a mechanism to alert a subscriber that a new message has arrived in a mailbox. This is typically achieved by turning on a light on the subscriber's IP phone.

Mobile Connect The ability to have a call for the user's enterprise IP phone number simultaneously ring up to ten remote devices, and switch seamlessly between IP phone and remote device while in the call. Also known as Single Number Reach.

Mobile Voice Access (MVA) The ability for a user to dial in from a remote device to an enterprise MVA access PSTN number, authenticate, and then place an outbound PSTN call that appears to come from their enterprise IP phone.

Monitor Mode/Watch Mode Line configuration that allows you to assign line instances to a Cisco IP phone that cannot be used for incoming or outgoing calls; rather, they can simply be used to check line status.

Network Time Protocol (NTP) Synchronizes the clock of a network device to a more accurate NTP server.

Nyquist theorem Describes the method of converting analog audio signals into digital format by sampling at twice the highest frequency of the audio.

other group pickup A method used with call pickup to answer a ringing phone from another group number, which must be specified after pressing the GPickUp softkey.

overlay line Allows shared line configurations by assigning multiple line instances to a single physical line button (overlay) on a Cisco IP phone; configured by using the o, c, or x separator.

packetization interval The amount of data (typically audio) placed into each packet. Cisco defaults to a 20 ms packetization interval for all codecs.

persistent chat Group chats that are preserved when a user rejoins a chat room after logging out.

phone user license A license belonging to each Cisco IP Phone that allows it to communicate with a Cisco Unified CME router or Cisco Unified Communications Manager server; most newly purchased IP Phones come with a phone user license.

power failover (PFO) The feature that allows the UC520 to complete calls out to the PSTN from a designated analog phone in the event of a power failure.

private branch exchange (PBX) A system that allows a company to run an internal, private voice network; PBX systems are usually used in larger companies and provide unique extensions to all devices.

private distribution lists A collection of subscribers created by a single subscriber for exclusive use by that subscriber.

Private Line Automatic Ringdown (PLAR) A configuration used to enable "immediate dial" applications, such as a phone that immediately dials an emergency number when a user lifts the handset.

public distribution list A collection of subscribers that is available to all Cisco Unity Express subscribers to use as a distribution list.

pulse-amplitude modulation (PAM) The process of sampling an analog waveform many times to determine numeric electric amplitude values for digital conversion; PAM is typically combined with pulse-code modulation (PCM).

pulse-code modulation (PCM) The process of converting pulse-amplitude modulation (PAM) values into binary number equivalents that voice equipment can transmit over digital circuits.

pulse dialing A type of address signaling in which the rotary-dial wheel of a phone connects and disconnects the local loop circuit as it rotates around to signal specific digits.

Q.931 A signaling protocol used by ISDN CCS implementations.

quantization The process of assigning analog signals a numeric value that can be transported over a digital network.

Real-Time Transport Control Protocol (RTCP) The UDP-based protocol responsible for transporting audio statistics; uses random, odd-numbered UDP ports from 16,384 to 32,767 for communication.

Real-Time Transport Protocol (RTP) The UDP-based protocol responsible for transporting audio packets; uses random, even-numbered UDP ports from 16,384 to 32,767 for communication.

resource May be an administrative web page, part of a web page, or a tool or interface within an application.

robbed bit signaling (RBS) An implementation of channel associated signaling (CAS) that steals the eighth bit of every sixth frame of a digital T1 circuit for signaling information.

role (CUCM) Defines a set of privileges to an application's resources. Privileges may be defined as No Access, Read, or Update.

router-on-a-stick An inter-VLAN routing configuration that allows a single router to move data between VLANs by using a FastEthernet or greater interface broken into subinterfaces connected to a switch trunk port.

service engine The interface that is created on Cisco Unified CME after the Cisco Unity Express hardware platform is installed and recognized by the router. CME will route calls through this interface to the service module for Cisco Unity Express to process.

service module The internal interface of Cisco Unity Express. Cisco Unity Express routes calls through this interface to the service engine for Cisco Unified CME to process.

Session Initiation Protocol (SIP) Voice-signaling protocol created by the IETF as a lightweight alternative to H.323.

Signaling System 7 (SS7) The protocol used within the telephony service provider network to provide inter-CO communication and call routing.

Single Number Reach A feature that allows users to have incoming calls ring another, preconfigured number after a defined time limit. Single number reach also allows for mobility, which allows the transfer of an active call to or from the same, preconfigured number.

SIP for Instant Messaging and Presence Leveraging (SIMPLE) Used for IM with third-party systems.

Skinny Client Control Protocol (SCCP) Cisco-proprietary voice-signaling protocol used to control Cisco IP phones.

smartports A CCA macro that aids in the configuration of roles for individual ports.

Spanning Tree Protocol (STP) A method designed to prevent loops in switched networks due to redundant inter-switch connections.

super frame (SF) An early T1 signaling method that sent 12 T1 frames at a time.

Survivable Remote Site Telephony (SRST) A configuration that allows a router to act as a failover device if an IP phone is unable to reach a Cisco Unified Communications Manager server.

switched virtual interface (SVI) A routed interface on a switch.

time-division multiplexing (TDM) A method of transmitting multiple channels of voice or data over a single digital connection by sending data in specific time slots.

troubleshooting A sequence of steps by which the possible causes of a problem, and the possible steps to correct it, are determined and executed.

trunk port A port on a Cisco switch specifically configured to transmit multiple VLANs. Trunks are typically used between switch devices and in router-on-a-stick configurations.

user group (CUCM) Associated with one or more roles. Users who members of a group inherit the privileges of the role(s) associated with the group. Membership in multiple groups may create conflicting privilege assignments; the enterprise parameter **Effective Access Privileges for Overlapping User Groups and Roles** defines whether the effective privilege is Maximum or Minimum.

virtual LAN (VLAN) A configuration used to break a switch into multiple broadcast domains and IP subnets.

VLAN Trunking Protocol (VTP) A Cisco-proprietary protocol that replicates VLAN database information to all switches participating in the same VTP domain.

voice expansion port The integrated voice/WAN interface card (VWIC) port in all models of the UC520 that allows for PSTN voice expansion. This port does not support WAN connectivity.

Voice Profile for Internet Mail (VPIM) A feature that allows one voicemail system to exchange messages with another voicemail system.

VoiceView Express An XML application that provides to subscribers GUI access to their voicemail messages via the Services button on an IP phone.

Index

Numbers

A

B

C

D

H

I-J

K-L

M

N

O

P

Q

R

S

T

U

V-Z